Classics in Mathematics

Richa[illegible]

Richard S. Ellis

Entropy, Large Deviations, and Statistical Mechanics

Reprint of the 1985 Edition

Springer

Richard S. Ellis
Department of Mathematics and Statistics
University of Massachusetts
Amherst, Massachusetts 01003
U.S.A.

Originally published as Vol. 271 in the series
Grundlehren der mathematischen Wissenschaften

Mathematics Subject Classification (2000): 82A05, 60K35

Library of Congress Control Number: 2005934786

ISSN 1431-0821
ISBN-10 3-540-29059-1 Springer Berlin Heidelberg New York
ISBN-13 978-3-540-29059-9 Springer Berlin Heidelberg New York

Springer is a part of Springer Science+Business Media

springeronline.com

Production: LE-TEX Jelonek, Schmidt & Vöckler GbR, Leipzig
Cover design: *design & production* GmbH, Heidelberg

Printed on acid-free paper   41/3142/YL - 5 4 3 2 1 0

Grundlehren der mathematischen Wissenschaften 271

A Series of Comprehensive Studies in Mathematics

Grundlehren der mathematischen Wissenschaften

A Series of Comprehensive Studies in Mathematics

A Selection

190. Faith: Algebra: Rings, Modules, and Categories I
191. Faith: Algebra II, Ring Theory
192. Mal'cev: Algebraic Systems
193. Pólya/Szegö: Problems and Theorems in Analysis I
194. Igusa: Theta Functions
195. Berberian: Baer*-Rings
196. Athreya/Ney: Branching Processes
197. Benz: Vorlesungen über Geometrie der Algebren
198. Gaal: Linear Analysis and Representation Theory
199. Nitsche: Vorlesungen über Minimalflächen
200. Dold: Lectures on Algebraic Topology
201. Beck: Continuous Flows in the Plane
202. Schmetterer: Introduction to Mathematical Statistics
203. Schoeneberg: Elliptic Modular Functions
204. Popov: Hyperstability of Control Systems
205. Nikol'skii: Approximation of Functions of Several Variables and Imbedding Theorems
206. André: Homologie des Algèbres Commutatives
207. Donoghue: Monotone Matrix Functions and Analytic Continuation
208. Lacey: The Isometric Theory of Classical Banach Spaces
209. Ringel: Map Color Theorem
210. Gihman/Skorohod: The Theory of Stochastic Processes I
211. Comfort/Negrepontis: The Theory of Ultrafilters
212. Switzer: Algebraic Topology—Homotopy and Homology
213. Shafarevich: Basic Algebraic Geometry
214. van der Waerden: Group Theory and Quantum Mechanics
215. Schaefer: Banach Lattices and Positive Operators
216. Pólya/Szegö: Problems and Theorems in Analysis II
217. Stenström: Rings of Quotients
218. Gihman/Skorohod: The Theory of Stochastic Process II
219. Duvant/Lions: Inequalities in Mechanics and Physics
220. Kirillov: Elements of the Theory of Representations
221. Mumford: Algebraic Geometry I: Complex Projective Varieties
222. Lang: Introduction to Modular Forms
223. Bergh/Löfström: Interpolation Spaces. An Introduction
224. Gilbarg/Trudinger: Elliptic Partial Differential Equations of Second Order
225. Schütte: Proof Theory
226. Karoubi: K-Theory, An Introduction
227. Grauert/Remmert: Theorie der Steinschen Räume
228. Segal/Kunze: Integrals and Operators
229. Hasse: Number Theory
230. Klingenberg: Lectures on Closed Geodesics
231. Lang: Elliptic Curves: Diophantine Analysis
232. Gihman/Skorohod: The Theory of Stochastic Processes III
233. Stroock/Varadhan: Multi-dimensional Diffusion Processes
234. Aigner: Combinatorial Theory

Continued after Index

Richard S. Ellis

Entropy, Large Deviations, and Statistical Mechanics

Springer-Verlag
New York Berlin Heidelberg Tokyo

Richard S. Ellis
Department of Mathematics and Statistics
University of Massachusetts
Amherst, Massachusetts 01003
U.S.A.

AMS Subject Classifications: 82A05, 60K35

Library of Congress Cataloging in Publication Data
Ellis, Richard S. (Richard Steven)
Entropy, large deviations, and statistical mechanics.
(Grundlehren der mathematischen Wissenschaften; 271)
Bibliography: p.
Includes index.
1. Statistical mechanics. 2. Entropy. 3. Large deviations. I. Title. II. Series.
QC174.8.E45 1985 530.1'3 84-13891

With 8 Illustrations

Typeset by Asco Trade Typesetting Ltd., Hong Kong.
Printed and bound by R. R. Donnelley and Sons, Harrisonburg, Virginia.
Printed in the United States of America.

9 8 7 6 5 4 3 2 1

ISBN 0-387-96052-X Springer-Verlag New York Berlin Heidelberg Tokyo
ISBN 3-540-96052-X Springer-Verlag Berlin Heidelberg New York Tokyo

For Alison, Melissa, and Michael
my wife and children
who make it all worthwhile

Preface

This book has two main topics: large deviations and equilibrium statistical mechanics. I hope to convince the reader that these topics have many points of contact and that in being treated together, they enrich each other. Entropy, in its various guises, is their common core.

The large deviation theory which is developed in this book focuses upon convergence properties of certain stochastic systems. An elementary example is the weak law of large numbers. For each positive ε, $P\{|S_n/n| \geq \varepsilon\}$ converges to zero as $n \to \infty$, where S_n is the nth partial sum of independent identically distributed random variables with zero mean. Large deviation theory shows that if the random variables are exponentially bounded, then the probabilities converge to zero exponentially fast as $n \to \infty$. The exponential decay allows one to prove the stronger property of almost sure convergence ($S_n/n \to 0$ a.s.). This example will be generalized extensively in the book. We will treat a large class of stochastic systems which involve both independent and dependent random variables and which have the following features: probabilities converge to zero exponentially fast as the size of the system increases; the exponential decay leads to strong convergence properties of the system. The most fascinating aspect of the theory is that the exponential decay rates are computable in terms of entropy functions. This identification between entropy and decay rates of large deviation probabilities enhances the theory significantly.

Entropy functions have their roots in statistical mechanics. They originated in the work of L. Boltzmann, who in the 1870's studied the relation between entropy and probability in physical systems. Thus statistical mechanics has a strong historical connection with large deviation theory. It also provides a natural context in which the theory can be applied. Applications of large deviations to models in equilibrium statistical mechanics are presented in Chapters III–V. These applications illustrate convincingly the power of the theory.

Equilibrium statistical mechanics is an exciting area of mathematical physics but one which remains inaccessible to many mathematicians. Some texts on the subject provide an introduction to the physics but do not develop the mathematics in much detail or with great rigor. Other texts treat mathematical problems in statistical mechanics with complete rigor but assume an extensive background in the physics. The uninitiated reader has difficulty understanding how concepts like ensemble, free energy, or entropy connect up with more familiar concepts in mathematics. My approach in this book is to emphasize strongly the connections between statistical mechanics on the one hand and probability and large deviations on the other. I hope that in so doing, I have succeeded in providing a readable treatment of statistical mechanics which is accessible to a general mathematical audience. My large deviation approach to statistical mechanics was inspired in part by the article of O. E. Lanford (1973).

In recent years, the scope of large deviations has been greatly expanded by M. D. Donsker and S. R. S. Varadhan. This book contains an introduction to their theory. I illustrate the main features in the context of independent identically distributed random vectors taking values in $\mathbb{R}^d$. I also present my own large deviation results, which are particularly suited for applications to statistical mechanics. Since readability rather than completeness has been my goal, the large deviation theorems are not stated in the greatest generality.

There are two parts to the book, Part I consisting of Chapters I–V. Chapter I introduces large deviations by means of elementary examples involving combinatorics and Stirling's formula. Chapter II presents the Donsker–Varadhan theory as well as my own large deviation results. The proofs of the theorems in this chapter are detailed and are postponed until Part II. Postponing proofs allows the reader to reach, as soon as possible, interesting applications of large deviations to statistical mechanics in Chapters III–V. Chapter III gives a large deviation analysis of a discrete gas model. Chapters IV–V discuss the Ising model of ferromagnetism and related spin systems. The emphasis in these two chapters is upon properties of Gibbs states. While large deviation theory provides a terminology and a set of results that are useful for treating Gibbs states, the book also develops other tools that are needed. These include convexity and moment inequalities.

Part II consists of Chapters VI–IX. Chapter VI is a summary of the theory of convex functions on $\mathbb{R}^d$. Chapters VII–IX prove the large deviation results stated in Chapter II without proof. The prerequisite for these chapters is a good working knowledge of probability and measure theory. The essential definitions and theorems in probability are listed in Appendix A. The appendix is intended to be a review or an outline for study rather than a detailed exposition.

This book can be used as a text. It contains over 100 problems, many of which have hints. Chapters I and II and VI–IX are a self-contained treatment of large deviations and convex functions. Readers primarily interested in spin systems can concentrate upon Chapters IV and V and refer to the statements and proofs of large deviation results as needed. Those portions of Chapters IV and V which do not rely on large deviations are self-contained. Chapters IV and V can be completely understood without reading Chapter III.

This book contains new results and new proofs of known theorems. These include the following: exponential convergence properties of Gibbs states

[Theorems IV.5.5, IV.6.6, and V.6.1]; a large deviation proof of the Gibbs variational formula [Theorem IV.7.3(a)]; a proof of the central limit theorem for spin systems [Theorem V.7.2(a)]; a level-3 large deviation theorem for i.i.d. random variables with a finite state space [Theorem IX.1.1]; a level-3 large deviation theorem for Markov chains with a finite state space [Problems IX.6.10–IX.6.15]; the solution of the Gibbs variational formula for finite-range interactions on $\mathbb{Z}$ via large deviations [Appendix C.6]. Many of the large deviation results and applications in the book depend upon my large deviation theorem, Theorem II.6.1. The proof of the level-3 theorem in Chapter IX was inspired by statistical mechanics [see Appendix C.6] and information theory.

I have had the good fortune of interacting with a number of special people. Todd Baker edited the manuscript with creativity and care. The book benefited greatly from his involvement. Peg Bombardier was my superb typist. She was always cheerful and patient, despite the numerous revisions, and was a pleasure to work with. Alan Sokal read portions of the manuscript and was a big help with the statistical mechanics. I owe a special debt of gratitude to Srinivasa Varadhan. He answered my many questions about large deviations patiently and with insight and showed a strong interest in the book. The encouragement of my family and friends was greatly appreciated. Above all, I thank my wife Alison. Her love is a blessing.

I am grateful to Alejandro de Acosta, Hans-Otto Georgii, Joseph Horowitz, Jonathan Machta, Charles Newman, and R. Tyrrell Rockafellar for reading portions of the manuscript and suggesting improvements. I am also indebted to the many other people, too numerous to mention by name, with whom I have consulted. While writing the book, I received support from the University of Massachusetts, the National Science Foundation, and the Lady Davis Fellowship Trust. Their support is gratefully acknowledged.

Richard S. Ellis

Contents

Part I

Large Deviations and Statistical Mechanics

Chapter I

Introduction to Large Deviations

I.1. Overview

One of the common themes of probability theory and statistical mechanics is the discovery of regularity in the midst of chaos. The laws of probability theory, which include laws of large numbers and central limit theorems, summarize the behavior of a stochastic system in terms of a few parameters (e.g., mean and variance). In statistical mechanics, one derives macroscopic properties of a substance from a probability distribution that describes the complicated interactions among the individual constituent particles. A central concept linking the two fields is entropy.[1] The term was introduced into thermodynamics by Clausius in 1865 after many years of intensive work by him and others on the second law of thermodynamics. An early important step in its development and enrichment was the discovery by Boltzmann of a statistical interpretaton of entropy. Boltzmann's discovery, which was published in 1877, has three parts. We have augmented part (c) to include the possibility of phase transitions.

(a) Entropy is a measure of randomness or disorder in a statistical mechanical system.
(b) If S is the entropy for a system in a given state and W is the "thermodynamical probability" of that state,* then $S = k \log W$, where k is a positive physical constant.
(c) The equilibrium states, which are the states of the system observed in nature, are those states with the largest thermodynamical probability and thus the largest entropy. By (a), they are the "most random" states of the system consistent with any constraints which the system must satisfy (e.g., conservation of energy). The existence of more than one equilibrium state corresponds to a phase transition.

All the notions of entropy discussed in this book are variations on the Boltzmann theme. In analyzing stochastic or statistical mechanical systems,

*The thermodynamical probability is defined to be the number of microstates compatible with the given state (see Wehrl (1978, page 223)).

one must extrapolate from a microscopic level, on which the system is defined, to a macroscopic level, on which the laws describing the behavior of the system are formulated. Boltzmann shows that entropy is the bridge between these two levels. We will illustrate these ideas and outline the main themes of this book in terms of a basic stochastic model. This discussion is intended for the reader who has a knowledge of probability theory consistent with Appendix A. Other readers may still perceive the global picture without turning to Appendix A at this time. That task may be postponed until the end of the first chapter.

Each of our systems is modeled microscopically by a collection of random variables $\{X_\alpha; \alpha \in \mathscr{A}\}$ which are defined on a probability space $(\Omega, \mathscr{F}, P)$ and which take values in a space Γ. Ω is a nonempty set, $\mathscr{F}$ is a σ-field of subsets of Ω, and P is a probability measure on $(\Omega, \mathscr{F})$. Γ is called the *state space* or the *outcome space*; in all of our applications, Γ is $\mathbb{R}^d$, $d \in \{1, 2, \ldots\}$, or a subset of $\mathbb{R}^d$. $\mathscr{A}$ is a suitable index set. Our results depend only on the distribution of the random variables $\{X_\alpha; \alpha \in \mathscr{A}\}$. Hence we may take Ω to be the product space $\Gamma^{\mathscr{A}}$ and the collection $\{X_\alpha; \alpha \in \mathscr{A}\}$ to be the coordinate representation process. That is, given a point $\omega = \{\omega_\alpha; \alpha \in \mathscr{A}\}$ in $\Gamma^{\mathscr{A}}$, we define $X_\alpha(\omega) = \omega_\alpha$, the αth coordinate of ω. Each $\omega \in \Gamma^{\mathscr{A}}$ represents a possible configuration or microstate of the system, and the entire space $\Gamma^{\mathscr{A}}$ is the set of all the configurations. The definition of the model is completed by specifying a probability measure P on configuration space. Here are some examples.

Example I.1.1. (a) Let $\mathscr{A}$ be the set of integers $\mathbb{Z} = \{\ldots, -1, 0, 1, \ldots\}$ and Γ a finite set of distinct real numbers $\{x_1, x_2, \ldots, x_r\}$. Define P to be an infinite product measure on $\Gamma^{\mathbb{Z}}$ with identical one-dimensional marginals, which we denote by ρ. Thus ρ is a probability measure on Γ, and it has the form $\sum_{i=1}^r \rho_i \delta_{x_i}$, where $\rho_i > 0$, $\sum_{i=1}^r \rho_i = 1$, and δ_{x_i} is the unit point measure at x_i. The variables $\{X_j; j \in \mathbb{Z}\}$ are independent and identically distributed (i.i.d.) and each has distribution ρ. We write P_ρ for P.

(b) A simple case of (a) is $\Gamma = \{0, 1\}$. The $\{X_j; j \in \mathbb{Z}\}$ are then Bernoulli trials and $\rho = \rho_0 \delta_0 + \rho_1 \delta_1$. The $\{X_j; j \in \mathbb{Z}\}$ may, for example, represent the successive outcomes of the toss of a coin in an infinite sequence of tosses separated by a constant time interval. Each outcome is recorded as 0 for a tail and 1 for a head. A fair coin corresponds to $\rho = \frac{1}{2}\delta_0 + \frac{1}{2}\delta_1$.

(c) In (a), the set $\Gamma = \{x_1, x_2, \ldots, x_r\}$ may represent a set of possible velocities of the molecules of an ideal gas which are constrained to move in an interval and which undergo elastic collisions at the endpoints. Then X_j denotes the velocity of the molecule labeled j. A configuration $\omega \in \Gamma^{\mathbb{Z}}$ is a specification of the velocities for each molecule. Independence means that the molecules do not interact. This model is treated in Chapter III.

(d) A ferromagnet is modeled by random configurations of spins (microscopic magnets) at sites in the D-dimensional integer lattice $\mathbb{Z}^D$, $D \in$

$\{1, 2, \ldots\}$. We set $\mathscr{A} = \mathbb{Z}^D$ and $\Gamma = \{1, -1\}$. The values 1 and -1 represent "spin-up" and "spin-down," respectively. Configuration space Ω is $\{1, -1\}^{\mathbb{Z}^D}$ and $X_j(\omega)$ is the spin at site $j \in \mathbb{Z}^D$ for the configuration ω. A product measure on Ω is not appropriate for this model since the spins at different sites interact. The ferromagnet is modeled by a probability measure P on Ω which is translation invariant; i.e., invariant with respect to spatial translations in $\mathbb{Z}^D$. Ferromagnetic models are the subject of Chapters IV and V.

As these examples show, a probability measure on a configuration space provides a microscopic definition of a stochastic or physical system. However, the laws describing the behavior of such a system are macroscopic descriptions which in contrast to the number of all configurations, involve many fewer variables. For each description, the range of possible values of these variables defines the set of macrostates. Each macrostate is compatible with, and hence is a summary of, many microstates. The entropy of a macrostate is a measure of this multiplicity. Those macrostates compatible with the most microstates—i.e., those with the largest entropy—are the ones observed in nature. Generally, a system will have several possible macroscopic descriptions, each differing in complexity and in choice of macrostate. For each description, there is a different entropy concept.

We return to the coin tossing model in order to explain these ideas. This model is represented by the infinite product measure P_ρ on the configuration space $\Omega = \{0, 1\}^{\mathbb{Z}}$ $(\rho = \rho_0 \delta_0 + \rho_1 \delta_1)$. Macroscopically, the behavior of the coin can be expressed by a single number, its mean value. The possible mean values are all numbers $z \in (0, 1)$, and there is no harm including the endpoints. We call the set of $z \in [0, 1]$ macrostates. The weak law of large numbers (WLLN) enables one to estimate the macrostate in terms of microstates. Define $S_n(\omega) = \sum_{j=1}^n X_j(\omega)$ for $n = 1, 2, \ldots$ and $\omega \in \Omega$. $S_n(\omega)/n$ is the average value of the tosses $\omega_1, \omega_2, \ldots, \omega_n$. The sum $S_n(\omega)/n$ is called a *microscopic sum* or *n-sum*. Let m_ρ be the mean value of the measure ρ $(m_\rho = 0 \cdot \rho_1 + 1 \cdot \rho_1 = \rho_1)$ and Q_n the distribution of S_n/n. The WLLN says that for any $\varepsilon > 0$,

$$Q_n\{(m_\rho - \varepsilon, m_\rho + \varepsilon)\} = P_\rho\{\omega \in \Omega : S_n(\omega)/n \in (m_\rho - \varepsilon, m_\rho + \varepsilon)\} \to 1 \quad \text{as } n \to \infty.$$

In other words, if n is large, then with respect to P_ρ essentially all microscopic n-sums are close to the macrostate m_ρ. The latter is called the *equilibrium state.*

Here is how entropy arises. Assume for simplicity that the coin is fair $(m_\rho = \frac{1}{2})$. For any $z \in \mathbb{R}$ and $\varepsilon > 0$, let $A_{z,\varepsilon}$ be the interval $(z - \varepsilon, z + \varepsilon)$. By the WLLN, $Q_n\{A_{m_\rho, \varepsilon}\} \to 1$ as $n \to \infty$ while if $z \neq m_\rho$ and $0 < \varepsilon < |z - m_\rho|$, then $Q_n\{A_{z,\varepsilon}\} \to 0$. In the latter case, it is not hard to refine the WLLN and to

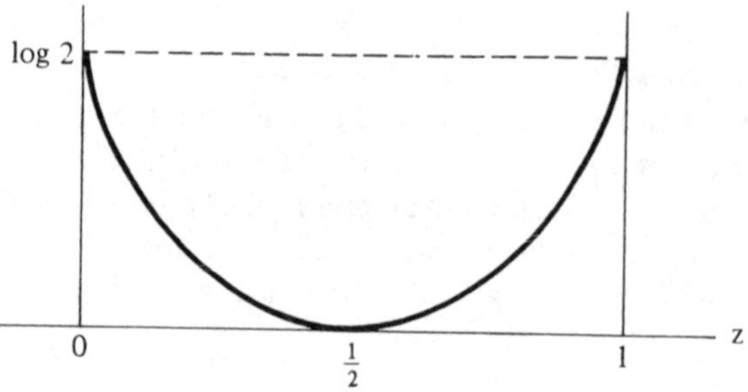

Figure I.1. The entropy function $I(z)$ for the coin tossing model.

prove that $Q_n\{A_{z,\varepsilon}\}$ decays to 0 exponentially fast. The exponential rate of decay is defined by $F(z,\varepsilon) = -\lim_{n\to\infty} n^{-1} \log Q_n\{A_{z,\varepsilon}\}$.* By a simple combinatoric argument (as in the proof of Theorem I.3.1), one shows that $F(z,\varepsilon)$ equals the infimum over $A_{z,\varepsilon}$ of the function

$$(1.1) \qquad I(z) = \begin{cases} z\log(2z) + (1-z)\log(2(1-z)) & \text{for } z\in[0,1], \\ \infty & \text{for } z\notin[0,1], \end{cases}$$

where $0\log 0 = 0$. The graph of the non-negative convex function $I(z)$ is shown in Figure I.1; clearly, $F(z,\varepsilon) > 0$. $I(z)$ is called the *entropy function* for the coin tossing model. We now interpret it in view of our earlier remarks on entropy.

For any $z\in\mathbb{R}$ and large n, $Q_n\{A_{z,\varepsilon}\}$ is approximately $\exp(-nF(z,\varepsilon))$. Since $F(z,\varepsilon)\to I(z)$ as $\varepsilon\to 0$, we may heuristically write

$$(1.2) \qquad Q_n\{A_{z,\varepsilon}\} \approx \exp(-nI(z))$$

for n large and ε small. If $z \neq m_\rho$, then $I(z)$ is positive and $\exp(-nI(z))\to 0$ as $n\to\infty$. This is consistent with the exponential decay of $Q_n\{A_{z,\varepsilon}\}$ for $0<\varepsilon<|z-m_\rho|$. The heuristic formula (1.2) shows that to a small value of $I(z)$ there corresponds a large probability $Q_n\{A_{z,\varepsilon}\}$ or, in other words, a high multiplicity of microstates. In this sense, $I(z)$ is a measure of the multiplicity of microstates compatible with the macrostate z. For another interpretation, given points z_1 and z_2 in $[0,1]$, it is reasonable to call z_1 more random than z_2 if $I(z_1) < I(z_2)$; that is, if there are more microstates compatible with z_1 than with z_2. Thus $I(z)$ also measures the randomness of the macrostate z. The equilibrium state $m_\rho = \frac{1}{2}$ is that macrostate which is compatible with the most microstates. In fact,

$$I(m_\rho) = 0 = \min\{I(z): z\in\mathbb{R}\} \quad \text{and} \quad I(z) > 0 \qquad \text{for } z \neq m_\rho.$$

Thus the equilibrium state, being the unique minimum point of I, is the most random macrostate. Points z outside of $[0,1]$ are forbidden values for S_n/n:

*If $Q_n\{A_{z,\varepsilon}\} = 0$, then set $\log Q_n\{A_{z,\varepsilon}\} = -\infty$.

if $A \cap [0,1]$ is empty, then $Q_n\{A\} = 0$ and $I(z) = \infty$ for each $z \in A$. For $z \neq m_\rho$ and $0 < \varepsilon < |z - m_\rho|$, $Q_n\{A_{z,\varepsilon}\}$ is called a *large deviation probability* since the event $\{\omega \in \Omega : S_n(\omega)/n \in A_{z,\varepsilon}\}$ corresponds to a fluctuation or deviation of S_n/n of order $|z - m_\rho|$ away from the limiting mean. It is a very rare event since $Q_n\{A_{z,\varepsilon}\} \to 0$ exponentially.

An equivalent statement of the WLLN is that the distributions $\{Q_n;\ n = 1, 2, \ldots\}$ converge weakly to the unit point measure at m_ρ (written $Q_n \Rightarrow \delta_{m_\rho}$). In this book, we will study much more general but analogous situations. A sequence of probability measures $\{Q_n; n = 1, 2, \ldots\}$ on a complete separable metric space $\mathscr{X}$ will converge weakly to the unit point measure at some $x_0 \in \mathscr{X}$. $\{Q_n\}$ will have a large deviation property in the sense that $Q_n\{K\}$ will decay exponentially for all closed sets K in $\mathscr{X}$ which do not contain x_0. The decay rate will be given by $-\inf_{x \in K} I(x)$, where $I(x)$ is some non-negative function on $\mathscr{X}$ with a unique minimum point at x_0 ($I(x_0) = 0$). $I(x)$ is called the *entropy function of the measures* $\{Q_n\}$.

In a series of important papers beginning in 1975, Donsker and Varadhan have identified three levels of large deviations which fit into the general framework just described. These levels will be treated in detail in this book. Let $\{X_j; j \in \mathbb{Z}\}$ be a sequence of i.i.d. random vectors taking values in $\mathbb{R}^d$. Let ρ be the distribution of X_1 and P_ρ the corresponding infinite product measure on $\Omega = (\mathbb{R}^d)^{\mathbb{Z}}$.

Level-1. Define $Q_n^{(1)}$ to be the distribution of $S_n(\omega)/n = \sum_{j=1}^n X_j(\omega)/n$ on $\mathbb{R}^d$ and assume that $\int_{\mathbb{R}^d} \|x\| \rho(dx)$ is finite. Then by the WLLN the sequence $\{Q_n^{(1)};\ n = 1, 2, \ldots\}$ converges weakly to δ_{m_ρ}, where m_ρ is the mean $\int_{\mathbb{R}^d} x\rho(dx)$ of ρ. If, furthermore, the moment generating function $\int_{\mathbb{R}^d} \exp\langle t, x\rangle \rho(dx)$ is finite for all $t \in \mathbb{R}^d$, then $Q_n^{(1)}\{K\}$ decays exponentially for all closed subsets K of $\mathbb{R}^d$ which do not contain m_ρ [Theorem II.4.1]. The decay rate is given in terms of a level-1 entropy function $I_\rho^{(1)}$ which is a non-negative convex function on $\mathbb{R}^d$ and which has a unique minimum point at m_ρ ($I_\rho^{(1)}(m_\rho) = 0$). The sums $\{S_n(\omega)/n\}$ are called *level*-1 *microscopic n-sums* and points $z \in \mathbb{R}^d$ are called *level*-1 *macrostates*. The mean m_ρ is the unique level-1 equilibrium state.

Level-2. For $\omega \in \Omega$, define the empirical measure $L_n(\omega, \cdot) = n^{-1} \sum_{j=1}^n \delta_{X_j(\omega)}(\cdot)$ (for $A \subseteq \mathbb{R}^d$ a Borel set, $L_n(\omega, A) = n^{-1} \sum_{j=1}^n \delta_{X_j(\omega)}\{A\}$). $L_n(\omega)$ takes values in the space $\mathscr{M}(\mathbb{R}^d)$ of probability measures on $\mathbb{R}^d$. $\mathscr{M}(\mathbb{R}^d)$ is a topological space with respect to the topology of weak convergence, and it is metrizable as a complete separable metric space. Let $Q_n^{(2)}$ denote the distribution of $L_n(\omega, \cdot)$ on $\mathscr{M}(\mathbb{R}^d)$. By the ergodic theorem, the sequence $\{L_n(\omega, \cdot); n = 1, 2, \ldots\}$ converges weakly to ρ (almost surely), and this implies that $\{Q_n^{(2)}; n = 1, 2, \ldots\}$ converges weakly to δ_ρ. In addition, $Q_n^{(2)}\{K\}$ decays exponentially for all closed subsets K of $\mathscr{M}(\mathbb{R}^d)$ which do not contain ρ [Theorem II.4.3]. The decay rate is given in terms of a level-2 entropy function $I_\rho^{(2)}$ which is a non-negative convex function on $\mathcal{M}(\mathbb{R}^d)$ and which has a unique minimum point at ρ ($I_\rho^{(2)}(\rho) = 0$). For $\nu \in \mathcal{M}(\mathbb{R}^d)$, $I_\rho^{(2)}(\nu)$ equals the relative entropy of ν with respect to ρ. The empirical measures $\{L_n(\omega, \cdot)\}$ are called *level-2 microscopic n-sums* and measures $\nu \in \mathcal{M}(\mathbb{R}^d)$ are called *level-2 macrostates*. The distribution ρ is the unique level-2 equilibrium state.

Level-3. Let $\mathcal{M}_s(\Omega)$ denote the set of strictly stationary probability measures on Ω. $\mathcal{M}_s(\Omega)$ is a topological space with respect to the topology of weak convergence, and it is metrizable as a complete separable metric space. For $\omega \in \Omega$, one defines the so-called empirical process $R_n(\omega, \cdot)$ which takes values in $\mathcal{M}_s(\Omega)$ [see page 22]. Let $Q_n^{(3)}$ denote the distribution of $R_n(\omega, \cdot)$ on $\mathcal{M}_s(\Omega)$. By the ergodic theorem, the sequence $\{R_n(\omega, \cdot); n = 1, 2, \ldots\}$ converges weakly to P_ρ (almost surely). Hence $\{Q_n^{(3)}; n = 1, 2, \ldots\}$ converges weakly to δ_{P_ρ}. In addition, $Q_n^{(3)}\{K\}$ decays exponentially for all closed subsets K of $\mathcal{M}_s(\Omega)$ which do not contain P_ρ [Theorem II.4.4]. The decay rate is given in terms of a level-3 entropy function $I_\rho^{(3)}$ which is a non-negative affine function on $\mathcal{M}_s(\Omega)$ and which has a unique minimum point at P_ρ ($I_\rho^{(3)}(P_\rho) = 0$). The empirical processes $\{R_n(\omega, \cdot)\}$ are called *level-3 microscopic n-sums* and measures $P \in \mathcal{M}_s(\Omega)$ are called *level-3 macrostates*. The measure P_ρ is the unique level-3 equilibrium state.

For each of the three levels, we may heuristically express the asymptotic behavior of the distributions $Q_n^{(i)}(dx)$ by the formula $\exp(-nI_\rho^{(i)}(x))dx$. $Q_n^{(i)}(dx)$ is a measure on the complete separable metric space $\mathcal{X} = \mathbb{R}^d$, $\mathcal{M}(\mathbb{R}^d)$, or $\mathcal{M}_s(\mathbb{R}^d)$ for $i = 1$, 2, or 3, respectively. By analogy with coin tossing, we may interpret each entropy function $I_\rho^{(i)}(x)$ as a measure of the multiplicity of microstates compatible with the macrostate $x \in \mathcal{X}$. In that sense, $I_\rho^{(i)}(x)$ is also a measure of the randomness of x.

Varadhan (1966) gave a useful application of the large deviation property to calculate the asymptotics of certain integrals. The heuristic formula $Q_n^{(i)}(dx) \approx \exp(-nI_\rho^{(i)}(x))dx$ suggests that if F is a bounded continuous function on $\mathcal{X}$, then

$$\begin{aligned}\lim_{n\to\infty} \frac{1}{n} \log \int_{\mathcal{X}} \exp(nF(x)) Q_n^{(i)}(dx) &= \lim_{n\to\infty} \frac{1}{n} \log \int_{\mathcal{X}} \exp[n(F(x) - I_\rho^{(i)}(x))]dx \\ &= \sup_{x\in\mathcal{X}} \{F(x) - I_\rho^{(i)}(x)\}.\end{aligned}$$

This limit is valid under suitable hypotheses [Theorem II.7.1] and will be applied a number of times in the book.

So far we have discussed large deviations for i.i.d. random vectors. Statistical mechanical systems have a similar three-level structure with one additional feature: there need not be a unique equilibrium state for a given level. This lack of a unique equilibrium state corresponds physically to a phase transition and probabilistically to a breakdown in the law of large numbers (or ergodic theorem) for the corresponding microscopic n-sums. But in general, whether or not there is a phase transition, we may consider the Legendre–Fenchel transform of the corresponding entropy function. This transform defines a convex function which in statistical mechanics is called the free energy. Free energy functions will play a central role in analyzing statistical mechanical systems in Chapters III–V.

In the remainder of this chapter we will introduce level-1, 2, and 3 large deviations by considering i.i.d. random variables with a finite state space. The corresponding entropy functions will be calculated by means of elementary combinatorics. In Chapter II, the three levels of large deviations will be generalized to i.i.d. random vectors taking values in $\mathbb{R}^d$. Section II.6 presents additional large deviation results, which are particularly suited for applications to statistical mechanics. The proofs of the theorems in Chapter II are detailed and will be postponed until Chapters VI–IX. In Chapters III–V the large deviation results will be applied to an ideal gas model and to ferromagnetic spin models in statistical mechanics.

I.2. Large Deviations for I.I.D. Random Variables with a Finite State Space

In its simplest form the theory of large deviations refines the classical law of large numbers. Let S_n be the nth partial sum of independent, identically distributed random variables $X_1, X_2, \ldots$. The strong law of large numbers states that if the expectation $E\{|X_1|\}$ is finite, then S_n/n converges to $E\{X_1\}$ almost surely. This was proved by Kolmogorov (1930). It implies the weak law, which states that S_n/n converges to $E\{X_1\}$ in probability. The first large deviation results were those of Cramér (1938) and Chernoff (1952). They showed that if X_1 has a finite moment generating function in a neighborhood of 0, then the probability that S_n/n deviates from $E\{X_1\}$ by a small amount $\varepsilon > 0$ is exponentially small as $n \to \infty$. After Chernoff, these results were applied and extended in statistics and probability by many people, and they have played a key role in information theory. In a series of papers starting in 1975, Donsker and Varadhan have generalized these results to Markov processes with general state spaces and have found many interesting new applications.

The Donsker–Varadhan theory identifies three levels of large deviations, which were mentioned in the previous section. An elementary way of introducing the three levels is by means of well-known but instructive examples involving i.i.d. random variables with a finite state space. The rest of this chapter focuses upon these examples. Later chapters will generalize the large deviation results beyond this elementary setting.

Let $r \geq 2$ be an integer and consider a finite set $\Gamma = \{x_1, x_2, \ldots, x_r\}$, where $x_1 < x_2 < \cdots < x_r$ are real numbers. Let $\mathscr{B}(\Gamma)$ denote the set of all subsets of Γ. We fix a probability measure ρ on $\mathscr{B}(\Gamma)$ for which $\rho_i = \rho\{x_i\} > 0$ for each $x_i \in \Gamma$. Thus ρ has the form $\sum_{i=1}^r \rho_i \delta_{x_i}$. For A a subset of Γ $\rho\{A\}$ equals $\sum_{i=1}^r \rho_i \delta_{x_i}\{A\}$, where $\delta_{x_i}\{A\}$ equals 1 if $x_i \in A$ and equals 0 if $x_i \notin A$. Denote by ω the doubly infinite sequence $(\ldots, \omega_{-2}, \omega_{-1}, \omega_0, \omega_1, \omega_2, \ldots)$ with each $\omega_j \in \Gamma$. Configuration space Ω is the set of all such sequences; thus $\Omega = \Gamma^{\mathbb{Z}}$. Let P_ρ be the infinite product measure on Ω with identical one-dimensional marginals ρ. To a cylinder set of the form

(1.3) $$\Sigma = \{\omega \in \Omega : \omega_{m+1} \in F_1, \ldots, \omega_{m+k} \in F_k\}, \quad F_1, \ldots, F_k \text{ subsets of } \Gamma,$$

P_ρ assigns the probability $P_\rho\{\Sigma\} = \prod_{j=1}^k \rho\{F_j\}$. P_ρ is uniquely determined by these probabilities.* For each integer j define the coordinate function X_j on Ω by $X_j(\omega) = \omega_j$. The functions $\{X_j; j \in \mathbb{Z}\}$ form a sequence of i.i.d. random variables with finite state space Γ and distribution ρ.

Until Section I.5, we will work with level-1 and level-2. These are defined by two random quantities, called level-1 and level-2 (microscopic) n-sums. The level-1 n-sum is the average value $S_n(\omega)/n = \sum_{j=1}^n X_j(\omega)/n$, $n = 1, 2, \ldots$. The level-2 n-sum is defined in terms of the empirical frequency $L_{n,i}(\omega)$ with which x_i appears in the sequence $X_1(\omega), \ldots, X_n(\omega)$:

(1.4) $$L_{n,i}(\omega) = \frac{1}{n} \sum_{j=1}^n \delta_{X_j(\omega)}\{x_i\}, \qquad n = 1, 2, \ldots, \quad i = 1, \ldots, r.$$

For each ω, the numbers $\{L_{n,i}(\omega); i = 1, \ldots, r\}$ define a probability measure $L_n(\omega, \cdot)$ on the set of all subsets of Γ. For A a subset of Γ,

(1.5) $$L_n(\omega, A) = \sum_{x_i \in A} L_{n,i}(\omega) = \frac{1}{n} \sum_{j=1}^n \delta_{X_j(\omega)}\{A\},^\dagger$$

where $\delta_{X_j(\omega)}\{A\}$ equals 1 if $X_j(\omega) \in A$ and equals 0 if $X_j(\omega) \notin A$. The measure $L_n(\omega, \cdot)$ is the level-2 n-sum. It is called the *empirical measure* corresponding to $X_1(\omega), \ldots, X_n(\omega)$. The average value $S_n(\omega)/n$ can be calculated by multiplying each x_i by the empirical frequency $L_{n,i}(\omega)$ and summing over Γ; i.e.,

(1.6) $$S_n(\omega)/n = \sum_{i=1}^r x_i L_{n,i}(\omega).$$

The right-hand side is the mean of the empirical measure.

With respect to the measure P_ρ, the asymptotic behavior of $S_n(\omega)/n$ and of $L_n(\omega, \cdot)$ follows from the law of large numbers. Indeed the summands in $S_n(\omega)/n$ are i.i.d. with mean

$$m_\rho = \int_\Omega X_j(\omega) P_\rho(d\omega) = \sum_{i=1}^r x_i \rho_i,$$

*Appendix A summarizes all the properties of probability measures that are needed in the text.

†The sum over an empty set is defined to be 0.

while the summands in $L_{n,i}(\omega)$ are i.i.d. with mean

$$\int_\Omega \delta_{X_j(\omega)}\{x_i\} P_\rho(d\omega) = P_\rho\{\omega \in \Omega: X_j(\omega) = x_i\} = \rho_i.$$

Hence for any $\varepsilon > 0$

$$(1.7) \qquad \begin{aligned} \lim_{n\to\infty} P_\rho\{\omega\in\Omega: |S_n(\omega)/n - m_\rho| \geq \varepsilon\} = 0, \\ \lim_{n\to\infty} P_\rho\{\omega\in\Omega: \max_{i=1,\ldots,r} |L_{n,i}(\omega) - \rho_i| \geq \varepsilon\} = 0. \end{aligned}$$

The vector $(\rho_1, \rho_2, \ldots, \rho_r)$ is the limiting mean of the random vector $(L_{n,i}(\omega), \ldots, L_{n,r}(\omega))$. The probabilities in (1.7) represent large deviations since they involve fluctuations of order ε of the respective n-sums away from the limiting means, and ε is fixed. Below we show by elementary combinatorial arguments that each of these probabilities decays exponentially.

I.3. Levels-1 and 2 for Coin Tossing

Coin tossing is defined by the state space $\Gamma = \{0, 1\}$ and the measure $\rho = \frac{1}{2}\delta_0 + \frac{1}{2}\delta_1$. The value 0 represents a tail and 1 a head. The proof for this simple case will set a pattern of proof for the more general large deviation results which follow. We have $\rho_1 = \rho_2 = \frac{1}{2}$, $m_\rho = \frac{1}{2}$, $L_{n,1}(\omega) = 1 - S_n(\omega)/n$, and $L_{n,2}(\omega) = S_n(\omega)/n$. Hence for each ω

$$|L_{n,1}(\omega) - \rho_1| = |L_{n,2}(\omega) - \rho_2| = |S_n(\omega)/n - m_\rho|,$$

and so the level-1 and 2 probabilities coincide:

$$(1.8) \qquad P_\rho\{|S_n/n - m_\rho| \geq \varepsilon\} = P_\rho\{\max_{i=1,2} |L_{n,i} - \rho_i| \geq \varepsilon\}.$$

Let $Q_n^{(1)}$ be the P_ρ-distribution of S_n/n on $\mathbb{R}$ and define the closed set

$$A = \{z \in \mathbb{R}: |z - m_\rho| \geq \varepsilon\}, \qquad \text{where } 0 < \varepsilon < \tfrac{1}{2}.$$

The set $A \cap [0, 1]$ is nonempty and $Q_n^{(1)}\{A\} = P_\rho\{|S_n/n - m_\rho| \geq \varepsilon\}$ is positive for all sufficiently large n. Since A does not contain m_ρ, $Q_n^{(1)}\{A\} \to 0$. According to the next theorem, $Q_n^{(1)}\{A\}$ decays exponentially, and the decay rate is given in terms of the entropy function

$$(1.9) \qquad I_\rho^{(1)}(z) = \begin{cases} z\log(2z) + (1-z)\log(2(1-z)) & \text{for } z \in [0,1], \\ \infty & \text{for } z \notin [0,1], \end{cases}$$

where $0 \log 0 = 0$. $I_\rho^{(1)}(z)$ is convex, is symmetric about $z = m_\rho = \frac{1}{2}$, and attains its minimum value of 0 at the unique point $z = m_\rho$. $I_\rho^{(1)}$ is depicted in Figure I.1.

Theorem I.3.1.

$$\lim_{n\to\infty}\frac{1}{n}\log Q_n^{(1)}\{A\} = \lim_{n\to\infty}\frac{1}{n}\log P_\rho\{|S_n/n - m_\rho| \geq \varepsilon\} = -\min_{z\in A} I_\rho^{(1)}(z).$$

Since the set A is closed and does not contain m_ρ, $\min_{z\in A} I_\rho^{(1)}(z) > I_\rho^{(1)}(m_\rho) = 0$. Hence $Q_n^{(1)}\{A\}$ converges to zero exponentially fast as $n\to\infty$.

Proof. Let Ω_n be the finite configuration space consisting of all sequences $\omega = (\omega_1, \omega_2, \ldots, \omega_n)$, with each $\omega_j \in \Gamma = \{0, 1\}$; thus $\Omega_n = \Gamma^n$. If $\pi_n P_\rho$ is the finite product measure on Ω_n with identical one-dimensional marginals ρ, then $Q_n^{(1)}\{A\} = \pi_n P_\rho\{\omega\in\Omega_n : S_n(\omega)/n \in A\}$. For fixed n and $\omega\in\Omega_n$, $S_n(\omega)$ may take any value $k\in\{0, 1, \ldots, n\}$. $S_n(\omega)/n$ is in A if and only if k is in the set $A_n = \{k\in\{0, 1, \ldots, n\} : |k/n - \frac{1}{2}| \geq \varepsilon\}$. For $k\in A_n$, define $C(n,k) = n!/(k!(n-k)!)$. There are $C(n,k)$ points ω in Ω_n for which $S_n(\omega) = k$, and $\pi_n P_\rho\{\omega\} = 2^{-n}$ for each $\omega\in\Omega_n$. Hence

$$(1.10)\quad Q_n^{(1)}\{A\} = \sum_{k\in A_n} \pi_n P_\rho\{\omega\in\Omega_n : S_n(\omega)/n = k/n\} = \sum_{k\in A_n} C(n,k)\frac{1}{2^n}.$$

Since there are no more than $n+1$ terms in the sum,

$$\max_{k\in A_n} C(n,k)\frac{1}{2^n} \leq Q_n^{(1)}\{A\} \leq (n+1)\max_{k\in A_n} C(n,k)\frac{1}{2^n},$$

and since log is an increasing function

$$(1.11)\quad \max_{k\in A_n}\left[\frac{1}{n}\log\left(C(n,k)\frac{1}{2^n}\right)\right] \leq \frac{1}{n}\log Q_n^{(1)}\{A\} \leq \frac{\log(n+1)}{n} + \max_{k\in A_n}\left[\frac{1}{n}\log\left(C(n,k)\frac{1}{2^n}\right)\right].$$

Thus the asymptotic behavior of $Q_n^{(1)}\{A\}$ is governed by the asymptotic behavior of the largest summand in (1.10). Entropy arises by the following lemma.

Lemma I.3.2. *Uniformly in $k\in\{0, 1, \ldots, n\}$,*

$$\frac{1}{n}\log C(n,k) = -\frac{k}{n}\log\frac{k}{n} - \left(1-\frac{k}{n}\right)\log\left(1-\frac{k}{n}\right) + O\left(\frac{\log n}{n}\right) \qquad \text{as } n\to\infty.$$

Proof. Since $C(n,0) = C(n,n) = 1$ and $C(n,1) = C(n,n-1) = n$, the lemma holds for all $n \geq 1$ and $k = 0, 1, n-1, n$. A weak form of Stirling's approximation states that for all $n \geq 2$, $\log(n!) = n\log n - n + \beta_n$, where $|\beta_n| = O(\log n)$ [Problem I.8.1]. Hence for $2 \leq k \leq n-2$,

$$\frac{1}{n}\log C(n,k) = \log n - \frac{k}{n}\log k - \frac{n-k}{n}\log(n-k) + \frac{1}{n}(\beta_n - \beta_k - \beta_{n-k}).$$

Write

$$\log n = -\frac{k}{n}\log\frac{1}{n} - \frac{n-k}{n}\log\frac{1}{n}$$

and combine these terms with the other log terms to give

$$\frac{1}{n}\log C(n,k) = -\frac{k}{n}\log\frac{k}{n} - \left(1-\frac{k}{n}\right)\log\left(1-\frac{k}{n}\right) + \frac{1}{n}(\beta_n - \beta_k - \beta_{n-k}).$$

For $2 \leq k \leq n-2$, the last term can be bounded by $O(n^{-1}\log n)$ uniformly in k. This completes the proof. □

The lemma shows that

$$\frac{1}{n}\log\left(C(n,k)\frac{1}{2^n}\right) = \log\frac{1}{2} - \frac{k}{n}\log\frac{k}{n} - \left(1-\frac{k}{n}\right)\log\left(1-\frac{k}{n}\right) + O\left(\frac{\log n}{n}\right).$$

The first three terms are exactly $-I_\rho^{(1)}(k/n)$, where $I_\rho^{(1)}$ is defined in (1.9). Thus

$$\text{(1.12)}\quad \frac{1}{n}\log Q_n^{(1)}\left\{\frac{k}{n}\right\} = \frac{1}{n}\log\left(C(n,k)\frac{1}{2^n}\right) = -I_\rho^{(1)}\left(\frac{k}{n}\right) + O\left(\frac{\log n}{n}\right).$$

Since $n^{-1}\log n$ and $n^{-1}\log(n+1)$ both tend to zero, we have by (1.11)

$$\text{(1.13)}\quad \lim_{n\to\infty}\frac{1}{n}\log Q_n^{(1)}\{A\} = \lim_{n\to\infty}\max_{k\in A_n}\left(-I_\rho^{(1)}\left(\frac{k}{n}\right)\right) = -\lim_{n\to\infty}\min_{k\in A_n} I_\rho^{(1)}\left(\frac{k}{n}\right).$$

For each n the set $\{z\in[0,1]: z = k/n \text{ for some } k\in A_n\}$ is a subset of $A\cap[0,1]$. Since $I_\rho^{(1)}(z) = \infty$ for $z\notin[0,1]$, we conclude using Problem I.8.2 that

$$\lim_{n\to\infty}\frac{1}{n}\log Q_n^{(1)}\{A\} = -\min_{z\in A\cap[0,1]} I_\rho^{(1)}(z) = -\min_{z\in A} I_\rho^{(1)}(z). \qquad \square$$

I.4. Levels-1 and 2 for I.I.D. Random Variables with a Finite State Space

In general, the state space Γ equals $\{x_1, x_2, \ldots, x_r\}$, where $x_1 < x_2 < \cdots < x_r$ are real numbers. The exponential decay rates of the two probabilities in (1.7) are expressed in terms of a function called the relative entropy. Let $\mathscr{B}(\Gamma)$ denote the set of all subsets of Γ and $\mathscr{M}(\Gamma)$ the set of probability measures on $\mathscr{B}(\Gamma)$. Each $\nu\in\mathscr{M}(\Gamma)$ has the form $\nu = \sum_{i=1}^r \nu_i\delta_{x_i}$, where $\nu_i \geq 0$ and $\sum_{i=1}^r \nu_i = 1$. $\mathscr{M}(\Gamma)$ may be identified with the compact convex subset of $\mathbb{R}^r$ consisting of all vectors $\nu = (\nu_1, \ldots, \nu_r)$ which satisfy $\nu_i \geq 0$ and $\sum_{i=1}^r \nu_i = 1$. The *relative entropy*[2] of ν with respect to the measure $\rho = \sum_{i=1}^r \rho_i\delta_{x_i}$ $(\rho_i > 0)$ is defined by

$$I_\rho^{(2)}(\nu) = \sum_{i=1}^r \nu_i \log\frac{\nu_i}{\rho_i} \qquad \text{where } 0\log 0 = 0.$$

We have the following properties.

Proposition I.4.1. (a) *$I_\rho^{(2)}(\nu)$ is a convex function of $\nu \in \mathscr{M}(\Gamma)$.*

(b) *$I_\rho^{(2)}(\nu)$ measures the discrepancy between ν and ρ in the sense that $I_\rho^{(2)}(\nu) \geq 0$ with equality if and only if $\nu = \rho$. Thus $I_\rho^{(2)}(\nu)$ attains its infimum over $\mathscr{M}(\Gamma)$ at the unique measure $\nu = \rho$.*

Proof. (a) $I_\rho^{(2)}(\nu)$ equals $\sum_{i=1}^r k(\nu_i/\rho_i)\rho_i$, where $h(x)$ is the convex function $x \log x$, $x \geq 0$. Let μ and ν be probability measures on $\mathscr{B}(\Gamma)$. Then for $0 \leq \lambda \leq 1$

$$\begin{aligned} I_\rho^{(2)}(\lambda\mu + (1-\lambda)\nu) &= \sum_{i=1}^r h(\lambda\mu_i/\rho_i + (1-\lambda)\nu_i/\rho_i)\rho_i \\ &\leq \lambda \sum_{i=1}^r h(\mu_i/\rho_i)\rho_i + (1-\lambda)\sum_{i=1}^r h(\nu_i/\rho_i)\rho_i \\ &= \lambda I_\rho^{(2)}(\mu) + (1-\lambda) I_\rho^{(2)}(\nu). \end{aligned}$$

(b) For any $x \geq 0$, $x \log x \geq x - 1$ with equality iff $x = 1$. Hence

$$\frac{\nu_i}{\rho_i}\log\frac{\nu_i}{\rho_i} \geq \frac{\nu_i}{\rho_i} - 1 \tag{1.14}$$

with equality iff $\nu_i = \rho_i$. Multiplying this inequality by ρ_i and summing over i yields

$$I_\rho^{(2)}(\nu) = \sum_{i=1}^r \nu_i \log\frac{\nu_i}{\rho_i} \geq 0.$$

$I_\rho^{(2)}(\nu)$ equals 0 iff equality holds in (1.14) for each i; ν_i equals ρ_i for each i iff ν equals ρ. □

We single out an important special case of relative entropy.

Example I.4.2. If ρ is the uniform measure on $\Gamma = \{x_1, x_2, \ldots, x_r\}$ ($\rho_i = 1/r$ for each i), then $I_\rho^{(2)}(\nu) = \log r + \sum_{i=1}^r \nu_i \log \nu_i$. The quantity $H(\nu) = -\sum_{i=1}^r \nu_i \log \nu_i$ is called the *Shannon entropy* of ν.[3] Since $-\nu_i \log \nu_i \geq 0$, $H(\nu)$ is nonnegative. We show that $H(\nu)$ is a measure of the randomness in ν. By Proposition I.4.1, $H(\nu) = \log r - I_\rho^{(2)}(\nu) \leq \log r$; $H(\nu) = \log r$ iff $I_\rho^{(2)}(\nu) = 0$ and this holds iff each $\nu_i = \rho_i = 1/r$. Hence $H(\nu)$ attains its maximum value of $\log r$ iff ν equals the uniform measure ρ. The measure ρ is in a sense the most random probability measure on $\mathscr{B}(\Gamma)$. At the other extreme, $H(\nu)$ equals 0 iff one of the ν_i's, say $\nu_{i'}$, is 1 and the other ν_i's, $i \neq i'$, are 0. The corresponding measures $\delta_{x_{i'}}$ are the least random probability measures on $\mathscr{B}(\Gamma)$.

We now turn to the large deviation results. For each $\omega \in \Omega$, the empirical measure $L_n(\omega, \cdot) = n^{-1}\sum_{j=1}^n \delta_{X_j(\omega)}(\cdot)$ is a probability measure on $\mathscr{B}(\Gamma)$. Hence $L_n(\omega, \cdot)$ takes values in $\mathscr{M}(\Gamma)$. Let $Q_n^{(2)}$ be the P_ρ-distribution of L_n on $\mathscr{M}(\Gamma)$ and define the closed set

(1.15) $$A_2 = \{\nu \in \mathscr{M}(\Gamma) : \max_{i=1,\ldots,r} |\nu_i - \rho_i| \geq \varepsilon\}$$
$$\text{where } 0 < \varepsilon < \min_{i=1,\ldots,r} \{\rho_i, 1 - \rho_i\}.$$

The set A_2 is nonempty and $Q_n^{(2)}\{A_2\} = P_\rho\{\max_{i=1,\ldots,r}|L_{n,i} - \rho_i| \geq \varepsilon\}$ is positive for all sufficiently large n. Since A_2 does not contain ρ, $Q_n^{(2)}\{A_2\} \to 0$. According to Theorem I.4.3, $Q_n^{(2)}\{A_2\}$ decays exponentially and the decay rate is given in terms of the relative entropy $I_\rho^{(2)}(\nu)$. For this reason $I_\rho^{(2)}(\nu)$ is called the *level-2 entropy function*. For level-1, let $Q_n^{(1)}$ be the P_ρ-distribution of S_n/n on $\mathbb{R}$ and define the closed set*

(1.16) $$A_1 = \{z \in \mathbb{R} : |z - m_\rho| \geq \varepsilon\} \quad \text{where } 0 < \varepsilon < \min\{m_\rho - x_1, x_r - m_\rho\}.$$

The set $A_1 \cap [x_1, x_r]$ is nonempty and $Q_n^{(1)}\{A_1\} = P_\rho\{|S_n/n - m_\rho| \geq \varepsilon\}$ is positive for all sufficiently large n. Since A_1 does not contain m_ρ, $Q_n^{(1)}\{A_1\} \to 0$. According to Theorem I.4.3, $Q_n^{(1)}\{A_1\}$ decays exponentially and the decay rate is given in terms of a function $I_\rho^{(1)}$ calculated from $I_\rho^{(2)}(\nu)$ by a variational formula

(1.17) $$I_\rho^{(1)}(z) = \begin{cases} \min\{I_\rho^{(2)}(\nu) : \nu \in \mathscr{M}(\Gamma), \sum_{i=1}^r x_i \nu_i = z\} & \text{for } z \in [x_1, x_r], \\ \infty & \text{for } z \notin [x_1, x_r]. \end{cases}$$

$I_\rho^{(1)}$ is called the *level-1 entropy function*. It is well-defined, and it measures the discrepancy between z and m_ρ in the sense that $I_\rho^{(1)}(z) \geq 0$ with equality if and only if $z = m_\rho$. Thus, the point m_ρ is the unique minimum point of $I_\rho^{(1)}(z)$. In addition $I_\rho^{(1)}(z)$ is a continuous convex function of $z \in [x_1, x_r]$. These properties are proved in Sections VII.5 and VIII.3. In Section II.4 we give another formula for $I_\rho^{(1)}$ in terms of a Legendre–Fenchel transform [see (2.14)].

For coin tossing ($\Gamma = \{0, 1\}$, $\rho = \frac{1}{2}\delta_0 + \frac{1}{2}\delta_1$), formula (1.17) for $I_\rho^{(1)}$ reduces to (1.9). Indeed the only measure $\nu \in \mathscr{M}(\Gamma)$ which satisfies the constraint $\sum_{i=1}^2 x_i \nu_i = z \in [0, 1]$ is $\nu = (1 - z)\delta_0 + z\delta_1$. Hence by (1.17)

$$I_\rho^{(1)}(z) = I_\rho^{(2)}((1 - z)\delta_0 + z\delta_1) = (1 - z)\log(2(1 - z)) + z\log(2z)$$
$$\text{for } z \in [0, 1].$$

Formula (1.17), which relates the level-1 and level-2 entropy functions, is called a *contraction principle*. It will be seen to follow directly from (1.6), which expresses S_n/n as the mean of the empirical measure L_n. Here is the large deviation theorem for levels-1 and 2.

Theorem I.4.3.

(1.18) $$\lim_{n\to\infty} \frac{1}{n} \log Q_n^{(1)}\{A_1\} = \lim_{n\to\infty} \frac{1}{n} \log P_\rho\{|S_n/n - m_\rho| \geq \varepsilon\} = -\min_{z \in A_1} I_\rho^{(1)}(z),$$

*The point $m_\rho = \sum_{i=1}^r x_i\rho_i$ is in the open interval (x_1, x_r) since $\rho_i > 0$, $\sum_{i=1}^r \rho_i = 1$.

$$\lim_{n\to\infty}\frac{1}{n}\log Q_n^{(2)}\{A_2\} = \lim_{n\to\infty}\frac{1}{n}\log P_\rho\{\max_{i=1,\ldots,r}|L_{n,i}-\rho_i|\geq\varepsilon\} = -\min_{\nu\in A_2} I_\rho^{(2)}(\nu). \tag{1.19}$$

Since the set A_1 is closed and does not contain m_ρ, $\min_{z\in A_1} I_\rho^{(1)}(z) > I_\rho^{(1)}(m_\rho) = 0$. Similarly the minimum of $I_\rho^{(2)}$ over A_2 is positive. Hence both $Q_n^{(1)}\{A_1\}$ and $Q_n^{(2)}\{A_2\}$ converge to zero exponentially fast.

It is instructive to interpret Theorem I.4.3 with reference to the discussion in Section I.1. Think of $\{X_j; j\in\mathbb{Z}\}$ as giving the successive outcomes of a gambling game in an infinite number of plays of the game separated by a constant time interval. For level-1 the macrostates are all real numbers $z\in[x_1, x_r]$. These correspond to a macroscopic description of the game in terms of the expected value of the outcome of a single play. The microscopic n-sums are $\{S_n(\omega)/n\}$. The P_ρ-probability that S_n/n is close to z behaves for large n like $\exp(-nI_\rho^{(1)}(z))$ [see (1.12)]. $I_\rho^{(1)}$ is the entropy function and the mean $m_\rho = \sum_{i=1}^r x_i\rho_i$ is the equilibrium state. For level-2 the macrostates are all probability measures $\nu\in\mathcal{M}(\Gamma)$. Each ν is a candidate for the distribution of the r outcomes $x_1, \ldots, x_r$ in each play of the game. The microscopic n-sums are $\{L_n(\omega,\cdot)\}$. The P_ρ-probability that L_n is close to ν behaves for large n like $\exp(-nI_\rho^{(2)}(\nu))$ [see (1.21)]. $I_\rho^{(2)}$ is the entropy function and the measure ρ is the equilibrium state.

Proof of Theorem I.4.3. First consider level-2. For fixed n and ω and $i\in\{1,\ldots,r\}$, let k_i be the number of times x_i appears in the sequence $X_1(\omega), \ldots, X_n(\omega)$. Then $L_{n,i}(\omega) = k_i/n$, and $L_n(\omega,\cdot)$ is in A_2 if and only if $\mathbf{k} = (k_1, \ldots, k_r)$ is in the set

$$A_{2,n} = \left\{\mathbf{k} = (k_1,\ldots,k_r): k_i\in\{0,1,\ldots,n\}, \sum_{i=1}^r k_i = n, \max_{i=1,\ldots,r}\left|\frac{k_i}{n}-\rho_i\right|\geq\varepsilon\right\}.$$

For fixed $\mathbf{k}\in A_{2,n}$, define

$$C(n,\mathbf{k}) = \frac{n!}{k_1!k_2!\ldots k_r!} \quad\text{and}\quad \rho^{\mathbf{k}} = \rho_1^{k_1}\rho_2^{k_2}\ldots\rho_r^{k_r}.$$

There are $C(n,\mathbf{k})$ points $\omega = (\omega_1, \omega_2, \ldots, \omega_n)$ in the finite configuration space $\Omega_n = \Gamma^n$ for which $L_{n,i}(\omega) = k_i/n$ for each i. Let $\pi_n P_\rho$ be the finite product measure on Ω_n with identical one-dimensional marginals ρ. We have

$$Q_n^{(2)}\{A_2\} = \sum_{\mathbf{k}\in A_{2,n}} \pi_n P_\rho\{\omega\in\Omega_n: L_{n,i}(\omega) = k_i/n \text{ for each } i\} = \sum_{\mathbf{k}\in A_{2,n}} C(n,\mathbf{k})\rho^{\mathbf{k}}. \tag{1.20}$$

The next lemma is proved like Lemma I.3.2 [Problem I.8.1(b)].

Lemma I.4.4. *Uniformly in $\mathbf{k} = (k_1, \ldots, k_r)$,*

$$\frac{1}{n}\log C(n,\mathbf{k}) = -\sum_{i=1}^r \frac{k_i}{n}\log\frac{k_i}{n} + O\left(\frac{\log n}{n}\right) \quad \textit{as } n\to\infty.$$

The lemma implies that for each $\mathbf{k}$,

$$\frac{1}{n}\log(C(n,\mathbf{k})\rho^k)) = \sum_{i=1}^{r} \frac{k_i}{n}\left(\log \rho_i - \log\frac{k_i}{n}\right) + O\left(\frac{\log n}{n}\right).$$

Define the measure $\nu_{\mathbf{k}/n} = \sum_{i=1}^{r} (k_i/n)\delta_{x_i} \in \mathscr{M}(\Gamma)$. The sum in the last display is exactly $-I_\rho^{(2)}(\nu_{\mathbf{k}/n})$, where $I_\rho^{(2)}(\nu_{\mathbf{k}/n})$ is the relative entropy of $\nu_{\mathbf{k}/n}$ with respect to ρ. Since $L_{n,i}(\omega) = k_i/n$ for each i if and only if $L_n(\omega, \cdot) = \nu_{\mathbf{k}/n}$, we see that

$$\frac{1}{n}\log Q_n^{(2)}\{\nu_{\mathbf{k}/n}\} = -I_\rho^{(2)}(\nu_{\mathbf{k}/n}) + O\left(\frac{\log n}{n}\right). \tag{1.21}$$

In the sum (1.20) for $Q_n^{(2)}\{A_2\}$ there are no more than $(n+1)^r$ terms. As in the proof for coin tossing, we conclude that

$$\frac{1}{n}\log Q_n^{(2)}\{A_2\} = \max_{\mathbf{k}\in A_{2,n}} \{-I_\rho^{(2)}(\nu_{\mathbf{k}/n})\} + O\left(\frac{\log n}{n}\right) + O\left(\frac{\log(n+1)^r}{n}\right).$$

For each n the set $\{\nu \in \mathscr{M}(\Gamma)\colon \nu = \nu_{\mathbf{k}/n} \text{ for some } \mathbf{k}\in A_{2,n}\}$ is a subset of A_2. Problem I.8.2 yields (1.19):

$$\lim_{n\to\infty}\frac{1}{n}\log Q_n^{(2)}\{A_2\} = -\lim_{n\to\infty}\min_{\mathbf{k}\in A_{2,n}} I_\rho^{(2)}(\nu_{\mathbf{k},n}) = -\min_{\nu\in A_2} I_\rho^{(2)}(\nu).$$

The level-1 limit is proved by expressing it in terms of a level-2 limit. Since S_n/n equals $\sum_{i=1}^{r} x_i L_{n,i}$, S_n/n is in the set A_1 if and only if L_n is in the set of measures $B_2 = \{\nu \in M(\Gamma)\colon |\sum_{i=1}^{r} x_i\nu_i - m_\rho| \geq \varepsilon\}$. Hence $Q_n^{(1)}\{A_1\}$ equals $Q_n^{(2)}\{B_2\}$. The level-2 argument just given for the set A_2 can be easily modified for the set B_2, and we find

$$\lim_{n\to\infty}\frac{1}{n}\log Q_n^{(1)}\{A_1\} = \lim_{n\to\infty}\frac{1}{n}\log Q_n^{(2)}\{B_2\} = -\min_{\nu\in B_2} I_\rho^{(2)}(\nu).$$

We evaluate this minimum in two steps:

$$\begin{aligned}\min_{\nu\in B_2} I_\rho^{(2)}(\nu) &= \min_{z\in A_1\cap[x_1,x_r]}\ \min\left\{I_\rho^{(2)}(\nu)\colon \nu\in\mathscr{M}(\Gamma), \sum_{i=1}^{r} x_i\nu_i = z\right\}\\ &= \min_{z\in A_1\cap[x_1,x_r]} I_\rho^{(1)}(z).\end{aligned}$$

Since $I_\rho^{(1)}(z) = \infty$ for $z\notin[x_1,x_r]$, it follows that

$$\lim_{n\to\infty}\frac{1}{n}\log Q_n\{A_1\} = -\min_{z\in A_1} I_\rho^{(1)}(z). \qquad \square$$

I.5. Level-3: Empirical Pair Measure

Level-2 focuses on the empirical measure $L_n(\omega, \cdot)$, which is defined in terms of the empirical frequencies $\{L_{n,i}(\omega)\}$. We can generalize level-2 by considering the empirical frequencies of pairs of outcomes. Fix $\omega\in\Omega$ and $n\in\{2,3,\ldots\}$ and define $Y_\beta^{(n)}(\omega)$ to be the ordered pair $(X_\beta(\omega), X_{\beta+1}(\omega))$ if

$\beta \in \{1, 2, \ldots, n-1\}$ and to be the cyclic ordered pair $(X_n(\omega), X_1(\omega))$ if $\beta = n$. For each subset $\{x_i, x_j\}$ of Γ^2, let $M_{n,i,j}(\omega)$ be $1/n$ times the number of pairs $\{Y_\beta^{(n)}(\omega)\}$ for which $Y_\beta^{(n)}(\omega) = (x_i, x_j)$; thus

$$M_{n,i,j}(\omega) = \frac{1}{n} \sum_{\beta=1}^{n} \delta_{Y_\beta^{(n)}(\omega)}\{x_i, x_j\}.$$

For each ω, the numbers $\{M_{n,i,j}(\omega)\}$ define a probability measure $M_n(\omega, \cdot)$ on the set of all subsets of Γ^2. For A a subset of Γ^2,

$$(1.22) \qquad M_n(\omega, A) = \sum_{\{x_i, x_j\} \in A} M_{n,i,j}(\omega) = \frac{1}{n} \sum_{\beta=1}^{n} \delta_{Y_\beta^{(n)}(\omega)}\{A\}.$$

The measure $M_n(\omega, \cdot)$ is called the *empirical pair measure* corresponding to $X_1(\omega), \ldots, X_n(\omega)$. This measure is consistent with the empirical measure $L_n(\omega, \cdot)$ in the sense that both of the one-dimensional marginals of $M_n(\omega, \cdot)$ equal $L_n(\omega, \cdot)$:

$$(1.23) \quad L_{n,i}(\omega) = \sum_{j=1}^{r} M_{n,i,j}(\omega) = \sum_{k=1}^{r} M_{n,k,i}(\omega) \qquad \text{for each } i = 1, \ldots, r.$$

In fact, since $M_n(\omega, \cdot)$ considers the cyclic pair $Y_n^{(n)}(\omega)$, the number of times x_i appears in the sequence $X_1(\omega), \ldots, X_n(\omega)$ equals the number of times x_i appears as a left-hand member of a pair $Y_\beta^{(n)}(\omega)$. This gives the first equality in (1.23), and the second is proved similarly.

With respect to P_ρ the asymptotic behavior of $M_{n,i,j}(\omega)$ is determined by the ergodic theorem [Theorem A.9.3]. Since

$$(1.24) \quad M_{n,i,j}(\omega) = \frac{1}{n} \sum_{\beta=1}^{n-1} \delta_{X_\beta(\omega)}\{x_i\} \cdot \delta_{X_{\beta+1}(\omega)}\{x_j\} + \frac{1}{n} \delta_{X_n(\omega)}\{x_i\} \cdot \delta_{X_1(\omega)}\{x_j\},$$

$\lim_{n\to\infty} M_{n,i,j}(\omega)$ equals the limit of the sum. Since P_ρ is ergodic,

$$(1.25) \quad \begin{aligned} \lim_{n\to\infty} M_{n,i,j}(\omega) &= \int_\Omega \delta_{X_1(\omega)}\{x_i\} \delta_{X_2(\omega)}\{x_j\} P_\rho(d\omega) \\ &= P_\rho\{\omega \in \Omega : X_1(\omega) = x_i, X_2(\omega) = x_j\} = \rho_i \rho_j \qquad P_\rho\text{-a.s.} \end{aligned}$$

We now formulate a large deviation problem connected with the limit (1.25). Let $\mathscr{B}(\Gamma^2)$ denote the set of all subsets of Γ^2 and $\mathscr{M}_s(\Gamma^2)$ the set of probability measures on $\mathscr{B}(\Gamma^2)$ with equal one-dimensional marginals. As we have seen, $M_n(\omega, \cdot)$ takes values $\mathscr{M}_s(\Gamma^2)$ for each ω. Any $\tau \in \mathscr{M}_s(\Gamma^2)$ has the form $\tau = \sum_{i,j=1}^r \tau_{ij} \delta_{\{x_i, x_j\}}$, where $\tau_{ij} \geq 0$, $\sum_{i,j=1}^r \tau_{ij} = 1$, and $\sum_{i=1}^r \tau_{ij} = \sum_{i=1}^r \tau_{ki}$ for each i. $\mathscr{M}_s(\Gamma^2)$ may be identified with the compact convex subset of $\mathbb{R}^{r^2}$ consisting of all vectors $\tau = \{\tau_{ij}; i,j = 1, \ldots, r\}$ which satisfy $\tau_{ij} \geq 0$, $\sum_{i,j=1}^r \tau_{ij} = 1$, and $\sum_{j=1}^r \tau_{ij} = \sum_{k=1}^r \tau_{ki}$ for each i.

Let $Q_{n,2}^{(3)}$ be the P_ρ-distribution of M_n on $\mathscr{M}_s(\Gamma^2)$ and define the closed set

$$(1.26) \quad \begin{aligned} A_3 = \{\tau \in \mathscr{M}_s(\Gamma^2) : \max_{i,j=1,\ldots,r} |\tau_{ij} - \rho_i \rho_j| \geq \varepsilon\}, \\ \text{where } 0 < \varepsilon < \min_{i,j=1,\ldots,r} \{\rho_i \rho_j, 1 - \rho_i \rho_j\}. \end{aligned}$$

The set A_3 is nonempty and $Q_{n,2}^{(3)}\{A_3\} = P_\rho\{\max_{i,j=1,\ldots,r}|M_{n,i,j} - \rho_i\rho_j| \geq \varepsilon\}$ is positive for all sufficiently large n. Let $\pi_2 P_\rho$ be the product measure in $\mathscr{M}_s(\Gamma^2)$ with one-dimensional marginals ρ $(\pi_2 P_\rho\{x_i, x_j\} = \rho_i\rho_j)$. Since A_3 does not contain $\pi_2 P_\rho$, the ergodic limit (1.25) implies that $Q_{n,2}^{(3)}\{A_3\} \to 0$. A level-3 large deviation problem is to determine the decay rate of this probability. In the next section, we will consider other level-3 problems which involve the empirical frequencies of strings $\{x_{i_1}, x_{i_2}, \ldots, x_{i_k}\}$ of arbitrary length $k = 3, 4, \ldots$.

The probabilities $Q_{n,2}^{(3)}\{A_3\}$ decay exponentially and the decay rate is given in terms of a function $I_{\rho,2}^{(3)}$ which is a natural extension of the level-2 function $I_\rho^{(2)}$. The latter determines the decay rate of level-2 probabilities. For $\tau \in \mathscr{M}_s(\Gamma^2)$, set $(v_\tau)_i = \sum_{j=1}^r \tau_{ij}$ and define

$$I_{\rho,2}^{(3)}(\tau) = \sum \tau_{ij} \log \frac{\tau_{ij}}{(v_\tau)_i \rho_j}, \tag{1.27}$$

where the sum runs over all i and j for which $(v_\tau)_i > 0$. $I_{\rho,2}^{(3)}(\tau)$ is well-defined $(0 \log 0 = 0)$ and equals the relative entropy of τ with respect to $\{(v_\tau)_i\rho_j\}$. We have the following properties [Problems IX.6.1–IX.6.3]. $I_{\rho,2}^{(3)}(\tau)$ is a convex function of τ. $I_{\rho,2}^{(3)}(\tau)$ measures the discrepancy between $\tau \in \mathscr{M}_s(\Gamma^2)$ and $\pi_2 P_\rho$ in the sense that $I_{\rho,2}^{(3)}(\tau) \geq 0$ with equality if and only if $\tau = \pi_2 P_\rho$. Thus the measure $\pi_2 P_\rho$ is the unique minimum point of $I_{\rho,2}^{(3)}$ on $\mathscr{M}_s(\Gamma^2)$.

Here is a level-3 large deviation theorem for the empirical pair measure. We sketch the proof in the case where Γ consists of two points $(r = 2)$.

Theorem I.5.1.

$$\begin{aligned} \lim_{n\to\infty} \frac{1}{n} \log Q_{n,2}^{(3)}\{A_3\} &= \lim_{n\to\infty} \frac{1}{n} \log P_\rho\{\max_{i,j=1,\ldots,r} |M_{n,i,j} - \rho_i\rho_j| \geq \varepsilon\} \\ &= -\min_{\tau\in A_3} I_{\rho,2}^{(3)}(\tau). \end{aligned} \tag{1.28}$$

Since the set A_3 is closed and does not contain $\pi_2 P_\rho$, $\min_{\tau\in A_3} I_{\rho,2}^{(3)}(\tau) > I_{\rho,2}^{(3)}(\pi_2 P_\rho) = 0$. Hence $Q_{n,2}^{(3)}\{A_3\}$ converges to zero exponentially fast.

Proof for $r = 2$. For fixed n and ω, define $N_{ij} = nM_{n,i,j}(\omega)$. $M_n(\omega, \cdot)$ is in A_3 if and only if $N = \{N_{ij}; i,j = 1, 2\}$ is in the set

$$\begin{aligned} A_{3,n} = \Bigg\{ & N = \{N_{ij}; i,j = 1,2\}\colon N_{ij} \in \{0, 1, \ldots, n\},\ \sum_{i,j=1}^{2} N_{ij} = n, \\ & \sum_{j=1}^{2} N_{ij} = \sum_{k=1}^{2} N_{ki} \quad \text{for each } i, \quad \max_{i,j=1,2} |N_{ij}/n - \rho_i\rho_j| \geq \varepsilon \Bigg\}. \end{aligned}$$

Now fix $N \in A_{3,n}$. Let $k_i = \sum_{j=1}^2 N_{ij}$ and define $\gamma(n, N)$ to be the number of points ω in the finite configuration space $\Omega_n = \Gamma^n$ for which $M_{n,i,j}(\omega) = N_{ij}/n$ for each i and j. Then

$$\begin{aligned} Q_{n,2}^{(3)}\{A_3\} &= \sum_{N\in A_{3,n}} \pi_n P_\rho\{\omega \in \Omega_n\colon M_{n,i,j}(\omega) = N_{ij}/n \quad \text{for each } i \text{ and } j\} \\ &= \sum_{N\in A_{3,n}} \gamma(n, N)\rho^{\mathbf{k}}, \end{aligned}$$

where $\rho^{\mathrm{k}} = \rho_1^{k_1}\rho_2^{k_2}$. The asymptotics of $\gamma(n, N)$ are given by the next lemma. Define $D_n = \{N: \gamma(n, N) > 0\}$.

Lemma I.5.2. *If* $N_{12} > 0$, *then* $N \in D_n$. *Uniformly in* $N \in D_n$,

$$(1.29)\quad \frac{1}{n}\log\gamma(n, N) = \sum_{i=1}^{2}\frac{k_i}{n}\log\frac{k_i}{n} - \sum_{i,j=1}^{2}\frac{N_{ij}}{n}\log\frac{N_{ij}}{n} + O\left(\frac{\log n}{n}\right) \quad \textit{as } n \to \infty.$$

Define the measure $\tau_{N/n} = \sum_{i,j=1}^{2}(N_{ij}/n)\delta_{\{x_i, x_j\}} \in \mathcal{M}_s(\Gamma^2)$. Since $M_{n,i,j}(\omega) = N_{ij}/n$ for each i and j if and only if $M_n(\omega, \cdot) = \tau_{N/n}$, the lemma implies that for $N \in D_n$

$$\frac{1}{n}\log Q_{n,2}^{(3)}\{\tau_{N/n}\} = \frac{1}{n}\log(\gamma(n, N)\rho^{\mathrm{k}}) = -I_{\rho,2}^{(3)}(\tau_{N/n}) + O\left(\frac{\log n}{n}\right).$$

As in the proof of Theorem I.4.3, we have

$$\lim_{n\to\infty}\frac{1}{n}\log Q_{n,2}^{(3)}\{A_3\} = -\lim_{n\to\infty}\min_{N\in A_{3,n}\cap D_n} I_{\rho,2}^{(3)}(\tau_{N/n}).$$

For each n the set $\{\tau \in \mathcal{M}_s(\Gamma^2): \tau = \tau_{N/n}$ for some $N \in A_{3,n} \cap D_n\}$ is a subset of A_3. Problem I.8.2 yields (1.28):

$$\lim_{n\to\infty}\frac{1}{n}\log Q_{n,2}^{(3)}\{A_3\} = -\min_{\tau\in A_3} I_{\rho,2}^{(3)}(\tau). \qquad \square$$

Formula (1.31) below is due to J. K. Percus (private communication).

Proof of Lemma I.5.2. Suppose that $\Gamma = \{x_1, x_2\}$. For fixed $\omega \in \Omega_n$, let $N_{ij} = nM_{n,i,j}(\omega)$ for $i, j = 1, 2$. The point ω has the form $(x_{i_1}, x_{i_2}, \ldots, x_{i_n})$, where each $i_j = 1$ or 2. We introduce the product $B(\omega) = B_{i_1 i_2}B_{i_2 i_3}\cdots B_{i_{n-1}i_n}B_{i_n i_1}$, where B_{11}, B_{12}, B_{21}, B_{22} are four positive real variables. In the product defining $B(\omega)$, each B_{ij} appears exactly N_{ij} times, so that $B(\omega) = B_{11}^{N_{11}}B_{12}^{N_{12}}B_{21}^{N_{21}}B_{22}^{N_{22}}$. The sum of all $B(\omega)$ for $\omega \in \Omega_n$ can be written in terms of the 2×2 matrix $B = \{B_{ij}; i,j = 1, 2\}$. In fact,*

$$(1.30)\quad \begin{aligned}\operatorname{Tr} B^n &= \sum_{i_1,\ldots,i_n=1}^{2} B_{i_1 i_2}B_{i_2 i_3}\cdots B_{i_{n-1}i_n}B_{i_n i_1}\\ &= \sum_{\omega\in\Omega_n} B(\omega) = \sum \gamma(n, N)B_{11}^{N_{11}}B_{12}^{N_{12}}B_{21}^{N_{21}}B_{22}^{N_{22}},\end{aligned}$$

where the last sum runs over all $N = \{N_{ij}; i,j = 1, 2\}$ such that $N_{ij} \in \{0, 1, \ldots, n\}$, $\sum_{i,j=1}^{2} N_{ij} = n$, and $\sum_{j=1}^{2} N_{ij} = \sum_{k=1}^{2} N_{ki}$ for each i. If $0 < B_{11} + B_{12} + B_{21} + B_{22} < 1$, then the sum $\sum_{n=0}^{\infty} B^n$ converges and equals the inverse matrix $(I - B)^{-1}$. We find

$$\begin{aligned}\sum_{n=0}^{\infty}\operatorname{Tr} B^n &= \operatorname{Tr}\sum_{n=0}^{\infty} B^n = \operatorname{Tr}(I - B)^{-1}\\ &= \frac{2 - B_{11} - B_{22}}{1 - (B_{11} + B_{22} + B_{12}B_{21} - B_{11}B_{22})}.\end{aligned}$$

*Tr denotes trace.

The last term can be expanded in a power series in B_{11}, B_{12}, B_{21}, B_{22}. Comparing this series with (1.30), one calculates [Problem I.8.5(a)]

$$\gamma(n, N) = \frac{k_1!k_2!}{N_{11}!N_{12}!N_{21}!N_{22}!}\frac{nN_{12}}{k_1k_2} \quad \text{for } k_i = N_{i1} + N_{i2} \geq 1 \ (i = 1, 2). \tag{1.31}$$

Formula (1.29) follows from Stirling's approximation. □

I.6. Level-3: Empirical Process

The previous sections have considered three levels of large deviations. Level-1 studies S_n/n, level-2 the empirical measure L_n, and level-3 the empirical pair measure M_n. For each level there is an entropy function $I_\rho^{(1)}$, $I_\rho^{(2)}$, and $I_{\rho,2}^{(3)}$, respectively. In this section we formulate another level-3 problem which includes as special cases all the results in the previous sections. Associated with this problem is another level-3 entropy function $I_\rho^{(3)}$. The three entropy functions encountered already can be obtained from $I_\rho^{(3)}$ by contraction principles.

A subset Σ of Ω is said to be a *cylinder set* if it has the form

$$\Sigma = \{\omega \in \Omega : (\omega_{m+1}, \ldots, \omega_{m+k}) \in F\}, \tag{1.32}$$

where m and k are integers with $k \geq 1$ and F is a subset of Γ^k. For now, we consider cylinder sets of the form

$$\Sigma = \{\omega \in \Omega : \omega_{m+1} = x_{i_1}, \omega_{m+2} = x_{i_2}, \ldots, \omega_{m+k} = x_{i_k}\},$$

where $x_{i_1}, x_{i_2}, \ldots, x_{i_k}$ are elements of Γ (not necessarily distinct). We will define a random quantity $R_n(\omega, \cdot)$ which for each ω is a strictly stationary probability measure on Ω. $R_n(\omega, \Sigma)$ gives the empirical frequency of the string $\{x_{i_1}, x_{i_2}, \ldots, x_{i_k}\}$ in the sequence $X_1(\omega), X_2(\omega), \ldots, X_n(\omega)$. If $k = 1$, then $R_n(\omega, \Sigma)$ reduces to the empirical frequency $L_{n,i_1}(\omega)$ while if $k = 2$, then it reduces to the empirical pair frequency $M_{n,i_1,i_2}(\omega)$. In general, with respect to the measure P_ρ, $R_n(\omega, \Sigma)$ converges almost surely to the value $P_\rho\{\Sigma\} = \rho_{i_1}\rho_{i_2} \cdots \rho_{i_k}$ as $n \to \infty$. A level-3 large deviation problem is to determine the decay rate of the P_ρ-probability that $R_n(\omega, \Sigma)$ differs from $P_\rho\{\Sigma\}$ for finitely many cylinder sets Σ. This level-3 problem extends the level-2 problem, which studies fluctuations of $L_n(\omega, \cdot)$ away from the measure ρ, as well as the level-3 pair problem, which studies fluctuations of $M_n(\omega, \cdot)$ away from the product measure $\pi_2 P_\rho$. We will see that level-3 large deviation probabilities decay exponentially and that the decay rate is determined by a function $I_\rho^{(3)}$ which is a natural extension of $I_\rho^{(2)}$ and $I_{\rho,2}^{(3)}$.

We first define strict stationarity. The set Γ is topologized by the discrete topology and the set $\Omega = \Gamma^{\mathbb{Z}}$ by the product topology. The σ-field generated by the open sets of the product topology is called the *Borel σ-field* of Ω

and is denoted by $\mathscr{B}(\Omega)$. $\mathscr{B}(\Omega)$ coincides with the σ-field generated by the cylinder sets Σ in (1.32) [Propositions A.3.2 and A.3.5(b)]. Let T be the mapping from Ω onto Ω defined by

$$(T\omega)_j = \omega_{j+1} \qquad \text{for } j \in \mathbb{Z}.$$

T is called the *shift mapping* on Ω. For each $B \in \mathscr{B}(\Omega)$, $T^{-1}B$ is also in $\mathscr{B}(\Omega)$ so that T is measurable. A probability measure P on $\mathscr{B}(\Omega)$ is said to be *strictly stationary*, or *translation invariant*, if the shift T preserves P: $P\{B\} = P\{T^{-1}B\}$ for all $B \in \mathscr{B}(\Omega)$ or equivalently $P\{\Sigma\} = P\{T^{-1}\Sigma\}$ for all cylinder sets Σ. We denote by $\mathscr{M}_s(\Omega)$ the set of strictly stationary probability measures on $\mathscr{B}(\Omega)$.

Example I.6.1. (a) [See Example A.7.3(a)] *Infinite product measure.* Let ν be a probability measure on the set $\Gamma = \{x_1, x_2, \ldots, x_r\}$. A cylinder set of the form $\Sigma = \{\omega \in \Omega : \omega_{m+1} \in F_1, \ldots, \omega_{m+k} \in F_k\}$, $F_1, \ldots, F_k$ subsets of Γ, is called a *product cylinder set*. The class of product cylinder sets generates $\mathscr{B}(\Omega)$. For such a set, we define $P\{\Sigma\} = \prod_{j=1}^{k} \nu\{F_j\}$. Clearly $P\{T^{-1}\Sigma\} = P\{\Sigma\}$. P can be extended uniquely to a strictly stationary probability measure on $\mathscr{B}(\Omega)$. The extension is the infinite product measure with identical one-dimensional marginals ν and is written P_ν.

(b) [See Example A.7.3(b)] *Markov chain.* Let ν be a probability measure on the set $\Gamma = \{x_1, x_2, \ldots, x_r\}$ and let $\{\gamma_{ij}\}$ be a non-negative $r \times r$ matrix with row sums 1. Assume that $\sum_{i=1}^{r} \nu_i \gamma_{ij} = \nu_j$ for each j. Let Σ be the cylinder set $\{\omega \in \Omega : (\omega_{m+1}, \ldots, \omega_{m+k}) \in F\}$, where F is a subset of Γ^k. We define $P\{\Sigma\} = \nu\{F\}$ for $k = 1$ and

$$P\{\Sigma\} = \sum_{(x_{i_1}, \ldots, x_{i_k}) \in F} \nu_{i_1} \gamma_{i_1 i_2} \cdots \gamma_{i_{k-1} i_k} \qquad \text{for } k \geq 2.$$

Clearly $P\{T^{-1}\Sigma\} = P\{\Sigma\}$. P can be extended uniquely to a strictly stationary probability measure on $\mathscr{B}(\Omega)$. The extension is a Markov chain with transition matrix $\{\gamma_{ij}\}$; ν is an invariant measure for the chain.

We now state a level-3 large deviation problem. Given a positive integer n, repeat the sequence $(X_1(\omega), X_2(\omega), \ldots, X_n(\omega))$ periodically into a doubly infinite sequence, obtaining a point

$$X(n, \omega) = (\ldots X_n(\omega), X_1(\omega), X_2(\omega), \ldots, X_n(\omega), \\ X_1(\omega), X_2(\omega), \ldots, X_n(\omega), X_1(\omega), \ldots)$$

in Ω; $(X(n, \omega))_1 = X_1(\omega)$, $(X(n, \omega))_2 = X_2(\omega)$, etc. For each $\omega \in \Omega$, define a probability measure on $\mathscr{B}(\Omega)$ by

$$R_n(\omega, \cdot) = \frac{1}{n} \sum_{k=0}^{n-1} \delta_{T^k X(n, \omega)}(\cdot), \tag{1.33}$$

where T^0 is the identity mapping and $T^k = T(T^{k-1})$ for $k = 2, \ldots, n-1$. For each Borel subset B of Ω, $R_n(\omega, B)$ is the relative frequency with which $X(n,\omega)$, $TX(n,\omega)$, $\ldots$, $T^{n-1}X(n,\omega)$ is in B. Since $X(n,\omega)$ is periodic of period n, $R(n,\cdot)$ is for each ω a strictly stationary probability measure. It is called the *empirical process* corresponding to $X_1(\omega)$, $\ldots$, $X_n(\omega)$.

Let k be a positive integer less than n. In order to interpret formula (1.33), define $Y_\beta^{(n)}(\omega)$ to be the k-tuple $(X_\beta(\omega), X_{\beta+1}(\omega), \ldots, X_{\beta+k-1}(\omega))$ if $\beta \in \{1, 2, \ldots, n-k+1\}$ and to be the cyclic k-tuple $(X_\beta(\omega), \ldots, X_n(\omega), X_1(\omega), \ldots, X_{\beta+k-1-n}(\omega))$ if $\beta \in \{n-k+2, \ldots, n\}$. If Σ is the cylinder set $\{\bar{\omega} \in \Omega: \bar{\omega}_1 = x_{i_1}, \bar{\omega}_2 = x_{i_2}, \ldots, \bar{\omega}_k = x_{i_k}\}$,* then

$$R_n(\omega, \Sigma) = \frac{1}{n} \sum_{\beta=1}^{n} \delta_{Y_\beta^{(n)}(\omega)}\{x_{i_1}, x_{i_2}, \ldots, x_{i_k}\}.$$

This is the empirical frequency of the string $\{x_{i_1}, x_{i_2}, \ldots, x_{i_k}\}$ in the sequence $X_1(\omega)$, $\ldots$, $X_n(\omega)$ (with cyclic counting). The contribution of the cyclic terms $Y_\beta^{(n)}(\omega)$, $\beta \in \{n-k+2, \ldots, n\}$, to $R_n(\omega, \Sigma)$ is $o(n)$ as $n \to \infty$.

$R_n(\omega, \cdot)$ is a natural generalization of the empirical measure $L_n(\omega, \cdot)$ and of the empirical pair measure $M_n(\omega, \cdot)$. If Σ is the one-dimensional cylinder set $\{\bar{\omega} \in \Omega: \bar{\omega}_1 = x_i\}$, then

$$R_n(\omega, \Sigma) = L_n(\omega, \{x_i\}) = L_{n,i}(\omega). \tag{1.34}$$

Thus $L_n(\omega, \cdot)$ is the one-dimensional marginal of $R_n(\omega, \cdot)$. If Σ is the two-dimensional cylinder set $\{\bar{\omega} \in \Omega: \bar{\omega}_1 = x_i, \bar{\omega}_2 = x_j\}$, then

$$R_n(\omega, \Sigma) = M_n(\omega, \{x_i, x_j\}) = M_{n,i,j}(\omega). \tag{1.35}$$

Thus $M_n(\omega, \cdot)$ is the two-dimensional marginal of $R_n(\omega, \cdot)$.

If Σ is any cylinder set, then the ergodic theorem implies that

$$\lim_{n\to\infty} R_n(\omega, \Sigma) = P_\rho\{\Sigma\} \qquad P_\rho\text{-a.s.}$$

[Theorems A.9.2(c) and Corollary A.9.8]. Let $\Sigma_1, \ldots, \Sigma_N$ be cylinder sets such that $0 < P_\rho\{\Sigma_k\} < 1$, $k = 1, \ldots, N$, and define the closed set†

$$B_3 = \{P \in \mathscr{M}_s(\Omega): \max_{k=1,\ldots,N} |P\{\Sigma_k\} - P_\rho\{\Sigma_k\}| \geq \varepsilon\}. \tag{1.36}$$

Let $Q_n^{(3)}$ be the P_ρ-distribution of R_n on $\mathscr{M}_s(\Omega)$. For all sufficiently small $\varepsilon > 0$ the set B_3 is nonempty, and for all sufficiently large n

$$Q_n^{(3)}\{B_3\} = P_\rho\{\omega \in \Omega: \max_{k=1,\ldots,N} |R_n(\omega, \Sigma_k) - P_\rho\{\Sigma_k\}| \geq \varepsilon\} > 0.$$

Since B_3 does not contain P_ρ, the ergodic theorem implies that $Q_n^{(3)}\{B_3\} \to 0$. In fact, the probabilities decay exponentially, and the decay rate is given in terms of a function $I_\rho^{(3)}$, which we now define.

Let α be a positive integer and π_α the projection of $\Gamma^{\mathbb{Z}}$ onto Γ^α defined by

* We use $\bar{\omega}$ since ω labels the empirical process.

† The topology on $M_s(\Omega)$ is the topology of weak convergence [Sections A.8–A.9].

$\pi_\alpha\omega = (\omega_1, \ldots, \omega_\alpha)$. If P is a measure in $\mathscr{M}_s(\Omega)$, then define a probability measure $\pi_\alpha P$ on $\mathscr{B}(\Gamma^\alpha)$ by requiring

$$\pi_\alpha P\{F\} = P\{\pi_\alpha^{-1}F\} = P\{\omega : (\omega_1, \ldots, \omega_\alpha) \in F\}$$

for subsets F of Γ^α. The measure $\pi_\alpha P$ is called the *α-dimensional marginal* of P. We consider the quantity

$$I^{(2)}_{\pi_\alpha P_\rho}(\pi_\alpha P) = \sum_{\omega \in \Gamma^\alpha} \pi_\alpha P\{\omega\} \log \frac{\pi_\alpha P\{\omega\}}{\pi_\alpha P_\rho\{\omega\}},$$

which is the relative entropy of $\pi_\alpha P$ with respect to $\pi_\alpha P_\rho$. In Chapter IX, we prove that the limit

$$I^{(3)}_\rho(P) = \lim_{\alpha \to \infty} \frac{1}{\alpha} I^{(2)}_{\pi_\alpha P_\rho}(\pi_\alpha P) \tag{1.37}$$

exists, that $I^{(3}_\rho (P)$ is an affine function of $P \in \mathscr{M}_s(\Omega)$, and that $I^{(3)}_\rho(P) \geq 0$ with equality if and only if $P = P_\rho$. Thus $I^{(3)}_\rho(P)$ measures the discrepancy between P and P_ρ. It is called the *mean relative entropy* of P with respect to P_ρ. Here are some examples of mean relative entropy.

Example I.6.2. Let ρ be the uniform measure on $\Gamma = \{x_1, x_2, \ldots, x_r\}$ ($\rho_i = 1/r$ for each i). For $P \in \mathscr{M}_s(\Omega)$

$$I^{(2)}_{\pi_\alpha P_\rho}(\pi_\alpha P) = \alpha \log r + \sum_{\omega \in \Gamma^\alpha} \pi_\alpha P\{\omega\} \cdot \log \pi_\alpha P\{\omega\}$$

since $\pi_\alpha P_\rho\{\omega\} = r^{-\alpha}$ for each $\omega \in \Gamma^\alpha$. According to Note 2 of Chapter IX, the limit

$$h(P) = -\lim_{\alpha \to \infty} \frac{1}{\alpha} \sum_{\omega \in \Gamma^\alpha} \pi_\alpha P\{\omega\} \log \pi_\alpha P\{\omega\}$$

exists; $h(P)$ is called the *mean entropy*[4] of P. It follows that the limit $I^{(3)}_\rho(P)$ in (1.37) exists and $I^{(3)}_\rho(P) = \log r - h(P)$. By properties of $I^{(3)}_\rho(P)$ mentioned above, $h(P) \leq \log r$ and $h(P) = \log r$ if and only if $P = P_\rho$. Mean entropy $h(P)$ generalizes Shannon entropy, which was discussed in Example I.4.2. Accordingly, $h(P)$ can be interpreted as a measure of the randomness in P per unit time.

(a) For the infinite product measure P_ν [Example I.6.1(a)], $I^{(3)}_\rho(P_\nu) = \log r - h(P_\nu)$, where $h(P_\nu) = -\sum_{i=r}^{r} \nu_i \log \nu_i$. The latter is the Shannon entropy of ν, $H(\nu)$.

(b) Let P be a Markov chain with transition matrix $\{\gamma_{ij}\}$ and invariant measure ν [Example I.6.1 (b)]. Then $I^{(3)}_\rho(P) = \log r - h(P)$, where $h(P) = -\sum_{i,j=1}^{r} \nu_i \gamma_{ij} \log \gamma_{ij}$.

For any measure ρ in $\mathscr{M}(\Gamma)$ with each $\rho_i > 0$ and for any $P \in \mathscr{M}_s(\Omega)$, one can express $I^{(3)}_\rho(P)$ as an expectation involving relative entropy [see (2.20)]. Here is the level-3 large deviation theorem. It will be proved in Chapter IX.

Theorem I.6.3. *For all sufficiently small* $\varepsilon > 0$

$$\lim_{n\to\infty}\frac{1}{n}\log Q_n^{(3)}\{B_3\} = \lim_{n\to\infty}\frac{1}{n}\log P_\rho\{\omega\in\Omega: \max_{k=1,\ldots,N}|R_n(\omega,\Sigma_k) - P_\rho\{\Sigma_k\}| \geq \varepsilon\}$$

$$(1.38) \qquad = -\min_{P\in B_3} I_\rho^{(3)}(P).$$

Since the set B_3 *is closed and does not contain* P_ρ, $\min_{P\in B_3} I_\rho^{(3)}(P) > I_\rho^{(3)}(P_\rho) = 0$. *Hence* $Q_n^{(3)}\{B_3\}$ *converges to zero exponentially fast.*

The theorem can be interpreted as follows. For level-3, the macrostates are all strictly stationary probability measures $P\in\mathscr{M}_s(\Omega)$. Each P is a candidate for describing the probabilistic structure of a gambling game on Ω [see page 16]. The microscopic n-sums are the empirical processes $\{R_n(\omega,\cdot)\}$. The P_ρ-probability that $R_n(\omega,\cdot)$ is close to $P\in\mathscr{M}_s(\Omega)$ behaves for large n like $\exp(-nI_\rho^{(3)}(P))$. $I_\rho^{(3)}$ is the entropy function and the measure P_ρ is the equilibrium state.

We end this chapter by discussing the relationship between Theorem I.6.3 and the previous large deviation theorems. Theorem I.5.1 treated the empirical pair measure $M_n(\omega,\cdot)$. If P is a measure in $\mathscr{M}_s(\Omega)$, then the two-dimensional marginal $\pi_2 P$ is a probability measure on Γ^2, and since P is strictly stationary, $\pi_2 P$ has equal one-dimensional marginals. Hence $\pi_2 P$ belongs to $\mathscr{M}_s(\Gamma^2)$. In the set B_3 in (1.36), let $\{\Sigma_k\}$ run through all two-dimensional cylinder sets $\{\bar\omega\in\Omega: \bar\omega_1 = x_i, \bar\omega_2 = x_j\}$, $i,j = 1,\ldots,r$. Then

$$\begin{aligned}\pi_2 B_3 &= \{\tau\in\mathscr{M}_s(\Gamma^2): \tau = \pi_2 P \text{ for some } P\in B_3\}\\ &= \{\tau\in\mathscr{M}_s(\Gamma^2): \max_{i,j=1,\cdots,r}|\tau_{ij} - \rho_i\rho_j| \geq \varepsilon\}.\end{aligned}$$

The latter is the set A_3 in (1.26). Since $\pi_2 R_n(\omega,\cdot) = M_n(\omega,\cdot)$ for each ω, it follows that

$$P_\rho\{\omega\in\Omega: M_n(\omega,\cdot)\in A_3\} = P_\rho\{\omega\in\Omega: R_n(\omega,\cdot)\in B_3\}.$$

By Theorem I.5.1 the decay rate of the first probability is $-\min_{\tau\in A_3} I_{\rho,2}^{(3)}(\tau)$. This must equal the decay rate of $P_\rho\{\omega\in\Omega: R_n(\omega,\cdot)\in B_3\}$, which by Theorem I.6.3 equals $-\min\{I_\rho^{(3)}(P): P\in B_3\}$. The latter can be rewritten as

$$-\min_{\tau\in B_3}\min\{I_\rho^{(3)}(P): P\in\mathscr{M}_s(\Omega), \pi_2 P = \tau\}.$$

Thus, one expects that for $\tau\in\mathscr{M}_s(\Gamma^2)$

$$(1.39) \quad I_{\rho,2}^{(3)}(\tau) = \sum \tau_{ij}\log\frac{\tau_{ij}}{(\nu_\tau)_i\rho_j} = \min\{I_\rho^{(3)}(P): P\in\mathscr{M}_s(\Omega), \pi_2 P = \tau\},$$

where $(\nu_\tau)_i = \sum_{j=1}^r \tau_{ij}$. We shall prove (1.39) in Chapter IX. It is the contraction principle relating $I_{\rho,2}^{(3)}$ and $I_\rho^{(3)}$.

The connection with Theorem I.4.3 is similar. If P is a measure in $\mathscr{M}_s(\Omega)$, then the one-dimensional marginal $\pi_1 P$ is an element of $\mathscr{M}(\Gamma)$. In the set B_3

in (1.36), let $\{\Sigma_k\}$ run through all one-dimensional cylinder sets $\{\bar{\omega}\in\Omega\colon \bar{\omega}_1 = x_i\}, i = 1, \ldots, r$. Then

$$\pi_1 B_3 = \{\nu\in\mathscr{M}(\Gamma)\colon \nu = \pi_1 P \text{ for some } P\in B_3\}$$
$$= \{\nu\in\mathscr{M}(\Gamma)\colon \max_{i=1,\ldots,r} |\nu_i - \rho_i| \geq \varepsilon\}.$$

The latter is the set A_2 in (1.15). Since $\pi_1 R_n(\omega, \cdot) = L_n(\omega, \cdot)$ for each ω, it follows that

$$P_\rho\{\omega\in\Omega\colon L_n(\omega, \cdot)\in A_2\} = P_\rho\{\omega\in\Omega\colon R_n(\omega, \cdot)\in B_3\}.$$

Again, comparing decay rates, we have

$$\min_{\nu\in A_2} I_\rho^{(2)}(\nu) = \min_{P\in B_3} I_\rho^{(3)}(P) = \min_{\nu\in A_2} \min\{I_\rho^{(3)}(P)\colon P\in\mathscr{M}_s(\Omega), \pi_1 P = \nu\}.$$

In Chapter IX, we prove for $\nu\in\mathscr{M}(\Gamma)$ the following contraction principle which is consistent with the last display:

$$(1.40) \qquad I_\rho^{(2)}(\nu) = \sum_{i=1}^{r} \nu_i \log\frac{\nu_i}{\rho_i} = \min\{I_\rho^{(3)}(P)\colon P\in\mathscr{M}_s(\Omega), \pi_1 P = \nu\}.$$

Finally, for level-1, recall the contraction principle (1.17) relating $I_\rho^{(1)}$ and $I_\rho^{(2)}$: for $z\in[x_1, x_r]$

$$I_\rho^{(1)}(z) = \min\left\{I_\rho^{(2)}(\nu)\colon \nu\in\mathscr{M}(\Gamma), \sum_{i=1}^{r} x_i\nu_i = z\right\}.$$

Comparing this with (1.40), we conclude that for $z\in[x_1, x_r]$,

$$(1.41) \qquad I_\rho^{(1)}(z) = \min\left\{I_\rho^{(3)}(P)\colon P\in\mathscr{M}_s(\Omega), \sum_{i=1}^{r} x_i(\pi_1 P)_i = z\right\}.$$

This completes our discussion of large deviations for i.i.d. random variables with a finite outcome space. An interesting feature of this chapter was the use of combinatorics to calculate explicit formulas for the entropy functions $I_\rho^{(1)}$, $I_\rho^{(2)}$, and $I_{\rho,2}^{(3)}$. In the next chapter, the three levels of large deviations will be generalized to random vectors taking values in $\mathbb{R}^d$. The theory will be applied to statistical mechanics in later parts of the book.

I.7. Notes

1 (page 3). Wehrl (1978) discusses the historical and physical backgrounds of entropy together with modern developments. He lists many references.

2 (page 13). Relative entropy $I_\rho^{(2)}(\nu)$, introduced by Kullback and Leibler in 1951, is also known as the Kullback–Leibler information number. It plays an important role in statistics, especially in large sample theories of estimation and testing [Kullback (1959), Bahadur (1967, 1971)]. $I_\rho^{(2)}(\nu)$ measures the statistical distance between ν and ρ. The smaller this distance, the harder

it is to discriminate between ν and ρ. Applications of large deviations to statistics are discussed by Bahadur (1971), Chernoff (1972), and individual articles in "Grandes deviations et applications statistiques," Astérisque *68*, Sociéte Mathématique de France, Paris, 1979.

3 (page 14). Shannon entropy $H(\nu)$ was first defined by Shannon (1948) and independently by Wiener (1948). The form of $H(\nu)$ can be derived from a set of axioms which a reasonable measure of randomness should satisfy; see, e.g., Khinchin (1957, pages 9–13).

4 (page 24). In information theory, the mean entropy $h(P)$, $P \in \mathscr{M}_s(\Omega)$, measures the amount of information per symbol in a message which is generated according to P [Khinchin (1957), MeEliece (1977)]. Large deviation bounds, known as Chernoff bounds [see Problem VII.8.9], are widely used [Wozencraft and Jacobs (1965), Gallager (1968)]. In ergodic theory, $h(P)$ is the Kolmogorov–Sinai invariant of the dynamical system $(\Omega, \mathscr{F}, P, T)$ [Martin and England (1981)].

I.8. Problems

I.8.1. (a) Given $n \in \{2, 3, \ldots,\}$, prove that $n \log n - n + 1 < \log(n!) < (n+1)\log(n+1) - (n+1) + 1$ by considering the area under the graph of $\log x$, $x \geq 1$. Deduce the weak form of Stirling's approximation: $\log(n!) = n \log n - n + O(\log n)$, $n \geq 2$.

(b) Prove Lemma I.4.4.

I.8.2. This problem shows how to complete the proofs of Theorems I.3.1, I.4.3, and I.5.1. Let A be a compact subset of $\mathbb{R}^d$ and f a real-valued function on A which is continuous relative to A. Let $\{A_n; n = 1, 2, \ldots\}$ be closed subsets of A such that for any $a \in A$ there exists a sequence $a_n \in A_n$ with $a_n \to a$ as $n \to \infty$. Prove that $\lim_{n\to\infty} \min_{x \in A_n} f(x) = \min_{x \in A} f(x)$.

I.8.3. Let $p = \{p_{ij}; i, j = 1, \ldots, r\}$ be a set of positive numbers such that $\sum_{i,j=1}^r p_{ij} = 1$. Define $\nu_i = \sum_{j=1}^r p_{ij}$ and $\mu_j = \sum_{i=1}^r p_{ij}$. Let $\rho = \{\rho_i; i = 1, \ldots, r\}$ be a sequence of positive numbers such that $\sum_{i=1}^r \rho_i = 1$. Prove that

$$I^{(2)}_{\{\rho_i\rho_j\}}(\{p_{ij}\}) = \sum_{i,j=1}^r p_{ij} \log \frac{p_{ij}}{\rho_i \rho_j} \geq I^{(2)}_\rho(\nu) + I^{(2)}_\rho(\mu)$$

$$= \sum_{i=1}^r \nu_i \log \frac{\nu_i}{\rho_i} + \sum_{j=1}^r \mu_j \log \frac{\mu_j}{\rho_j}$$

with equality if and only if $p_{ij} = \nu_i \mu_j$ for each i and j.

If each $\rho_i = 1/r$, then we conclude that the randomness in p (as measured by Shannon entropy) is no greater than the sum of the randomness in ν and the randomness in μ and that equality holds if and only if p is product measure.

I.8.4. (a) Prove Jensen's inequality: if $w_1, w_2, \ldots, w_r$ and $y_1, y_2, \ldots, y_r$ are nonnegative numbers such that $\sum_{j=1}^r w_j = 1$, then

$$\left(\sum_{j=1}^r w_j y_j\right) \log\left(\sum_{j=1}^r w_j y_j\right) \leq \sum_{j=1}^r w_j y_j \log y_j.$$

[*Hint:* Let $h(x) = x \log x$. If $a = \sum_{j=1}^r w_j y_j > 0$, then $h(y_j) \geq h(a) + h'(a) \cdot (y_j - a)$.]

(b) [Rényi (1970b, page 556)]. Let $\nu = \{\nu_j; j = 1, \ldots, r\}$ be a sequence of non-negative numbers such that $\sum_{j=1}^r \nu_j = 1$. Let $\gamma = \{\gamma_{jk}\}$ be an $r \times r$ doubly stochastic matrix ($\gamma_{jk} \geq 0, \sum_{j=1}^r \gamma_{jk} = 1$ for each k, $\sum_{k=1}^r \gamma_{jk} = 1$ for each j) and define $\mu_k = \sum_{j=1}^r \nu_j \gamma_{jk}$. Set $\rho_i = 1/r$ for $i = 1, \ldots, r$. Prove that

$$\sum_{k=1}^r \mu_k \log \mu_k \leq \sum_{j=1}^r \nu_j \log \nu_j \quad \text{and} \quad I_\rho^{(2)}(\mu) \leq I_\rho^{(2)}(\nu).$$

See Voight (1981) for generalizations.

I.8.5 [J. K. Percus (private communication)]. (a) Derive formula (1.31). [*Hint:* Expand

$$\frac{2 - B_{11} - B_{22}}{(1-c)(1 - B_{12}B_{21}/(1-c))} = (1 - B_{11} + 1 - B_{22}) \sum_{n \geq 0} (B_{12}B_{21})^n (1-c)^{-n-1},$$

where $1 - c = 1 - (B_{11} + B_{22} - B_{11}B_{22}) = (1 - B_{11})(1 - B_{22})$.]

(b) For $r \geq 2$, let $k_i = \sum_{j=1}^r N_{ij}$. Show that

$$\gamma(n, N) = \frac{\prod_{i=1}^r k_i!}{\prod_{i,j=1}^r N_{ij}!} \cdot \sum_{\alpha=1}^r \left[\delta_{ij} - \frac{N_{ij}}{k_j}\right]_{\alpha\alpha} \qquad \text{for all } N_{ij} \geq 1,$$

where $[\delta_{ij} - N_{ij}/k_j]_{\alpha\alpha}$ is the $\alpha\alpha$-cofactor of the $r \times r$ matrix $\{\delta_{ij} - N_{ij}/k_j\}$. [*Hint:* If $B = \{B_{ij}\}$ is an $r \times r$ positive matrix with $0 < \sum_{i,j=1}^r B_{ij} < 1$, then $\operatorname{Tr}(I - B)^{-1} = \sum_{\alpha=1}^r [I - B]_{\alpha\alpha}/\det(I - B)$, where $[I - B]_{\alpha\alpha}$ is the $\alpha\alpha$-cofactor of $I - B$. Use the formula (which holds in the sense of distribution theory)

$$\frac{1}{\det(I - B)} = \int_{\mathbb{R}^r} \int_{\mathbb{R}^r} \exp[i(\langle x, y\rangle - \langle x, By\rangle)]\,dx\,dy \cdot \frac{1}{(2\pi)^r}.]$$

I.8.6. Let $B = \{B_{ij}\}$ be a positive 2×2 matrix and $\lambda(B)$ the largest eigenvalue of B in absolute value. By Lemma IX.4.1, $\lambda(B)$ is positive and $\log \lambda(B) = \lim_{n \to \infty} n^{-1} \log \operatorname{Tr} B^n$. As in the proof of Lemma I.5.2,

$$\operatorname{Tr} B^n = \Sigma \gamma(n, N) \prod_{i,j=1}^2 B_{ij}^{N_{ij}}. \tag{1.42}$$

where the sum runs over all $N = \{N_{ij}; i, j = 1, 2\}$ such that $N_{ij} \in \{0, 1, \ldots, n\}$, $\sum_{i,j=1}^{2} N_{ij} = n$, and $\sum_{j=1}^{2} N_{ij} = \sum_{k=1}^{2} N_{ki}$ for each i. Using Lemma I.5.2, prove that

$$\log \lambda(B) = \max \left\{ \sum \tau_{ij} \log \frac{B_{ij}(v_\tau)_i}{\tau_{ij}} : \tau \in \mathcal{M}_{s,2} \right\}, \tag{1.43}$$

where $(v_\tau)_i = \sum_{j=1}^{2} \tau_{ij}$ and $\mathcal{M}_{s,2}$ is the set of $\tau = \{\tau_{ij}; i, j = 1, 2\}$ satisfying $\tau_{ij} \geq 0$, $\sum_{i,j=1}^{2} \tau_{ij} = 1$, and $\sum_{j=1}^{2} \tau_{ij} = \sum_{k=1}^{2} \tau_{ki}$ for each i. The sum in (1.43) runs over all i and j for which $(v_\tau)_i > 0$. Theorem IX.4.4 and Problem IX.6.4 are generalizations.

I.8.7. Verify the calculations in Examples I.6.2(a), (b).

Chapter II

Large Deviation Property and Asymptotics of Integrals

II.1. Introduction

The previous chapter treated three levels of large deviations for independent, identically distributed (i.i.d.) random variables with a finite state space. The main results show the exponential decay of large deviation probabilities. A level-1 example is $P_\rho\{|S_n/n - m_\rho| \geq \varepsilon\}$, where S_n is the nth partial sum of the random variables and m_ρ is their common mean. Levels-2 and 3 treat analogous probabilities for the empirical measures $\{L_n\}$ and the empirical processes $\{R_n\}$, respectively. One purpose of the present chapter is to expand the scope of large deviations and to prepare the groundwork for applications to statistical mechanics later in the book.[1] This chapter extends the ideas of Chapter I by considering random vectors taking values in $\mathbb{R}^d$, where $d \geq 1$ is a fixed integer. $\mathbb{R}^d$ is a natural state space for stochastic models, including models in statistical mechanics. The theory, which is somewhat technical and detailed, is presented in a manner that closely parallels the development in Chapter I, where elementary proofs based upon combinatorics were possible because of the finite state space. The reader who followed that development should find the new theorems familiar looking and plausible. In order to reach illuminating applications of the theory to statistical mechanical models, the proofs of the theorems will be postponed until Chapters VI–IX.

The last section of this chapter shows how one can apply large deviations to evaluate the asymptotics of certain configuration space integrals that depend on a parameter. The main theorem in this section is due to Varadhan (1966). In the next few chapters we will apply Varadhan's theorem to study integrals associated with the statistical mechanical models (partition functions).

II.2. Levels-1, 2, and 3 Large Deviations for I.I.D. Random Vectors

The presentation of the three levels of large deviations for vector-valued random variables involves some topology. We next outline the information we need, saving a more thorough treatment for Appendix A.

The basic underlying model in this chapter is a sequence $\{X_j; j \in \mathbb{Z}\}$ of i.i.d. random vectors defined on a probability space $(\Omega, \mathscr{F}, P)$ and taking values in $\mathbb{R}^d$, where d is a fixed positive integer. Since all of our results depend only on the distribution of the random vectors, we may suppose $(\Omega, \mathscr{F}, P)$ to be the following standard model.

Ω is the space of all sequences $\omega = \{\omega_j; j \in \mathbb{Z}\}$ with each $\omega_j \in \mathbb{R}^d$. That is, Ω is the infinite product space $\prod_{j \in \mathbb{Z}} \mathbb{R}^d$ or simply $(\mathbb{R}^d)^{\mathbb{Z}}$. We endow Ω with the product topology. An open base for this topology consists of all sets $\prod_{j \in \mathbb{Z}} U_j$, where all but a finite number of the factors equal $\mathbb{R}^d$ and the remainder, say $U_{j_1}, \ldots, U_{j_k}$, are open sets in $\mathbb{R}^d$. Convergence in this topology is equivalent to coordinatewise convergence in $\mathbb{R}^d$.

$\mathscr{F}$ is the Borel σ-field of Ω. The Borel σ-field, denoted by $\mathscr{B}(\Omega)$, is defined as the σ-field generated by the open subsets of Ω. $\mathscr{B}(\Omega)$ coincides with the σ-field generated by the product cylinder sets

(2.1) $$\Sigma_{m,k} = \{\omega \in \Omega: \omega_{m+1} \in F_1, \ldots, \omega_{m+k} \in F_k\},$$

where m and k are integers with $k \geq 1$ and $F_1, \ldots, F_k$ are Borel subsets of $\mathbb{R}^d$.

P is an infinite product measure on $\mathscr{B}(\Omega)$ with identical one-dimensional marginals, which we denote by ρ. Thus, for example,

$$P\{\Sigma_{m,k}\} = \prod_{j=1}^{k} \rho\{F_j\}.$$

We write P_ρ for P. $X_j, j \in \mathbb{Z}$, is the jth coordinate function on Ω defined by $X_j(\omega) = \omega_j$. The set $\{X_j; j \in \mathbb{Z}\}$ is a sequence of i.i.d. random vectors taking values in $\mathbb{R}^d$. This sequence is called the *coordinate representation process*. Each X_j has distribution ρ:

$$P_\rho\{\omega \in \Omega: X_j(\omega) \in B\} = \rho\{B\} \qquad \text{for Borel subsets } B \text{ of } \mathbb{R}^d.$$

Ω can serve as a configuration space for physical systems. In statistical mechanical models, Ω may be the configuration space for the velocities of a system of particles [Chapter III] or for the spins at sites of a ferromagnet [Chapters IV and V]. A microscopic model for such a system is completed by choice of a probability measure P on $(\Omega, \mathscr{F})$. The measure P, called an *ensemble*, expresses the microscopic interactions among the particles or spins. Product measure corresponds to the absence of interactions. It models a system of independent particles (ideal gas) or independent spins. When interactions are present, an ensemble which is not product measure must be chosen.

We now define the three levels for a sequence of i.i.d. random vectors $\{X_j; j \in \mathbb{Z}\}$ on a probability space $\{\Omega, \mathscr{F}, P\}$. Let ρ be the distribution of X_1. Level-1 large deviations treat the microscopic n-sums $S_n(\omega)/n = \sum_{j=1}^n X_j(\omega)/n$, $n = 1, 2, \ldots$. The law of large numbers states that if $\int_{\mathbb{R}^d} \|x\| \, \rho(dx)$ is finite,* then

* $\|x\| = (\sum_{i=1}^d x_i^2)^{1/2}$, the Euclidean norm of x.

$$(2.2)\qquad S_n(\omega)/n \to m_\rho = \int_{\mathbb{R}^d} x\rho(dx) \qquad P\text{-a.s.}$$

Denote by $Q_n^{(1)}$ the distribution of S_n/n on $\mathbb{R}^d$. The limit (2.2) implies that if A is any Borel set whose closure does not contain m_ρ, then $Q_n^{(1)}\{A\} \to 0$ as $n \to \infty$. Hence the distributions $\{Q_n^{(1)}; n = 1, 2, \ldots\}$ converge weakly to the unit point measure δ_{m_ρ} on $\mathbb{R}^d$ [Theorem A.8.2]. Level-1 large deviations study the exponential decay of the probabilities $Q_n^{(1)}\{A\}$.

Level-2 large deviations treat the empirical measures $L_n(\omega, \cdot) = n^{-1}\sum_{j=1}^n \delta_{X_j(\omega)}(\cdot)$, $n = 1, 2, \ldots$, which are level-2 microscopic n-sums. For each Borel subset B of $\mathbb{R}^d$, $L_n(\omega, B)$ is the relative frequency with which $X_1(\omega), \ldots, X_n(\omega)$ is in B. Therefore, for each ω, $L_n(\omega, \cdot)$ is an element of $\mathscr{M}(\mathbb{R}^d)$, the set of probability measures on $\mathscr{B}(\mathbb{R}^d)$. The mean of the empirical measure is $\int_{\mathbb{R}^d} xL_n(\omega, dx) = S_n(\omega)/n$. The set $\mathscr{M}(\mathbb{R}^d)$ with the topology of weak convergence is a complete separable metric space [Example A.8.4(a)]. It is easily checked that if A is a Borel subset of $\mathscr{M}(\mathbb{R}^d)$, then the set $L_n^{-1}A = \{\omega \in \Omega: L_n(\omega, \cdot) \in A\}$ belongs to the σ-field $\mathscr{F}$. The distribution of L_n is the measure $Q_n^{(2)}$ defined by $Q_n^{(2)}\{A\} = P_\rho\{L_n^{-1}A\}$ for A a Borel subset of $\mathscr{M}(\mathbb{R}^d)$.

For level-2, the limit (2.2) is replaced by

$$(2.3)\qquad L_n(\omega, \cdot) \Rightarrow \rho \qquad P\text{-a.s.}$$

This limit states that the sequence $\{L_n(\omega, \cdot); n = 1, 2, \ldots\}$ converges weakly to ρ P-a.s. It is a consequence of the ergodic theorem [Corollary A.9.8]. The limit (2.3) implies that if A is any Borel subset of $\mathscr{M}(\mathbb{R}^d)$ whose closure does not contain ρ, then $Q_n^{(2)}\{A\} \to 0$ as $n \to \infty$. Hence the distributions $\{Q_n^{(2)}; n = 1, 2, \ldots\}$ converge weakly to the unit point measure δ_ρ on $\mathscr{M}(\mathbb{R}^d)$. Level-2 large deviations study the exponential decay of the probabilities $Q_n^{(2)}\{A\}$.

Level-3 large deviations treat the empirical processes

$$(2.4)\qquad R_n(\omega, \cdot) = \frac{1}{n}\sum_{k=0}^{n-1} \delta_{T^k X(n,\omega)}(\cdot), \qquad n = 1, 2, \ldots,$$

which are level-3 microscopic n-sums. T is the shift mapping on $(\mathbb{R}^d)^{\mathbb{Z}}$ and $X(n, \omega)$ is the periodic point in $(\mathbb{R}^d)^{\mathbb{Z}}$ obtained by repeating $(X_1(\omega), X_2(\omega), \ldots, X_n(\omega))$ periodically [cf., Section I.6]. For each Borel subset B of $(\mathbb{R}^d)^{\mathbb{Z}}$, $R_n(\omega, B)$ is the relative frequency with which $X(n, \omega), TX(n, \omega), \ldots, T^{n-1}X(n, \omega)$ is in B. Therefore, for each ω, $R_n(\omega, \cdot)$ is an element of $\mathscr{M}_s((\mathbb{R}^d)^{\mathbb{Z}})$, the set of strictly stationary probability measures on $\mathscr{B}((\mathbb{R}^d)^{\mathbb{Z}})$. All the one-dimensional marginals of $R_n(\omega, \cdot)$ equal $L_n(\omega, \cdot)$. The set $\mathscr{M}_s((\mathbb{R}^d)^{\mathbb{Z}})$ with the topology of weak convergence is a complete separable metric space [Theorem A.9.2(a)]. It is easily checked that if A is a Borel subset of $\mathscr{M}_s(\mathbb{R}^d)^{\mathbb{Z}}$, then the set $R_n^{-1}A = \{\omega \in \Omega: R_n(\omega, \cdot) \in A\}$ belongs to the σ-field $\mathscr{F}$. The distribution of R_n is the measure $Q_n^{(3)}$ defined by $Q_n^{(3)}\{A\} = P_\rho\{R_n^{-1}A\}$ for A a Borel subset of $\mathscr{M}_s((\mathbb{R}^d)^{\mathbb{Z}})$.

For level-3, the limits (2.2) and (2.3) are replaced by

$$(2.5)\qquad R_n(\omega, \cdot) \Rightarrow P_\rho \qquad P\text{-a.s.}$$

Table II.1. Definitions of Levels-1, 2, and 3

Level	Random quantity	A.s. limit	Metric space $\mathscr{X}$	Distribution Q_n on $\mathscr{B}(\mathscr{X})$	Weak limit of $\{Q_n\}$
1	S_n/n	m_ρ	$\mathbb{R}^d$	$Q_n^{(1)}\{B\} = P\left\{\frac{S_n}{n}\in B\right\}$	$Q_n^{(1)} \Rightarrow \delta_{m_\rho}$
2	L_n	ρ	$\mathscr{M}(\mathbb{R}^d)$	$Q_n^{(2)}\{B\} = P\{L_n \in B\}$	$Q_n^{(2)} \Rightarrow \delta_\rho$
3	R_n	P_ρ	$\mathscr{M}_s((\mathbb{R}^d)^{\mathbb{Z}})$	$Q_n^{(3)}\{B\} = P\{R_n \in B\}$	$Q_n^{(3)} \Rightarrow \delta_{P_\rho}$

This limit states that the sequence $\{R_n(\omega,\cdot); n = 1, 2, \ldots\}$ converges weakly to P_ρ P-a.s. It is a consequence of the ergodic theorem [Corollary A.9.8]. The limit (2.5) implies that if A is any Borel subset of $\mathscr{M}_s((\mathbb{R}^d)^{\mathbb{Z}})$ whose closure does not contain P_ρ, then $Q_n^{(3)}\{A\} \to 0$ as $n \to \infty$. Hence the distributions $\{Q_n^{(3)}; n = 1, 2, \ldots\}$ converge weakly to the unit point measure δ_{P_ρ} on $\mathscr{M}_s((\mathbb{R}^d)^{\mathbb{Z}})$. Level-3 large deviations study the exponential decay of the probabilities $Q_n^{(3)}\{A\}$.

Thus for each level there is a random quantity S_n/n, L_n, or R_n. Each quantity takes values in a complete separable metric space $\mathscr{X} = \mathbb{R}^d$, $\mathscr{M}(\mathbb{R}^d)$, or $\mathscr{M}_s((\mathbb{R}^d)^{\mathbb{Z}})$, and has an a.s. limit which is a point x_0 in $\mathscr{X}$ ($x_0 = m_\rho$, ρ, or P_ρ). The distributions of these random quantities are measures $Q_n^{(1)}$, $Q_n^{(2)}$, or $Q_n^{(3)}$ on the Borel subsets of the respective space $\mathscr{X}$, and they converge weakly to the corresponding unit point measure δ_{x_0} on $\mathscr{X}$. Large deviations study the asymptotic behaviors of the distributions $Q_n^{(1)}$, $Q_n^{(2)}$, and $Q_n^{(3)}$. This information is summarized in Table II.1.

In the next section, we give an abstract formulation of a class of large deviation problems, which includes the three levels of large deviations for i.i.d. random vectors. We return to the three levels in Section II.4.

II.3. The Definition of Large Deviation Property

A central theme of Chapter I was to express the exponential decay of large deviation probabilities in terms of entropy functions. For each of the three levels of large deviations we had limits of the form

$$\text{(2.6)} \qquad \lim_{n\to\infty} \frac{1}{n} \log Q_n^{(i)}\{A_i\} = -\min_{x\in A_i} I_\rho^{(i)}(x), \qquad i = 1, 2, 3.$$

The functions $I_\rho^{(1)}$, $I_\rho^{(2)}$, and $I_\rho^{(3)}$ are the entropy functions corresponding to the three levels and the sets A_i are closed sets not containing the unique minimum point of the respective function $I_\rho^{(i)}$. In each case, the minimum of $I_\rho^{(i)}$ over A_i is positive. Hence the limit (2.6) implies that the large deviation probabilities $Q_n^{(i)}\{A_i\}$ decay exponentially. The right-hand side is the decay rate.

For example, in Chapter I, A_1 is the set $\{z \in \mathbb{R} : |z - m_\rho| \geq \varepsilon\}$ and $I_\rho^{(1)}$ is defined by the contraction principle (1.17). The point m_ρ is the unique minimum point of $I_\rho^{(1)}$. The limit (2.6) for $i = 1$ can be used to show the exponential decay of probabilities of other sets. Let K be any closed set in $\mathbb{R}$ which does not contain the point m_ρ. Since K is a subset of A_1 for sufficiently small ε, the limit (2.6) implies that there exists a number $N = N(K) > 0$ such that

$$(2.7) \qquad Q_n^{(1)}\{K\} \leq e^{-nN} \qquad \text{for all sufficiently large } n.$$

In applications, these bounds for closed sets are often as important as limits of the form (2.6). Our first task is a formulation of such bounds for distributions on $\mathbb{R}^d$. The bounds will be expressed directly in terms of entropy functions on $\mathbb{R}^d$ which for now we assume, and later will prove, to exist. This formulation will suggest how to prove limits like (2.6) and will motivate the general concept of large deviation property. Large deviation property will be defined later in the section.

Let $\{Q_n; n = 1, 2, \dots\}$ be a sequence of Borel probability measures on $\mathbb{R}^d$. Suppose that there exists a non-negative real-valued function $I(x)$, $x \in \mathbb{R}^d$, such that $\inf_{x \in \mathbb{R}^d} I(x) = 0$ and

$$(2.8) \qquad \limsup_{n \to \infty} \frac{1}{n} \log Q_n\{K\} \leq -\inf_{x \in K} I(x)$$

for each nonempty closed set K in $\mathbb{R}^d$. Assume also that $I(x)$ attains its infimum of 0 at some point in $\mathbb{R}^d$. Such a point is called a *minimum point* of I; a minimum point need not be unique. What additional properties of $I(x)$ guarantee that a bound of the form (2.7) holds whenever K does not contain a minimum point of I? It suffices if the following hold.

(a) $I(x)$ is lower semicontinuous on $\mathbb{R}^d$: $x_n \to x$ implies $\liminf_{n\to\infty} I(x_n) \geq I(x)$.

(b) $I(x)$ has compact level sets: the set $\{x \in \mathbb{R}^d : I(x) \leq b\}$ is compact for each real number b.

Indeed, if $I(K)$ denotes the infimum of $I(x)$ over K, then (2.8) implies that

$$(2.9) \qquad Q_n\{K\} \leq \exp(-nI(K)/2) \qquad \text{for all sufficiently large } n.$$

If we prove that $I(K)$ is positive, then we will have a bound of the form (2.7). The set

$$K_1 = K \cap \{x \in \mathbb{R}^d : I(x) \leq I(K) + 1\}$$

is nonempty and since $I(x)$ has compact level sets, K_1 is compact. According to a standard result, a lower semicontinuous function attains its infimum over a nonempty compact set [Problem II.9.2]. Since $I(K_1) = I(K)$, there exists a point $z \in K_1 \subseteq K$ such that $I(z) = I(K)$. If $I(K) = 0$, then z must equal a minimum point of I. This contradicts the assumption that K contains no minimum point of I. Thus $I(K)$ is positive, and we are done.

In many cases, functions $I(x)$ arise which take the value ∞. Thus it may happen that in the argument just given, the infimum $I(K)$ equals ∞. But then (2.8) implies that for any number $N > 0$, $Q_n\{K\} \leq e^{-nN}$ for all sufficiently large n.

Along with the upper bound (2.8), it is reasonable to consider the complementary lower bound for each nonempty open set G in $\mathbb{R}^d$:

(2.10) $$\liminf_{n\to\infty} \frac{1}{n} \log Q_n\{G\} \geq -\inf_{x \in G} I(x).$$

Let us deduce a simple criterion on a nonempty Borel subset A of $\mathbb{R}^d$ which implies that if both bounds (2.8) and (2.10) are valid, then

(2.11) $$\lim_{n\to\infty} \frac{1}{n} \log Q_n\{A\} = -\inf_{x \in A} I(x).$$

Denote the closure of A by $\operatorname{cl} A$ and the interior of A by $\operatorname{int} A$. Since $\operatorname{cl} A \supseteq A \supseteq \operatorname{int} A$, it follows that

(2.12) $$\begin{aligned} -\inf_{x \in \operatorname{cl} A} I(x) &\geq \limsup_{n\to\infty} \frac{1}{n} \log Q_n\{\operatorname{cl} A\} \geq \limsup_{n\to\infty} \frac{1}{n} \log Q_n\{A\} \\ &\geq \liminf_{n\to\infty} \frac{1}{n} \log Q_n\{A\} \geq \liminf_{n\to\infty} \frac{1}{n} \log Q_n\{\operatorname{int} A\} \geq -\inf_{x \in \operatorname{int} A} I(x). \end{aligned}$$

If the two extreme terms are equal, then the limit (2.11) follows.

We return to the level-1 problem studied in Chapter I. Let $Q_n = Q_n^{(1)}$ be the distribution of $n^{-1} \sum_{j=1}^n X_j$, where $X_1, X_2, \ldots$ are i.i.d. random vectors taking values in a finite set $\Gamma = \{x_1, x_2, \ldots, x_r\}$ with $x_1 < x_2 < \cdots < x_r$. According to Theorems II.4.1 and II.5.1 below, the bounds (2.8) and (2.10) are valid for $Q_n^{(1)}$; I equals the entropy function $I_\rho^{(1)}$ defined in (1.17). Let A_1 be the closed set $\{z \in \mathbb{R} : |z - m_\rho| \geq \varepsilon\}$, where $0 < \varepsilon < \min\{m_\rho - x_1, x_r - m_\rho\}$. Since the infimum of $I_\rho^{(1)}(z)$ over A_1 equals the infimum of $I_\rho^{(1)}(z)$ over $\operatorname{int} A_1$, the bounds (2.8) and (2.10) yield

$$\lim_{n\to\infty} \frac{1}{n} \log Q_n^{(1)}\{A_1\} = -\inf_{z \in A_1} I_\rho^{(1)}(z).$$

In addition, since $I_\rho^{(1)}$ is a continuous function on the compact set $[x_1, x_r]$ [Theorem II.4.1(c)], $I_\rho^{(1)}$ is lower semicontinuous on $\mathbb{R}$ and has compact level sets.

This discussion leads us to the definition of large deviation property. We want the definition to be general enough to cover the three levels of large deviations summarized in Table II.1. Let $\mathscr{X}$ be a complete separable metric space and $\{Q_n; n = 1, 2, \ldots\}$ a sequence of probability measures on the Borel subsets of $\mathscr{X}$. Suppose that as $n \to \infty$, $\{Q_n\}$ converges weakly to the unit point measure δ_{x_0} at some point x_0 in $\mathscr{X}$. Thus for most Borel sets A, $Q_n\{A\} \to 0$ as $n \to \infty$. In the examples we consider, $Q_n\{A\}$ converges to zero exponentially fast as $n \to \infty$ with an exponential rate depending on the set A. We allow for the possibility of other scaling constants besides n. The situation is abstracted in the following definition.[2]

Definition II.3.1. Let $\mathscr{X}$ be a complete separable metric space, $\mathscr{B}(\mathscr{X})$ the Borel σ-field of $\mathscr{X}$, and $\{Q_n; n = 1, 2, \ldots\}$ a sequence of probability measures on $\mathscr{B}(\mathscr{X})$. $\{Q_n\}$ is said to have a *large deviation property* if there exist a sequence of positive numbers $\{a_n; n = 1, 2, \ldots\}$ which tend to ∞ and a function $I(x)$ which maps $\mathscr{X}$ into $[0, \infty]$ such that the following hypotheses hold.

(a) $I(x)$ is lower semicontinuous on $\mathscr{X}$.
(b) $I(x)$ has compact level sets.
(c) $\limsup_{n\to\infty} a_n^{-1} \log Q_n\{K\} \le -\inf_{x\in K} I(x)$ for each closed set K in $\mathscr{X}$.*
(d) $\liminf_{n\to\infty} a_n^{-1} \log Q_n\{G\} \ge -\inf_{x\in G} I(x)$ for each open set G in $\mathscr{X}$.

$I(x)$ is called an *entropy function* of $\{Q_n\}$.

We note several consequences of the definition.

The infimum of $I(x)$ over $\mathscr{X}$ equals 0. This follows from the upper and lower large deviation bounds (c)–(d) with $K = G = \mathscr{X}$.

$I(x)$ attains its infimum over any nonempty closed set (the infimum may be ∞). This follows from hypotheses (a) and (b) by the same argument given earlier in the case $\mathscr{X} = \mathbb{R}^d$ [page 34].

According to the following theorem, if a large deviation property holds, then the entropy function is unique.[3]

Theorem II.3.2. *Let $\mathscr{X}$ be a complete separable metric space and $\{Q_n; n = 1, 2, \ldots\}$ a sequence of probability measures on $\mathscr{B}(\mathscr{X})$. If for a fixed sequence $\{a_n; n = 1, 2, \ldots\}$ $\{Q_n\}$ has a large deviation property with entropy function I and with entropy function J, then $I(x) = J(x)$ for all $x \in \mathscr{X}$.*

Proof. Fix a point $x \in \mathscr{X}$ and define K_j and G_j to be the closed ball and the open ball, respectively, with radius $1/j$ and center x ($j = 1, 2, \ldots$). Since $K_{j+1} \subset G_j \subset K_j$,

$$-I(K_{j+1}) \le -I(G_j) \le \liminf_{n\to\infty} \frac{1}{a_n} \log Q_n\{G_j\}$$

$$\le \limsup_{n\to\infty} \frac{1}{a_n} \log Q_n\{K_j\} \le -J(K_j),$$

where if A is a subset of $\mathscr{X}$, then $I(A)$ and $J(A)$ denote the infimum of I and of J over A, respectively. Repeating the argument with I and J interchanged, we see that

$$J(K_j) \le I(K_{j+1}) \le J(K_{j+2}) \le I(K_{j+3}) \le \cdots.$$

Hence to prove that $I(x) = J(x)$, it suffices to show that $I(K_j) \to I(x)$ and $J(K_j) \to J(x)$ as $j \to \infty$. We prove the first. The second is proved the same way. Since $\{I(K_j); j = 1, 2, \ldots\}$ is an increasing sequence and x is a point in each set K_j, $\lim_{j\to\infty} I(K_j)$ exists as an extended real-valued number and $\lim_{j\to\infty} I(K_j) \le I(x)$. I attains its infimum over the closed set K_j. Hence there exists a point x_j in K_j such that $I(x_j) = I(K_j)$. The sequence $\{x_j; j = 1, 2, \ldots\}$ converges to x and by lower semicontinuity

*If K or G equals the empty set ϕ, then (c)–(d) hold trivially if we set $\log Q_n(\phi) = \log 0 = -\infty$ and $\inf_{x\in\phi} I(x) = \infty$.

$$I(x) \leq \lim_{j\to\infty} I(x_j) = \lim_{j\to\infty} I(K_j) \leq I(x).$$

Thus $I(K_j) \to I(x)$, and the proof is complete. □

We remarked above that if $I(x)$ is an entropy function, then the infimum of $I(x)$ over the whole space $\mathscr{X}$ equals 0. Since $\mathscr{X}$ is closed, $I(x)$ attains its infimum of 0 at some point of $\mathscr{X}$. Such a point is called a *minimum point* of I. Recall the case $\mathscr{X} = \mathbb{R}^d$ which was considered at the beginning of this section. We proved using hypotheses (a)–(c) of large deviation property that $Q_n\{K\}$ decays exponentially for any closed set K not containing a minimum point of I. The same proof works for any complete separable metric space $\mathscr{X}$, and we have the following result.

Theorem II.3.3. *Let $\mathscr{X}$ be a complete separable metric space, $\{Q_n; n = 1, 2, \ldots\}$ a sequence of probability measures on $\mathscr{B}(\mathscr{X})$, and $I(x)$ a function which maps $\mathscr{X}$ into $[0, \infty]$ such that $\inf_{x\in\mathscr{X}} I(x) = 0$. Suppose that hypotheses* (a)–(c) *of large deviation property are valid for some positive sequence $a_n \to \infty$. Then for any closed set K not containing a minimum point of I, there exists a number $N = N(K) > 0$ such that $Q_n\{K\} \leq e^{-a_n N}$ for all sufficiently large n.*

We end this section with a condition on a Borel set A which guarantees that $\lim_{n\to\infty} a_n^{-1} \log Q_n\{A\} = -\inf_{x\in A} I(x)$. The theorem is proved as in (2.12).

Theorem II.3.4. *Let $\mathscr{X}$ be a complete separable metric space and $\{Q_n; n = 1, 2, \ldots\}$ a sequence of probability measures on $\mathscr{B}(\mathscr{X})$. Suppose that $\{Q_n\}$ has a large deviation property with constants $\{a_n; n = 1, 2, \ldots\}$ and entropy function I. Call a Borel subset A of $\mathscr{X}$ an I-continuity set if*

$$\inf_{x\in \mathrm{cl}\, A} I(x) = \inf_{x\in \mathrm{int}\, A} I(x).$$

If A is an I-continuity set, then $\lim_{n\to\infty} a_n^{-1} \log Q_n(A) = -\inf_{x\in A} I(x)$.

II.4. Statement of Large Deviation Properties for Levels-1, 2, and 3

The microscopic n-sums and distributions determining the three levels were summarized in Table II.1. This section states the large deviation property for each level, then summarizes the results in Table II.2.

Level-1. We denote by $\langle \cdot, \cdot \rangle$ the Euclidean inner product on $\mathbb{R}^d$ and by $\|\cdot\|$ the Euclidean norm.

Theorem II.4.1.* *Let $X_1, X_2, \ldots$ be a sequence of* i.i.d. *random vectors taking values in $\mathbb{R}^d$ and let $\rho \in \mathscr{M}(\mathbb{R}^d)$ be the distribution of X_1. For $t \in \mathbb{R}^d$, define*

*Parts (a) and (b) of Theorem II.4.1 are proved in Section VII.5; part (c) is proved in Theorem VIII.3.3(c).

(2.13) $$c_\rho(t) = \log E\{\exp\langle t, X_1\rangle\} = \log \int_{\mathbb{R}^d} \exp\langle t, x\rangle \rho(dx).$$

Assume that $c_\rho(t)$ *is finite for all* t *and define*

(2.14) $$I_\rho^{(1)}(z) = \sup_{t \in \mathbb{R}^d} \{\langle t, z\rangle - c_\rho(t)\} \quad \text{for} \quad z \in \mathbb{R}^d.$$

Let S_n *denote the* nth *partial sum* $\sum_{j=1}^n X_j$, $n = 1, 2, \ldots$. *Then the following conclusions hold.*

(a) $\{Q_n^{(1)}\}$, *the distributions of* $\{S_n/n\}$ *on* $\mathbb{R}^d$, *have a large deviation property with* $a_n = n$ *and entropy function* $I_\rho^{(1)}$.

(b) $I_\rho^{(1)}(z)$ *is a convex function of* z. $I_\rho^{(1)}(z)$ *measures the discrepancy between* z *and the mean** $m_\rho = \int_{\mathbb{R}^d} x\rho(dx)$ *in the sense that* $I_\rho^{(1)}(z) \geq 0$ *with equality if and only if* $z = m_\rho$.

(c) *If* ρ *is supported on a finite set* $\Gamma = \{x_1, x_2, \ldots, x_r\} \subseteq \mathbb{R}$ *with* $x_1 < x_2 < \cdots < x_r$, *then* $I_\rho^{(1)}$ *is finite and continuous on the interval* $[x_1, x_r]$ *and* $I_\rho^{(1)}(z) = \infty$ *for* $z \notin [x_1, x_r]$.

The hypothesis that $c_\rho(t)$ be finite for all t is satisfied if, for example, X_1 is bounded.

In Chapter I, we defined the level-1 entropy function $I_\rho^{(1)}$ by the contraction principle (1.17). A new aspect of Theorem II.4.1 is formula (2.14) for $I_\rho^{(1)}$. Although the proof of the theorem is postponed, we will motivate this formula by assuming that $\{Q_n^{(1)}\}$ has a large deviation property with some entropy function $I_\rho^{(1)}$ and then deducing (2.14) as a consequence. For the contraction principle, see Theorem II.5.1.

If $\{Q_n^{(1)}\}$ has a large deviation property with entropy function $I_\rho^{(1)}$, then it is tempting to think of $Q_n^{(1)}(dz)$ as having the asymptotic form $\exp(-nI_\rho^{(1)}(z))dz$. Since $X_1, X_2, \ldots$ are i.i.d.,

$$\frac{1}{n}\log E\{\exp\langle t, S_n\rangle\} = \frac{1}{n}\log \prod_{j=1}^n E\{\exp\langle t, X_j\rangle\} = \log E\{\exp\langle t, X_1\rangle\} = c_\rho(t).$$

Using the heuristic formula $Q_n^{(1)}(dz) \approx \exp(-nI_\rho^{(1)}(z))dz$, we may write

$$\begin{aligned} c_\rho(t) &= \frac{1}{n}\log E\{\exp(n\langle t, S_n/n\rangle)\} = \frac{1}{n}\log \int_{\mathbb{R}^d} \exp(n\langle t, z\rangle) Q_n^{(1)}(dz) \\ &\approx \frac{1}{n}\log \int_{\mathbb{R}^d} \exp[n(\langle t, z\rangle - I_\rho^{(1)}(z))]dz. \end{aligned}$$

As $n \to \infty$, the main contribution to the integral comes from where the integrand is a maximum (Laplace's method). Hence it is plausible that

(2.15) $$c_\rho(t) = \sup_{z \in \mathbb{R}^d} \{\langle t, z\rangle - I_\rho^{(1)}(z)\}.$$

The function $c_\rho(t)$ is a convex function on $\mathbb{R}^d$ [Example VII.1.2]. Formula

* $\int_{\mathbb{R}^d} \|x\| \rho(dx) < \infty$ since $c_\rho(t) < \infty$ for all t.

(2.15) expresses this convex function as the Legendre–Fenchel transform of $I_\rho^{(1)}(z)$. When the theory of these transforms is developed, we will see that formula (2.15) can be inverted to yield formula (2.14) for $I_\rho^{(1)}$. In Section II.7, we will justify the manipulations leading to (2.15). The function $c_\rho(t)$ is well-known in probability theory as the logarithm of the moment generating function of ρ (also called the cumulant generating function of ρ). By analogy with statistical mechanics, we call $c_\rho(t)$ the *free energy function* of ρ. The reason for this choice of terminology will be clear in Section IV.5.

Here are two examples of entropy functions calculated by (2.14).

Example II.4.2. (a) For coin tossing, $\rho = \frac{1}{2}\delta_0 + \frac{1}{2}\delta_1$ and $c_\rho(t) = \log[\frac{1}{2}(1 + e^t)]$. Fix z real. $c_\rho(t)$ is convex and $tz - c_\rho(t)$ is concave. Hence $tz - c_\rho(t)$ attains its supremum at t if and only if

$$0 = \frac{d}{dt}(tz - c_\rho(t)) = z - c_\rho'(t).$$

The derivative $c_\rho'(t) = e^t/(1 + e^t)$ has range $(0, 1)$ as t runs through $(-\infty, \infty)$. For $z \in (0, 1)$ $c_\rho'(t)$ equals z at the unique value $t = t(z) = \log(z/(1-z))$. By (2.14)

$$I_\rho^{(1)}(z) = t(z)\cdot z - c_\rho(t(z)) = z\log(2z) + (1-z)\log(2(1-z)).$$

For $z = 0$ or 1, the supremum in (2.14) is not attained but $I_\rho^{(1)}(0) = I_\rho^{(1)}(1) = \log 2$; for $z \notin [0, 1]$, $I_\rho^{(1)}(z) = \infty$ [Theorem VIII.3.3(b)]. Hence we obtain the formula for $I_\rho^{(1)}$ in (1.9).

(b) Given m real and $\sigma^2 > 0$, consider the Gaussian probability measure on $\mathbb{R}$

$$\rho(dz) = N(m, \sigma^2)(dz) = \frac{1}{(2\pi\sigma^2)^{1/2}}\exp\left[-\frac{(z-m)^2}{2\sigma^2}\right] dz;$$

$m = \int_{\mathbb{R}} x\rho(dx)$ and $\sigma^2 = \int_{\mathbb{R}}(x-m)^2\rho(dx)$. A standard calculation shows that $c_\rho(t) = \frac{1}{2}\sigma^2 t^2 + mt$. For any z real the supremum in (2.14) is attained at $t = (z-m)/\sigma^2$ and $I_\rho^{(1)}(z) = (z-m)^2/2\sigma^2$.

Theorem II.4.1 has an interesting implication concerning the convergence of $\{S_n/n\}$. For any $\varepsilon > 0$, let K be the closed set $\{z \in \mathbb{R}^d : \|z - m_\rho\| \geq \varepsilon\}$. The distributions $\{Q_n^{(1)}\}$ of $\{S_n/n\}$ have a large deviation property and the entropy function $I_\rho^{(1)}(z)$ attains its infimum of 0 at the unique point $z = m_\rho$. By Theorem II.3.3, there exists a number $N = N(\varepsilon) > 0$ such that

(2.16) $\quad Q_n^{(1)}\{K\} = P\{\|S_n/n - m_\rho\| \geq \varepsilon\} \leq e^{-nN}$ for all sufficiently large n.

Since $\sum_{n=1}^\infty \exp(-nN)$ is finite, the Borel–Cantelli lemma implies the strong law of large numbers: $S_n/n \to m_\rho$ P-a.s. [Theorem A.5.3]. We may summarize (2.16) by saying that S_n/n converges exponentially to m_ρ. Exponential convergence will be considered in Section II.6 for other random vectors.

Level-2. Let ρ be a Borel probability measure on $\mathbb{R}$ whose support is a finite set $\Gamma = \{x_1, x_2, \ldots, x_r\}$, with $x_1 < x_2 < \cdots < x_r$. Thus if $\rho_i = \rho\{x_i\}$, then each ρ_i is positive and $\Sigma_{i=1}^r \rho_i = 1$. $\mathcal{M}(\Gamma)$ denotes the closed subset of $\mathcal{M}(\mathbb{R})$ consisting of probability measures ν for which $\nu\{\Gamma\} = 1$. Let A_2 be the set

$$\{\nu \in \mathcal{M}(\Gamma) \colon \max_{i=1,\ldots,r} |\nu_i - \rho_i| \geq \varepsilon\} \qquad \text{where } 0 < \varepsilon < \min_{i=1,\ldots,r} \{\rho_i, 1 - \rho_i\}.$$

In Theorem I.4.3, we proved that

$$\lim_{n\to\infty} \frac{1}{n} \log Q_n^{(2)}\{A_2\} = \lim_{n\to\infty} \frac{1}{n} \log P_\rho\{L_n \in A_2\} = -\min_{\nu \in A_2} I_\rho^{(2)}(\nu), \tag{2.17}$$

where $I_\rho^{(2)}(\nu) = \sum_{i=1}^r \nu_i \log(\nu_i/\rho_i)$. We now state a generalization. If $\nu \in \mathcal{M}(\mathbb{R}^d)$ is absolutely continuous with respect to $\rho \in \mathcal{M}(\mathbb{R}^d)$, then we write $\nu \ll \rho$ and denote by $d\nu/d\rho$ the Radon–Nikodym derivative of ν with respect to ρ.

Theorem II.4.3. *Let $X_1, X_2, \ldots$ be a sequence of* i.i.d. *random vectors taking values in $\mathbb{R}^d$ and let $\rho \in \mathcal{M}(\mathbb{R}^d)$ be the distribution of X_1. For $\nu \in \mathcal{M}(\mathbb{R}^d)$, define the relative entropy of ν with respect to ρ by the formula*[4]

$$I_\rho^{(2)}(\nu) = \begin{cases} \displaystyle\int_{\mathbb{R}^d} \log \frac{d\nu}{d\rho}(x)\, \nu(dx) & \text{if } \nu \ll \rho \text{ and } \displaystyle\int_{\mathbb{R}^d} \left| \log \frac{d\nu}{d\rho} \right| d\nu < \infty, \\ \infty & \text{otherwise.} \end{cases} \tag{2.18}$$

Let L_n denote the empirical measure $n^{-1} \sum_{j=1}^n \delta_{X_j}$, $n = 1, 2, \ldots$. Then the following conclusions hold.

(a) *$\{Q_n^{(2)}\}$, the distributions of $\{L_n\}$ on $\mathcal{M}(\mathbb{R}^d)$, have a large deviation property with $a_n = n$ and entropy function $I_\rho^{(2)}$.*

(b) *$I_\rho^{(2)}(\nu)$ is a convex function of ν. $I_\rho^{(2)}(\nu)$ measures the discrepancy between ν and ρ in the sense that $I_\rho^{(2)}(\nu) \geq 0$ with equality if and only if $\nu = \rho$.*

This theorem is proved in Section VIII.2 in the case of a finite state space.

Theorem II.4.3 implies the limit (2.17). Indeed, if the support of ρ is $\Gamma = \{x_1, x_2, \ldots, x_r\}$, then $Q_n^{(2)}\{\mathcal{M}(\Gamma)\} = 1$. For $\nu \in \mathcal{M}(\Gamma)$, $I_\rho^{(2)}(\nu)$ in (2.18) reduces to the sum $\sum_{i=1}^r \nu_i \log(\nu_i/\rho_i)$. The space $\mathcal{M}(\Gamma)$ with the topology of weak convergence is homeomorphic to the compact convex subset of $\mathbb{R}^r$ consisting of all vectors $\nu = (\nu_1, \ldots, \nu_r)$ which satisfy $\nu_i \geq 0$ and $\sum_{i=1}^r \nu_i = 1$ [Example A.8.4(b)]. Since $I_\rho^{(2)}(\nu)$ is a continuous function on this compact set, the set A_2 in (2.17) is an $I_\rho^{(2)}$-continuity set relative to $\mathcal{M}(\Gamma)$. The limit (2.17) follows from Theorem II.3.4.

Level-3. Let ρ be a Borel probability measure on $\mathbb{R}$ whose support is a finite set $\Gamma = \{x_1, x_2, \ldots, x_r\}$ with $x_1 < x_2 < \cdots < x_r$. $\mathcal{M}_s(\Gamma^{\mathbb{Z}})$ denotes the closed subset of $\mathcal{M}_s(\mathbb{R}^{\mathbb{Z}})$ consisting of probability measures P for which $P\{\Gamma^{\mathbb{Z}}\} = 1$. Let $\Sigma_1, \ldots, \Sigma_N$ be cylinder sets of $\Gamma^{\mathbb{Z}}$ such that $0 < P_\rho\{\Sigma_k\} < 1, k = 1, \ldots, N$, and define the closed set

$$B_3 = \{P \in \mathscr{M}_s(\Gamma^{\mathbb{Z}}): \max_{k=1,\cdots,N} |P\{\Sigma_k\} - P_\rho\{\Sigma_k\}| \geq \varepsilon\}.$$

In Theorem I.6.3, we stated that for all sufficiently small $\varepsilon > 0$

$$(2.19) \quad \lim_{n\to\infty} \frac{1}{n} \log Q_n^{(3)}\{B_3\} = \lim_{n\to\infty} \frac{1}{n} \log P_\rho\{R_n \in B_3\} = -\min_{P \in B_3} I_\rho^{(3)}(P).$$

$I_\rho^{(3)}(P)$ is defined by the limit (1.37). Theorem II.4.4 is a generalization. A new theme is a second formula for $I_\rho^{(3)}$, stated in (2.20) below.

Let ρ be a Borel probability measure on $\mathbb{R}^d$ and $\{X_j; j \in \mathbb{Z}\}$ the coordinate functions on $((\mathbb{R}^d)^{\mathbb{Z}}, \mathscr{B}((\mathbb{R}^d)^{\mathbb{Z}}), P_\rho)$. $\{X_j, j \in \mathbb{Z}\}$ is a sequence of i.i.d. random vectors with distribution ρ. If P is a measure in $\mathscr{M}_s((\mathbb{R}^d)^{\mathbb{Z}})$, then let $\tilde{P}_\omega$ be a regular conditional distribution, with respect to P, of X_1 given the σ-field $\mathscr{F}(\{X_j, j \leq 0\})$ [Theorem A.6.3]. For each fixed ω, $\tilde{P}_\omega$ is a probability measure on $\mathscr{B}(\mathbb{R}^d)$. For each Borel subset B of $\mathbb{R}^d$, $\tilde{P}_\omega\{B\}$ is a version of the conditional probability $P\{X_1 \in B | \{X_j, j \leq 0\}\}$ and so is a measurable function of $(\ldots, X_{-2}(\omega), X_{-1}(\omega), X_0(\omega))$. There exists a version of $I_\rho^{(2)}(\tilde{P}_\omega)$, the relative entropy of $\tilde{P}_\omega$ with respect to ρ, which is a non-negative measurable function of $(\ldots, X_{-2}(\omega), X_{-1}(\omega), X_0(\omega))$ [Theorem A.9.9]. Formula (2.20) for $I_\rho^{(3)}(P)$ is then well-defined.

Theorem II.4.4. *For $P \in \mathscr{M}_s((\mathbb{R}^d)^{\mathbb{Z}})$, define the mean relative entropy of P with respect to P_ρ by*

$$(2.20) \qquad I_\rho^{(3)}(P) = \int_{(\mathbb{R}^d)^{\mathbb{Z}}} I_\rho^{(2)}(\tilde{P}_\omega) P(d\omega),$$

where $I_\rho^{(2)}$ is defined in (2.18). *Define the empirical processes $\{R_n; n = 1, 2, \ldots\}$ by* (2.4). *Then the following conclusions hold.*

(a) *$\{Q_n^{(3)}\}$, the distributions of $\{R_n\}$ on $\mathscr{M}_s((\mathbb{R}^d)^{\mathbb{Z}})$, have a large deviation property with $a_n = n$ and entropy function $I_\rho^{(3)}$.*

(b) *$I_\rho^{(3)}(P)$ is an affine function of P. $I_\rho^{(3)}(P)$ measures the discrepancy between P and P_ρ in the sense that $I_\rho^{(3)}(P) \geq 0$ with equality if and only if $P = P_\rho$.*

The theorem will be proved in Chapter IX in the case of a finite state space. There, we will show that the limit (1.37) defining $I_\rho^{(3)}$ coincides with formula (2.20).

Theorem II.4.4 implies the limit (2.19). Indeed, if the support of ρ is $\Gamma = \{x_1, x_2, \ldots, x_r\}$, then $Q_n^{(3)}\{\mathscr{M}_s(\Gamma^{\mathbb{Z}})\} = 1$. Thus we may deduce the limit (2.19) by proving that the set B_3 is an $I_\rho^{(3)}$-continuity set relative to $\mathscr{M}_s(\Gamma^{\mathbb{Z}})$. Being an entropy function, $I_\rho^{(3)}$ attains its infimum over the closed set B_3 at some measure $\bar{P}$. $\bar{P}$ satisfies $|\bar{P}\{\Sigma_k\} - P_\rho\{\Sigma_k\}| \geq \varepsilon$ for some $k \in \{1, \ldots, N\}$. Assume that $\bar{P}\{\Sigma_k\} - P_\rho\{\Sigma_k\} \geq \varepsilon$. If $\bar{P}\{\Sigma_k\} - P_\rho\{\Sigma_k\} \leq -\varepsilon$, then the proof is similar. We define the closed convex set

$$C = \{P \in \mathscr{M}_s(\Gamma^{\mathbb{Z}}): P\{\Sigma_k\} - P_\rho\{\Sigma_k\} \geq \varepsilon\}.$$

Table II.2. Large Deviation Properties for Levels-1, 2, and 3

Level	Random quantity	A.s. limit	Properties of the entropy function	Where large deviation property is proved
1	S_n/n	m_ρ	$I_\rho^{(1)}$ convex $I_\rho^{(1)}(z) > I_\rho^{(1)}(m_\rho) = 0,\ z \neq m_\rho$	Section VII.5
2	L_n	ρ	$I_\rho^{(2)}$ convex $I_\rho^{(2)}(\nu) > I_\rho^{(2)}(\rho) = 0,\ \nu \neq \rho$	Section VIII.2 (finite state space)
3	R_n	P_ρ	$I_\rho^{(3)}$ affine $I_\rho^{(3)}(P) > I_\rho^{(3)}(P_\rho) = 0,\ P \neq P_\rho$	Chapter IX (finite state space)

The interior of C is the set of $P \in \mathscr{M}_s(\Gamma^{\mathbb{Z}})$ satisfying $P\{\Sigma_k\} - P_\rho\{\Sigma_k\} > \varepsilon$. If P_0 is any measure in int C, then $I_\rho^{(3)}(P_0)$ is finite [Theorem IX.2.3], and since $I_\rho^{(3)}$ is affine

$$\inf_{P \in B_3} I_\rho^{(3)}(P) = I_\rho^{(3)}(\bar{P}) = \lim_{n\to\infty}\left(\frac{1}{n} I_\rho^{(3)}(P_0) + \left(1 - \frac{1}{n}\right) I_\rho^{(3)}(\bar{P})\right)$$
$$= \lim_{n\to\infty} I_\rho^{(3)}\left(\frac{1}{n} P_0 + \left(1 - \frac{1}{n}\right)\bar{P}\right).$$

Since $n^{-1}P_0 + (1 - n^{-1})\bar{P}$ belongs to int C, it follows that $\inf_{P \in \operatorname{int} C} I_\rho^{(3)}(P) = \inf_{P \in B_3} I_\rho^{(3)}(P)$. Since int C is a subset of int B_3, we conclude that B_3 is an $I_\rho^{(3)}$-continuity set relative to $\mathscr{M}_s(\Gamma^{\mathbb{Z}})$.

The large deviation properties for the three levels are summarized in Table II.2.

II.5. Contraction Principles

In Chapter I, we stated contraction principles relating levels-1 and 2 [see (1.17)] and levels-2 and 3 [see (1.40)]. These generalize essentially without change. But they also have additional features that are worth emphasizing. The first theorem states the contraction principle relating levels-1 and 2.

Theorem II.5.1.* *Let ρ be a Borel probability measure on $\mathbb{R}^d$ such that $c_\rho(t) = \log \int_{\mathbb{R}^d} \exp\langle t, x\rangle \rho(dx)$ is finite for all $t \in \mathbb{R}^d$. Define $I_\rho^{(1)}$ by (2.14) and $I_\rho^{(2)}$ by (2.18). Then for all points $z \in \mathbb{R}^d$*

$$I_\rho^{(1)}(z) = \inf\left\{ I_\rho^{(2)}(\nu) : \nu \in \mathscr{M}(\mathbb{R}^d), \int_{\mathbb{R}} x\nu(dx) = z \right\}. \tag{2.21}$$

For $z = m_\rho$, (2.21) is certainly valid since $I_\rho^{(1)}(m_\rho)$ equals 0 and the in-

*Theorem II.5.1 is proved in Donsker–Varadhan (1976a, page 425). It will be proved in Chapter VIII [Theorems VIII.3.1, VIII.4.1] under additional hypotheses on ρ.

fimum on the right-hand side is attained at $\nu = \rho$ ($I_\rho^{(2)}(\rho) = 0$). For other z the right-hand side may be ∞ unless there exists a probability measure ν on $\mathbb{R}^d$ which is absolutely continuous with respect to ρ and satisfies $\int_{\mathbb{R}^d} x\nu(dx) = z$. In order to determine those points z for which the terms in (2.21) are finite, consider again the definition (2.14) of $I_\rho^{(1)}$. The free energy function $c_\rho(t)$ is convex and differentiable [Theorem VII.5.1]. Hence the supremum in (2.14) is attained at $t \in \mathbb{R}^d$ if and only if

$$\nabla_t(\langle t, z\rangle - c_\rho(t)) = 0 \qquad \text{or } \nabla c_\rho(t) = z.$$

Suppose that this equation has a solution $t = t(z)$. The gradient of c_ρ can be calculated by differentiating under the integral sign; that is,

$$\frac{\partial}{\partial t_i}\int_{\mathbb{R}^d} \exp\langle t, x\rangle \rho(dx) = \int_{\mathbb{R}^d} x_i \exp\langle t, x\rangle \rho(dx)$$

and so

$$\nabla c_\rho(t) = z = \int_{\mathbb{R}^d} x \exp\langle t, x\rangle \rho(dx) \cdot \frac{1}{\int_{\mathbb{R}^d} \exp\langle t, x\rangle \rho(dx)}.$$

Define ρ_t to be the Borel probability measure on $\mathbb{R}^d$ which is absolutely continuous with respect to ρ and which has Radon–Nikodym derivative[5]

(2.22) $$\frac{d\rho_t}{d\rho}(x) = \exp\langle t, x\rangle \cdot \frac{1}{\int_{\mathbb{R}^d} \exp\langle t, x\rangle \rho(dx)}.$$

We see that $\nabla c_\rho(t)$ equals the mean of ρ_t. Since $\nabla c_\rho(t) = z$, we conclude that the right-hand side of (2.21) is finite for z in the range of ∇c_ρ; e.g., the point $z = m_\rho$ corresponds to $t = 0$ and $\rho_0 = \rho$. In fact, we will see that $I_\rho^{(2)}(\nu)$ attains its infimum over the set $\{\nu \in \mathscr{M}(\mathbb{R}^d): \int_{\mathbb{R}^d} x\nu(dx) = z\}$ at the unique measure $\nu = \rho_t$ and $I_\rho^{(1)}(z)$ equals $I_\rho^{(2)}(\rho_t)$. The range of ∇c_ρ is closely related to the support of ρ.

For coin tossing, where $\rho = \frac{1}{2}\delta_0 + \frac{1}{2}\delta_1$ and $I_\rho^{(1)}$ is given by (1.9), these assertions are easy to verify. By Example II.4.2(a), the equation $c_\rho'(t) = z$ has for $z \in (0, 1)$ a unique solution $t(z) = \log(z/(1-z))$. Hence

$$\rho_{t(z)} = \frac{1}{2}\frac{1}{\int_{\mathbb{R}} \exp(t(z)x)\rho(dx)}\delta_0 + \frac{1}{2}\frac{\exp(t(z))}{\int_{\mathbb{R}} \exp(t(z)x)\rho(dx)}\delta_1 = (1-z)\delta_0 + z\delta_1.$$

Since $\rho_{t(z)}$ is the only probability measure which is absolutely continuous with respect to ρ and which has mean z, the right-hand side of (2.21) indeed attains its infimum at the unique measure $\nu = \rho_{t(z)}$. A short calculation shows that $I_\rho^{(2)}(\rho_{t(z)}) = I_\rho^{(1)}(z)$. This equality also holds for $z = 0$ (resp., $z = 1$) with $\rho_{t(z)}$ replaced by δ_0 (resp., δ_1), and the infimum in (2.21) is attained at this unique measure. Finally, for $z \notin [0, 1]$ $I_\rho^{(1)}(z) = \infty$; if ν has mean z, then $\nu \ll \rho$ cannot hold and $I_\rho^{(2)}(\nu) = \infty$. Hence both sides of (2.21) equal ∞.

If ρ is a Borel probability measure on $\mathbb{R}$ whose support is a finite set Γ, then the situation is analogous to coin tossing. $I_\rho^{(2)}(\nu)$ equals ∞ unless $\nu \ll \rho$, and $\nu \ll \rho$ is equivalent to ν belonging to $\mathcal{M}(\Gamma)$. Hence any ν in $\mathcal{M}(\mathbb{R})$ but not in $\mathcal{M}(\Gamma)$ contributes $I_\rho^{(2)}(\nu) = \infty$ to the infimum in (2.21). The next theorem will be proved in Section VIII.3 [Theorem VIII.3.1].

Theorem II.5.2. *Let ρ be a Borel probability measure on $\mathbb{R}$ whose support is a finite set $\Gamma = \{x_1, x_2, \ldots, x_r\}$ with $x_1 < x_2 < \cdots < x_r$. For t real, define*

$$\rho_{t,i} = \frac{\exp(tx_i)\rho_i}{\int_{\mathbb{R}} \exp(tx)\rho(dx)} = \frac{\exp(tx_i)\rho_i}{\sum_{j=1}^{r} \exp(tx_j)\rho_j} \tag{2.23}$$

and let ρ_t be the probability measure $\sum_{i=1}^r \rho_{t,i}\delta_{x_i}$ in $\mathcal{M}(\Gamma)$. Define

$$c_\rho(t) = \log \int_\Gamma \exp(tx)\rho(dx) = \log \sum_{i=1}^{r} \exp(tx_i)\rho_i.$$

The following conclusions hold.

(a) *For each point $z \in (x_1, x_r)$, there exists a unique value $t = t(z)$ such that $c'_\rho(t) = \int_{\mathbb{R}} x\rho_t(dx) = z$.*

(b) *For each point $z \in (x_1, x_r)$, $I_\rho^{(2)}(\nu)$ attains its infimum over the set $\{\nu \in \mathcal{M}(\Gamma): m_\nu = z\}$ at the unique measure $\rho_{t(z)}$ and*

$$I_\rho^{(1)}(z) = I_\rho^{(2)}(\rho_{t(z)}) = \inf\left\{ I_\rho^{(2)}(\nu): \nu \in \mathcal{M}(\Gamma), \int_{\mathbb{R}^d} x\nu(dx) = z \right\} < \infty. \tag{2.24}$$

For $z = x_1$ (resp., $z = x_r$) the previous sentence is true with $\rho_{t(z)}$ replaced by δ_{x_1} (resp., δ_{x_r}). For $z \notin [x_1, x_r]$ $I_\rho^{(1)}(z)$ equals ∞, and since there is no $\nu \in \mathcal{M}(\Gamma)$ with mean z, the infimum in (2.24) is over an empty set.*

The contraction principle stated in part (b) has an interesting interpretation as a principle for inferring a distribution under insufficient knowledge. Assume that ρ is the uniform measure on the set $\Gamma = \{x_1, x_2, \ldots, x_r\}$ with $x_1 < x_2 < \cdots < x_r$ ($\rho_i = 1/r$ for each i). Then for $\nu \in \mathcal{M}(\Gamma)$ $I_\rho^{(2)}(\nu)$ equals $\log r - H(\nu)$, where $H(\nu) = -\sum_{i=1}^r \nu_i \log \nu_i$ is the Shannon entropy of ν. As we saw in Example I.4.2, $H(\nu)$ is a measure of the randomness of ν. The measure $\rho_{t(z)}$ in Theorem II.5.2 gives the infimum of $I_\rho^{(2)}(\nu)$, or the supremum of $H(\nu)$, over the set of ν in $\mathcal{M}(\Gamma)$ satisfying $\int_{\mathbb{R}} x\nu(dx) = z$. Hence $\rho_{t(z)}$ is the most random measure in $\mathcal{M}(\Gamma)$ satisfying this constraint. Consider now a sequence of i.i.d. random variables with state space Γ and distribution ρ. Suppose that ρ is changed to another nondegenerate measure in $\mathcal{M}(\Gamma)$ and that the only information supplied is that the mean of the new distribution is z. Then $\rho_{t(z)}$, being the most random measure in $\mathcal{M}(\Gamma)$ with mean z, is the most reasonable choice for the new distribution. This interpretation of the contraction principle has been advanced by Jaynes in a series of papers beginning in 1957.

*For $z = x_1$ (resp., $z = x_r$), the only measure $\nu \in \mathcal{M}(\Gamma)$ with mean z is δ_{x_1} (resp., δ_{x_r}).

The contraction principle relating levels-2 and 3 is given next. As in the contraction principle relating levels-1 and 2, there is a unique measure which gives the infimum provided $I_\rho^{(2)}(\nu)$ is finite. Ω denotes the space $(\mathbb{R}^d)^{\mathbb{Z}}$.

Theorem II.5.3.* *Given ρ a Borel probability measure on $\mathbb{R}^d$, define $I_\rho^{(2)}$ by (2.18) and $I_\rho^{(3)}$ by (2.20). For $P \in \mathcal{M}_s(\Omega)$, define $\pi_1 P$ to be its one-dimensional marginal. Let ν be a Borel probability measure on $\mathbb{R}^d$. Then the following conclusions hold.*

(a) $I_\rho^{(2)}(\nu) = \inf\{I_\rho^{(3)}(P) : P \in \mathcal{M}_s(\Omega), \pi_1 P = \nu\}$.

(b) *Let P_ν be the infinite product measure with $\pi_1 P_\nu = \nu$. If $I_\rho^{(2)}(\nu)$ is finite, then $I_\rho^{(3)}(P)$ attains its infimum over the set $\{P \in \mathcal{M}_s(\Omega) : \pi_1 P = \nu\}$ at the unique measure P_ν and*

$$I_\rho^{(2)}(\nu) = I_\rho^{(3)}(P_\nu) = \inf\{I_\rho^{(3)}(P) : P \in \mathcal{M}_s(\Omega), \pi_1 P = \nu\}. \tag{2.25}$$

When $\Gamma = \{x_1, x_2, \ldots, x_r\}$ and ρ is the uniform measure on Γ, we recall from Example I.6.2 that $I_\rho^{(3)}(P)$ equals $\log r - h(P)$, where the mean entropy $h(P)$ is a measure of the randomness of P. In this case, the measure P_ν in Theorem II.5.3 is the most random strictly stationary probability measure with one-dimensional marginal ν.

For applications to statistical mechanics, it is useful to combine the two contraction principles already stated and obtain a new one relating levels-1 and 3. We formulate it for a finite state space $\Gamma = \{x_1, x_2, \ldots, x_r\}$ with $x_1 < x_2 < \cdots < x_r$ and consider only points $z \in [x_1, x_r]$.

Theorem II.5.4. *Assume the hypotheses of Theorem* II.5.2. *Fix $z \in (x_1, x_r)$. Then $I_\rho^{(3)}(P)$ attains its infimum over the set $\{P \in \mathcal{M}_s(\Gamma^{\mathbb{Z}}) : \int_\Gamma x \pi_1 P(dx) = z\}$ at the unique measure $P_{\rho_{t(z)}}$ and*

$$I_\rho^{(1)}(z) = I_\rho^{(3)}(P_{\rho_{t(z)}}) = \inf\left\{I_\rho^{(3)}(P) : P \in \mathcal{M}_s(\Gamma^{\mathbb{Z}}), \int_\Gamma x \pi_1 P(dx) = z\right\} < \infty. \tag{2.26}$$

If $z = x_1$ (resp., $z = x_r$), then the previous sentence is true with $P_{\rho_{t(z)}}$ replaced by $P_{\delta_{x_1}}$ (resp. $P_{\delta_{x_r}}$).

Sketch of Proof. Calculate the infimum in (2.26) in two steps: over all $P \in \mathcal{M}_s(\Gamma^{\mathbb{Z}})$ with fixed one-dimensional marginal $\pi_1 P = \nu \in \mathcal{M}(\Gamma)$, then over all $\nu \in \mathcal{M}(\Gamma)$ with mean z. The infimum over P is attained at the unique measure P_ν. Since $I_\rho^{(3)}(P_\nu) = I_\rho^{(2)}(\nu)$, the theorem follows from Theorem II.5.2. □

The three contraction principles are summarized in Table II.3 for measures $\rho \in \mathcal{M}(\mathbb{R}^d)$.

* Theorem II.5.3 is proved in Chapter IX in the case where the support of ρ is a finite set.

Table II.3. Contraction Principles for $\rho \in \mathcal{M}(\mathbb{R}^d)$

Levels	Contraction principle	Unique minimizing measure	Where proved
1 and 2	$I_\rho^{(1)}(z) = \inf\{I_\rho^{(2)}(v) : \int_{\mathbb{R}^d} xv(dx) = z\}$	For $v = \rho_{t(z)}$, $I_\rho^{(2)}(v) = I_\rho^{(1)}(z)$.	Theorems VIII.3.1 and VIII.4.1
2 and 3	$I_\rho^{(2)}(v) = \inf\{I_\rho^{(3)}(P) : \pi_1 P = v\}$	For $P = P_v$, $I_\rho^{(3)}(P) = I_\rho^{(2)}(v)$.	Theorem IX.3.1 (finite state space)
1 and 3	$I_\rho^{(1)}(z) = \inf\{I_\rho^{(3)}(P) : \int_{\mathbb{R}^d} x\, \pi_1 P(dx) = z\}$	For $P = P_{\rho_{t(z)}}$, $I_\rho^{(3)}(P) = I_\rho^{(1)}(z)$.	Combine previous two principles

II.6. Large Deviation Property for Random Vectors and Exponential Convergence

This is a good place to survey our progress. In Chapter I we considered large deviation results for i.i.d. random variables with a finite state space. The theorems in Sections II.2–II.5 generalized the Chapter I results to i.i.d. random vectors taking values in $\mathbb{R}^d$. These theorems included three levels of large deviations and contraction principles relating the levels. It would also be useful to have a large deviation result for dependent random vectors since these arise naturally in statistical mechanics and other applications. In this section, we state such a result. It can handle either independent or dependent random vectors, and it generalizes the level-1 theorem, Theorem II.4.1.

Let $\mathscr{W} = \{W_n ; n = 1, 2, \ldots\}$ be a sequence of random vectors which are defined on probability spaces $\{(\Omega_n, \mathscr{F}_n, P_n) ; n = 1, 2, \ldots\}$ and which take values in $\mathbb{R}^d$. For example, W_n may be the nth partial sum of a sequence of random vectors, but these vectors need not be i.i.d. Also, the probability spaces need not be equal. We define functions

$$(2.27) \qquad c_n(t) = \frac{1}{a_n} \log E_n\{\exp\langle t, W_n\rangle\}, \qquad n = 1, 2, \ldots, \quad t \in \mathbb{R}^d,$$

where $\{a_n ; n = 1, 2, \ldots\}$ is a sequence of positive real numbers tending to infinity, E_n denotes expectation with respect to P_n, and $\langle -, - \rangle$ is the Euclidean inner product on $\mathbb{R}^d$. The following hypotheses are assumed to hold.

(a) Each function $c_n(t)$ is finite for all $t \in \mathbb{R}^d$.

(b) $c_{\mathscr{W}}(t) = \lim_{n \to \infty} c_n(t)$ exists for all $t \in \mathbb{R}^d$ and is finite.

Hypothesis (b) is natural for statistical mechanical applications since $c_{\mathscr{W}}$ is closely related to the concept of free energy. We call $c_{\mathscr{W}}$ the *free energy function* of $\mathscr{W}$. A basic fact is that $c_n(t)$ and $c_{\mathscr{W}}(t)$ are convex functions on $\mathbb{R}^d$ [Proposition VII.1.1].

The hypotheses on $c_n(t)$ and $c_{\mathscr{W}}(t)$ are satisfied if, for example, W_n is the nth partial sum of i.i.d. random vectors $X_1, X_2, \ldots$ and $E\{\exp\langle t, X_1\rangle\}$ is finite for all $t \in \mathbb{R}^d$. In this case, with $a_n = n$

$$c_n(t) = \frac{1}{n} \log E\{\exp\langle t, W_n \rangle\} = \log E\{\exp\langle t, X_1 \rangle\} = c_{\mathscr{W}}(t).$$

This is the free energy function $c_\rho(t)$ in Theorem II.4.1.

Theorem II.6.1 states a large deviation property for the distributions of W_n/a_n. A natural candidate for entropy function is the Legendre–Fenchel transform

$$I_{\mathscr{W}}(z) = \sup_{t \in \mathbb{R}^d} \{\langle t, z \rangle - c_{\mathscr{W}}(t)\} \qquad \text{for } z \in \mathbb{R}^d. \tag{2.28}$$

Like the level-1 entropy function $I_\rho^{(1)}$, $I_{\mathscr{W}}$ is a convex, lower semicontinuous function, and it has compact level sets. The infimum of $I_{\mathscr{W}}$ over $\mathbb{R}^d$ is 0, and the infimum is attained at some point [Theorem VII.2.1]. However, Example II.6.2 shows that in contrast to $I_\rho^{(1)}$ the minimum point need not be unique. Whether or not $I_{\mathscr{W}}$ attains its infimum at a unique point has interesting consequences which will be explored below. We prove the next theorem in Chapter VII.

Theorem II.6.1. *Assume hypotheses* (a) *and* (b) *on page* 46. *Let Q_n be the distribution of W_n/a_n on $\mathbb{R}^d$. Then the following conclusions hold.*

(a) *$I_{\mathscr{W}}(z)$ is convex, lower semicontinuous, and non-negative. $I_{\mathscr{W}}(z)$ has compact level sets and* $\inf_{z \in \mathbb{R}^d} I_{\mathscr{W}}(z) = 0$.

(b) *The upper large deviation bound is valid:*

$$\limsup_{n \to \infty} \frac{1}{a_n} \log Q_n\{K\} \le -\inf_{z \in K} I_{\mathscr{W}}(z) \qquad \textit{for each closed set } K \textit{ in } \mathbb{R}^d.* \tag{2.29}$$

(c) *Assume in addition that $c_{\mathscr{W}}(t)$ is differentiable for all t. Then the lower large deviation bound is valid:*

$$\liminf_{n \to \infty} \frac{1}{a_n} \log Q_n\{G\} \ge -\inf_{z \in G} I_{\mathscr{W}}(z) \qquad \textit{for each open set } G \textit{ in } \mathbb{R}^d. \tag{2.30}$$

Hence, if $c_{\mathscr{W}}(t)$ is differentiable for all t, then (2.30) *and parts* (a) *and* (b) *imply that $\{Q_n; n = 1, 2, \ldots\}$ has a large deviation property with entropy function $I_{\mathscr{W}}$.*

This theorem generalizes Theorem II.4.1 since the free energy function $c_\rho(t)$ is differentiable for all t [Theorem VII.5.1].

Part (a) of Theorem II.6.1 states that the upper bound (2.29) is valid for all closed sets K. In the following simple example, the free energy function $c_{\mathscr{W}}$ is not differentiable at a single point and the lower bound (2.30) fails for a whole class of open sets G. Nevertheless, a large deviation property holds (with a nonconvex entropy function).

Example II.6.2. Let W_n have the distribution $\frac{1}{2}\delta_n + \frac{1}{2}\delta_{-n}$. Then with $a_n = n$

$$c_{\mathscr{W}}(t) = \lim_{n \to \infty} \frac{1}{n} \log \frac{1}{2}\{e^{nt} + e^{-nt}\} = \begin{cases} t & \text{if } t \ge 0, \\ -t & \text{if } t \le 0. \end{cases}$$

*See footnote on page 36.

Thus, $c_{\mathscr{W}}(t) = |t|$, which is not differentiable at 0, and

$$I_{\mathscr{W}}(z) = \sup_{t \in \mathbb{R}} \{tz - |t|\} = \begin{cases} 0 & \text{if } |z| \leq 1, \\ \infty & \text{if } |z| > 1. \end{cases}$$

$I_{\mathscr{W}}$ attains its infimum not at a single point, but on the whole interval $[-1, 1]$. If G is any open subset of $(-1, 1)$, then $Q_n\{G\} = 0$ and $\log Q_n\{G\} = -\infty$, while the infimum of $I_{\mathscr{W}}$ over G equals 0. Thus the lower bound (2.30) fails. On the other hand, it is easy to see that the distributions $\{Q_n\}$ of $\{W_n/n\}$ have a large deviation property with entropy function $I(z) = 0$ for $z = 1$ and $z = -1$, $I(z) = \infty$ for $z \notin \{1, -1\}$.

In general, if $c_{\mathscr{W}}(t)$ exists but is not differentiable for all t, then the validity of the large deviation property for the distributions of $\{W_n/a_n\}$ is an unsolved problem. If the large deviation property does hold with some entropy function $I(z)$, then the function $I_{\mathscr{W}}(z)$ defined by (2.28) equals the closed convex hull of $I(z)$; the latter is defined as the largest lower semicontinuous, convex function majorized by $I(z)$ [Problem VII.8.2].

We next introduce a notion of stochastic convergence which is useful in statistical mechanics. Let $\{W_n; n = 1, 2, \ldots\}$ be a sequence of random vectors on probability spaces $\{(\Omega_n, \mathscr{F}_n, P_n); n = 1, 2, \ldots\}$ and $\{a_n; n = 1, 2, \ldots\}$ a sequence of positive real numbers which tend to infinity. We say that W_n/a_n *converges exponentially* to a constant z_0, and write $W_n/a_n \xrightarrow{\text{exp}} z_0$, if for any $\varepsilon > 0$ there exists a number $N = N(\varepsilon) > 0$ such that

$$(2.31) \quad P_n\{\| W_n/a_n - z_0\| \geq \varepsilon\} \leq e^{-a_n N} \qquad \text{for all sufficiently large } n.$$

An example of exponential convergence was given in (2.16) ($S_n/n \xrightarrow{\text{exp}} m_\rho$).[6]

In statistical mechanics, the parameters $\{a_n\}$ may represent the numbers of particles in a sequence of systems indexed by $\{n\}$. These particles may assume different microstates; e.g., positions and velocities of molecules in a gas or spins in a magnet. The $\{W_n\}$ represent microscopic sums (e.g., total energy or total magnetization) which are proportional to $\{a_n\}$. The inequality (2.31) states that in the limit $n \to \infty$ all but an exponentially small set of configurations have essentially the same value z_0 of W_n per particle; z_0 is the equilibrium value.

We may apply Theorems II.3.3 and II.6.1 to give a simple criterion for the exponential convergence of W_n/a_n to a constant z_0. According to Theorem II.6.1, if the free energy function $c_{\mathscr{W}}(t)$ exists, then the Legendre–Fenchel transform $I_{\mathscr{W}}(z)$ is non-negative and lower semicontinuous, and it has compact level sets. Also the upper large deviation bound (2.29) is valid for all closed sets K. These are hypotheses (a)–(c) of large deviation property. Now suppose that $I_{\mathscr{W}}(z)$ attains its infimum of 0 at a unique point z_0 of $\mathbb{R}^d$. By Theorem II.3.3 there exists for any $\varepsilon > 0$ a number $N = N(\varepsilon) > 0$ such that

$$P_n\{\| W_n/a_n - z_0\| \geq \varepsilon\} \leq e^{-a_n N} \qquad \text{for all sufficiently large } n.$$

Thus W_n/a_n converges exponentially to z_0.

The next theorem strengthens this result considerably. It states that W_n/a_n converges exponentially to a constant if and only if the free energy function $c_{\mathscr{W}}(t)$ is differentiable at $t = 0$. Furthermore, the differentiability is equivalent to $I_{\mathscr{W}}(z)$ attaining its infimum at a unique point.

Theorem II.6.3. *Assume hypotheses* (a) *and* (b) *on page* 46. *Then the following statements are equivalent.*

(a) $W_n/a_n \xrightarrow{\text{exp}} z_0$.
(b) *$c_{\mathscr{W}}(t)$ is differentiable at $t = 0$ and $\nabla c_{\mathscr{W}}(0) = z_0$.*
(c) *$I_{\mathscr{W}}(z)$ attains its infimum over $\mathbb{R}^d$ at the unique point $z = z_0$.*

The theorem will be proved in Section VII.6. For now, we prove only that if $c_{\mathscr{W}}(t)$ is differentiable at $t = 0$, then

$$E\{W_n/a_n\} \to \nabla c_{\mathscr{W}}(0) \qquad \text{as } n \to \infty. \tag{2.32}$$

This shows that $z_0 = \nabla c_{\mathscr{W}}(0)$ is a plausible value for the exponential limit of W_n/a_n. The proof of (2.32) is a nice application of convexity. By definition, $c_{\mathscr{W}}(t) = \lim_{n\to\infty} c_n(t)$, where $c_n(t) = a_n^{-1} \log E_n\{\exp\langle t, W_n\rangle\}$. The function $c_n(t)$ is convex and is differentiable at $t = 0$, and

$$\nabla c_n(0) = E_n\{W_n/a_n\}$$

[Proposition VII.1.1 and Remark VII.5.4]. By Lemma IV.6.3, we may interchange derivatives and limits to deduce that $\nabla c_n(0) \to \nabla c_{\mathscr{W}}(0)$ as $n \to \infty$. This gives (2.32).

Our final theorem shows that if the $\{W_n\}$ are all defined on the same space, then exponential convergence implies almost sure convergence provided $\sum_{n=1}^{\infty} \exp(-a_n N)$ is finite for all $N > 0$. This extra condition is needed in order to apply the Borel–Cantelli lemma [see Theorem A.5.3]. When used together, Theorems II.6.3 and II.6.4 are a handy tool for proving strong laws of large numbers.

Theorem II.6.4. *Assume that the random vectors $\{W_n; n = 1, 2, \ldots\}$ are all defined on the same probability space $(\Omega, \mathscr{F}, P)$. If $\sum_{n=1}^{\infty} \exp(-a_n N) < \infty$ for all $N > 0$, then $W_n/a_n \xrightarrow{\text{exp}} z_0$ implies $W_n/a_n \xrightarrow{\text{a.s.}} z_0$.*

The theorems in this section were motivated by large deviation results in statistical mechanics proved by Lanford (1973). In our formulation of these results for random vectors $\{W_n\}$, the free energy function $c_{\mathscr{W}}(t)$ plays a central role. If this function is differentiable for all t, then the distributions of $\{W_n/a_n\}$ have a large deviation property [Theorem II.6.1]. The differentiability of $c_{\mathscr{W}}(t)$ at $t = 0$ is equivalent to the exponential convergence of $\{W_n/a_n\}$ to a constant [Theorem II.6.3]. These theorems will have many applications in subsequent chapters. Theorem II.6.1 will be used to derive the level-1 theorem for i.i.d. random vectors [Theorem II.4.1] and the level-2 and 3 theorems for i.i.d. random variables with a finite state space

[Theorems II.4.3 and II.4.4]. Theorem II.6.3 will be applied to the study of phase transitions in statistical mechanics.

II.7. Varadhan's Theorem on the Asymptotics of Integrals

The basic problems and techniques in the subject of large deviations have been inspired by a number of different areas. These include probability, analysis, statistics, information theory, and statistical mechanics. Its roots in analysis go back to Laplace, who devised a method for studying the asymptotic behavior of real integrals of the form $\int_{\mathbb{R}} \exp(nF(x))dx$ as $n \to \infty$.[7]

For example, consider the integral $n! = \int_0^\infty x^n e^{-x} dx$, where n is a positive integer. Change variables to $z = x/n$, obtaining

$$n! = n^{n+1} \int_0^\infty \exp[-n(z - \log z)]dz.$$

Laplace says that the leading order asymptotic behavior of the integral is determined by the largest value of the integrand, in this case e^{-n} since $z - \log z$ has a minimum value of 1 at $z = 1$. In fact, it is not hard to prove that $n! \sim n^{n+1}e^{-n}$, in the sense that $n^{-1}\log(n!/n^{n+1}) \to -1$ as $n \to \infty$. By making a quadratic approximation about $z = 1$ and suitably expanding, one derives Stirling's formula $n! = n^n e^{-n}\sqrt{2\pi n}(1 + \varepsilon_n)$, where ε_n is an asymptotic series in powers of $1/n$ which converges to 0 as $n \to \infty$.

Similar problems arise in probability theory. Let ρ be a Borel probability measure on $\mathbb{R}^d$ and $X_1, X_2, \ldots$ i.i.d. random vectors with distribution ρ. Set $S_n = \sum_{j=1}^n X_j$. What is the leading order asymptotic behavior of the expectation

(2.33) $$Z_n^{(1)} = E\{\exp[nF(S_n(\omega)/n)]\},$$

where F is a bounded continuous function from $\mathbb{R}$ to $\mathbb{R}$? The special case where $X_1, X_2, \ldots$ are Gaussian $N(0, 1)$ random variables is easy. For then S_n/n is Gaussian $N(0, n^{-1})$, and so

$$Z_n^{(1)} = \left(\frac{n}{2\pi}\right)^{1/2} \int_{\mathbb{R}} \exp\left[n\left(F(z) - \frac{1}{2}z^2\right)\right] dz.$$

Since $n^{1/2n} \to 1$, Laplace tells us that

(2.34) $$\lim_{n\to\infty} \frac{1}{n} \log Z_n^{(1)} = \sup_{z\in\mathbb{R}} \{F(z) - \tfrac{1}{2}z^2\}.$$

Thus $Z_n^{(1)}$ grows or decays exponentially depending on the sign of the supremum.

One may determine the leading order asymptotic behavior of $Z_n^{(1)}$ for arbitrary i.i.d. random vectors $X_1, X_2, \ldots$ by applying the level-1 large deviation property. Indeed, if $Q_n^{(1)}$ is the distribution of S_n/n, then

(2.35) $$Z_n^{(1)} = \int_{\mathbb{R}^d} \exp(nF(z)) Q_n^{(1)}(dz).$$

Assume that ρ satisfies the hypothesis of Theorem II.4.1 ($c_\rho(t) < \infty$ for all $t \in \mathbb{R}^d$). Then $\{Q_n^{(1)}\}$ has a large deviation property, which may be summarized by the heuristic formula

$$dQ_n^{(1)}(z) \approx \exp(-nI_\rho^{(1)}(z))dz.$$

Inserting this in (2.35), we would like to conclude that by analogy with Laplace's method

(2.36) $$\lim_{n\to\infty} \frac{1}{n} \log Z_n^{(1)} = \sup_{z\in\mathbb{R}} \{F(z) - I_\rho^{(1)}(z)\}.$$

In order to check this against (2.34), recall from Example II.4.2(b) that for an $N(0, 1)$ distribution ρ, $I_\rho^{(1)}(z)$ equals $\frac{1}{2}z^2$. Hence in the Gaussian case (2.34) and (2.36) agree. The limit (2.36) is a consequence of Theorem II.7.1 below.

The asymptotic evaluation of $Z_n^{(1)}$ gives another way of interpreting large deviations. Suppose that the supremum in (2.36) is attained at a unique point z_0. According to (2.33), the configurations ω which determine the exponential growth or decay of $Z_n^{(1)}$ are those for which $S_n(\omega)/n$ is close to z_0 for all large n. Since in general $z_0 \neq E\{X_1\}$, these ω represent large deviations for S_n/n.

The following theorem gives the leading order asymptotic behavior of a class of integrals on a complete separable metric space. The theorem is due to Varadhan (1966). It is proved in Appendix B.1.[8]

Theorem II.7.1. *Let $\mathscr{X}$ be a complete separable metric space, $\mathscr{B}(\mathscr{X})$ the Borel σ-field of $\mathscr{X}$, and $\{Q_n; n = 1, 2, \ldots\}$ a sequence of probability measures on $\mathscr{B}(\mathscr{X})$. Assume that $\{Q_n\}$ has a large deviation property with constants $\{a_n\}$ and entropy function I. Let F be a continuous function from $\mathscr{X}$ to $\mathbb{R}$.*

(a) *Assume that* $\sup_{x\in\mathscr{X}} F(x)$ *is finite. Then* $\sup_{x\in\mathscr{X}}\{F(x) - I(x)\}$ *is finite and*

(2.37) $$\lim_{n\to\infty} \frac{1}{a_n} \log \int_{\mathscr{X}} \exp(a_n F(x)) Q_n(dx) = \sup_{x\in\mathscr{X}} \{F(x) - I(x)\}.$$

(b) *More generally, assume that F satisfies*

(2.38) $$\lim_{L\to\infty} \limsup_{n\to\infty} \frac{1}{a_n} \log \int_{\{F\geq L\}} \exp(a_n F(x)) Q_n(dx) = -\infty.$$

Then the limit (2.37) *holds and is finite. In particular, if F is bounded above on the union of the supports of the $\{Q_n\}$, then* (2.38) *is satisfied and thus the limit* (2.37) *holds and is finite.*

Let $X_1, X_2, \ldots$ be a sequence of i.i.d. random vectors with distribution $\rho \in \mathscr{M}(\mathbb{R}^d)$. As an application of Theorem II.7.1, we derive the formula

(2.39) $$c_\rho(t) = \sup_{z \in \mathbb{R}^d} \{\langle t, z\rangle - I_\rho^{(1)}(z)\},$$

where $I_\rho^{(1)}$ is the level-1 entropy function and $c_\rho(t) = \log E\{\exp\langle t, X_1\rangle\}$ [see (2.15)]. Assume for simplicity that the random vector X_1 is bounded.* Since $X_1, X_2, \ldots$ are i.i.d., $c_\rho(t)$ equals $n^{-1}\log E\{\exp\langle t, S_n\rangle\}$ and so

$$c_\rho(t) = \lim_{n\to\infty} \frac{1}{n} \log \int_{\mathbb{R}^d} \exp(n\langle t, z\rangle) Q_n^{(1)}(dz).$$

Since X_1 and thus S_n/n is bounded, the function $z \to \langle t, z\rangle$ is bounded on the union of the supports of the $\{Q_n^{(1)}\}$. Hence (2.39) follows from the previous theorem and Theorem II.4.1.

Theorem II.7.1 allows us to analyze the asymptotic behavior of a class of measures, examples of which will arise in our study of statistical mechanical models.

Theorem II.7.2. *Let $\mathscr{X}$ be a complete separable metric space and $\{Q_n; n = 1, 2, \ldots\}$ a sequence of probability measures on $\mathscr{B}(\mathscr{X})$. Assume that $\{Q_n\}$ has a large deviation property with constants $\{a_n\}$ and entropy function I. Let F be a continuous function from $\mathscr{X}$ to $\mathbb{R}$ such that $\int_{\mathscr{X}} \exp(a_n F(x)) Q_n(dx)$ is finite for each n and condition* (2.38) *is satisfied. For $n = 1, 2, \ldots$ and $A \in \mathscr{B}(\mathscr{X})$, define probability measures*

$$Q_{n,F}\{A\} = \int_A \exp(a_n F(x)) Q_n(dx) \cdot \frac{1}{\int_{\mathscr{X}} \exp(a_n F(x)) Q_n(dx)}.$$

The following conclusions hold.

(a) *The sequence $\{Q_{n,F}; n = 1, 2, \ldots\}$ has a large deviation property with constants $\{a_n\}$ and entropy function*

$$I_F(x) = I(x) - F(x) - \inf_{x \in \mathscr{X}} \{I(x) - F(x)\}.$$

(b) *Let K be a closed set in $\mathscr{X}$ which does not contain a minimum point of I_F. Then there exists a number $N = N(K) > 0$ such that*

$$Q_{n,F}\{K\} \le e^{-a_n N} \qquad \textit{for all sufficiently large } n.$$

In this sense, the asymptotic behavior of $\{Q_{n,F}\}$ is determined by the minimum points of I_F. In particular, if I_F has a unique minimum point x, then $Q_{n,F} \Rightarrow \delta_x$.

Sketch of Proof. (a) By the previous theorem,

$$\lim_{n\to\infty} \frac{1}{a_n} \log \int_{\mathscr{X}} \exp(a_n F(x)) Q_n(dx) = \sup_{x \in \mathscr{X}} \{F(x) - I(x)\} = -\inf_{x \in \mathscr{X}} \{I(x) - F(x)\}.$$

*See Problem II.9.9 for a more general case.

In Appendix B.2 we will prove that for each nonempty closed set K in $\mathscr{X}$

$$\limsup_{n\to\infty} \frac{1}{a_n} \log \int_K \exp(a_n F(x)) Q_n(dx) \le \sup_{x\in K} \{F(x) - I(x)\}$$
$$= -\inf_{x\in K} \{I(x) - F(x)\}$$

and that for each nonempty open set G in $\mathscr{X}$

$$\liminf_{n\to\infty} \frac{1}{a_n} \log \int_G \exp(a_n F(x)) Q_n(dx) \ge \sup_{x\in G} \{F(x) - I(x)\}$$
$$= -\inf_{x\in G} \{I(x) - F(x)\}.$$

The last three displays yield the upper and lower large deviation bounds for $\{Q_{n,F}\}$ with entropy function I_F. The function I_F is lower semicontinuous since I is lower semicontinuous and F is continuous. That I_F has compact level sets will be shown in the Appendix B.2.

(b) Theorems II.3.3 and A.8.2. □

The next theorem gives three different integrals that can be evaluated by Theorem II.7.1. Each evaluation uses one of the levels of large deviations. In the next three chapters, we will see that these integrals are associated with some basic statistical mechanical models.

Theorem II.7.3. *Let $X_1, X_2, \ldots$ be a sequence of* i.i.d. *random vectors with distribution $\rho \in \mathscr{M}(\mathbb{R}^d)$. Suppose that* $c_\rho(t) = \log E\{\exp\langle t, X_1\rangle\}$ is finite for all $t \in \mathbb{R}^d$. Let $F: \mathbb{R}^d \to \mathbb{R}$ and $G: \mathbb{R}^d \times \mathbb{R}^d \to \mathbb{R}$ be bounded continuous functions.*

(a) *Define $Z_n^{(1)} = E\{\exp(nF(\sum_{j=1}^n X_j/n)\}$. Then*

$$\lim_{n\to\infty} \frac{1}{n} \log Z_n^{(1)} = \sup_{z\in\mathbb{R}^d} \{F(z) - I_\rho^{(1)}(z)\}. \tag{2.40}$$

(b) *Define $Z_n^{(2)} = E\{\exp \sum_{j=1}^n F(X_j)\}$. Then*

$$\lim_{n\to\infty} \frac{1}{n} \log Z_n^{(2)} = \sup_{\nu\in\mathscr{M}(\mathbb{R}^d)} \left\{ \int_{\mathbb{R}^d} F(z)\nu(dz) - I_\rho^{(2)}(\nu) \right\}. \tag{2.41}$$

(c) *Define $Z_n^{(3)} = E\{\exp \sum_{j=1}^{n-1} G(X_j, X_{j+1})\}$. Then*

$$\lim_{n\to\infty} \frac{1}{n} \log Z_n^{(3)} = \sup_{P\in\mathscr{M}_s((\mathbb{R}^d)^{\mathbb{Z}})} \left\{ \int_{(\mathbb{R}^d)^{\mathbb{Z}}} G(\bar\omega_1, \bar\omega_2) P(d\bar\omega) - I_\rho^{(3)}(P) \right\}. \tag{2.42}$$

The proof of part (c) involves a comparison between $Z_n^{(3)}$ and a related expectation. Similar comparisons will recur in later chapters, and so we state the elementary fact as a lemma.

*This hypothesis is needed only to derive (2.40); (2.41)–(2.42) hold without any extra hypothesis on ρ.

Lemma II.7.4. *Let* $(\Omega, \mathscr{F}, P)$ *be a probability space and* f *and* g *bounded measurable functions from* Ω *to* $\mathbb{R}$. *Then*

$$\left| \log \int_\Omega e^f \, dP - \log \int_\Omega e^g \, dP \right| \le \|f - g\|_\infty .$$

Proof. Almost surely, we have

$$\exp(g) \cdot \exp(-\|f-g\|_\infty) \le \exp(f) \le \exp(g) \cdot \exp(\|f-g\|_\infty).$$

Integrating and taking logarithms completes the proof. □

Proof of Theorem II.7.3. (a) We have already seen $Z_n^{(1)}$ in (2.33). The limit (2.40) follows from Theorems II.4.1 (level-1 large deviation property) and Theorem II.7.1.

(b) For each ω

$$\int_{\mathbb{R}^d} F(z)\, L_n(\omega, dz) = \frac{1}{n} \sum_{j=1}^n \int_{\mathbb{R}^d} F(z) \delta_{X_j(\omega)}(dz) = \frac{1}{n} \sum_{j=1}^n F(X_j(\omega)).$$

Hence $Z_n^{(2)}$ can be written as

$$Z_n^{(2)} = E\left\{ \exp\left(n \int_{\mathbb{R}^d} F(z) L_n(\omega, dz) \right) \right\} = \int_{\mathscr{M}(\mathbb{R}^d)} \exp\left(n \int_{\mathbb{R}^d} F(z) \nu(dz) \right) Q_n^{(2)}(d\nu),$$

where $Q_n^{(2)}$ is the distribution of L_n on $\mathscr{M}(\mathbb{R}^d)$. The function $\nu \to \int_{\mathbb{R}^d} F(z)\nu(dz)$ is a bounded continuous function on $\mathscr{M}(\mathbb{R}^d)$. The limit (2.41) follows from Theorem II.4.3 (level-2 large deviation property) and Theorem II.7.1.

(c) We may suppose that $\{X_j; j \in \mathbb{Z}\}$ is the coordinate representation process on $\Omega = (\mathbb{R}^d)^{\mathbb{Z}}$. Lemma II.7.4 implies that since G is bounded, $Z_n^{(3)}$ has the same leading order asymptotic behavior as $\bar{Z}_n^{(3)} = E\{\exp(\sum_{j=1}^{n-1} G(X_j, X_{j+1}) + G(X_n, X_1))\}$. For each ω,

$$\begin{aligned} \int_\Omega G(\bar\omega_1, \bar\omega_2) R_n(\omega, d\bar\omega) &= \frac{1}{n} \sum_{j=1}^{n-1} \int_{\mathbb{R}^d \times \mathbb{R}^d} G(\bar\omega_1, \bar\omega_2)\, \delta_{X_j(\omega)}(d\bar\omega_1) \delta_{X_{j+1}(\omega)}(d\bar\omega_2) \\ &\quad + \frac{1}{n} \int_{\mathbb{R}^d \times \mathbb{R}^d} G(\bar\omega_1, \bar\omega_2)\, \delta_{X_n(\omega)}(d\bar\omega_1) \delta_{X_1(\omega)}(d\bar\omega_2) \\ &= \frac{1}{n} \sum_{j=1}^{n-1} G(X_j(\omega), X_{j+1}(\omega)) + \frac{1}{n} G(X_n(\omega), X_1(\omega)). \end{aligned}$$

Thus

$$\bar{Z}_n^{(3)} = \int_{\mathscr{M}_s(\Omega)} \exp\left(n \int_\Omega G(\bar\omega_1, \bar\omega_2) P(d\bar\omega) \right) Q_n^{(3)}(dP),$$

where $Q_n^{(3)}$ is the distribution of R_n on $\mathscr{M}_s(\Omega)$. The function $P \to \int_\Omega G(\bar\omega_1, \bar\omega_2) P(d\bar\omega)$ is a bounded continuous function on $\mathscr{M}_s(\Omega)$. The limit (2.42) follows from Theorem II.4.4 (level-3 large deviation property) and Theorem II.7.1.
□

The main results in the first part of this chapter concerned three levels of large deviations for i.i.d. random vectors taking values in $\mathbb{R}^d$. In Section II.6 the level-1 theorem was generalized to random vectors for which a free energy function exists. In that section we also introduced the notion of exponential convergence and related it to the large deviation theorem. In the next three chapters, large deviations will be applied to models in statistical mechanics.

II.8. Notes

1 (page 30) Azencott (1980), Stroock (1984), and Varadhan (1984) treat large deviations for discrete time processes and continuous time processes with applications.

2 (page 35). An extended real-valued function on a complete separable metric space is lower semicontinuous if and only if it has closed level sets. Hence hypothesis (b) in Definition II.3.1 of large deviation property includes hypothesis (a). But it is useful to state separately the hypothesis of lower semicontinuity. The definition of large deviation property is taken from Varadhan (1966, Section 3). Theorem II.3.2 is due to S.R.S. Varadhan (private communication). Theorem II.3.4 is given in Donsker and Varadhan (1977a).

3 (page 36) Orey (1985) has a different proof of Theorem II.3.2 (uniqueness of the entropy function). His proof does not use hypothesis (b) of large deviation property (compact level sets).

4 (page 40) In the definition (2.18) of $I_\rho^{(2)}(\nu)$, suppose that $\nu \ll \rho$ and $d\nu/d\rho = f(x)$. Since $f(x)\log f(x)$ is bounded for small $f(x)$, $f(x)(\log f(x))^-$ is always integrable with respect to ρ. Thus $f(x)\log f(x) \in L^1(\mathbb{R}^d, \rho)$ if and only if $f(x)(\log f(x))^+ \in L^1(\mathbb{R}^d, \rho)$, and then

$$I_\rho^{(2)}(\nu) = \int_{\mathbb{R}^d} \log f(x)\, \nu(dx) = \int_{\mathbb{R}^d} f(x)\log f(x)\, \rho(dx).$$

5 (page 43). The measure ρ_t with Radon–Nikodym derivative $d\rho_t/d\rho$ given by (2.22) is said to be in the exponential family generated by ρ. These families are a basic technical tool in many branches of statistics [Barndorff–Nielsen (1978), Johansen (1979), Brown (1985)].

6 (page 48). Exponential convergence for random variables has been studied by Baum, Katz, and Read (1962).

7 (page 50). Laplace's method is discussed in Erdélyi (1956) and Henrici (1977), for example.

8 (page 51). Special cases of Theorems II.7.1 and II.7.2 are given in Schilder (1966) and Pincus (1968). Varadhan (1966, Section 3) generalizes Theorem II.7.1 to sequences of functions $\{F_n; n = 1, 2, \ldots\}$.

II.9. Problems

II.9.1. (a) (page 32) L_n maps Ω into $\mathscr{M}(\mathbb{R}^d)$. Prove that if A is a Borel subset of $\mathscr{M}(\mathbb{R}^d)$, then $L_n^{-1}A$ belongs to the σ-field $\mathscr{F}$ of Ω. [*Hint:* Consider a basic open set A of $\mathscr{M}(\mathbb{R}^d)$.]

(b) (page 32) R_n maps Ω into $\mathscr{M}_s((\mathbb{R}^d)^{\mathbb{Z}})$. Prove that if A is a Borel subset of $\mathscr{M}_s((\mathbb{R}^d)^{\mathbb{Z}})$, then $R_n^{-1}A$ belongs to the σ-field $\mathscr{F}$ of Ω.

II.9.2. Let F be an extended real-valued function on a complete separable metric space $\mathscr{X}$.

(a) Prove that F is lower semicontinuous (l.s.c.) if and only if all its level sets $\{x\in\mathscr{X}: F(x)\leq b\}$, b real, are closed.

(b) Assume that F does not take the value $-\infty$. Let K be a nonempty compact subset of $\mathscr{X}$. Prove that if F is l.s.c., then $\inf_{x\in K}F(x) > -\infty$ and F attains its infimum over K.

II.9.3. Let $\mathscr{X}$ be a complete separable metric space and $\{Q_n; n=1,2,\ldots\}$ a sequence of probability measures on $\mathscr{B}(\mathscr{X})$. Suppose that $\{Q_n\}$ has a large deviation property with $a_n = n$ and entropy function I. Given a Borel subset A of $\mathscr{X}$ and $0<\theta<1$, let $N(A,\theta)$ be the smallest number n_0 such that $Q_n\{A\}\leq\theta$ for all $n\geq n_0$. Define $I(A)=\inf_{x\in A}I(x)$. Prove that for any closed set K and open set G with $0 < I(K), I(G) < \infty$

$$\limsup_{\theta\to 0^+}\frac{N(K,\theta)}{\log 1/\theta}\leq\frac{1}{I(K)},\qquad \liminf_{\theta\to 0^+}\frac{N(G,\theta)}{\log 1/\theta}\geq\frac{1}{I(G)}.$$

II.9.4. Let $\mathscr{X}$ be a complete separable metric space and $\{Q_n; n=1,2,\ldots\}$ a sequence of probability measures on $\mathscr{B}(\mathscr{X})$. Suppose that $\{Q_n\}$ has a large deviation property with constants $\{a_n\}$ and entropy function I. Suppose that there exists a point x_0 in $\mathscr{X}$ with the property that if K is any closed set which does not contain x_0, then for some $N=N(K)>0$, $Q_n\{K\}\leq e^{-a_nN}$ for all sufficiently large n. Prove that

$$I(x) > I(x_0) = 0 \qquad \text{for all } x\neq x_0.$$

II.9.5. Let ρ be a Borel probability measure on $\mathbb{R}^d$, f a bounded continuous function from $\mathbb{R}^d$ to $\mathbb{R}$, and A a closed interval in $\mathbb{R}$ with nonempty interior. Define the set $C=\{\nu\in\mathscr{M}(\mathbb{R}^d): \int_{\mathbb{R}^d}f(x)\nu(dx)\in A\}$.

(a) Prove that C is a closed set in $\mathscr{M}(\mathbb{R}^d)$ and that the interior of C contains the set $\{\nu\in\mathscr{M}(\mathbb{R}^d): \int_{\mathbb{R}^d}f(x)\nu(dx)\in\operatorname{int}A\}$.

(b) Assume that $\inf_{\nu\in C}I_\rho^{(2)}(\nu)$ is finite. Prove that $I_\rho^{(2)}$ attains its infimum over C at some measure $\bar{\nu}$. [*Hint:* $I_\rho^{(2)}$ is lower semicontinuous and has compact level sets.]

(c) Assume that $\inf_{\nu\in C}I_\rho^{(2)}(\nu)$ is finite and that there exists a measure ν_0 for which $\int_{\mathbb{R}^d}f(x)\nu_0(dx)$ belongs to $\operatorname{int}A$ and $I_\rho^{(2)}(\nu_0)$ is finite. Prove that C is an $I_\rho^{(2)}$-continuity set. [*Hint:* Consider $I_\rho^{(2)}(n^{-1}\nu_0+(1-n^{-1})\bar{\nu})$, $n\to\infty$.]

II.9.6. [Donsker and Varadhan (1976a, page 431)] Parts (a) and (b) show how to derive the level-1 large deviation bounds from the level-2 large deviation bounds and the contraction principle (2.21). Let ρ be a Borel probability measure on $\mathbb{R}^d$ whose support is a bounded set Γ.

(a) Prove that the mapping $\nu \to m_\nu = \int_\Gamma x\nu(dx)$ is a continuous mapping of $\mathscr{M}(\Gamma)$ into $\mathbb{R}^d$. What is the image of $L_n(\omega, \cdot)$ under this mapping?

(b) Using part (a) and the large deviation bounds for L_n, prove that for each closed set K in $\mathbb{R}^d$,

$$\limsup_{n\to\infty} \frac{1}{n} \log P\{S_n/n \in K\} \leq -\inf_{z\in K} \inf\{I_\rho^{(2)}(\nu): \nu \in \mathscr{M}(\Gamma), m_\nu = z\}$$

and that for each open set G in $\mathbb{R}^d$

$$\liminf_{n\to\infty} \frac{1}{n} \log P\{S_n/n \in G\} \geq -\inf_{z\in G} \inf\{I_\rho^{(2)}(\nu): \nu \in \mathscr{M}(\Gamma), m_\nu = z\}.$$

Using (2.21), deduce the level-1 large deviation bounds with entropy function $I_\rho^{(1)}$.

(c) Let ρ be any Borel probability measure on $\mathbb{R}^d$. Derive the level-2 large deviation bounds from the level-3 large deviation bounds and the contraction principle (2.25).

II.9.7. Let ρ be a Borel probability measure on $\mathbb{R}^d$ with bounded support. Using Theorem II.3.2 (uniqueness of entropy function), derive the contraction principle (2.21) from the level-1 and 2 large deviation properties [Theorems II.4.1 and II.4.3]. [*Hint*: $J(z) = \inf\{I_\rho^{(2)}(\nu): \nu \in \mathscr{M}(\mathbb{R}^d), m_\nu = z\}$ is a lower semicontinuous function of $z \in \mathbb{R}^d$ and $J(z)$ has compact level sets [Donsker and Varadhan (1976a, page 425)].]

II.9.8. Fill in the details in the proof of Theorem II.5.4.

II.9.9. Let ρ be a Borel probability measure on $\mathbb{R}^d$ such that $c_\rho(t)$ is finite for all $t \in \mathbb{R}^d$. Using Theorem II.7.1(b), prove formula (2.39). [*Hint*: Verify (2.38). If $\langle t, z\rangle \geq L$, then $e^{-nL}e^{n\langle t, z\rangle} \geq 1$.]

II.9.10. Assume that in Theorem II.6.1, $c_{\mathscr{W}}(t)$ is finite and differentiable for all $t \in \mathbb{R}^d$. Using Theorem II.7.1, prove that $c_{\mathscr{W}}(t) = \sup_{z\in\mathbb{R}^d}\{\langle t, z\rangle - I_{\mathscr{W}}(z)\}$.

II.9.11. Let A be the closure of a nonempty bounded open set in $\mathbb{R}^d$ and F a continuous function from A to $\mathbb{R}$. Using Theorem II.7.1, prove that $\lim_{n\to\infty} n^{-1} \log \int_A \exp(nF(x))dx = \max_{x\in A} F(x)$.

II.9.12. Let F be a continuous function from $\mathbb{R}$ to $\mathbb{R}$ which satisfies $\limsup_{|x|\to\infty} F(x)/(x^2/2) < 1$. Using Theorem II.7.1, prove that

$$\lim_{n\to\infty} \frac{1}{n} \log \int_{\mathbb{R}} \exp[n(F(x) - \tfrac{1}{2}x^2)]dx = \sup_{x\in\mathbb{R}} \{F(x) - \tfrac{1}{2}x^2\} < \infty.$$

II.9.13. Let F be a bounded continuous function from $\mathbb{R}^d$ to $\mathbb{R}$. In the notation of Theorem II.7.3, prove that

$$\lim_{n\to\infty}\frac{1}{n}\log E\left\{\exp\left[nF\left(\sum_{j=1}^{n}X_j/n\right)\right]\right\}=\sup_{P\in\mathscr{M}_s(\Omega)}\left\{F\left(\int_\Omega \bar{\omega}_1 P(d\bar{\omega})\right)-I_\rho^{(3)}(P)\right\},$$

$$\lim_{n\to\infty}\frac{1}{n}\log E\left\{\exp\sum_{j=1}^{n}F(X_j)\right\}=\sup_{P\in\mathscr{M}_s(\Omega)}\left\{\int_\Omega F(\bar{\omega}_1)P(d\bar{\omega})-I_\rho^{(3)}(P)\right\},$$

where $\Omega=(\mathbb{R}^d)^{\mathbb{Z}}$.

Chapter III

Large Deviations and the Discrete Ideal Gas

III.1. Introduction

In the next three chapters we apply the theory of large deviations to analyze some basic models in equilibrium statistical mechanics.[1] This branch of physics applies probability theory to study equilibrium properties of systems consisting of a large number of particles. The systems fall into two groups: continuous systems, which include the solids, liquids, and gases common to everyday experience; and lattice systems, of which ferromagnets are the main example. This chapter introduces the continuous theory by treating a simple model called a discrete ideal gas. This model, which has no interactions, is a physical analog of i.i.d. random variables.

The macroscopic description of a physical system such as an ideal gas is given by thermodynamics. Thermodynamics summarizes the properties of the gas in terms of macroscopic variables such as pressure, volume, temperature, and internal energy. But this theory takes no account of the fact that the gas is composed of molecules. The main aim of statistical mechanics is to derive properties of the gas from a probability distribution which describes its microscopic (i.e., molecular) behavior. This distribution is called an ensemble.

Statistical mechanics identifies certain macroscopic variables with ensemble averages of appropriate microscopic sums; that is, as averages of sums depending upon the microstate of the system. An example is the internal energy of a gas. The corresponding microscopic sums are proportional to the sums of the energies of the individual molecules. What approximation is involved in replacing the microscopic sums by the ensemble averages? For the discrete ideal gas, we prove that the microscopic sums converge exponentially to their ensemble averages as the number n of particles in the gas increases to infinity. Thus, for all sufficiently large n the sums are close to the ensemble averages except for an exponentially small set of microstates. One may therefore replace the sums by the averages with a very high degree of accuracy. Our key tool for proving the exponential convergence will be the large deviation theory developed in Chapter II.

The set of all ensemble averages of microscopic sums defines a level-1 equilibrium state of the discrete ideal gas. In the last section of this chapter, we will consider the mathematically convenient system consisting of infinitely many particles. We will define a level-3 equilibrium state, which is a strictly stationary probability measure on the infinite-particle configuration space. For the discrete ideal gas, the level-3 equilibrium state turns out to be product measure. This physically uninteresting measure arises because of the absence of interactions in the model. When we study phase transitions in the next two chapters, it will be clear that the level-3 equilibrium state gives information not available in the level-1 description.

While the central topic of this chapter is the statistical mechanics of the discrete ideal gas, it is important to understand its relationship with the complementary macroscopic theory, thermodynamics. An overview of thermodynamics is given in the next section.[2]

III.2. Physics Prelude: Thermodynamics

We consider a closed cylinder filled with a homogeneous gas which consists of only one kind of molecule (e.g., pure helium). The gas is called *simple*, and the cylinder and gas together are called a *simple system*. When the system suffers a change in its surroundings, its observable properties will usually change. For example, if the cylinder is placed in a tub of hot water, then its temperature will rise. After a time, however, the system will reach a state where no further change occurs. It is then said to have come to thermodynamic equilibrium. The equilibrium state is completely determined by the volume V of the cylinder and the pressure P exerted by the gas on the cylinder walls. Relationships among V, P, and all the other equilibrium properties of the system are derived from the four laws of thermodynamics plus empirical observations. The zeroth, first, and second laws define the basic concepts of temperature, internal energy, and entropy, respectively. These laws are explained below. We will not discuss the third law, which is concerned with the properties of entropy at low temperatures.

One can change the equilibrium state of the simple system by a thermodynamic process. Such a process may involve mechanical interactions or thermal interactions, according to the nature of the cylinder walls. A movable wall (e.g., a piston fitted to one end of the cylinder) allows one to change the volume of the system by exerting a force which moves the wall. A diathermal wall permits the flow of heat and so allows thermal interactions. Two systems separated by a diathermal wall are said to be in *thermal contact*.

Experimentally, all simple gases behave in a universal way when they are sufficiently dilute. The *ideal gas* is an idealization of this limiting behavior. The microscopic picture of the ideal gas is a gas of molecules which do not interact with each other and whose energy is all kinetic.

We now turn to the laws of thermodynamics. Two systems in thermal contact can influence each other by an interchange of heat. We know from experience that the heat flows from the warmer system to the cooler system until their temperatures equalize. The two systems are said to be in thermal equilibrium when the interchange of heat between them has ceased. The zeroth law states that if two systems A and B are separately in thermal equilibrium with a third system, then they must also be in thermal equilibrium with each other. The pressures as well as the volumes of systems A and B will in general have different values. But a simple argument based on the zeroth law shows there is a function of the pressure P and volume V—call it $\theta(P, V)$—which takes a common value for the two systems [Adkins (1975, Section 2.2)]. This common value is called the temperature T, and the equation $\theta(P, V) = T$ is called the *equation of state*. The precise form of this equation must be determined experimentally. For the ideal gas, it is found to be $PV = nkT$, when n is the number of molecules in the gas (assumed fixed) and k is a physical constant, called Boltzmann's constant. From now on we set k to 1 in order to simplify the notation. In the absence of phase transitions, the equation of state can be used to express either P or V uniquely in terms of the other variable and T. Thus the equilibrium state of the simple system can be defined equivalently by specifying (P, V) or (P, T) or (V, T). Once the volume and temperature, for example, of the system are chosen, the value of P at equilibrium is determined uniquely by nature.

The fundamental measure of temperature in thermodynamics is the *absolute temperature* T, which for the systems under discussion is always a positive number. The limit $T \to 0^+$ defines a point called absolute zero (-273.16°C) at which no substance has any molecular motion or heat.

The equilibrium state of the simple system can be changed either by performing work on the system or by adding heat. The first law extends the principle of the conservation of energy to include both such processes. According to the first law, there exists a function U of the equilibrium state (V, T) with the property that if the state is changed from (V_1, T_1) to (V_2, T_2), then the change in the total energy of the system is given by $\Delta U = U(V_2, T_2) - U(V_1, T_1) = Q - W$. Q is the amount of heat added to the system and W is the amount of work performed by the system.* The function U is called the *internal energy*. For the ideal gas, Joule found that $U(V, T)$ is a function of T only. If, for example, the gas is monatomic,† then further experiments, confirmed by microscopic theories, have shown that $U = \frac{3}{2}nT$. For a nonideal simple gas, U depends on both V and T, and one may invert the formula $U = U(V, T)$ to express T uniquely as a function of U and V. The variables (U, V) are a useful equivalent label, replacing (V, T), for the equilibrium state of the simple system.

There are processes which, although they conserve energy, do not occur in nature. For example, it would not violate the first law if the temperatures

* $Q < 0$ means that the system supplies heat. $W < 0$ means that work is done on the system.

† Its molecules are single atoms (e.g., helium, but not diatomic oxygen).

of two systems placed in thermal contact were to diverge rather then equalize. The second law expresses a natural direction for thermodynamic processes, implying in particular that the diverging of the temperatures is impossible. The second law applies most generally to a *composite system*, which is a set of several interacting simple systems. For example, two simple gases contained within a closed cylinder and separated from each other by an internal piston form a composite system. We assume that the composite system is *completely isolated*; i.e., surrounded by a wall which prevents both mechanical and thermal interactions between the system and its surroundings. Walls within the composite system which prevent interactions among the individual subsystems are called *internal constraints*. If in the above example the internal piston is in a rigidly fixed position, then it is an internal constraint. Suppose that the composite system is in equilibrium with respect to certain internal constraints and that a process occurs which alters some of these constraints (in the above example, the internal piston moves). Eventually the system reaches a new equilibrium state. The second law allows one to determine it.

We label the equilibrium state of each subsystem by (U, V). According to the second law, there exists a function of the equilibrium state known as the *entropy* and written $S(U, V)$. The entropy of the composite system is defined to be the sum of the entropies of the individual subsystems. The entropy has the following properties.

III.2.1. Properties of the Entropy. (a) The entropy of a completely isolated composite system cannot decrease during any thermodynamic process.

(b) The new equilibrium state of the system corresponds to the maximum value of the entropy consistent with the altered internal constraints.

Thermodynamics classifies all processes as being either reversible or irreversible, the latter encompassing most of the familiar processes in nature. For irreversible processes, Property III.2.1(a) can be strengthened: the entropy of the system must increase. Thus this property defines a natural direction for change. Property III.2.1(b) is consistent with Boltzmann's interpretation of entropy [page 3].

To see the physical meaning of entropy, consider a gas in a cylinder which has a rigid, diathermal wall. A standard thermodynamic relation [Callen (1960, Chapter 2)] is that if the internal energy U of the gas is increased by adding heat, then

$$\frac{\partial S(U, V)}{\partial U} = \frac{1}{T} > 0. \tag{3.1}$$

Thus, an increase in entropy is associated with adding heat to the system at constant volume.

In order to illustrate Properties III.2.1, we present an elementary application to an irreversible process. A monatomic ideal gas is contained in a completely isolated cylinder of volume V_2. The gas is confined by a thin membrane to a volume V_1 of the container ($V_1 < V_2$). The unoccupied part of the container is a vacuum. Let U_1 be the internal energy of the system. Suddenly the membrane ruptures, and the gas expands to fill the entire container. During the expansion no work is done on the system and no heat is added. Hence by the first law, $\Delta U = 0$. We use the second law to characterize the new equilibrium state (U_1, V_2). Let (U, V) be a candidate for the new equilibrium state after the membrane ruptures $(U = U_1, 0 < V \leq V_2)$. For a monatomic ideal gas, $S(U, V)$ is given by

$$S(U. V) = \tfrac{3}{2}n \log U + n \log V + S_0, \tag{3.2}$$

where S_0 is a constant [Callen (1960, Section 3.4)]. Hence, $\Delta S = S(U_1, V) - S(U_1, V_1) = n \log(V/V_1)$. By Property III.2.1(a), ΔS must be non-negative. Thus, V must exceed V_1 and the gas must expand. The actual equilibrium state (U_1, V_2) gives the unique maximum in $\max\{S(U, V): U = U_1, 0 < V \leq V_2\}$, and this is consistent with Property III.2.1(b).

After the second law had been formulated, much effort was spent in trying to interpret it microscopically.[3] As the work of Maxwell and of Boltzmann showed, such an interpretation requires a statistical description of matter. Given this, the increase in entropy in the previous example, which corresponds to the expansion of the gas into the bigger volume, measures an observer's increased ignorance about the location of a specific molecule. One says that the disorder of the gas has increased. Equation (3.1), which shows that the entropy of a system increases when heat is added at constant volume, has a similar statistical interpretation. Adding heat increases the chaotic, thermal motions of the molecules. The rate of increase of this disorder per unit of energy added is $1/T$.

The second law enables one to determine the equilibrium state of a completely isolated composite system. However, most processes in the real world involve a composite system which is not completely isolated but which is in thermal contact with *a heat reservoir*. The latter is a system which is so large that its temperature does not change with the gain or loss of any finite amount of heat. Hence the reservoir maintains a fixed temperature in each subsystem of the composite system. We consider two subclasses of thermodynamic processes: those which do not change the volume of the composite system and those which do not change its pressure. The Helmholtz free energy, rather than entropy, is suited to finding the equilibrium state of a system subject to the first kind of process. We label the equilibrium state of each subsystem by (V, T). Let $U(V, T)$ be the internal energy. Substituting $U = U(V, T)$ into the formula for the entropy S of the subsystem gives the entropy as $S(V, T)$. The *Helmholtz free energy*[4] is defined by $F(V, T) = U(V, T) - TS(V, T)$. The Helmholtz free energy of the composite system is defined as the sum of the Helmholtz free energies of each subsystem. Suppose that the composite system is in equilibrium with respect to certain internal constraints and that a process occurs which alters some of these constraints without changing the total volume of the system. Eventually the system reaches a new equilibrium state. The following properties enable one to find the new state.

III.2.2. Properties of the Helmholtz Free Energy. (a) The Helmholtz free energy of a composite system of fixed volume which is in thermal contact with a heat reservoir cannot decrease during any thermodynamic process.

(b) The new equilibrium state of the system corresponds to the minimum value of the Helmholtz free energy (at constant temperature equal to that of the heat reservoir) consistent with the altered internal constraints.

There is another important thermodynamic function called the Gibbs free energy. Properties of this function analogous to those in III.2.2 enable one to find the equilibrium state of a composite system which is in thermal contact with a heat reservoir and which is subject to a process that does not change its pressure. In the statistical mechanics to be developed later in this chapter, the Helmholtz free energy rather than the Gibbs free energy will play a central role. The importance of the Gibbs free energy will be seen in Chapters IV and V when we study the statistical mechanics of ferromagnetic systems.

This completes our overview of thermodynamics. We now turn to the discrete ideal gas, for which we will see analogs of many of the concepts just discussed.

III.3. The Discrete Ideal Gas and the Microcanonical Ensemble

For the standard ideal gas,[5] the possible particle velocities are all vectors in $\mathbb{R}^3$. Our discrete ideal gas model restricts the possible velocities to a finite subset of $\mathbb{R}$. This restriction will allow us to use the large deviation results in Chapter II developed for random variables with a finite state space. The model is first defined dynamically, but for the purpose of studying equilibrium properties this definition is too detailed. Instead, the properties are derived from the microcanonical ensemble and the canonical ensemble, two probability distributions on configuration space which give equivalent results.

Let Δ be a closed bounded interval in $\mathbb{R}$ with nonempty interior. The interval Δ may be thought of as a completely isolated, long, thin tube containing a gas. The discrete ideal gas is defined by the motions in Δ of a large number n of noninteracting point particles. The mass of each particle is one. Fixing a positive integer r, let Γ be a set of $2r$ numbers

$$\Gamma = \{v_i; i = -r, \ldots, -1, 1, \ldots, r\}$$

which are to be the possible velocities of the particles. Arrange the $\{v_i\}$ in increasing order and assume that $v_{-i} = -v_i$; thus

$$v_{-r} = -v_r < \cdots < v_{-1} = -v_1 < 0 < v_1 < \cdots < v_r. \tag{3.3}$$

At time $t = 0$, the jth particle has an initial position $x_j(0)$ in the interior of Δ, $\operatorname{int}\Delta$, and an initial velocity $y_j(0) = v \in \Gamma$. The particle moves with constant velocity v until it first reaches the boundary of Δ, $\operatorname{bd}\Delta$, at which time it is reflected elastically, moving back into $\operatorname{int}\Delta$ with velocity $-v$. The motion continues at velocity $-v$ until the particle again reaches $\operatorname{bd}\Delta$, undergoes elastic reflection, then moves back into $\operatorname{int}\Delta$ with velocity $-(-v) = v$, etc. We may define the velocity $y_j(t)$ at time $t \geq 0$ to be a piecewise constant, left-continuous function taking values in Γ. The position at time $t \geq 0$ is defined to be $x_j(t) = x_j(0) + \int_0^t y_j(s)ds$.

The motion of the system of n particles is the superposition of all these individual motions. Particles may pass through each other. The system at time t is represented by the microstate $\omega(t) = (\mathbf{x}(t), \mathbf{y}(t))$, where $\mathbf{x}(t) = (x_1(t), \ldots, x_n(t))$ and $\mathbf{y}(t) = (y_1(t), \ldots, y_n(t))$. For each $t \geq 0$, $\omega(t)$ is a point in n-particle configuration space, which is the set

$$\Omega_n = \{\omega\colon \omega = (x_1, \ldots, x_n, y_1, \ldots, y_n), x_j \in \Delta, y_j \in \Gamma\} = \Delta^n \times \Gamma^n.$$

Mathematically, this dynamical definition is perfectly valid. But there are obvious practical problems with it. The definition requires one to know the initial positions and velocities of all n particles and to follow these motions for all time. Since n is typically of the order of 10^{24}, this is of course impossible. Even if it could be defined, the microstate $\omega(t)$ carries much more information than should be needed in order to deduce equilibrium properties, which are determined by only a few thermodynamic variables. We now consider an alternate approach based upon probability theory.

It is natural to treat as random the positions and velocities of the n particles since they cannot be measured precisely. To do this, we introduce a probability measure P_n on the configuration space Ω_n in terms of which the discrete ideal gas system will be analyzed. Before defining P_n, we return for a moment to the dynamics. Conservation of energy restricts the possible motions of the particles. Since each particle has mass one, the kinetic energy of the jth particle at time t is $\frac{1}{2}(y_j(t))^2$ and the total kinetic energy is $U_n(t) = \sum_{j=1}^n \frac{1}{2}(y_t(t))^2$. Since $|y_t(t)| = |y_j(0)|$, we have $U_n(t) = U_n(0)$. Ideally we should assume that $U_n(0)$ or, equivalently, the energy per particle $U_n(0)/n$ is a fixed number. But it is mathematically more convenient to restrict $U_n(0)/n$ to lie in an interval A. The range of $U_n(0)/n$ is the interval $[\frac{1}{2}v_1^2, \frac{1}{2}v_r^2]$. Choose A to be a closed subinterval of $(\frac{1}{2}v_1^2, \frac{1}{2}v_r^2)$ with nonempty interior. A can be as small as we like. Restricting $U_n(0)/n$ to lie in A means that the microstate $\omega(t)$ stays in the "energy sandwich" $D_n(A) = \{\omega \in \Omega_n\colon \sum_{j=1}^n \frac{1}{2} y_j^2/n \in A\}$.

For actual physical systems, macroscopic observations cannot possibly give a precise value for the energy per particle but only determine the true value up to some small error. Hence the introduction of A is physically justifiable. One should think of A as a small interval which contains the true value of the energy per particle for the discrete ideal gas. The mathematics supports this interpretation [see Theorem III.4.2].

Since the speeds $|y_j(t)|$ are each constant in time, there are actually n conservation laws, one for each particle. But the individual speeds cannot be measured, while the total energy can. Conservation of energy is the only physically meaningful conservation law.

We are given the information that the total energy per particle is in the set A. The microstate of the system could be any point in the energy sandwich

$D_n(A)$. Therefore it is reasonable to give all microstates in $D_n(A)$ an equal probability. This leads to the microcanonical ensemble, which is the universally accepted choice of measure for the discrete ideal gas system. This ensemble is defined next. We topologize Δ by the relative topology as a subset of $\mathbb{R}$, Γ by the discrete topology, and Ω_n by the product topology. $\mathscr{B}(\Delta)$, $\mathscr{B}(\Gamma)$, and $\mathscr{B}(\Omega_n)$ denote the respective Borel σ-fields.

Definition III.3.1. Let λ be normalized Lebesgue measure on $\mathscr{B}(\Delta)$ with $\lambda\{\Delta\} = 1$ and ρ the uniform measure on $\mathscr{B}(\Gamma)$ defined by $\rho = \sum_{1 \leq |i| \leq r} (2r)^{-1} \delta_{v_i}$. Let $\pi_n P_{\lambda,\rho}$ be the product measure on $\mathscr{B}(\Omega_n)$ defined by

$$\pi_n P_{\lambda,\rho}(d\omega) = \lambda(dx_1) \ldots \lambda(dx_n)\rho(dy_1) \ldots \rho(dy_n).$$

The *microcanonical ensemble* is the (normalized) restriction of $\pi_n P_{\lambda,\rho}$ to the energy sandwich $D_n(A)$. This restriction is the conditional measure

$$P_n(d\omega) = \pi_n P_{\lambda,\rho}\{d\omega \mid U_n/n \in A\},$$

where $U_n = \sum_{j=1}^n \frac{1}{2} y_j^2$. We write $P_n = P_{n,\Delta,A}$ to emphasize the dependence upon Δ and A.

With respect to $P_{n,\Delta,A}$, the positions and velocities of the n particles are random. Given $\omega_0 \in \Omega_n$ and $t > 0$, define the microstate

$$\omega(t, \omega_0) = (x_1(t), \ldots, x_n(t), y_1(t), \ldots, y_n(t))$$

with initial value $\omega(0, \omega_0) = \omega_0$. If B is any Borel subset of Ω_n, then for each $t > 0$

$$P_{n,\Delta,A}\{\omega_0 \in \Omega_n : \omega(t, \omega_0) \in B\} = P_{n,\Delta,A}\{\omega_0 \in B\}; \tag{3.4}$$

i.e., $\omega(t, \omega_0)$ is also distributed by $P_{n,\Delta,A}$ [Problem III.10.1]. Because of (3.4), we drop t from the notation for the microstate. Let $\{X_j; j = 1, \ldots, n\}$ and $\{Y_j; j = 1, \ldots, n\}$ be the coordinate functions on Ω_n defined by $X_j(\omega) = x_j$ and $Y_j(\omega) = y_j$. $X_j(\omega)$ and $Y_j(\omega)$ are, respectively, the random position and velocity of the jth particle.

III.4. Thermodynamic Limit, Exponential Convergence, and Equilibrium Values

In the laboratory, a gas is studied by measuring *observables*, which are functions of the positions and velocities of the molecules. For the discrete ideal gas, we consider observables of the form $F_n(\omega) = \sum_{j=1}^n f(X_j(\omega), Y_j(\omega))$, $n = 1, 2, \ldots$, where f is a bounded measurable real-valued function on $\Delta \times \Gamma$. In some cases we must assume that f is also continuous on $\Delta \times \Gamma$. The corresponding observable is called a *continuous observable*.

Examples III.4.1. (a) For $x \in \Delta$ and $y \in \Gamma$, define $f(x, y) = \frac{1}{2} y^2$. Then $F_n(\omega) = \sum_{j=1}^n \frac{1}{2} Y_j(\omega)^2$. This sum is denoted by $U_n(\omega)$ and is called the (total) energy in ω. The observable $U_n(\omega)$ corresponds to the internal energy U in thermodynamics.

(b) Let J be a subinterval of Δ and v_i a point in Γ. For $x \in \Delta$ and $y \in \Gamma$, define $f(x, y) = \chi_J(x)\chi_{\{v_i\}}(y)$.* Then $F_n(\omega)$ is the number of particles in ω with position in J and velocity v_i. $F_n(\omega)/n$ is the fraction of such particles. A continuous observable approximating $F_n(\omega)$ is obtained by choosing $f(x, y) = g(x)\chi_{\{v_i\}}(y)$, where $g(x)$ is a continuous function on Δ which approximates $\chi_J(x)$.

The aim of this section is to study the asymptotics of continuous observables $F_n(\omega)$ with respect to the microcanonical ensemble $P_{n,\Delta,A}$ as $n \to \infty$.[6] The limit $n \to \infty$ is called the *infinite-particle limit* or the *thermodynamic limit*. The relevant notion of convergence is exponential convergence, defined in Section II.6. We prove that there exists a number $\langle f \rangle$ such that F_n/n converges exponentially to $\langle f \rangle$ (written $F_n/n \xrightarrow{\exp} \langle f \rangle$. That is, for any $\varepsilon > 0$ there exists a number $N = N(\varepsilon) > 0$ such that

$$P_{n,\Delta,A}\{\omega \in \Omega_n : |F_n(\omega)/n - \langle f \rangle| \geq \varepsilon\} \leq e^{-nN} \qquad \text{for all sufficiently large } n. \tag{3.5}$$

Thus for large n all but an exponentially small set of microstates have essentially the same value $\langle f \rangle$ of the observable per particle. $F_n(\omega)/n$ is a microscopic n-sum which approximates the equilibrium value $\langle f \rangle$. In Example III.4.1(a) this equilibrium value is called the *specific energy*, and in Example III.4.1(b) it is called the *density* of particles in the subset J with velocity v_i.

As a first case, we consider Example III.4.1(a), where F_n is the energy U_n. Some definitions are needed. We have denoted by ρ the probability measure $\sum_{1 \leq |i| \leq r} (2r)^{-1}\delta_{v_i}$ on $\mathscr{B}(\Gamma)$. Define

$$\alpha = \int_\Gamma \frac{1}{2} y^2 \rho(dy) = \frac{1}{2r} \sum_{1 \leq |i| \leq r} \frac{1}{2} v_i^2.$$

The set A is closed subinterval of $[\frac{1}{2}v_i^2, \frac{1}{2}v_r^2]$ with nonempty interior. We define a new number $u(A)$. If α is in A, let $u(A) = \alpha$. If α is not in A, let $u(A)$ be the point in A closest to α. In general $u(A) \in (\frac{1}{2}v_1^2, \frac{1}{2}v_r^2)$ [see Figure III.1.]. The next theorem shows the physical importance of $u(A)$.

Theorem III.4.2. *Let* $U_n = \sum_{j=1}^n \frac{1}{2} Y_j^2$, $n = 1, 2, \ldots$. *Then as* $n \to \infty$,

$$U_n/n \xrightarrow{\exp} u(A) \text{ with respect to the microcanonical ensembles } \{P_{n,\Delta,A}\}.$$

The quantity $u(A)$ *represents the equilibrium value of the energy per particle and is called the specific (microcanonical) energy of the discrete ideal gas.*

Proof. Define the closed set $K = \{u \in \mathbb{R} : |u - u(A)| \geq \varepsilon\}$, where ε is positive. Then $P_{n,\Delta,A}\{|U_n/n - u(A)| \geq \varepsilon\} = P_{n,\Delta,A}\{U_n/n \in K\}$. We prove that there exists a number $N = N(\varepsilon) > 0$ such that

* χ_C denotes the characteristic function of a set C.

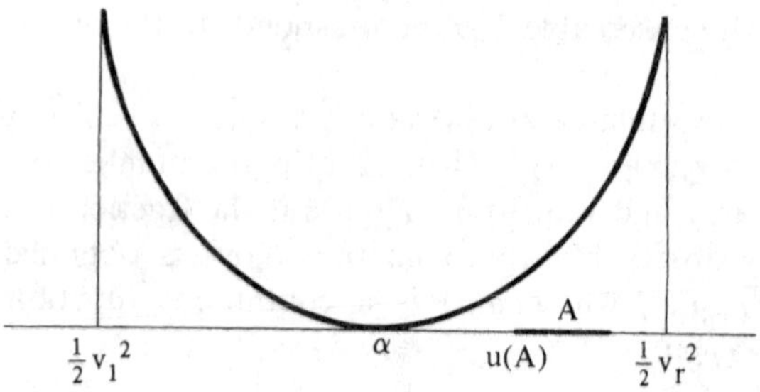

Figure III.1. The entropy function $\bar{I}_\rho^{(1)}(u)$ and the set A.

$$P_{n,\Delta,A}\{U_n/n \in K\} \le e^{-nN} \quad \text{for all sufficiently large } n. \tag{3.6}$$

Assume that $\varepsilon > 0$ is so small that $K \cap A$ has nonempty interior. For sufficiently large n, $\pi_n P_{\lambda,\rho}\{U_n/n \in A\}$ is positive and

$$P_{n,\Delta,A}\{U_n/n \in K\} = \pi_n P_{\lambda,\rho}\{U_n/n \in K \cap A\} \cdot \frac{1}{\pi_n P_{\lambda,\rho}\{U_n/n \in A\}}.$$

Since U_n is a function of velocities only, the λ-integrals on the right-hand side can be done. Let $\pi_n P_\rho$ denote the product measure on $\mathscr{B}(\Gamma^n)$ with identical one-dimensional marginals ρ. The last display reduces to

$$P_{n,\Delta,A}\{U_n/n \in K\} = \pi_n P_\rho\{U_n/n \in K \cap A\} \cdot \frac{1}{\pi_n P_\rho\{U_n/n \in A\}}.$$

With respect to $\pi_n P_\rho$, the summands $\frac{1}{2}Y_j^2$ are i.i.d. and their state space is the finite set $\{\frac{1}{2}v_1^2, \ldots, \frac{1}{2}v_r^2\}$.* In order to complete the proof of the theorem, we use the following level-1 large deviation result. It is a consequence of Theorem II.4.1. □

Lemma III.4.3. *For t real, define*

$$\bar{c}_\rho(t) = \log \int_{\Gamma^n} \exp(\tfrac{1}{2}tY_1^2)\,d(\pi_n P_\rho) = \log \int_\Gamma \exp(\tfrac{1}{2}ty^2)\rho(dy). \tag{3.7}$$

The following conclusions hold.

(a) *The $\pi_n P_\rho$-distributions of $\{U_n/n\}$ have a large deviation property with $a_n = n$ and entropy function*

$$\bar{I}_\rho^{(1)}(u) = \sup_{t \in \mathbb{R}} \{tu - \bar{c}_\rho(t)\}, \quad u \in \mathbb{R}. \tag{3.8}$$

(b) *$\bar{I}_\rho^{(1)}(u)$ is a convex function of u. $\bar{I}_\rho^{(1)}(u)$ measures the discrepancy between u and $\alpha = \int_\Gamma \frac{1}{2}y^2\rho(dy) = \int_{\Gamma^n} \frac{1}{2}Y_1^2\,d(\pi_n P_\rho)$ in the sense that $\bar{I}_\rho^{(1)}(u) \ge 0$ with equality if and only if $u = \alpha$ [see Figure* III.1].

(c) *$\bar{I}_\rho^{(1)}$ is finite and continuous on the interval $[\frac{1}{2}v_1^2, \frac{1}{2}v_r^2]$ and $\bar{I}_\rho^{(1)}(u) = \infty$ for $u \notin [\frac{1}{2}v_1^2, \frac{1}{2}v_r^2]$.*

We return to the proof of the theorem. Since the intervals $K \cap A$ and

*By (3.3), $\frac{1}{2}v_{-i}^2 = \frac{1}{2}v_i^2$.

A are $I_\rho^{(1)}$-continuity sets [Theorem II.3.4], the lemma implies that

$$\lim_{n\to\infty}\frac{1}{n}\log P_{n,\Delta,A}\{U_n/n\in K\} = \lim_{n\to\infty}\frac{1}{n}\log\left[\pi_n P_\rho\{U_n/n\in K\cap A\}\cdot\frac{1}{\pi_n P_\rho\{U_n/n\in A\}}\right]$$

$$(3.9) \qquad = -\left[\inf_{u\in K\cap A}\bar{I}_\rho^{(1)}(u) - \inf_{u\in A}\bar{I}_\rho^{(1)}(u)\right].$$

We will prove that

$$(3.10) \qquad \inf_{u\in A}\bar{I}_\rho^{(1)}(u) \text{ is attained at the unique point } u = u(A).$$

Since $K\cap A$ is a closed interval which does not contain $u(A)$, (3.10) will imply that $\inf_{u\in K\cap A}\bar{I}_\rho^{(1)}(u) > \inf_{u\in A}\bar{I}_\rho^{(1)}(u)$ and thus that the last term in (3.9) is negative. The exponential bound (3.6) follows.

If $\alpha\in A$, then $u(A)$ equals α, $\bar{I}_\rho^{(1)}(\alpha)$ equals 0, and (3.10) holds. The case $\alpha\notin A$ is shown in Figure III.1. Since $\bar{I}_\rho^{(1)}$ is convex with a unique minimum point at α, $\bar{I}_\rho^{(1)}$ attains its infimum over A at a unique point, which is the point in A closest to α. By definition, the closest point is $u(A)$. This proves (3.10). □

Property III.2.1(b) of the thermodynamic entropy states that for a completely isolated system the equilibrium state is the state of maximum entropy S consistent with any internal constraints. The function $s(u) = -\bar{I}_\rho^{(1)}(u)$ is a natural microcanonical analog. It is called the *specific microcanonical entropy*. Any point $u\in A$ is a candidate for the equilibrium value of energy per particle. For sufficiently small $\varepsilon > 0$ and $u\in\operatorname{int}A$, $P_{n,\Delta,A}\{|U_n/n - u| < \varepsilon\}$ equals $\pi_n P_\rho\{|U_n/n - u| < \varepsilon\}$. For $u = u(A)$, $P_{n,\Delta,A}\{|U_n/n - u(A)| < \varepsilon\}$ equals $\pi_n P_\rho\{u(A) - \varepsilon < U_n/n < u(A)\}$ or $\pi_n P_\rho\{u(A) < U_n/n < u(A) + \varepsilon\}$ according to whether $u(A)$ is a right-hand end point or a left-hand end point of A. Hence by Lemma III.4.3, $s(u)$ measures, with respect to the microcanonical ensemble, the multiplicity of microstates ω for which $U_n(\omega)/n$ is close to u. By (3.10), the equilibrium value $u(A)$ is the unique point at which $s(u)$ attains its supremum over the constraint set A. It is the point in A consistent with the most microstates.

The next theorem states that for any continuous observable $F_n = \sum_{j=1}^n f(X_j, Y_j)$, F_n/n converges exponentially to a number $\langle f\rangle$. This number is the integral of f with respect to a new measure which we now define. For β real, define probabilities

$$(3.11) \qquad \bar{\rho}_{\beta,i} = \frac{\exp(-\frac{1}{2}\beta v_i^2)\cdot(1/2r)}{\int_\Gamma \exp(-\frac{1}{2}\beta y^2)\rho(dy)} = \frac{\exp(-\frac{1}{2}\beta v_i^2)\cdot(1/2r)}{\sum_{1\le|j|\le r}\exp(-\frac{1}{2}\beta v_j^2)\cdot(1/2r)}$$

and let $\bar{\rho}_\beta$ be the probability measure $\sum_{1\le|i|\le r}\bar{\rho}_{\beta,i}\delta_{v_i}$ on $\mathscr{B}(\Gamma)$. This measure is closely related to the measure ρ_t in Section II.5. If in the definition (2.23) of $\rho_{t,i}$ one replaces t by $-\beta$ and x_i by $\frac{1}{2}v_i^2$, then $\rho_{t,i}$ becomes $\bar{\rho}_{\beta,i}$. The parameter β and the measure $\bar{\rho}_\beta$ will be interpreted physically in the next section.

Theorem III.4.4.[7] *Let $u(A)$ be the specific energy of the discrete ideal gas [see previous theorem].*

(a) *Then there exists a unique real number* $\beta = \beta(u(A))$ *such that*

(3.12)
$$\int_\Gamma \tfrac{1}{2}y^2 \bar{\rho}_\beta(dy) = u(A).$$

(b) *Let* f *be a bounded continuous real-valued function on* $\Delta \times \Gamma$ *and define*

(3.13)
$$\langle f \rangle_{\Delta, u(A)} = \int_{\Delta \times \Gamma} f(x, y)\lambda(dx)\bar{\rho}_\beta(dy),$$

where $\beta = \beta(u(A))$. *Then*

(3.14)
$$\frac{1}{n}\sum_{j=1}^{n} f(X_j, Y_j) \overset{\exp}{\to} \langle f \rangle_{\Delta, u(A)}$$

with respect to the microcanonical ensembles $\{P_{n,\Delta,A}\}$.

This theorem is consistent with Theorem III.4.2. If $f(x, y) = \frac{1}{2}y^2$, then by (3.13), $\langle \frac{1}{2}y^2 \rangle_{\Delta, u(A)} = u(A)$. Thus (3.14) reduces to $U_n/n \overset{\exp}{\to} u(A)$.

While Theorem III.4.2 was proved by level-1 large deviations, the proof of Theorem III.4.4 requires the level-2 theory. Define $L_n(\omega, \cdot) = n^{-1}\sum_{j=1}^n \delta_{Y_j(\omega)}(\cdot)$, $n = 1, 2, \ldots$. This is the empirical measure corresponding to $Y_1, \ldots, Y_n$, and it takes values in the set $\mathscr{M}(\Gamma)$ of probability measures on $\mathscr{B}(\Gamma)$. Each $\nu \in \mathscr{M}(\Gamma)$ has the form $\sum_{1 \le |i| \le r} \nu_i \delta_{v_i}$, where $\nu_i \ge 0$ and $\sum_{1 \le |i| \le r} \nu_i = 1$. $I_\rho^{(2)}(\nu)$ denotes the relative entropy of ν with respect to the measure $\rho = \sum_{1 \le |i| \le r} (2r)^{-1}\delta_{v_i}$:

$$I_\rho^{(2)}(\nu) = \int_\Gamma \log \frac{d\nu}{d\rho}(y)\nu(dy) = \sum_{1 \le |i| \le r} \nu_i \log(2r\nu_i).$$

The next lemma states the level-2 large deviation property together with a contraction principle relating $I_\rho^{(2)}$ and the function $\bar{I}_\rho^{(1)}$ in Lemma III.4.3. Theorem III.4.4 is proved afterwards.

Lemma III.4.5. (a) *The* $\pi_n P_\rho$*-distributions of* $\{L_n\}$ *have a large deviation property with* $a_n = n$ *and entropy function* $I_\rho^{(2)}$.

(b) *For each point* $u \in (\frac{1}{2}v_1^2, \frac{1}{2}v_r^2)$, *there exists a unique value* $\beta = \beta(u)$ *such that* $\int_\Gamma \frac{1}{2}y^2 \bar{\rho}_\beta(dy) = u$.

(c) *For* $u \in (\frac{1}{2}v_1^2, \frac{1}{2}v_r^2)$, *define the set* $\Psi(u) = \{\nu \in \mathscr{M}(\Gamma) : \int_\Gamma \frac{1}{2}y^2 \nu(dy) = u\}$. *Then* $I_\rho^{(2)}$ *attains its infimum over* $\Psi(u)$ *at the unique measure* $\bar{\rho}_{\beta(u)}$ *and*

(3.15)
$$\bar{I}_\rho^{(1)}(u) = I_\rho^{(2)}(\bar{\rho}_{\beta(u)}) = \inf\{I_\rho^{(2)}(\nu) : \nu \in \Psi(u)\} < \infty.$$

Proof. (a) Since $\pi_n P_\rho\{L_n \in \mathscr{M}(\Gamma)\} = 1$, part (a) follows from Theorem II.4.3(a) [see Theorem VIII.2.1].

(b) For t real, define

$$\bar{c}_\rho(t) = \log \int_{\Gamma^n} \exp(\tfrac{1}{2}tY_1^2)d(\pi_n P_\rho) = \log \int_\Gamma \exp(\tfrac{1}{2}ty^2)\rho(dy).$$

We have

$$\bar{c}_\rho'(t) = \frac{\int_\Gamma \frac{1}{2}y^2 \exp(\frac{1}{2}ty^2)\rho(dy)}{\int_\Gamma \exp(\frac{1}{2}ty^2)\rho(dy)}$$

and

$$\bar{c}_\rho'(-\beta) = \int_\Gamma \tfrac{1}{2}y^2 \bar{\rho}_\beta(dy).$$

If $\bar{\rho}$ denotes the $\pi_n P_\rho$-distribution of $\frac{1}{2}Y_1^2$, then $\bar{c}_\rho(t) = \log \int_{\mathbb{R}} e^{tx}\bar{\rho}(dx)$. This is the free energy function of $\bar{\rho}$, which was introduced in Section II.4. The support of $\bar{\rho}$ is the finite set $\{\frac{1}{2}v_1^2, \ldots, \frac{1}{2}v_r^2\}$. By Theorem II.5.2(a), for $u \in (\frac{1}{2}v_1^2, \frac{1}{2}v_r^2)$ there exists a unique value $\beta = \beta(u)$ such that

$$\bar{c}_\rho'(-\beta) = \int_\Gamma \tfrac{1}{2}y^2 \bar{\rho}_\beta(dy) = u.$$

(c) While it is possible to apply Theorem II.5.2, we prefer a direct proof. By definition

$$\bar{I}_\rho^{(1)}(u) = \sup_{t \in \mathbb{R}} \{tu - \bar{c}_\rho(t)\} = \sup_{\beta \in \mathbb{R}} \{-\beta u - \bar{c}_\rho(-\beta)\}.$$

By part (b), for $u \in (\frac{1}{2}v_1^2, \frac{1}{2}v_r^2)$, there exists a unique value $\beta = \beta(u)$ such that $\bar{c}_\rho'(-\beta) = u$. Since $\bar{c}_\rho$ is convex [Example VII.1.2], this value of β gives the supremum in the last display, and so

$$\bar{I}_\rho^{(1)}(u) = -\beta(u)\cdot u - \bar{c}_\rho(-\beta(u)).$$

Now consider $I_\rho^{(2)}(v)$ for $v \in \Psi(u)$. If $\beta = \beta(u)$, then

$$\begin{aligned} I_\rho^{(2)}(\bar{\rho}_\beta) &= \int_\Gamma \left[-\tfrac{1}{2}\beta y^2 - \log \int_\Gamma \exp(-\tfrac{1}{2}\beta y^2)\rho(dy)\right] \bar{\rho}_\beta(dy) \\ &= -\beta u - \bar{c}_\rho(-\beta) = \bar{I}_\rho^{(1)}(\beta(u)). \end{aligned}$$

If $v \neq \bar{\rho}_\beta$ is any other measure in $\Psi(u)$, then

$$\begin{aligned} I_\rho^{(2)}(v) &= \int_\Gamma \log \frac{dv}{d\rho}(y)\, v(dy) = \int_\Gamma \log \frac{dv}{d\bar{\rho}_\beta}(y)\, v(dy) + \int_\Gamma \log \frac{d\bar{\rho}_\beta}{d\rho}(y)\, v(dy) \\ &= I_{\bar{\rho}_\beta}^{(2)}(v) + \int_\Gamma \left[-\tfrac{1}{2}\beta y^2 - \log \int_\Gamma \exp(-\tfrac{1}{2}\beta y^2)\rho(dy)\right] v(dy) \\ &= I_{\bar{\rho}_\beta}^{(2)}(v) - \beta u - \bar{c}_\rho(-\beta) = I_{\bar{\rho}_\beta}^{(2)}(v) + \bar{I}_\rho^{(1)}(u). \end{aligned}$$

Since $v \neq \bar{\rho}_\beta$, $I_{\bar{\rho}_\beta}^{(2)}(v)$ is positive [Proposition I.4.1(b)] and so $I_\rho^{(2)}(v) > \bar{I}_\rho^{(1)}(u)$. We conclude that $I_\rho^{(2)}(v) \geq \bar{I}_\rho^{(1)}(u) = I_\rho^{(2)}(\bar{\rho}_\beta)$ and that equality holds if and only if $v = \bar{\rho}_\beta$. This completes the proof. □

Proof of Theorem III.4.4. (a) Since $u(A) \in (\frac{1}{2}v_1^2, \frac{1}{2}v_r^2)$, Lemma III.4.5 shows that there exists a unique real number $\beta = \beta(u(A))$ solving (3.12).

(b) We prove the limit (3.14) first for $f(x, y)$ which are functions only of velocity y. The strategy is to prove the exponential bound (3.5). If K denotes the closed set $\{u \in \mathbb{R}: |u - \langle f \rangle| \geq \varepsilon\}$, $\varepsilon > 0$, then for sufficiently large n

$$P_{n,\Delta,A}\left\{\left|\frac{1}{n}\sum_{j=1}^{n} f(Y_j) - \langle f \rangle\right| \geq \varepsilon\right\}$$

$$(3.16) \qquad = P_{n,\Delta,A}\left\{\frac{1}{n}\sum_{j=1}^{n} f(Y_j) \in K\right\}$$

$$= \frac{\pi_n P_\rho\{n^{-1}\sum_{j=1}^n f(Y_j) \in K, n^{-1}\sum_{j=1}^n \frac{1}{2} Y_j^2 \in A\}}{\pi_n P_\rho\{n^{-1}\sum_{j=1}^n \frac{1}{2} Y_j^2 \in A\}}.$$

Each of the sums in this display can be written in terms of the empirical measure $L_n(\omega, \cdot)$. We have

$$(3.17) \qquad \begin{aligned} \frac{1}{n}\sum_{j=1}^{n} f(Y_j(\omega)) &= \int_\Gamma f(y) L_n(\omega, dy), \\ \frac{1}{n}\sum_{j=1}^{n} \tfrac{1}{2}(Y_j(\omega))^2 &= \int_\Gamma \tfrac{1}{2} y^2 L_n(\omega, dy). \end{aligned}$$

Now define two closed subsets of $\mathcal{M}(\Gamma)$,

$$\Psi_1 = \left\{\nu \in \mathcal{M}(\Gamma): \int_\Gamma f(y)\nu(dy) \in K\right\}$$

and

$$\Psi_2 = \left\{\nu \in \mathcal{M}(\Gamma): \int_\Gamma \tfrac{1}{2} y^2 \nu(dy) \in A\right\}.$$

The ratio in (3.16) becomes $\pi_n P_\rho\{L_n \in \Psi_1 \cap \Psi_2\}/\pi_n P_\rho\{L_n \in \Psi_2\}$, and so by the level-2 large deviation property

$$\limsup_{n\to\infty} \frac{1}{n} \log P_{n,\Delta,A}\left\{\frac{1}{n}\sum_{j=1}^{n} f(Y_j) \in K\right\} \leq -\left\{\inf_{\nu \in \Psi_1 \cap \Psi_2} I_\rho^{(2)}(\nu) - \inf_{\nu \in \Psi_2} I_\rho^{(2)}(\nu)\right\}.$$

We have used the fact that Ψ_2 is an $I_\rho^{(2)}$-continuity set [Problem II.9.5]. In order to complete the proof that $n^{-1}\sum_{j=1}^n f(Y_j) \overset{\exp}{\to} \langle f \rangle$, we must show that

$$(3.18) \qquad \inf_{\nu \in \Psi_1 \cap \Psi_2} I_\rho^{(2)}(\nu) > \inf_{\nu \in \Psi_2} I_\rho^{(2)}(\nu).$$

Bring in the measure $\bar{\rho}_\beta$, where $\beta = \beta(u(A))$. Since $u(A)$ is in A and by (3.12) $\int_\Gamma \frac{1}{2} y^2 \bar{\rho}_\beta(dy) = u(A)$, $\bar{\rho}_\beta$ is in the set Ψ_2. The key is to prove that

(3.19) $\inf_{\nu \in \Psi_2} I_\rho^{(2)}(\nu)$ is attained at the unique measure $\nu = \bar{\rho}_{\beta(u(A))}$.

Say (3.19) is proved. By definition $\int_\Gamma f(y)\bar{\rho}_\beta(dy)$ equals $\langle f \rangle$. Since $\langle f \rangle$ is not in K, $\bar{\rho}_\beta$ is not in the set $\Psi_1 \cap \Psi_2$. Since this set is closed, (3.19) will yield (3.18). We now prove (3.19). According to Lemma III.4.5

$$I_\rho^{(2)}(\bar{\rho}_{\beta(u(A))}) = \bar{I}_\rho^{(1)}(u(A)),$$

and $\bar{\rho}_{\beta(u(A))}$ is the unique measure in $\Psi(u(A))$ with this property. If v belongs to $\Psi(u)$ for u in A, $u \neq u(A)$, then by (3.10)

$$I_\rho^{(2)}(v) \geq \inf_{v \in \Psi(u)} I_\rho^{(2)}(v) = \bar{I}_\rho^{(1)}(u) > \bar{I}_\rho^{(1)}(u(A)).$$

The last two displays prove (3.19) and complete the proof of the limit (3.14) for functions f of velocity.

We now prove the limit (3.14) for a bounded continuous function $f(x, y)$ on $\Delta \times \Gamma$. Define

$$L_n(\omega, dxdy) = n^{-1} \sum_{j=1}^{n} \delta_{(X_j(\omega), Y_j(\omega))}(dxdy), \qquad n = 1, 2, \ldots .$$

This is the empirical measure corresponding to $(X_1, Y_1), \ldots, (X_n, Y_n)$. It takes values in the set $\mathcal{M}(\Delta \times \Gamma)$ of probability measures on $\mathscr{B}(\Delta \times \Gamma)$). Each measure τ in $\mathcal{M}(\Delta \times \Gamma)$ has the form $\sum_{1 \leq |i| \leq r} \tau_i(dx)\delta_{v_i}(dy)$, where $\tau_i(dx)$ is a measure on $\mathscr{B}(\Delta)$ $(0 \leq \tau_i(\Delta) \leq 1)$. By Theorem II.4.3, the $\pi_n P_{\lambda,\rho}$-distributions of $L_n(\omega, dxdy)$ on $\mathcal{M}(\Delta \times \Gamma)$ have a large deviation property with $a_n = n$ and entropy function

$$I_{\lambda,\rho}^{(2)}(\tau) = \begin{cases} \displaystyle\sum_{1 \leq |i| \leq r} \int_\Delta \log\left(2r \frac{d\tau_i}{d\lambda}(x)\right) \tau_i(dx) & \text{if each } \tau_i \ll \lambda \text{ and } \displaystyle\int_\Delta \left|\log \frac{d\tau_i}{d\lambda}\right| d\tau_i < \infty, \\ \infty & \text{otherwise.} \end{cases}$$

Define sets $K = \{u \in \mathbb{R}: |u - \langle f \rangle| \geq \varepsilon\}$, $\varepsilon > 0$,

$$\Psi_1 = \left\{\tau \in \mathcal{M}(\Delta \times \Gamma): \int_{\Delta \times \Gamma} f(x, y)\tau(dx\,dy) \in K\right\}$$

and

$$\Psi_2 = \left\{\tau \in \mathcal{M}(\Delta \times \Gamma): \int_{\Delta \times \Gamma} \tfrac{1}{2}y^2\tau(dx\,dy) \in A\right\}.$$

The set Ψ_2 is closed and since f is continuous, Ψ_1 is closed. Also, Ψ_2 is an $I_{\lambda,\rho}^{(2)}$-continuity set [Problem II.9.5]. Therefore

$$\limsup_{n \to \infty} \frac{1}{n} \log P_{n,\Delta,A}\left\{\frac{1}{n}\sum_{j=1}^{n} f(X_j, Y_j) \in K\right\}$$

$$= \limsup_{n \to \infty} \frac{1}{n} \log\left[\pi_n P_{\lambda,\rho}\{L_n \in \Psi_1 \cap \Psi_2\} \cdot \frac{1}{\pi_n P_{\lambda,\rho}\{L_n \in \Psi_2\}}\right] \tag{3.20}$$

$$\leq -\left\{\inf_{\tau \in \Psi_1 \cap \Psi_2} I_{\lambda,\rho}^{(2)}(\tau) - \inf_{\tau \in \Psi_2} I_{\lambda,\rho}^{(2)}(\tau)\right\}.$$

Let $\bar{\tau}(dxdy)$ denote the measure $\lambda(dx)\bar{\rho}_{\beta(u(A))}(dy) \in \mathscr{M}(\Delta \times \Gamma)$. Since $\int_{\Delta\times\Gamma} \frac{1}{2}y^2\, d\bar{\tau} = u(A)$, $\bar{\tau}$ is in the set Ψ_2. If we prove that

$$\text{(3.21)} \qquad \inf_{\tau\in\Psi_2} I_{\lambda,\rho}^{(2)}(\tau) \qquad \text{is attained at the unique measure } \tau = \bar{\tau},$$

then since $\bar{\tau}$ is not in the set $\Psi_1 \cap \Psi_2$, it will follow from (3.20) that $n^{-1}\sum_{j=1}^n f(X_j, Y_j) \overset{\exp}{\to} \langle f \rangle$. Clearly

$$I_{\lambda,\rho}^{(2)}(\bar{\tau}) = I_\rho^{(2)}(\bar{\rho}_{\beta(u(A))}) = \bar{I}_\rho^{(1)}(u(A)).$$

We leave it to Problem III.10.4 to show that if $\tau \neq \bar{\tau}$ is any other measure in Ψ_2, then $I_{\lambda,\rho}^{(2)}(\tau) > \bar{I}_\rho^{(1)}(u(A))$. This will prove (3.21) and complete the proof of the theorem. □

III.5. The Maxwell–Boltzmann Distribution and Temperature

We recall the definition of the quantity $\langle f \rangle_{\Delta,u(A)}$ in Theorem III.4.4. For β real, $\bar{\rho}_\beta$ denotes the probability measure $\sum_{1\le|i|\le r} \bar{\rho}_{\beta,i}\delta_{v_i}$ on $\mathscr{B}(\Gamma)$, where

$$\bar{\rho}_{\beta,i} = \frac{\exp(-\frac{1}{2}\beta v_i^2)\cdot(1/2r)}{\int_\Gamma \exp(-\frac{1}{2}\beta y^2)\rho(dy)}.$$

$u(A)$ denotes the specific energy, which is a number in the interval $(\frac{1}{2}v_1^2, \frac{1}{2}v_r^2)$ [Theorem III.4.2]. There exists a unique real number $\beta = \beta(u(A))$ which satisfies $\int_\Gamma \frac{1}{2}y^2\bar{\rho}_\beta(dy) = u(A)$. We defined $\langle f \rangle_{\Delta,u(A)} = \int_{\Delta\times\Gamma} f(x,y)\lambda(dx)\bar{\rho}_\beta(dy)$, where $\beta = \beta(u(A))$. This integral gives the equilibrium value per particle of a corresponding observable. The set of averages $\langle f \rangle_{\Delta,u(A)}$ for all bounded continuous real-valued functions f on $\Delta \times \Gamma$ defines a level-1 equilibrium state of the discrete ideal gas. We first interpret the measure $\bar{\rho}_\beta$ physically and then argue that the parameter β should be identified with inverse absolute temperature.

Write the specific energy $u(A)$ as u. Theorem III.4.4 implies that given $\varepsilon > 0$ and $\theta > 0$, then for all sufficiently large n

$$P_{n,\Delta,A}\left\{\left|\frac{1}{n}\sum_{j=1}^n f(X_j(\omega), Y_j(\omega)) - \langle f \rangle_{\Delta,u}\right| \ge \varepsilon\right\} \le \theta$$

Since f is bounded, we conclude that

$$\text{(3.22)} \quad \lim_{n\to\infty} \int_{\Omega_n} f(X_1, Y_1)dP_{n,\Delta,A} = \lim_{n\to\infty} \int_{\Omega_n} \frac{1}{n}\sum_{j=1}^n f(X_j, Y_j)dP_{n,\Delta,A} = \langle f \rangle_{\Delta,u}.$$

Now tag particle number 1 in the gas of n particles. Since (3.22) holds for all bounded, continuous functions f, we conclude that the measure $\lambda(dx)\bar{\rho}_{\beta(u)}(dy)$ on $\Delta \times \Gamma$ is the limiting marginal distribution of the position and velocity of the tagged particle with respect to the microcanonical ensemble as $n \to \infty$. In other words, requiring the energy per particle to be close to u for all large n imposes the distribution $\bar{\rho}_\beta$ on the velocities $\{v_i\}$ of a single particle. The measure $\bar{\rho}_{\beta(u)}$ is the equilibrium distribution of the velocities $\{v_i\}$. It is called the *Maxwell–Boltzmann distribution*. With respect to velocities, it is the unique level-2 equilibrium state.

We can characterize $\bar{\rho}_{\beta(u)}$ in terms of entropy. Given a probability measure $\nu \in \mathscr{M}(\Gamma)$, define the energy in ν to be $\int_\Gamma \frac{1}{2}y^2\nu(dy)$. Let Ψ_2 be the set of $\nu \in \mathscr{M}(\Gamma)$ which satisfy the microcanonical constraint $\int_\Gamma \frac{1}{2}y^2\nu(dy) \in A$. Any $\nu \in \Psi_2$ is a candidate for the level-2 equilibrium state with respect to velocities. By (3.19) the actual equilibrium state $\bar{\rho}_{\beta(u)}$ is the unique minimum point of the relative entropy $I_\rho^{(2)}$ over Ψ_2. Hence the Maxwell–Boltzmann distribution can be regarded as the most random probability measure on $\mathscr{B}(\Gamma)$ which satisfies the microcanonical constraint.

The parameter β in the Maxwell–Boltzmann distribution has an important physical interpretation. It is identified with the absolute temperature T by the formula $\beta = 1/T$; $\beta = \beta(u)$ is called the *inverse (absolute) temperature* corresponding to specific energy u. According to this identification, since T is positive, β should also be positive.[8] For fixed $\beta > 0$, the sequence of equilibrium probabilities $\{\bar{\rho}_{\beta,i}; i = 1, 2, \ldots, r\}$ of the speeds $0 < v_1 < v_2 < \cdots < v_r$ is monotonically decreasing: $\bar{\rho}_{\beta,1} > \bar{\rho}_{\beta,2} > \cdots > \bar{\rho}_{\beta,r}$. As one intuitively expects, higher speeds, which correspond to higher energies, should have lower probabilities. If we allowed $\beta < 0$, then we would be in the reverse situation in which higher energy states would be more likely than lower energy states. But for the kind of system which we are modeling, this is nonphysical.[9] As $\beta \to \infty$, $(\bar{\rho}_\beta)_1 \to 1$ and for $i \in \{2, \ldots, r\}$, $(\bar{\rho}_\beta)_i \to 0$. As $\beta \to 0^+$, each $(\bar{\rho}_\beta)_i \to (2r)^{-1} = \rho_i$. In other words, as $\beta \to \infty$ or $T \to 0^+$, the higher energy states corresponding to $\bar{v}_2, \ldots, \bar{v}_r$ get "frozen out," while as $\beta \to 0^+$ or $T \to \infty$, all energy states become equally probable.

The identification of β and $1/T$ can also be justified by means of entropy. A microcanonical analog of thermodynamic entropy $S(U, V)$ is $-\bar{I}_\rho^{(1)}(u)$, defined in (3.8). One can show [Problem III.10.5] that for $u \in (\frac{1}{2}v_1^2, \frac{1}{2}v_r^2)$, $d(-\bar{I}_\rho^{(1)}(u))/du = \beta(u)$. If we identify β with $1/T$, then this formula is the analog of the thermodynamic formula $\partial S(U, V)/\partial U = 1/T$ [see (3.1)].

III.6. The Canonical Ensemble and Its Equivalence with the Microcanonical Ensemble

The canonical ensemble is a second ensemble with respect to which one can calculate equilibrium values of observables per particle. We present the form of the ensemble as a definition and motivate it afterwards. In Theorem III.6.2, we show that for the purpose of calculating equilibrium values, the microcanonical ensemble and the canonical ensemble give equivalent answers. The definition of the canonical ensemble involves the parameter β. In order to be consistent with the physical interpretation of β as inverse temperature, we will restrict β to positive values. However, all of the mathematical results which follow are valid for any β real.

Definition III.6.1. Let λ be normalized Lebesgue measure on $\mathscr{B}(\Delta)$ with $\lambda\{\Delta\} = 1$. Given $\beta > 0$, define

$$\bar{\rho}_{\beta,i} = \frac{\exp(-\frac{1}{2}\beta v_i^2)\cdot(1/2r)}{\int_\Gamma \exp(-\frac{1}{2}\beta y^2)\rho(dy)} = \frac{\exp(-\frac{1}{2}\beta v_i^2)\cdot(1/2r)}{\sum_{1\le|j|\le r}\exp(-\frac{1}{2}\beta v_j^2)\cdot(1/2r)}$$

and let $\bar{\rho}_\beta$ be the probability measure $\sum_{1\le|i|\le r}\bar{\rho}_{\beta,i}\delta_{v_i}$ on $\mathscr{B}(\Gamma)$ (Maxwell–Boltzmann distribution). Let $P_{n,\Delta,\beta}$ be the product measure on n-particle configuration space Ω_n defined by

$$P_{n,\Delta,\beta}(d\omega) = \lambda(dx_1)\cdots\lambda(dx_n)\bar{\rho}_\beta(dy_1)\cdots\bar{\rho}_\beta(dy_n),$$

where $\omega = (x_1, \ldots, x_n, y_1, \ldots, y_n)\in\Omega_n$. $P_{n,\Delta,\beta}$ is called the *canonical ensemble*.

The canonical ensemble has an interesting relation to the microcanonical ensemble. Fix a positive integer m, let n exceed m, and tag particles $1, 2, \ldots, m$ in a discrete ideal gas consisting of n particles. Let $u = u(A)$ be the specific energy of the discrete ideal gas and $\beta(u)$ the corresponding inverse temperature [see Note 8]. Then the canonical ensemble $P_{m,\Delta,\beta(u)}$ equals the limiting marginal distribution of the positions and velocities of the m tagged particles with respect to the n-particle microcanonical ensemble as $n\to\infty$. While we have a proof of this fact, the following motivation should convince the reader, and so the proof is omitted.[10] The microcanonical ensemble was defined by conditioning the energy per particle to lie in the interval A. But as n increases, the velocities of the m tagged particles depend less and less upon this conditioning. Since the particles do not interact, in the limit $n\to\infty$ the m position-velocity vectors should be independent. Further, for $m = 1$ the limiting marginal distribution was found in (3.22) to be $\lambda(dx)\bar{\rho}_{\beta(u)}(dy)$. It follows that the limiting marginal distribution of the m tagged particles should be product measure with identical one-dimensional marginals $\lambda(dx)\bar{\rho}_{\beta(u)}(dy)$. This product measure is $P_{m,\Delta,\beta(u)}$.

In Definition III.6.1, we defined the canonical ensemble by fixing β rather than by fixing u and setting $\beta = \beta(u)$. Fixing β corresponds more closely to physical situations since in experiments temperature rather than energy is the controlling parameter. Physically, the microcanonical ensemble describes a completely isolated system. By contrast, the canonical ensemble describes a system which is in diathermal contact with a heat reservoir that maintains a fixed temperature $T = 1/\beta$. In the discussion of the relation between the two ensembles in the previous paragraph, the role of the heat reservoir is played by the $n - m$ untagged particles in the discrete ideal gas where eventually $n\to\infty$.

The next theorem is an exponential law of large numbers which determines equilibrium values with respect to the canonical ensemble. We consider observables $\sum_{j=1}^{n} f(X_j, Y_j)$, $n = 1, 2, \ldots$, where f is bounded and measurable. The continuity condition on f required in Theorem III.4.4(b) is not needed.

Theorem III.6.2. *Let f be a bounded measurable real-valued function on $\Delta \times \Gamma$ and define $\langle f \rangle_{\Delta,\beta}^{(\text{can.})} = \int_{\Delta\times\Gamma} f(x,y)\lambda(dx)\bar{\rho}_\beta(dy)$, where the inverse temperature β is given. Then as $n \to \infty$*

$$\frac{1}{n}\sum_{j=1}^{n} f(X_j, Y_j) \overset{\exp}{\to} \langle f \rangle_{\Delta,\beta}^{(\text{can.})} \qquad \textit{with respect to the canonical ensembles } \{P_{n,\Delta,\beta}\}.$$

Proof. With respect to $\{P_{n,\Delta,\beta}\}$ the summands $\{f(X_j, Y_j)\}$ are bounded and i.i.d. and each has mean $\langle f \rangle_{\Delta,\beta}^{(\text{can.})}$. Theorem II.4.1 implies that the distributions of $\{n^{-1}\sum_{j=1}^{n} f(X_j, Y_j)\}$ have a large deviation property and that if I denotes the entropy function, then I has a unique minimum point at $\langle f \rangle_{\Delta,\beta}^{(\text{can.})}$. This property of the entropy function implies exponential convergence [Theorem II.6.3]. □

If $f(x, y) = \frac{1}{2}y^2$, then $\langle \frac{1}{2}y^2 \rangle_{\Delta,\beta}^{(\text{can.})}$ equals $\int_\Gamma \frac{1}{2}y^2 \bar{\rho}_\beta(dy)$. The latter is called the *specific canonical energy* and is written $u_c(\beta)$. We compare Theorem III.6.2 with the microcanonical analog, Theorem III.4.4. The latter says that for continuous f

$$\frac{1}{n}\sum_{j=1}^{n} f(X_j, Y_j) \overset{\exp}{\to} \langle f \rangle_{\Delta,u} = \int_{\Delta\times\Gamma} f(x,y)\lambda(dx)\bar{\rho}_{\beta(u)}(dy).$$

If we evaluate the limit in Theorem III.6.2 at $\beta = \beta(u)$, then clearly $\langle f \rangle_{\Delta,u} = \langle f \rangle_{\Delta,\beta(u)}^{(\text{can.})}$. In other words, for the purpose of calculating equilibrium values, the two ensembles give the same answer. The microcanonical ensemble with specific energy u is equivalent to the canonical ensemble with inverse temperature $\beta(u)$.

This *equivalence of ensembles* motivates the definition of canonical entropy. Microcanonical entropy was defined to be $s(u) = -\bar{I}_\rho^{(1)}(u)$. The contraction principle (3.15) shows that $\bar{I}_\rho^{(1)}(u) = I_\rho^{(2)}(\bar{\rho}_{\beta(u)})$, where $I_\rho^{(2)}$ is relative entropy with respect to $\rho = \sum_{1\le|i|\le r}(2r)^{-1}\delta_{v_i}$. If we define the *specific canonical entropy* to be $s_c(\beta) = -\bar{I}_\rho^{(2)}(\bar{\rho}_\beta)$, then by (3.15) $s(u) = s_c(\beta(u))$.

The canonical ensemble, which for the discrete ideal gas is product measure, is much easier to work with than the microcanonical ensemble, which is conditional measure. However, the canonical ensemble is defined for a fixed number of particles while experimentally the exact number is never known precisely. A third ensemble, called the grand canonical ensemble, treats systems in which the number of particles is arbitrary. For the purpose of calculating equilibrium values, this ensemble is equivalent to the first two.[11]

For future reference, we note an alternate way of writing the canonical ensemble $P_{n,\Delta,\beta}$. Let $\pi_n P_{\lambda,\rho}$ be the product measure on $\mathscr{B}(\Omega_n)$ with identical one-dimensional marginals $\lambda(dx)\rho(dy)$.

Proposition III.6.3. *The canonical ensemble can be written as*

$$P_{n,\Delta,\beta}(d\omega) = \frac{\exp[-\beta U_n(\omega)]\pi_n P_{\lambda,\rho}(d\omega)}{\int_{\Omega_n} \exp[-\beta U_n(\omega)]\pi_n P_{\lambda,\rho}(d\omega)}, \tag{3.23}$$

where $U_n(\omega) = \sum_{j=1}^n \frac{1}{2} Y_j(\omega)^2$ *is the energy in the configuration* $\omega = (x_1, \ldots, x_n, y_1, \ldots, y_n)$.

Proof. If $y_j = v_{i_j}$ is the velocity of the jth particle, then $U_n(\omega) = \sum_{j=1}^n \frac{1}{2} v_{i_j}^2$ and

$$\exp(-\beta U_n(\omega)) \prod_{j=1}^n \rho\{v_{i_j}\} = \prod_{j=1}^n \left[\exp(-\tfrac{1}{2}\beta v_{i_j}^2) \cdot \frac{1}{2r}\right].$$

Since $\bar{\rho}_\beta\{v_{i_j}\} = \exp(-\frac{1}{2}\beta v_{i_j}^2) \cdot (1/2r) / \int_\Gamma \exp(-\frac{1}{2}\beta y^2)\rho(dy)$, the proposition follows. □

The proposition shows that the canonical ensemble is absolutely continuous with respect to the product measure $\pi_n P_{\lambda,\rho}$ with a density that depends only on the energy.

III.7. A Derivation of a Thermodynamic Equation

A monatomic ideal gas of n particles fills a cylinder of volume V at absolute temperature T. The equilibrium values of the pressure P and the internal energy U are determined from the equations $PV = nT$ (ideal gas law) and $U = \frac{3}{2}nT$ [see page 61]. Eliminating T between the two equations yields $PV = \frac{2}{3}U$, which is called Boyle's law.[12] Using the canonical ensemble, we will derive the analog of Boyle's law for the discrete ideal gas.

First, the definition of pressure must be modified since Δ has no surface. Let b denote the left-hand endpoint of Δ. The *linear pressure* Pl of the discrete ideal gas is a thermodynamic variable, defined as the force at b exerted by the gas. The analog of Boyle's law for our one-dimensional model is $\text{Pl} \cdot |\Delta| = 2U$, where $|\Delta|$ is the length of Δ and U is the internal energy. The factor $\frac{1}{3}$, which appears in Boyle's law for the ideal gas because the latter is a three-dimensional system, is absent.

For our derivation we place the thermodynamic variables Pl and U by observables. Given a microstate ω in the n-particle configuration space Ω_n, define $U_n(\omega) = \sum_{j=1}^n \frac{1}{2} Y_j(\omega)^2$. This is the energy in ω. We now derive a formula for the observable $\text{Pl}_n(\omega)$, which is the force at b exerted by the gas in the microstate ω. Newton's second law implies that the force exerted by the gas at b equals the change in the total momentum of the gas at b per unit time. Let ε be the duration of a short time interval. For all sufficiently small ε, a particle in the interior of Δ with velocity v_i will strike b during the time interval if and only if v_i is negative and its position is within $(b, b + |v_i|\varepsilon)$. Since the mass of the particle is one, the change in momentum is $-v_i - v_i = 2|v_i|$. Thus, for all sufficiently small ε,

$$\frac{\mathrm{Pl}_n(\omega)}{n} = \frac{1}{\varepsilon}\sum_{v_i<0} 2|v_i|\frac{1}{n}\sum_{j=1}^{n} \chi_{(b,b+|v_i|\varepsilon)}(X_j(\omega))\cdot\chi_{\{v_i\}}(Y_j(\omega)).$$

This is a sum (over $v_i < 0$) of observables as in Example III.4.1(b). Both $\mathrm{Pl}_n(\omega)$ and $U_n(\omega)$ grow with n, and so we will compare Pl_n/n with $2U_n/n$. By Theorem III.6.2, with respect to the canonical ensembles $\{P_{n,\Delta,\beta}\}$,

$$\frac{U_n}{n} \overset{\exp}{\to} \langle \tfrac{1}{2}y^2\rangle_{\Delta,\beta}^{(\text{can.})} = \sum_{1\le|i|\le r} \frac{1}{2}v_i^2\bar{\rho}_{\beta,i} \qquad \text{as } n\to\infty, \tag{3.24}$$

$$\begin{aligned}\frac{\mathrm{Pl}_n}{n} &\overset{\exp}{\to} \frac{1}{\varepsilon}\sum_{v_i<0} 2|v_i|\langle\chi_{(b,b+|v_i|\varepsilon)}(x)\cdot\chi_{\{v_i\}}(y)\rangle_{\Delta,\beta}^{(\text{can.})}\\ &= \frac{1}{\varepsilon}\sum_{v_i<0} 2|v_i|\cdot\frac{|v_i|\varepsilon}{|\Delta|}\bar{\rho}_{\beta,i} = \frac{1}{|\Delta|}\sum_{1\le|i|\le r} v_i^2\bar{\rho}_{\beta,i} \qquad \text{as } n\to\infty.\end{aligned} \tag{3.25}$$

The last equality follows from the symmetry of the $\{v_i\}$ [see (3.3)]. Let u_c and pl denote the equilibrium values of energy per particle and linear pressure per particle defined by the limits in (3.24) and (3.25), respectively:

$$u_c = \sum_{1\le|i|\le r} \tfrac{1}{2}v_i^2\bar{\rho}_{\beta,i}, \qquad \mathrm{pl} = \frac{1}{|\Delta|}\sum_{1\le|i|\le r} v_i^2\bar{\rho}_{\beta,i}.$$

We have $\mathrm{pl}\cdot|\Delta| = 2u_c$. This completes our derivation of Boyle's law.

III.8. The Gibbs Variational Formula and Principle

In this section we introduce the magic statistical mechanical function called the free energy. This function is an analog of the Helmholtz free energy discussed in Section III.2. Our first theorem evaluates the free energy two different ways by means of Varadhan's theorem. An evaluation based on level-1 large deviations is related to the exponential law of large numbers in Theorem III.6.2. A level-3 evaluation gives a supremum over strictly stationary probability measures which is called the Gibbs variational formula. The supremum in the Gibbs variational formula is attained at the unique strictly stationary probability measure which is the natural extension of the canonical ensemble to an infinite particle system. This extension is called the level-3 equilibrium state of the discrete ideal gas. In our study of ferromagnetic systems in the next two chapters, the level-3 evaluation of the free energy will be generalized. This will lead to a level-3 notion of phase transition.

From now on, with respect to the canonical ensemble we consider only the marginal distribution of velocities. Γ^n, which denotes the set of all sequences $\omega = (\omega_1, \ldots, \omega_n)$ with each $\omega_j \in \Gamma$, is the velocity configuration space of n particles. Define $\pi_n P_{\bar{\rho}_\beta}$ to be the product measure on $\mathscr{B}(\Gamma^n)$ with identical one-dimensional marginals $\bar{\rho}_\beta$. By Definition III.6.1, $\pi_n P_{\bar{\rho}_\beta}$ is the marginal distribution of the velocities of n particles with respect to the canonical ensemble. It is called the *reduced-canonical ensemble*. As in (3.23), this measure can be written as

$$\pi_n P_{\rho_\beta}(d\omega) = \frac{\exp[-\beta U_n(\omega)]\pi_n P_\rho(d\omega)}{\int_{\Gamma^n} \exp[-\beta U_n(\omega)]\pi_n P_\rho(d\omega)}, \tag{3.26}$$

where $U_n(\omega) = \sum_{j=1}^n \frac{1}{2} Y_j(\omega)^2$ is the energy in ω.

In order to describe the infinite-particle discrete ideal gas, we let $n \to \infty$ in the above definitions. The velocity configuration space becomes the set $\hat{\Omega}$ of all infinite sequences $\omega = (\omega_1, \omega_2, \ldots)$ with each $\omega_j \in \Gamma$; thus $\hat{\Omega} = \Gamma^{\mathbb{Z}^+}$. In order to use the definitions from Chapter II concerning level-3 large deviations, we imbed $\hat{\Omega}$ in the set $\Omega = \Gamma^{\mathbb{Z}}$ of all doubly infinite sequences $\omega = (\ldots, \omega_{-1}, \omega_0, \omega_1, \ldots)$ with each $\omega_j \in \Gamma$. We topologize Γ by the discrete topology and Ω by the product topology and let $\mathscr{M}_s(\Omega)$ be the set of strictly stationary probability measures on the Borel subsets of Ω. One maps Ω onto $\hat{\Omega}$ by projecting the point $(\ldots, \omega_{-1}, \omega_0, \omega_1, \omega_2, \ldots)$ onto $(\omega_1, \omega_2, \ldots)$. Given $P \in \mathscr{M}_s(\Omega)$, this projection defines a probability measure $\hat{P}$ on $\hat{\Omega}$. Let $\mathscr{M}_s(\hat{\Omega})$ be the set $\{\hat{P} : P \in \mathscr{M}_s(\Omega)\}$. For the discrete ideal gas, a special measure in $\mathscr{M}_s(\hat{\Omega})$ is the infinite product measure $\hat{P}_{\bar{\rho}_\beta}$ with identical one-dimensional marginals $\bar{\rho}_\beta$. The measure $\hat{P}_{\bar{\rho}_\beta}$ is the infinite-particle analog of the reduced-canonical ensemble $\pi_n P_{\rho_\beta}$ in (3.26). Our main theorem, Theorem III.8.1, characterizes $P_{\bar{\rho}_\beta} \in \mathscr{M}_s(\Omega)$ as the unique solution of a variational formula involving the free energy. We now define the latter function.

We recall from Section III.2 that for a simple system in the equilibrium state (V, T), the Helmholtz free energy is defined by $F(V, T) = U(V, T) - TS(V, T)$, where U and S denote the internal energy and entropy, respectively. Analogs of U and S with respect to the canonical ensemble and the reduced canonical ensemble are the quantities* $u_c(\beta) = \int_\Gamma \frac{1}{2} y^2 \bar{\rho}_\beta(dy)$ and $s_c(\beta) = -I_\rho^{(2)}(\bar{\rho}_\beta)$. Hence an analog of $F = T(T^{-1}U - S)$ with respect to the reduced-canonical ensemble is

$$\phi(\beta) = \beta^{-1}(\beta u_c(\beta) - s_c(\beta)) = \beta^{-1}(\beta u_c(\beta) + I_\rho^{(2)}(\bar{\rho}_\beta)). \tag{3.27}$$

The function $\phi(\beta)$ is called the *specific free energy*. We have the simple formula

* $u_c(\beta)$ and $s_c(\beta)$ were defined after Theorem III.6.2.

(3.28) $$\phi(\beta) = -\beta^{-1} \log \int_\Gamma \exp(-\tfrac{1}{2}\beta y^2)\rho(dy).$$

Indeed,

$$I_\rho^{(2)}(\bar{\rho}_\beta) = \int_\Gamma \log \frac{d\bar{\rho}_\beta}{d\rho}(y)\bar{\rho}_\beta(dy) \qquad \text{where} \qquad \bar{\rho}_{\beta,i} = \frac{\exp(-\frac{1}{2}\beta v_i^2)\cdot(1/2r)}{\int_\Gamma \exp(-\frac{1}{2}\beta y^2)\rho(dy)}.$$

Hence

$$\begin{aligned} I_\rho^{(2)}(\bar{\rho}_\beta) &= \int_\Gamma \left[-\tfrac{1}{2}\beta y^2 - \log \int_\Gamma \exp(-\tfrac{1}{2}\beta y^2)\rho(dy)\right] \bar{\rho}_\beta(dy) \\ &= -\beta u_c(\beta) - \log \int_\Gamma \exp(-\tfrac{1}{2}\beta y^2)\rho(dy), \end{aligned}$$

and (3.28) follows. For t real define $\bar{c}_\rho(t) = \log \int_\Gamma \exp(\frac{1}{2}ty^2)\rho(dy)$. This function is the free energy function of the probability measure $\sum_{1 \le |i| \le r} (2r)^{-1} \delta_{v_i^2/2}$, and we can write the specific free energy as

(3.29) $$\phi(\beta) = -\beta^{-1}\bar{c}_\rho(-\beta).$$

The function $\bar{c}_\rho$ was used earlier in this chapter [Lemmas III.4.3 and III.4.5].

Let $Y(\omega) = \omega_j$ be the coordinate functions on $\Omega = \Gamma^{\mathbb{Z}}$ and P_ρ the infinite product measure on $\mathscr{B}(\Omega)$ with identical one-dimensional marginals ρ. E_ρ denotes expectation with respect to P_ρ. The function $-\beta\phi(\beta)$ equals $\log E_\rho\{\exp(-\frac{1}{2}\beta Y_1^2)\}$ or equivalently

(3.30) $$-\beta\phi(\beta) = \frac{1}{n}\log Z_n(\beta) \qquad \text{where} \quad Z_n(\beta) = E_\rho\left\{\exp\left(-\tfrac{1}{2}\beta \sum_{j=1}^n Y_j^2\right)\right\}$$

and n is any positive integer. $Z_n(\beta)$ is called the *reduced-canonical partition function*. It is the numerator in the formula (3.26) for the reduced-canonical ensemble. In the next theorem, $-\beta\phi(\beta) = \lim_{n\to\infty} n^{-1} \log Z_n(\beta)$ is evaluated by level-1 and by level-3 large deviations. Formula (3.32) is called the *Gibbs variational formula.*[13]

Theorem III.8.1. *Let $\bar{I}_\rho^{(1)}(u)$ and $I_\rho^{(3)}(P)$ be the level-1 and level-3 entropy functions defined in Lemma III.4.3 and Theorem II.4.4, respectively. Then the following conclusions hold.*

(a) *For each $\beta > 0$*

(3.31) $$-\beta\phi(\beta) = \lim_{n\to\infty} \frac{1}{n}\log Z_n(\beta) = \sup_{u\in\mathbb{R}}\{-\beta u - \bar{I}_\rho^{(1)}(u)\}.$$

The supremum is attained at the unique point $u = u_c(\beta) = \int_\Gamma \frac{1}{2}y^2\bar{\rho}_\beta(dy)$, which is the specific canonical energy.

(b) *For each $\beta > 0$*

(3.32) $$-\beta\phi(\beta) = \lim_{n\to\infty} \frac{1}{n}\log Z_n(\beta) = \sup_{P\in\mathscr{M}_s(\Omega)} \left\{-\beta\int_\Omega \tfrac{1}{2}\omega_1^2 P(d\omega) - I_\rho^{(3)}(P)\right\}.$$

The supremum is attained at the unique measure $P = P_{\bar{\rho}_\beta}$, which is the infinite product measure on $\mathscr{B}(\Omega)$ with identical one-dimensional marginals $\bar{\rho}_\beta$.

Proof. (a) According to Lemma III.4.3, the P_ρ-distributions of $\{\sum_{j=1}^n \frac{1}{2} Y_j^2/n; n = 1, 2, \ldots\}$ have a large deviation property with entropy function $\bar{I}_\rho^{(1)}$. Theorem II.7.3(a) implies that

$$-\beta\phi(\beta) = \lim_{n\to\infty} \frac{1}{n} \log Z_n(\beta) = \sup_{u\in\mathbb{R}} \{-\beta u - \bar{I}_\rho^{(1)}(u)\}.$$

The point $u = u_c(\beta)$ gives the supremum in this display. Indeed $u_c(\beta)$ equals $\int_\Gamma \frac{1}{2} y^2 \bar{\rho}_\beta(dy)$, and thus $\beta(u_c(\beta))$, the inverse temperature* corresponding to specific energy $u_c(\beta)$, is exactly β. By (3.27) and (3.15),

$$(3.33) \qquad -\beta\phi(\beta) = -\beta u_c(\beta) - I_\rho^{(2)}(\bar{\rho}_{\beta(u_c(\beta))}) = -\beta u_c(\beta) - \bar{I}_\rho^{(1)}(u_c(\beta)),$$

as we claimed. To show that the point $u_c(\beta)$ is the unique point at which the supremum in (3.31) is attained, we use the fact that $\bar{I}_\rho^{(1)}(u)$ is strictly convex and thus $-\beta u - \bar{I}_\rho^{(1)}(u)$ is strictly concave [Theorem VII.5.5(h)]. A strictly concave function attains its supremum at a unique point (if at all).

(b) Let $R_n(\omega, \cdot)$ be the empirical process corresponding to $Y_1, \ldots, Y_n$ (defined by (2.4) with $X_1, \ldots, X_n$ replaced by $Y_1, \ldots, Y_n$). By Theorem II.4.4 the P_ρ-distributions of $\{R_n\}$ have a large deviation property with entropy function $I_\rho^{(3)}$. Since

$$\frac{n-1}{n} \log Z_n(\beta) = \log E_\rho\left\{\exp\left(-\tfrac{1}{2}\beta \sum_{j=1}^{n-1} Y_j^2\right)\right\},$$

Theorem II.7.3(c) implies that

$$-\beta\phi(\beta) = \lim_{n\to\infty} \frac{1}{n} \log Z_n(\beta) = \sup_{P\in\mathscr{M}_s(\Omega)} \left\{-\beta \int_\Omega \tfrac{1}{2}\omega_1^2 P(d\omega) - I_\rho^{(3)}(P)\right\}.$$

We now locate where the supremum in this display is attained. For fixed $u \in [\frac{1}{2}v_1^2, \frac{1}{2}v_r^2]$, let $\Phi(u)$ be the set of $P \in \mathscr{M}_s(\Omega)$ for which $\int_\Omega \frac{1}{2}\omega_1^2 P(d\omega) = u$. In Theorem II.5.4, we stated a contraction principle relating levels-1 and 3. By analogy with this result, one can show that

$$(3.34) \qquad \inf_{P\in\Phi(u)} I_\rho^{(3)}(P) = \bar{I}_\rho^{(1)}(u)$$

and that if $u = u_c(\beta)$, then $I_\rho^{(3)}(P)$ attains its infimum over $\Phi(u_c(\beta))$ at the unique measure $P_{\bar{\rho}_\beta}$ [Problem III.10.7]. Any $P \in \mathscr{M}_s(\Omega)$ belongs to $\Phi(u)$ for some $u \in [\frac{1}{2}v_1^2, \frac{1}{2}v_r^2]$. If $u = u_c(\beta)$, then

$$\begin{aligned}
-\beta \int_\Omega \tfrac{1}{2}\omega_1^2 P(d\omega) - I_\rho^{(3)}(P) &\leq -\beta u_c(\beta) - \inf_{P\in\Phi(u_c(\beta))} I_\rho^{(3)}(P) \\
&= -u_c(\beta) - \bar{I}_\rho^{(1)}(u_c(\beta)) = -\beta\phi(\beta).
\end{aligned}$$

* See page 75.

Equality holds in the first line of the display if and only if $P = P_{\bar{\rho}_\beta}$. If $u \neq u_c(\beta)$, then by the proof of part (a) of the present theorem

$$-\beta \int_\Omega \tfrac{1}{2}\omega_1^2 P(d\omega) - I_\rho^{(3)}(P) \leq -\beta u - \inf_{P \in \Phi(u)} I_\rho^{(3)}(P) = -\beta u - \bar{I}_\rho^{(1)}(u)$$
$$< -\beta u_c(\beta) - \bar{I}_\rho^{(1)}(u_c(\beta)) = -\beta\phi(\beta).$$

It follows that the supremum in (3.32) is attained at the unique measure $P = P_{\bar{\rho}_\beta}$.

We now interpret Theorem III.8.1 physically. The level-1 interpretation involves the energy observable $U_n(\omega) = \sum_{j=1}^n \frac{1}{2} Y_j(\omega)^2$. Each number $u \in (\frac{1}{2}v_1^2, \frac{1}{2}v_r^2)$ is a candidate for the equilibrium value of the energy per particle $U_n(\omega)/n$. By Theorem III.6.2, U_n/n tends exponentially to $u_c(\beta)$, and so $u_c(\beta)$ is the actual equilibrium value. Part (a) of Theorem III.8.1 characterizes this number as the unique point at which the function $-\beta u - \bar{I}_\rho^{(1)}(u)$ attains its supremum over $u \in \mathbb{R}$. The quantity $-\beta^{-1}[\beta u_c(\beta) - I_\rho^{(1)}(u_c(\beta))]$ equals the specific free energy $\phi(\beta)$.

For level-3, each measure $P \in \mathcal{M}_s(\Omega)$ defines a measure $\hat{P} \in \mathcal{M}_s(\hat{\Omega})$ which is a candidate for the infinite-particle description of the gas. We define the specific energy in P to be $u(P) = \int_\Omega \frac{1}{2}\omega_1^2 P(d\omega)$, the specific entropy in P to be $s(P) = -I_\rho^{(3)}(P)$, and the specific free energy in P to be

$$\phi(\beta; P) = \beta^{-1}(\beta u(P) - s(P)) = \beta^{-1}\left(\beta \int_\Omega \tfrac{1}{2}\omega_1^2 P(d\omega) + I_\rho^{(3)}(P)\right). \tag{3.35}$$

Any measure $P \in \mathcal{M}_s(\Omega)$ at which $-\beta\phi(\beta; P)$ attains its supremum over $\mathcal{M}_s(\Omega)$ is called a *level-3 equilibrium state*. The same term applies to the corresponding measure $\hat{P} \in \mathcal{M}_s(\hat{\Omega})$. Part (b) of Theorem III.8.1 implies the following result, known as the *Gibbs variational principle*.

Theorem III.8.2. *$P_{\bar{\rho}_\beta}$ is the unique measure in $\mathcal{M}_s(\Omega)$ at which $-\beta\phi(\beta; P)$ attains its supremum. Thus $P_{\bar{\rho}_\beta}$ is the unique level-3 equilibrium state for the discrete ideal gas. The quantity $\phi(\beta; P_{\bar{\rho}_\beta})$ equals the specific free energy $\phi(\beta)$.*

Since β is positive, $P_{\bar{\rho}_\beta}$ is also the unique measure in $\mathcal{M}_s(\Omega)$ at which $\phi(\beta; P)$ attains its infimum. That the equilibrium state gives the unique infimum is the precise analog of Property III.2.2(b) of the Helmholtz free energy. This property is a minimum principle which specifies the thermodynamic equilibrium state of a composite system of fixed volume in diathermal contact with a heat reservoir.

An obvious connection between the level-1 and level-3 interpretations of the theorem is that $u_c(\beta)$ equals the specific energy in the equilibrium state $P_{\bar{\rho}_\beta}$:

$$u_c(\beta) = u(P_{\bar{\rho}_\beta}) = \int_\Omega \tfrac{1}{2}\omega_1^2 P_{\bar{\rho}_\beta}(d\omega) = \int_\Gamma \tfrac{1}{2}y^2 \bar{\rho}_\beta(dy).$$

Another important aspect of the theorem is the entropy characterization of $P_{\bar{\rho}\beta}$ used in the proof of Theorem III.8.1(b); namely,

$$\inf\left\{I_\rho^{(3)}(P): P \in \mathcal{M}_s(\Omega), \int_\Omega \tfrac{1}{2}\omega_1^2 P(d\omega) = u_c(\beta)\right\} \tag{3.36}$$

is attained at the unique measure $P_{\bar{\rho}\beta}$. Thus, $P_{\bar{\rho}\beta}$ can be interpreted as the most random strictly stationary measure which satisfies the energy constraint $\int_\Omega \frac{1}{2}\omega_1^2 P(d\omega) = u_c(\beta)$.

With (3.36), we have come full circle. The microcanonical ensemble was defined by conditioning upon the energy per particle. This conditioning has been transmuted into an energy constraint in the entropy principle that characterizes the level-3 equilibrium state. As we will see in the next chapter, Theorems III.8.1 and II.8.2 have natural extensions to systems with interactions. By contrast with the discrete ideal gas, these systems need not have unique level-3 equilibrium states, and microscopic sums associated with the systems need not converge exponentially to a constant. This different behavior is associated with phase transitions.

III.9. Notes

1 (page 59). For an overview of the statistical mechanics of continuous systems, Uhlenbeck and Ford (1963) is recommended. Huang (1963), Thompson (1972), and Reichl (1980) are also good. All of these texts derive the laws of thermodynamics from statistical mechanics. Ruelle (1969) is an advanced reference. Lanford (1973) and Martin-Löf (1979) study the statistical mechanics of continuous systems and derive the laws of thermodynamics using large deviations. Our treatment of the discrete ideal gas is a special case. Of historical interest is Khinchin (1949), which was the first extensive study of statistical mechanics from the viewpoint of probability theory. Also see Kac (1959b) and Mackey (1974). The classic treatment of the foundations of statistical mechanics is Ehrenfest and Ehrenfest (1959). Penrose (1979) reviews modern developments. Truesdell (1961), Farquhar (1964), and Lebowitz and Penrose (1973) discuss one of the central foundational issues, which is the role of the ergodic theorem in justifying the microcanonical ensemble. For nonequilibrium theories, see for example Prigogine (1973, 1978), Lanford (1975, 1976), Cercignani (1975), Lebowitz (1978), and Reichl (1980).

2 (page 60). For thermodynamics, Callen (1960) is recommended because of his clear treatment of entropy. Pippard (1957), Zemansky (1968), and Adkins (1975) are also good. Mendelssohn (1966) is a nice treatment of low temperature physics.

3 *Historical Sketch* (page 63). The laws of thermodynamics had been definitively formulated by Clausius in 1850. But the need for a statistical

theory of matter was shown by the very limited success of Clausius and others (including, initially, Boltzmann) in explaining the second law of thermodynamics on the basis of the mechanics of particles alone. Maxwell became interested in the molecular theory of gases when he read Clausius's papers on the subject. While Clausius simply used the average molecular velocity, Maxwell saw that an analysis of how molecular velocities were distributed would be of fundamental importance. This analysis was given by Maxwell in his work on the kinetic theory of gases (1860, 1867), which was the first application of probability theory to physics. Maxwell then turned to the second law. He invented his famous demon (1867) [see Klein (1970)] to show that the second law cannot be derived from mechanics alone, but rather that it expresses the statistical regularity of systems composed of very large numbers of particles. Spurred in part by Maxwell's work, Boltzmann after 1867 began to create a systematic statistical theory of gases. The Boltzmann equation (1872) enabled him to study how a gas not in equilibrium can reach equilibrium as a result of collisions among its molecules. The paper of 1877 expressed his statistical interpretation of entropy and of the second law. Gibbs, working in isolation in the United States, had two important achievements. His papers of 1873–1878 invented modern thermodynamics; his 1902 book was the first systematic exposition of the entire formalism of statistical mechanics. This work presents a mathematical structure for computations which is essentially the one used today. Another important early figure was Einstein. His first papers in statistical mechanics (1902–1904) also dealt with the general formalism, which he developed independently of Gibbs. He applied the theory to explain Brownian motion as well as many phenomena in quantum physics (see references in Pais (1982)).

4 (page 63). Callen (1960, page 109) explains why the term "free" is used in the Helmholtz free energy.

5 (page 64). Sinai (1977, Lecture 8) treats the ideal gas by means of ergodic theory.

6 (page 67). Our study of the asymptotics of the discrete ideal gas simplifies the standard statistical mechanical approach to classical continuous systems in a number of ways. The statistical mechanics of classical continuous systems is developed in detail in Ruelle (1969) and in Lanford (1973).

7 (page 69). Theorem III.4.4 has been proved by Lanford (1973) by a different method. A related form of the theorem goes back to Khinchin (1949). Blanc-LaPierre and Tortrat (1956, pages 161–162) give a heuristic large deviation proof suggested by K. Ito.

8 (page 75). The physical requirement $\beta > 0$ means that the only physically meaningful values of the specific energy u are those in the interval $(\frac{1}{2}v_1^2, \alpha)$, where $\alpha = \int_\Gamma \frac{1}{2}y^2 \rho(dy)$. Only for such u is $\beta(u) > 0$.

9 (page 75). In this chapter we are modeling systems for which the only energy is vibrational energy and for which the number of energy levels is infinite. For such systems the absolute temperature cannot be negative. However, there are other systems (spin systems) which have a finite number

of energy levels and for which it is possible to have more nuclei in the higher energy levels than in the lower energy levels. This situation corresponds to negative absolute (spin) temperatures. Zemansky (1968, Chapter 14) and Proctor (1978) are introductions to this matter.

10 (page 76). Spitzer (1974) and Zabell (1980) each derive the canonical ensemble as a limiting marginal distribution, starting from a slightly different definition of the microcanonical ensemble (conditioning on $U_n/n = u + z_n$, where $z_n \to 0$, instead of on $U_n/n \in A$). Also see van Campenhout and Cover (1981) and Csiszár (1984). Lenard (1978) has a self-contained justification of the canonical ensemble in quantum statistical mechanics.

11 (page 77). The equivalence of ensembles for continuous systems is treated by Aizenmann, Goldstein, and Lebowitz (1978) and for lattice systems by Dobrushin and Tirozzi (1977).

12 (page 78). Boyle's law is discussed in Adkins (1975, Section 8.2). A similar derivation of Boyle's law for the ideal gas goes back to Maxwell (1860).

13 (page 81). The specific free energy $\phi(\beta)$ equals $-\beta^{-1}\bar{c}_\rho(-\beta)$, where $\bar{c}_\rho(t) = \int_\Gamma \exp(\frac{1}{2}ty^2)\rho(dy)$ [see (3.29)]. Hence (3.31) can be derived from (3.8) by an inverse Legendre–Fenchel transform.

III.10. Problems

III.10.1. Prove equation (3.4). This is related to Liouville's theorem (see Thompson (1972, Appendix A)). [Hint: First prove (3.4) for Borel sets of the form

$$B = J_1 \times \cdots \times J_n \times \{v_{i_1}\} \times \cdots \times \{v_{i_n}\},$$

where $J_1, \ldots, J_n$ are subintervals of Δ and $v_{i_1}, \ldots, v_{i_n}$ are velocities in Γ.]

III.10.2. Prove that the distributions of $U_n/n = \frac{1}{2}\sum_{j=1}^n Y_j^2/n$, $n = 1, 2, \ldots$, with respect to the microcanonical ensembles $\{P_{n,\Delta,A}\}$ have a large deviation property with entropy function $I(u) = \bar{I}_\rho^{(1)}(u) - \inf_{z\in A}\bar{I}_\rho^{(1)}(z)$ for $u \in A$, $I(u) = \infty$ for $u \notin A$.

III.10.3. Let f be a real-valued function on Γ. Prove that the $P_{n,\Delta,A}$-distributions of $n^{-1}\sum_{j=1}^n f(Y_j)$, $n = 1, 2, \ldots$, have a large deviation property and determine the entropy function.

III.10.4. Complete the proof of (3.21) by showing that if τ is any measure in the set $\Psi_2 = \{\tau \in \mathcal{M}(\Delta \times \Gamma): \int_{\Delta\times\Gamma} \frac{1}{2}y^2\tau(dxdy) \in A\}$ which differs from $\bar{\tau}(dxdy) = \lambda(dx)\bar{\rho}_{\beta(u(A))}(dy)$, then

$$I_{\lambda,\rho}^{(2)}(\tau) > I_{\lambda,\rho}^{(2)}(\bar{\tau}) = \bar{I}_\rho^{(1)}(u(A)).$$

[*Hint*: Given $\tau \in \Psi_2$, define $\tilde{\tau}(dxdy) = \sum_{1\le|i|\le r}\lambda(dx)\nu_i\delta_{v_i}$, where $\nu_i = \int_\Delta \tau_i(dx)$. Show that if $I_{\lambda,\rho}^{(2)}(\tau)$ is finite, then

$$I^{(2)}_{\lambda,\rho}(\tau) = I^{(2)}_{\tilde{\tau}}(\tau) + I^{(2)}_{\rho}(\nu), \quad \text{where} \quad \nu = \sum_{1\le|i|\le r} \nu_i \delta_{v_i},$$

and use (3.19).]

III.10.5. (page 75). Prove that $d(-\bar{I}^{(1)}_{\rho}(u))/du = \beta(u)$, where $\bar{I}^{(1)}_{\rho}$ is defined in (3.8). [*Hint:* See the proof of Lemma III.4.5(b)–(c).]

III.10.6. We consider sequences of velocities $\{v_i^{(r)}; 1 \le |i| \le r\}$, $r = 1, 2, \ldots,$ of the discrete ideal gas. Suppose that there exists a positive sequence $\{b(r); r = 1, 2, \ldots\}$ such that $v_i^{(r)} - v_{i-1}^{(r)} = b(r)$, $b(r) \to 0$, and $rb(r) \to \infty$ as $r \to \infty$. Thus as $r \to \infty$ the velocities become dense in $\mathbb{R}$. Let

$$u_c^{(r)}(\beta) = \sum_{1\le|i|\le r} \tfrac{1}{2}(v_i^{(r)})^2 \bar{\rho}^{(r)}_{\beta,i},$$

where

$$\bar{\rho}^{(r)}_{\beta,i} = \frac{\exp[-\frac{1}{2}\beta(v_i^{(r)})^2]\cdot(1/2r)}{\sum_{1\le|j|\le r}\exp[-\frac{1}{2}\beta(v_j^{(r)})^2]\cdot(1/2r)}.$$

Prove that $\lim_{r\to\infty} u_c^{(r)}(\beta) = \frac{1}{2}\beta^{-1}$.

For large r, $u_c^{(r)}(\beta) \sim \frac{1}{2}\beta^{-1}$ is the analog for the discrete ideal gas of the thermodynamic relation $U = \frac{3}{2}nT$ for a monatomic ideal gas [see page 61].

III.10.7. (page 82). Complete the proof that the supremum in (3.32) is attained at the unique measure $P = P_{\bar{\rho}_\beta}$. [*Hint:* Combine Theorem II.5.3 and Lemma III.4.5(c). Extend the latter to $u = \frac{1}{2}v_1^2$ and $u = \frac{1}{2}v_r^2$.]

Chapter IV

Ferromagnetic Models on $\mathbb{Z}$

IV.1. Introduction

Phase transitions are a familiar aspect of nature. Water boils, becoming water vapor, or water vapor, under compression, liquefies. These are examples of a liquid–gas phase transition. The liquid and the gas are said to be two phases of the same substance. One of the most interesting problems in equilibrium statistical mechanics is to explain phase transitions in terms of the probability distributions on configuration space which describe the microscopic behavior of physical systems. The simplest systems for which this is possible are ferromagnetic models on a lattice. The present chapter introduces these models.

The phase transition for ferromagnetic systems has many similarities with the more common liquid–gas transition, although each is described by different variables.[1] Both phase transitions arise as a result of two competing microscopic effects. The first effect tends to order the system. It is caused by attractive forces of interaction and is measured by energy. The second effect tends to randomize the system. It is caused by thermal excitations and is measured by entropy. At sufficiently low temperatures, the energy effect predominates and a phase transition becomes possible.

This chapter develops the statistical mechanics of ferromagnetic models on the one-dimensional integer lattice $\mathbb{Z}$. Many of the results for models on $\mathbb{Z}$ generalize to ferromagnetic models on the D-dimensional integer lattice $\mathbb{Z}^D$, $D \in \{2, 3, \ldots\}$. These models will be treated in Chapter V. The next section discusses qualitatively the main features of ferromagnetic models as established by the theorems of this chapter and the next.

IV.2. An Overview of Ferromagnetic Models

The ultimate source of ferromagnetism is the quantum mechanical spinning of electrons. Because a small magnetic dipole moment is associated with the spin, the electron acts like a magnet with one north pole and one south pole. Both the spin and the magnetic moment can be represented by an arrow which defines the direction of the electron's magnetic field. The spin can point up (spin value 1) or down (spin value -1), and it flips between the two orientations. Ferromagnetic models were invented in order to represent, in simplified form, the interaction of electron spins in real ferromagnets. In this section we discuss the most popular ferromagnetic model which is the Ising model. Its properties are qualitatively the same as those of more complicated models to be discussed later in the book.[2]

Let Λ be a symmetric hypercube of the D-dimensional integer lattice $\mathbb{Z}^D$. To each site j of Λ there is assigned a variable ω_j which takes the value 1 (spin-up) or -1 (spin-down). Fix a number $\mathscr{J} > 0$. Associated with each configuration $\omega = \{\omega_j; j \in \Lambda\}$ of spins is a *Hamiltonian* or *interaction energy*

$$H_{\Lambda,h}(\omega) = -\frac{1}{2}\sum_{i,j\in\Lambda} J(i-j)\omega_i\omega_j - h\sum_{j\in\Lambda}\omega_j,$$

where $J(i-j)$ equals $\mathscr{J}$ if $\|i-j\| = 1$ and equals 0 if $\|i-j\| \neq 1$. Thus the first sum extends over all nearest neighbor pairs of sites in Λ. The number $\mathscr{J}$ is the strength of the nearest neighbor coupling and h is a real number which is the strength of an externally applied magnetic field. The configuration space is the set Ω_Λ of all sequences $\omega = \{\omega_j; j \in \Lambda\}$; thus $\Omega_\Lambda = \{1, -1\}^\Lambda$. Define $\mathscr{B}(\Omega_\Lambda)$ to be the set of all subsets of Ω_Λ. Let ρ be the measure $\frac{1}{2}\delta_1 + \frac{1}{2}\delta_{-1}$ and $\pi_\Lambda P_\rho$ the product measure on $\mathscr{B}(\Omega_\Lambda)$ with identical one-dimensional marginals ρ. The Ising model is defined by the probability measure $P_{\Lambda,\beta,h}$ on $\mathscr{B}(\Omega_\Lambda)$ which assigns to each $\{\omega\}$, $\omega \in \Omega_\Lambda$, the probability

$$P_{\Lambda,\beta,h}\{\omega\} = \exp[-\beta H_{\Lambda,h}(\omega)]\pi_\Lambda P_\rho\{\omega\}\cdot\frac{1}{Z}.$$

The parameter β represents the inverse absolute temperature $1/T$ and is positive. Z is the normalization $\int_{\Omega_\Lambda}\exp[-\beta H_{\Lambda,h}(\omega)]\pi_\Lambda P_\rho(d\omega)$. We call $P_{\Lambda,\beta,h}$ a *finite-volume Gibbs state* or a *finite-volume equilibrium state*. Z is called a *partition function*. Suppose that the external field h is nonzero and consider the configuration $\bar{\omega}$ whose spins are all aligned in the same direction as h. Since $\mathscr{J}$ is positive, this configuration has the smallest interaction energy, hence the largest probability $P_{\Lambda,\beta,h}\{\bar{\omega}\}$, of all configurations in Ω_Λ. Hence the positivity of $\mathscr{J}$ induces an alignment effect in the finite-volume Gibbs state. The effect becomes weaker as β is decreased. At $\beta = 0$ it disappears entirely as $P_{\Lambda,\beta,h}$ reduces to the product measure $\pi_\Lambda P_\rho$ which assigns equal probability to each configuration.

We now proceed to explain, without technicalities, the main properties of the Ising model. These themes are developed in detail in subsequent sections.

Magnetization. Let $S_\Lambda(\omega)$ be the observable $\sum_{j\in\Lambda}\omega_j$ which gives the total spin in Λ. We introduce two quantities

$$M(\Lambda,\beta,h) = \int_{\Omega_\Lambda} S_\Lambda(\omega)P_{\Lambda,\beta,h}(d\omega) \quad \text{and} \quad m(\beta,h) = \lim_{\Lambda\uparrow\mathbb{Z}^D}\frac{1}{|\Lambda|}M(\Lambda,\beta,h).$$

$M(\Lambda,\beta,h)$ is the average value of the total spin in Λ and is called the *magnetization*; $m(\beta,h)$ is the magnetization per site in the limit where Λ expands to fill $\mathbb{Z}^D$ and is called the *specific magnetization*. For $h > 0$, $M(\Lambda, \beta, h)$ is positive and because of the alignment effect built into the finite-volume Gibbs state, $M(\Lambda, \beta, h)$ is an increasing function of h. These properties persist for the specific magnetization after the limit $\Lambda \uparrow \mathbb{Z}^D$. However, the alignment effect becomes weaker as β is decreased. This is reflected in dramatically different behavior of the limit $\lim_{h \to 0^+} m(\beta, h)$ for different values of β.

Spontaneous magnetization. By symmetry, $M(\Lambda, \beta, 0)$ equals 0, and so $m(\beta, 0) = \lim_{\Lambda \uparrow \mathbb{Z}^D} |\Lambda|^{-1} M(\Lambda, \beta, 0)$ equals 0. There exists a critical value of β, called the *critical inverse temperature* and denoted by β_c, which has the following properties. If $0 < \beta < \beta_c$, then $m(\beta, h)$ converges to $m(\beta, 0) = 0$ as $h \to 0^+$. If $\beta > \beta_c$, then $m(\beta, h)$ converges to a positive number $m(\beta, +)$ as $h \to 0^+$. Thus for $\beta > \beta_c$, the system remains permanently magnetized after the external field is removed; $m(\beta, +)$ is known as the *spontaneous magnetization*. For $h < 0$, $m(\beta, h)$ behaves similarly: $m(\beta, h)$ is negative and as $h \to 0^-$, $m(\beta, h)$ converges to $m(\beta, 0) = 0$ or to the negative number $m(\beta, -) = -m(\beta, +)$ according to whether $0 < \beta < \beta_c$ or $\beta > \beta_c$. The value of the critical inverse temperature β_c depends upon $\mathscr{J}$ and upon the dimension D of the lattice. If $D = 1$, then β_c is infinite and spontaneous magnetization does not occur. By contrast, for any $D \geq 2$, β_c is finite. We have not discussed the value $m(\beta, +) = \lim_{h \to 0^+} m(\beta, h)$ at $\beta = \beta_c$. For the Ising model on $\mathbb{Z}^2$, $m(\beta_c, +)$ is 0. While $m(\beta_c, +)$ is believed to be 0 for any Ising model on $\mathbb{Z}^D$, $D \geq 3$, this has not been proved.

Curves showing m as a function of h for fixed β are depicted in Figure IV.1 for the Ising model on $\mathbb{Z}^2$. These curves are called *isotherms*, and the point $(\beta, h) = (\beta_c, 0)$ is called the *critical point*. Notice that $m(\beta, h)$ is an increasing, concave function of $h \geq 0$. The concavity represents a saturation effect. An increment $\Delta h > 0$ causes a change $\Delta m(\beta, h) = m(\beta, h + \Delta h) - m(\beta, h)$ in m. The larger the value of h, the smaller is $\Delta m(\beta, h)$. The quantity $\partial m(\beta, h)/\partial h$, which gives the slope of the isotherms, is called the *specific magnetic susceptibility* and is denoted by $\chi(\beta, h)$.

Infinite-volume Gibbs states. We now consider the limiting behavior of the finite-volume Gibbs states as $\Lambda \uparrow \mathbb{Z}^D$. First, we modify these states by means of *external conditions* or *boundary conditions*. An external condition is defined by fixing the values of the spins $\tilde{\omega}_j$ at each site j which is in $\Lambda^c = \mathbb{Z}^D \backslash \Lambda$ and has a nearest neighbor in Λ. The external condition $\tilde{\omega} = \{\tilde{\omega}_j\}$ changes the Hamiltonian of a configuration ω from $H_{\Lambda,h}(\omega)$ to

$$H_{\Lambda,h,\tilde{\omega}}(\omega) = -\frac{1}{2} \sum_{i,j \in \Lambda} J(i-j)\omega_i\omega_j - \sum_{i \in \Lambda} \left(h + \sum_{j \in \Lambda^c} J(i-j)\tilde{\omega}_j \right) \omega_i.$$

The corresponding finite-volume Gibbs state is defined by the probability measure $P_{\Lambda,\beta,h,\tilde{\omega}}$ on $\mathscr{B}(\Omega_\Lambda)$ which assigns to each $\{\omega\}$, $\omega \in \Omega_\Lambda$, the probability

$$P_{\Lambda,\beta,h,\tilde{\omega}}\{\omega\} = \exp[-\beta H_{\Lambda,h,\tilde{\omega}}(\omega)] \pi_\Lambda P_\rho\{\omega\} \cdot \frac{1}{Z},$$

where Z is a normalization. If each $\tilde{\omega}_j$ equals 1 (resp., -1), then the external

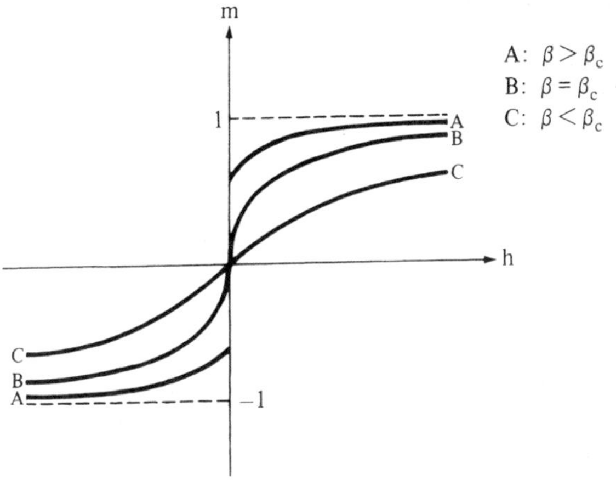

Figure IV.1. Isotherms for the Ising model on $\mathbb{Z}^2$. (Adapted from Figure 4.29 in L. E. Reichl, *A Modern Course in Statistical Physics*, University of Texas Press, Austin, 1980. Copyright © 1980 by the University of Texas Press.)

condition is called *plus* (resp., *minus*) and the measure is written as $P_{\Lambda,\beta,h,+}$ (resp., $P_{\Lambda,\beta,h,-}$). The external condition $\tilde{\omega}$ depends on Λ, and so we write $\tilde{\omega} = \tilde{\omega}(\Lambda)$. For fixed $\beta > 0$ and h real, let us consider the set of all weak limits

$$P = w\text{-}\lim_{\Lambda' \uparrow \mathbb{Z}^D} P_{\Lambda',\beta,h,\tilde{\omega}(\Lambda')}, \tag{4.1}$$

where $\{\Lambda'\}$ is any increasing sequence of symmetric hypercubes whose union is $\mathbb{Z}^D$ and $\tilde{\omega}(\Lambda')$ is any external condition for Λ'. Each weak limit P is a probability measure on the infinite-volume configuration space $\Omega = \{1, -1\}^{\mathbb{Z}^D}$. We call a probability measure P on Ω an *infinite-volume Gibbs state* or an *infinite-volume equilibrium state* if P belongs to the closed convex hull of the set of weak limits of the form (4.1).

Classification of infinite-volume Gibbs states. For the Ising model, various infinite-volume Gibbs states with different properties arise. They are listed in Table IV.1. The set of infinite-volume Gibbs states is denoted by $\mathscr{G}_{\beta,h}$. We now describe the structure of $\mathscr{G}_{\beta,h}$ for different values of β and h. Let β_c be the critical inverse temperature. For all $\beta > 0$, $h \neq 0$ and $0 < \beta < \beta_c$, $h = 0$, the finite-volume Gibbs states $\{P_{\Lambda,\beta,h,\tilde{\omega}(\Lambda)}\}$ have a unique weak limit for any choice of external conditions $\{\tilde{\omega}(\Lambda)\}$. Thus $\mathscr{G}_{\beta,h}$ consists of a unique measure $P_{\beta,h}$. The measure $P_{\beta,h}$ is translation invariant (strictly stationary with respect to the shift mappings on Ω) and ergodic. The ergodic phases are characterized among all translation invariant states by the property that macroscopic quantities are given definite values. For example, an experimenter might measure the spin per site $S_\Lambda(\omega)/|\Lambda| = \sum_{j\in\Lambda}\omega_j/|\Lambda|$ in a sample ω drawn from the magnet. The ergodic theorem implies that with respect to an ergodic phase such as $P_{\beta,h}$, $S_\Lambda(\omega)/|\Lambda|$ tends to a constant which is almost surely independent of the sample chosen. This justifies calling an ergodic phase *pure*.

Table IV.1. States of the Ferromagnet

Name	Definition
Infinite-volume Gibbs state	A member of the closed convex hull of the set of weak limits of $\{P_{\Lambda',\beta,h,\bar{\omega}(\Lambda')};\Lambda'\uparrow\mathbb{Z}^D\}$.
Phase	Translation invariant infinite-volume Gibbs state.
Pure phase	Ergodic, translation invariant infinite-volume Gibbs state.
Mixed phase	Nontrivial convex combination of pure phases.

For $\beta > \beta_c$ and $h = 0$, the situation is radically different. $\mathscr{G}_{\beta,0}$ contains two distinct pure phases $P_{\beta,0,+}$ and $P_{\beta,0,-}$, which arise from the finite-volume Gibbs states $P_{\Lambda,\beta,0,+}$ and $P_{\Lambda,\beta,0,-}$ with plus and minus external conditions, respectively. The average value of the spin at any site j with respect to each of these measures is given by

$$\int_\Omega \omega_j P_{\beta,0,+}(d\omega) = m(\beta,+) > 0,$$

(4.2)

$$\int_\Omega \omega_j P_{\beta,0,-}(d\omega) = -m(\beta,+) < 0,$$

where $m(\beta,+)$ is the spontaneous magnetization. $P_{\beta,0,+}$ is called a *pure plus phase* and $P_{\beta,0,-}$ a *pure minus phase*. In addition, $\mathscr{G}_{\beta,0}$ contains all convex combinations $P^{(\lambda)}_{\beta,0} = \lambda P_{\beta,0,+} + (1-\lambda)P_{\beta,0,-}$, $0<\lambda<1$. These measures are translation invariant but not ergodic and are called *mixed phases*. Such a phase corresponds physically to the situation where an experimenter has an a priori choice of external condition. With probability λ he prepares each finite-volume Gibbs state to have the plus external condition and with probability $1-\lambda$ to have the minus external condition. The existence of more than one pure phase for $\beta > \beta_c$ and $h = 0$ corresponds to a *phase transition*.

The phase transition reflects a crucial instability in the model. Choosing the plus or minus external condition outside Λ induces a slight preference for spin-up or spin-down at any fixed site inside Λ. For $\beta > \beta_c$ and $h = 0$, even this slight preference is strong enough, in the limit $\Lambda\uparrow\mathbb{Z}^D$, to push the infinite-volume system into a phase with net positive or negative specific magnetization. The phase transition is related to the notion of symmetry breaking [see page 116].

More is known about the structure of $\mathscr{G}_{\beta,0}$ for $\beta \geq \beta_c$. For the Ising model on $\mathbb{Z}^2$, $\mathscr{G}_{\beta_c,0}$ consists of a unique measure which is a pure phase. For $\beta > \beta_c$ and $h = 0$, $\mathscr{G}_{\beta,0}$ consists precisely of the measures $P_{\beta,0,+}$, $P_{\beta,0,-}$, and convex combinations [Aizenman (1979, 1980), Higuchi (1982)]. Thus all measures in $\mathscr{G}_{\beta,0}$ are translation invariant. For the Ising model on $\mathbb{Z}^D$ with $D \geq 3$, $\mathscr{G}_{\beta,0}$ contains nontranslation invariant states for any sufficiently large $\beta > \beta_c$ [Dobrushin (1972), van Beijeren (1975)].

The pure phases of the Ising model on $\mathbb{Z}^2$ are shown in Figure IV.2. The interval $h = 0$, $\beta > \beta_c$ is called the *coexistence interval* for the pure plus

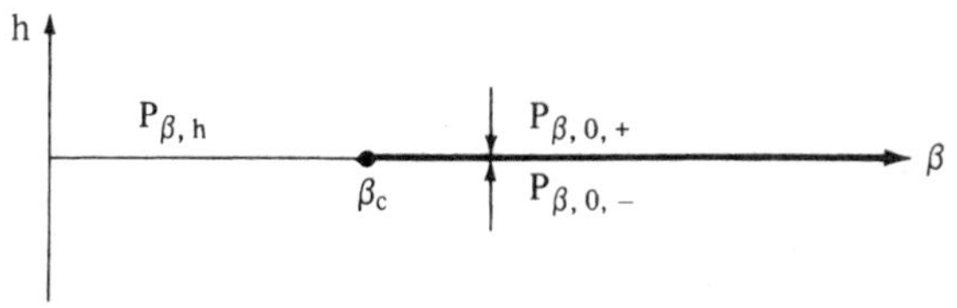

$P_{\beta, h}$: unique pure phase ($\beta > 0$, $h \neq 0$ and $0 < \beta \leq \beta_c$, $h \neq 0$)
$P_{\beta, 0, +}$: pure plus phase ($\beta > \beta_c$)
$P_{\beta, 0, -}$: pure minus phase ($\beta > \beta_c$)

Figure IV.2. The phase diagram of the Ising model on $\mathbb{Z}^2$.

phase and pure minus phase. Crossing the interval at constant β by decreasing h through 0 gives an abrupt transition between phases characterized by the discontinuity in the specific magnetization ($m(\beta, h)$ jumps from $m(\beta, +)$ to $m(\beta, -)$).

Correlations.[3] Correlations in the Ising model are related to the phase transition just discussed. We consider the model on $\mathbb{Z}^2$. For $D \geq 3$ similar behavior is expected. The following discussion is heuristic; not all the statements have been proved. Fix $h = 0$. At infinite temperature ($\beta = 0$), the finite-volume Gibbs state $P_{\Lambda, \beta, 0, \tilde{\omega}}$ reduces to the product measure $\pi_\Lambda P_\rho$. The corresponding infinite-volume Gibbs state is the product measure P_ρ on Ω, with respect to which the spins are independent and thus uncorrelated. At small but nonzero β, there is a unique infinite-volume Gibbs state $P_{\beta,0}$. Spins begin to be positively correlated with their nearest neighbors which in turn begin to be positively correlated with the second nearest neighbors, and so on. Since $\langle \omega_i \rangle_{\beta,0} = \int_\Omega \omega_i P_{\beta,0}(d\omega)$ equals 0 for each i, the convariance (or as we will call it, the pair correlation) equals $\langle \omega_i \omega_j \rangle_{\beta,0} = \int_\Omega \omega_i \omega_j P_{\beta,0}(d\omega)$. The correlations decrease with distance, and in fact $\langle \omega_i \omega_j \rangle_{\beta,0}$ has roughly an exponential decay when the Euclidean distance $\|i - j\|$ is large. We write

$$\langle \omega_i \omega_j \rangle_{\beta,0} \sim \exp[-\|i - j\|/\xi(\beta, 0)] \qquad \text{as } \|i - j\| \to \infty.$$

This relation defines the (exponential) *correlation length* $\xi(\beta, 0)$. The number $\xi(\beta, 0)$ is a rough measure of the distance over which correlations between spins are significant. As β increases, $\xi(\beta, 0)$ increases, and correlations begin to extend over larger and larger distances. These correlations take the form of spin fluctuations, which are islands of a few spins each that mostly point in the same direction. As β approaches the critical inverse temperature β_c, the correlation length grows rapidly, but the smaller fluctuations are not suppressed. They become contained in areas of larger fluctuations which themselves can be distinguished only because they have an overall excess of one spin orientation. When β equals β_c, the correlation length is infinite. Spin fluctuations persist at all scales of length and are extremely sensitive to small perturbations in h. The infinite correlation length is reflected in the fact that $\langle \omega_i \omega_j \rangle_{\beta_c,0}$ no longer decays exponentially but decays like a power of $\|i-j\|^{-1}$:

$$\langle \omega_i \omega_j \rangle_{\beta_c,0} \sim \|i-j\|^{-z} \qquad \text{as } \|i-j\| \to \infty,$$

where z is some positive number ($z = \frac{1}{4}$ for the Ising model on $\mathbb{Z}^2$).

For β larger than β_c, we enter the region of positive spontaneous magnetization $m(\beta, +)$. We consider the pair correlation with respect to the pure plus phase $P_{\beta,0,+}$. By (4.2), $\langle \omega_i \rangle_{\beta,0,+} = \langle \omega_j \rangle_{\beta,0,+} = m(\beta,+)$. As in the case $\beta < \beta_c$, the pair correlation decays exponentially:

$$\begin{aligned}&\langle \omega_i \omega_j \rangle_{\beta,0,+} - \langle \omega_i \rangle_{\beta,0,+} \langle \omega_j \rangle_{\beta,0,+} \\ &= \langle \omega_i \omega_j \rangle_{\beta,0,+} - [m(\beta,+)]^2 \sim \exp[-\|i-j\|/\xi(\beta,0)] \qquad \text{as } \|i-j\| \to \infty,\end{aligned}$$

where $0 < \xi(\beta,0) < \infty$. As β increases, $m(\beta,+)$ increases and the alignment effect becomes more rigid. Within a region of up-spins, islands of down-spins become, on the average, smaller and $\xi(\beta,0)$ decreases. As $\beta \to \infty$, $m(\beta,+)$ converges to 1, $\xi(\beta,0)$ converges to 0, and $P_{\beta,0,+}$ converges weakly to the state where all spins are oriented up. A similar discussion holds for the pure minus phase $P_{\beta,0,-}$.

The infinite correlation length at $\beta = \beta_c$ is related to the behavior of the specific magnetic susceptibility $\chi(\beta,h) = \partial m(\beta,h)/\partial h$ at $h = 0$. In Chapter V, we will prove that for $0 < \beta \leq \beta_c$

$$\chi(\beta,0) = \beta \sum_{k \in \mathbb{Z}^D} \langle \omega_0 \omega_k \rangle_{\beta,0}$$

($\langle \omega_0 \rangle_{\beta_c,0} = \langle \omega_k \rangle_{\beta_c,0} = 0$ for the model on $\mathbb{Z}^2$). For $0 < \beta < \beta_c$, $\langle \omega_0 \omega_k \rangle_{\beta,0}$ decays exponentially and $\chi(\beta,0)$ is finite. By contrast, at $\beta = \beta_c$, $\langle \omega_0 \omega_k \rangle_{\beta_c,0}$ decays like $\|k\|^{-1/4}$ and $\chi(\beta_c,0)$ is infinite. These conclusions are confirmed by Figure IV.1, in which $\chi(\beta,0)$ is the slope at $h = 0$ of the isotherm $m(\beta,h)$. The infinite slope at $h = 0$ of the critical isotherm $m(\beta_c,h)$ anticipates the onset of spontaneous magnetization for $\beta > \beta_c$ ($m(\beta,h)$ discontinuous at $h = 0$). By definition, a large value of $\chi(\beta,h)$ implies a dramatic response of the magnetization to a small change in external field. The divergence of the specific magnetic susceptibility at the critical point is one way in which the extreme sensitivity of the spin fluctuations to small perturbations in h shows up in the macroscopic behavior of the ferromagnet.[4]

This completes our qualitative discussion of the Ising model. The study of ferromagnetic models on $\mathbb{Z}$ begins in the next section.

IV.3. Finite-Volume Gibbs States on $\mathbb{Z}$

The models will be defined on the symmetric intervals $\Lambda = \{j \in \mathbb{Z} : |j| \leq N\}$, where N is a non-negative integer. To each site $j \in \Lambda$ there is assigned a spin ω_j which takes the value 1 (spin-up) or -1 (spin-down). The configuration space is the set Ω_Λ of all sequences $\omega = \{\omega_j; j \in \Lambda\}$; thus, $\Omega_\Lambda = \{1, -1\}^\Lambda$. The coordinate functions on Ω_Λ, defined by $Y_j(\omega) = \omega_j$, are called the *spin random variables* at the sites j. The presence of interactions distinguishes these models from the discrete ideal gas. The *Hamiltonian* or *interaction energy* of a spin configuration $\omega \in \Omega_\Lambda$ is defined as

(4.3)
$$H_{\Lambda,h}(\omega) = -\frac{1}{2} \sum_{i,j \in \Lambda} J(i-j)\omega_i\omega_j - h \sum_{j \in \Lambda} \omega_j.$$

We assume that J is a non-negative function on $\mathbb{Z}$ which is symmetric; i.e., satisfies $J(i-j) = J(j-i)$ for each i and j. J is called a *ferromagnetic interaction.** The parameter h is a real number which gives the strength of an external magnetic field acting at each site in Λ. The term $-J(i-j)\omega_i\omega_j$ in the first sum in (4.2) gives the interaction energy between the spins at sites i and j. The interaction strength $J(i-j)$ is translation invariant; i.e., $J((i+k)-(j+k)) = J(i-j)$ for each k. The factor $\frac{1}{2}$ is included in (4.3) because each pair i, j with $i \neq j$ appears twice with equal weight $J(i-j) = J(j-i)$. The term $-h\omega_j$ in the second sum in (4.3) gives the interaction energy between the external field and the spin at site j. An interaction J is said to have *finite-range* if $J(k)$ equals 0 for all sufficiently large k. The *range* is the smallest number L such that $J(k) = 0$ whenever $|k| > L$.

We denote by $\mathscr{B}(\Omega_\Lambda)$ the set of all subsets of Ω_Λ. Let ρ be the measure $\frac{1}{2}\delta_1 + \frac{1}{2}\delta_{-1}$ and define $\pi_\Lambda P_\rho$ to be the product measure on $\mathscr{B}(\Omega_\Lambda)$ with identical one-dimensional marginals ρ. For each $\omega \in \Omega_\Lambda$, $\pi_\Lambda P_\rho\{\omega\}$ equals $2^{-|\Lambda|}$, where $|\Lambda| = 2N+1$ is the number of sites in Λ. Let $\beta = 1/T > 0$ be the inverse absolute temperature. The ferromagnetic model is defined by the probability measure $P_{\Lambda,\beta,h}$ on $\mathscr{B}(\Omega_\Lambda)$ which assigns to each $\{\omega\}$, $\omega \in \Omega_\Lambda$, the probability

(4.4)
$$P_{\Lambda,\beta,h}\{\omega\} = \exp[-\beta H_{\Lambda,h}(\omega)]\pi_\Lambda P_\rho\{\omega\} \cdot \frac{1}{Z(\Lambda,\beta,h)}.$$

$Z(\Lambda,\beta,h)$ is a normalization which is picked so that $\sum_{\omega \in \Omega_\Lambda} P_{\Lambda,\beta,h}\{\omega\} = 1$:

$$Z(\Lambda,\beta,h) = \int_{\Omega_\Lambda} \exp[-\beta H_{\Lambda,h}(\omega)]\pi_\Lambda P_\rho(d\omega) = \sum_{\omega \in \Omega_\Lambda} \exp[-\beta H_{\Lambda,h}(\omega)]\frac{1}{2^{|\Lambda|}}.$$
(4.5)

For A a subset of Ω_Λ, we have

$$P_{\Lambda,\beta,h}\{A\} = \sum_{\omega \in A} P_{\Lambda,\beta,h}\{\omega\}.$$

The measure $P_{\Lambda,\beta,h}$ is called a *finite-volume Gibbs state* on Λ. $Z(\Lambda,\beta,h)$ is called the *partition function.*[5] Here are some examples of ferromagnetic models.

Example IV.3.1. (a) *The Ising model on* $\mathbb{Z}$. Fix a number $\mathscr{J} > 0$. Define $J(i-j)$ to be $\mathscr{J}$ if $|i-j| = 1$ and to be 0 if $|i-j| \neq 1$. This interaction, which couples only nearest neighbors, has range 1.

* More general interactions are discussed in Appendix C.3.

(b) *The Curie–Weiss model.* Fix a number $\mathscr{J}_0 > 0$. Define $J(i-j)$ to be $\mathscr{J}_0/|\Lambda|$ if both i and j are in Λ and to be 0 if either i or j is not in Λ. This interaction, which depends on the set Λ, couples all pairs of spins in Λ with equal strength. The range tends to ∞ as $|\Lambda| \to \infty$.

(c) *Infinite-range models.* Fix a number $\alpha > 1$. Define $J(0) = 0$ and $J(i-j) = |i-j|^{-\alpha}$ for all $i \neq j$ in $\mathbb{Z}$. Since $\alpha > 1$, this interaction is summable: $\sum_{k \in \mathbb{Z}} J(k) < \infty$ [see Section IV.5].

In order to exclude trivial cases, we assume that the interaction $J(i-j)$ in (4.3) is positive for at least one distinct pair i, j in Λ. The finite-volume Gibbs state $P_{\Lambda,\beta,h}$ has the same form as the reduced canonical ensemble for the discrete ideal gas, which is defined in (3.26). The difference is that the kinetic energy $U_n(\omega)$ in (3.26) is replaced by the interaction energy $H_{\Lambda,h}(\omega)$.* While (3.26) can be written as a product measure (with identical one-dimensional marginals $\bar{\rho}_\beta$), $P_{\Lambda,\beta,h}$ cannot be written as a product measure because of the positivity assumption on J.

Let us examine the form of $P_{\Lambda,\beta,h}$ for different values of β and h. As $\beta \to 0$, $P_{\Lambda,\beta,h}$ converges weakly to the product measure $\pi_\Lambda P_\rho$ which gives equal probability to each ω. Thus, at $\beta = 0$ (infinite temperature) the magnet is completely random. For $\beta > 0$ an alignment effect comes into play. Since $P_{\Lambda,\beta,h}\{\bar{\omega}\} > P_{\Lambda,\beta,h}\{\omega\}$ if and only if $H_{\Lambda,h}\{\bar{\omega}\} < H_{\Lambda,h}\{\omega\}$, the smaller the energy of a configuration, the more probable it is. Thus the most likely configurations are those that minimize $H_{\Lambda,h}$ over Ω_Λ. These minimizing configurations, called *ground states*, are not hard to identify. Let $\bar{\omega}_+$ (resp., $\bar{\omega}_-$) be the configuration with $\bar{\omega}_{+,j} = 1$ (resp., $\bar{\omega}_{-,j} = -1$) for each $j \in \Lambda$. If A is a subset of $\mathbb{Z}$, then the interaction J is said to be *irreducible on A* if for each pair of sites i, j in A either $J(i-j) > 0$ or there exists a finite sequence $i_1 = i, i_2, \ldots, i_{r-1}, i_r = j$ in A such that $J(i_{\alpha+1} - i_\alpha) > 0$, $\alpha = 1, 2, \ldots, r-1$. The next result is a consequence of the non-negativity assumption on J [Problem IV.9.1].

Proposition IV.3.2. (a) *For $h > 0$, $\bar{\omega}_+$ is the unique ground state.*

(b) *For $h < 0$, $\bar{\omega}_-$ is the unique ground state.*

(c) *For $h = 0$, the ground states include $\bar{\omega}_+$ and $\bar{\omega}_-$. These are the unique ground states if and only if J is irreducible on Λ.*

For $h > 0$, the measures $\{P_{\Lambda,\beta,h}; \beta > 0\}$ converge weakly to the unit point measure $\delta_{\bar{\omega}_+}$ as $\beta \to \infty$ [Problem IV.9.2]. That is, for $h > 0$ and 0 temperature the totally aligned ground state $\bar{\omega}_+$ is the only possible configuration. No randomness at all is left in the ferromagnet. A similar situation holds for $h < 0$.

The remarks in the previous paragraphs show that the finite-volume state $P_{\Lambda,\beta,h}$ defines a reasonable model for a ferromagnet. The form of $P_{\Lambda,\beta,h}$ can also be justified by means of an entropy principle. This principle

* $U_n(\omega) = \frac{1}{2}\sum_{j=1}^n y_j^2$ has the same form as $H_{\Lambda,0}(\omega)$ but with $J(i-j) = 0$ for all $i \neq j$.

will be generalized later in the chapter. Given a probability measure P on $\mathscr{B}(\Omega_\Lambda)$, we define the energy in P to be $U(\Lambda, h; P) = \int_{\Omega_\Lambda} H_{\Lambda,h}(\omega) P(d\omega)$. Let $U_{\min}$ and $U_{\max}$ denote the minimum and maximum, respectively, of $H_{\Lambda,h}(\omega)$ over Ω_Λ. Fix a number $U \in (U_{\min}, U_{\max})$. We prove that there exists a unique value β such that the finite-volume Gibbs state $P_{\Lambda,\beta,h}$ is the most random probability measure P on $\mathscr{B}(\Omega_\Lambda)$ which satisfies the energy constraint $U(\Lambda, h; P) = U$. The randomness in P, relative to the fixed measure $\pi_\Lambda P_\rho$, is measured by the negative of the relative entropy

$$(4.6) \qquad I^{(2)}_{\pi_\Lambda P_\rho}(P) = \int_{\Omega_\Lambda} \log \frac{dP}{d(\pi_\Lambda P_\rho)}(\omega) P(d\omega) = \sum_{\omega \in \Omega_\Lambda} \log \frac{P\{\omega\}}{\pi_\Lambda P_\rho\{\omega\}} \cdot P\{\omega\}.$$

Theorem IV.3.3.[6] *Let h real and $U \in (U_{\min}, U_{\max})$ be given. Then the following conclusions hold.*

(a) *There exists a unique value β such that $U(\Lambda, h; P_{\Lambda,\beta,h}) = U$.*

(b) *$I^{(2)}_{\pi_\Lambda P_\rho}(P)$ attains its infimum over the set $\{P \in \mathscr{M}(\Omega_\Lambda): U(\Lambda, h; P_{\Lambda,\beta,h}) = U\}$ at the unique measure $P = P_{\Lambda,\beta,h}$.*

Proof. (a) Define $c(t) = \log \int_{\Omega_\Lambda} \exp(tH_{\Lambda,h}(\omega)) \pi_\Lambda P_\rho(d\omega)$ for t real.* We have

$$c'(t) = \frac{\int_{\Omega_\Lambda} H_{\Lambda,h}(\omega) \exp[tH_{\Lambda,h}(\omega)] \pi_\Lambda P_\rho(d\omega)}{\int_{\Omega_\Lambda} \exp[tH_{\Lambda,h}(\omega)] \pi_\Lambda P_\rho(d\omega)},$$

$$c'(-\beta) = \int_{\Omega_\Lambda} H_{\Lambda,h}(\omega) P_{\Lambda,\beta,h}(d\omega) = U(P_{\Lambda,\beta,h}).$$

If ν denotes the $\pi_\Lambda P_\rho$-distribution of $H_{\Lambda,h}$, then $c(t) = \log \int_{\mathbb{R}} e^{tx} \nu(dx)$. This is the free energy function of ν, which was introduced in Section II.4. The support of ν is a finite set which equals the range of $H_{\Lambda,h}(\omega)$, $\omega \in \Omega_\Lambda$. By Theorem II.5.2(a), for $U \in (U_{\min}, U_{\max})$, there exists a unique value β such that $c'(-\beta) = U(\Lambda, h; P_{\Lambda,\beta,h}) = U$.

(b) $I^{(2)}_{\pi_\Lambda P_\rho}(P_{\Lambda,\beta,h}) = \int_{\Omega_\Lambda} [-\beta H_{\Lambda,h}(\omega) - \log Z] P_{\Lambda,\beta,h}(d\omega) = -\beta U - \log Z$, where $Z = Z(\Lambda, \beta, h)$. If $P \neq P_{\Lambda,\beta,h}$ is any other measure in $\mathscr{M}(\Omega_\Lambda)$ for which $U(\Lambda, h; P_{\Lambda,\beta,h}) = U$, then

$$\begin{aligned} I^{(2)}_{\pi_\Lambda P_\rho}(P) &= \int_{\Omega_\Lambda} \log \frac{dP}{d(\pi_\Lambda P_\rho)}(\omega) P(d\omega) \\ &= \int_{\Omega_\Lambda} \log \frac{dP}{dP_{\Lambda,\beta,h}}(\omega) P(d\omega) + \int_{\Omega_\Lambda} \log \frac{dP_{\Lambda,\beta,h}}{d(\pi_\Lambda P_\rho)}(\omega) P(d\omega) \\ &= I^{(2)}_{P_{\Lambda,\beta,h}}(P) + \int_{\Omega_\Lambda} [-\beta H_{\Lambda,h}(\omega) - \log Z] P(d\omega) \\ &= I^{(2)}_{P_{\Lambda,\beta,h}}(P) - \beta U - \log Z. \end{aligned}$$

* $c(t) = \log Z(\Lambda, -t, h)$.

Since $P \neq P_{\Lambda,\beta,h}$, $I^{(2)}_{P_{\Lambda,\beta,h}}(P)$ is positive [Proposition I.4.1(b)], and so $I^{(2)}_{\pi_\Lambda P_\rho}(P) > -\beta U - \log Z$. We conclude that

$$I^{(2)}_{\pi_\Lambda P_\rho}(P) \geq -\beta U - \log Z = I^{(2)}_{\pi_\Lambda P_\rho}(P_{\Lambda,\beta,h})$$

with equality if and only if $P = P_{\Lambda,\beta,h}$. □

Let $S_\Lambda(\omega)$ be the random sum $\sum_{j \in \Lambda} Y_j(\omega)$, called the (total) *spin in* Λ. The *magnetization* is defined as the expectation

$$\begin{aligned} M(\Lambda,\beta,h) &= \int_{\Omega_\Lambda} S_\Lambda(\omega) P_{\Lambda,\beta,h}(d\omega) \\ &= \sum_{j\in\Lambda}\sum_{\omega\in\Omega_\Lambda} \omega_j \exp[-\beta H_{\Lambda,h}(\omega)] 2^{-|\Lambda|} \frac{1}{Z(\Lambda,\beta,h)}. \end{aligned} \tag{4.7}$$

The next theorem states properties of $M(\Lambda,\beta,h)$ which will be proved in Theorem V.4.2.

Theorem IV.3.4. (a) *For each* $\beta > 0$, $M(\Lambda,\beta,0) = 0$; $M(\Lambda,\beta,h)$ *is a non-negative concave function of* $h \geq 0$ *and a nondecreasing function of* h *real. It satisfies* $M(\Lambda,\beta,-h) = -M(\Lambda,\beta,h)$ *and* $|M(\Lambda,\beta,h)| \leq |\Lambda|$.

(b) *For each* $h \geq 0$, $M(\Lambda,\beta,h)$ *is a non-negative, nondecreasing function of* $\beta > 0$.

By (4.7), $M(\Lambda,\beta,h)$ is a continuous function of h. Since $M(\Lambda,\beta,0) = 0$, $M(\Lambda,\beta,h)$ converges to 0 as $h \to 0$ for *any* value of $\beta > 0$.

We will study the magnetization per site in the limit as the symmetric intervals Λ expand to fill $\mathbb{Z}$. This limit is called the *infinite-volume limit* or the *thermodynamic limit* and is denoted by $\Lambda \uparrow \mathbb{Z}$. We define the *specific magnetization* as

$$m(\beta,h) = \lim_{\Lambda\uparrow\mathbb{Z}} \frac{1}{|\Lambda|} M(\Lambda,\beta,h). \tag{4.8}$$

Since $M(\Lambda,\beta,0)$ equals 0, $m(\beta,0)$ equals 0. We will see that if the interaction is summable, then $m(\beta,h)$ exists, $m(\beta,+) = \lim_{h\to 0^+} m(\beta,h)$ exists, and $m(\beta,+)$ is non-negative. The model is said to exhibit *spontaneous magnetization* at inverse temperature β if $m(\beta,+) > 0 = m(\beta,0)$. The quantities $m(\beta,h)$ and $m(\beta,+)$ are studied in the next two sections.

IV.4. Spontaneous Magnetization for the Curie–Weiss Model

In order to see how spontaneous magnetization can result from the microscopic alignment effects built into $P_{\Lambda,\beta,h}$, we first consider the Curie–Weiss model.[7] This model is ideal for doing exact calculations, and the analysis of it involves interesting applications of large deviation theory. The Curie–Weiss model is defined in Example IV.3.1(b). To ease the notation, we replace Λ by the set $\{1, 2, \ldots, n\}$, where n is a positive integer. All quantities are indexed by n instead of by Λ. Thus the finite-volume Gibbs state is given by the formula

$$(4.9) \qquad P_{n,\beta,h}\{\omega\} = \exp[-\beta H_{n,h}(\omega)]\pi_n P_\rho\{\omega\} \cdot \frac{1}{Z(n,\beta,h)}, \qquad \omega \in \Omega_n,$$

where $\Omega_n = \{1, -1\}^n$,

$$H_{n,h}(\omega) = -\tfrac{1}{2}(\mathscr{J}_0/n) \sum_{i,j=1}^{n} \omega_i \omega_j - h \sum_{j=1}^{n} \omega_j,$$

$$Z(n,\beta,h) = \int_{\Omega_n} \exp[-\beta H_{n,h}(\omega)]\pi_n P_\rho(d\omega).$$

We write the Hamiltonian as a function of the sum $\sum_{j=1}^n \omega_j/n$:

$$H_{n,h}(\omega) = -n\left[\frac{1}{2}\mathscr{J}_0\left(\frac{\sum_{j=1}^n \omega_j}{n}\right)^2 + h\frac{\sum_{j=1}^n \omega_j}{n}\right].$$

This simple form makes the model easy to handle.

By (4.7) and (4.8), $m(\beta,h)$ equals $\lim_{n\to\infty}\langle S_n/n\rangle_{n,\beta,h}$, where $\langle - \rangle_{n,\beta,h}$ denotes expectation with respect to $P_{n,\beta,h}$. We prove that $m(\beta,+) = \lim_{h\to 0^+} m(\beta,h)$ equals 0 for all $0 < \beta \le \mathscr{J}_0^{-1}$ and is positive for all $\beta > \mathscr{J}_0^{-1}$. Thus spontaneous magnetization occurs at all $\beta > \mathscr{J}_0^{-1}$. The number $\mathscr{J}_0^{-1}$ is called the *critical inverse temperature* for the Curie–Weiss model and is denoted by β_c^{CW}.

We will determine the limit of $\langle f(S_n/n)\rangle_{n,\beta,h}$ as $n \to \infty$ for any function $f \in \mathscr{C}(\mathbb{R})$. By doing this we will not only prove spontaneous magnetization but also determine the distribution limit of S_n/n as $n \to \infty$. Let $Q_n^{(1)}(dz)$ denote the distribution of the sum $\sum_{j=1}^n \omega_j/n$ with respect to the product measure $\pi_n P_\rho$. Then

$$(4.10) \qquad \begin{aligned} \left\langle f\left(\frac{S_n}{n}\right)\right\rangle_{n,\beta,h} &= \int_{\Omega_n} f\left(\frac{\sum_{j=1}^n \omega_j}{n}\right)\exp[-\beta H_{n,h}(\omega)]\pi_n P_\rho(d\omega)\cdot\frac{1}{Z(n,\beta,h)} \\ &= \int_{\mathbb{R}} f(z)\exp[n(\tfrac{1}{2}\beta\mathscr{J}_0 z^2 + \beta h z)]Q_n^{(1)}(dz)\cdot\frac{1}{Z(n,\beta,h)}, \end{aligned}$$

and the partition function $Z(n,\beta,h)$ can be written as

$$(4.11) \qquad Z(n,\beta,h) = \int_{\mathbb{R}} \exp[n(\tfrac{1}{2}\beta\mathscr{J}_0 z^2 + \beta h z)]Q_n^{(1)}(dz).$$

By Theorem II.4.1, the distributions $\{Q_n^{(1)}; n = 1, 2, \ldots\}$ have a large deviation property with $a_n = n$ and entropy function $I_\rho^{(1)}(z) = \sup_{t\in\mathbb{R}}\{tz - c_\rho(t)\} = \sup_{t\in\mathbb{R}}\{tz - \log\cosh t\}$. A simple calculation shows that

$$(4.12) \quad I_\rho^{(1)}(z) = \begin{cases} \dfrac{1-z}{2}\log(1-z) + \dfrac{1+z}{2}\log(1+z) & \text{for } |z| \le 1, \\ \infty & \text{for } |z| > 1. \end{cases}$$

We define $i_{\beta,h}(z) = -(\frac{1}{2}\beta\mathscr{J}_0 z^2 + \beta h z) + I_\rho^{(1)}(z)$. For large n (4.10) and (4.11) suggest the heuristic formula

$$(4.13)\quad \left\langle f\left(\frac{S_n}{n}\right)\right\rangle_{n,\beta,h} \approx \int_{\mathbb{R}} f(z)\exp[-n i_{\beta,h}(z)]dz \cdot \frac{1}{\int_{\mathbb{R}}\exp[-n i_{\beta,h}(z)]dz}.$$

According to this formula, the limit of $\langle f(S_n/n)\rangle_{n,\beta,h}$ as $n \to \infty$ should be determined by the points z at which the function $i_{\beta,h}(z)$ attains its infimum. In fact, this statement is true because of the large deviation result stated in Theorem II.7.2. We will locate the minimum points, then deduce the limit of $\langle f(S_n/n)\rangle_{n,\beta,h}$.

Minimum points of $i_{\beta,h}(z)$ satisfy the equation

$$(4.14)\quad \frac{\partial i_{\beta,h}(z)}{\partial z} = 0 \qquad \text{or} \qquad \beta\mathscr{J}_0 z + \beta h = (I_\rho^{(1)})'(z) = \frac{1}{2}\log\frac{1+z}{1-z}.$$

$(I_\rho^{(1)})'(z)$ is an odd function of $z \in [-1, 1]$, is concave for $z \geq 0$, and has slope $(I_\rho^{(1)})''(0) = 1$ at $z = 0$. Also, $|(I_\rho^{(1)})'(z)| \to \infty$ as $|z| \to 1$. Since the slope of the affine function $z \to \beta\mathscr{J}_0 z + \beta h$ is $\beta\mathscr{J}_0$, the nature of the solutions of (4.14) depends on whether $0 < \beta\mathscr{J}_0 \leq 1$ or $\beta\mathscr{J}_0 > 1$. For $0 < \beta \leq \mathscr{J}_0^{-1}$ and any real h, (4.14) has a unique solution $z(\beta, h)$, and $z(\beta, h)$ is the unique minimum point of $i_{\beta,h}(z)$. As $h \to 0$, $z(\beta, h) \to z(\beta, 0) = 0$ [Figure IV.3(a), (b)]. For $\beta > \mathscr{J}_0^{-1}$, Figures IV.3(c), (d) show the solutions of (4.14) for small $h > 0$ and for $h = 0$. For $\beta > \mathscr{J}_0^{-1}$ and $h \neq 0$, (4.14) has a unique solution $z(\beta, h)$ that has the same sign as h, and $z(\beta, h)$ is the unique minimum point of $i_{\beta,h}(z)$. For $\beta > \mathscr{J}_0^{-1}$ and $h = 0$, the minimum points are the nonzero solutions $z(\beta, +)$ and $z(\beta, -) = -z(\beta, +)$ of (4.14). As $h \to 0^+$, $z(\beta, h) \to z(\beta, +) > 0$ and as $h \to 0^-$, $z(\beta, h) \to z(\beta, -) < 0$.

Part (a) of the next theorem gives the limit of $\langle f(S_n/n)\rangle_{n,\beta,h}$ as $n \to \infty$. Part (b) shows that spontaneous magnetization occurs at all $\beta > \beta_c^{CW} = \mathscr{J}_0^{-1}$. Part (c) states that S_n/n converges exponentially to $m(\beta, h)$ for $\beta > 0$, $h \neq 0$ and $0 < \beta \leq \mathscr{J}_0^{-1}$, $h = 0$ but that exponential convergence to a constant fails for $\beta > \mathscr{J}_0^{-1}$ and $h = 0$.

Theorem IV.4.1. (a) *Let f be a bounded continuous function from $\mathbb{R}$ to $\mathbb{R}$. Then*

$$\lim_{n\to\infty} \langle f(S_n/n)\rangle_{n,\beta,h}$$

$$= \begin{cases} f(z(\beta, h)) & \text{for } \beta > 0, h \neq 0 \text{ and } 0 < \beta \leq \mathscr{J}_0^{-1}, h = 0, \\ \frac{1}{2}f(z(\beta, +)) + \frac{1}{2}f(z(\beta, -)) & \text{for } \beta > \mathscr{J}_0^{-1}, h = 0, \end{cases}$$

(4.15)

(b) *Let $m(\beta, h)$ be the specific magnetization for the Curie–Weiss model. Then $m(\beta, h)$ equals $z(\beta, h)$ for $\beta > 0$, $h \neq 0$ and $0 < \beta \leq \mathscr{J}_0^{-1}$, $h = 0$, and for each choice of sign*

$$(4.16)\quad m(\beta, \pm) = \lim_{h\to 0^\pm} m(\beta, h) = \begin{cases} z(\beta, 0) = 0 & \text{for } 0 < \beta \leq \mathscr{J}_0^{-1}, \\ z(\beta, \pm) \neq 0 & \text{for } \beta > \mathscr{J}_0^{-1}. \end{cases}$$

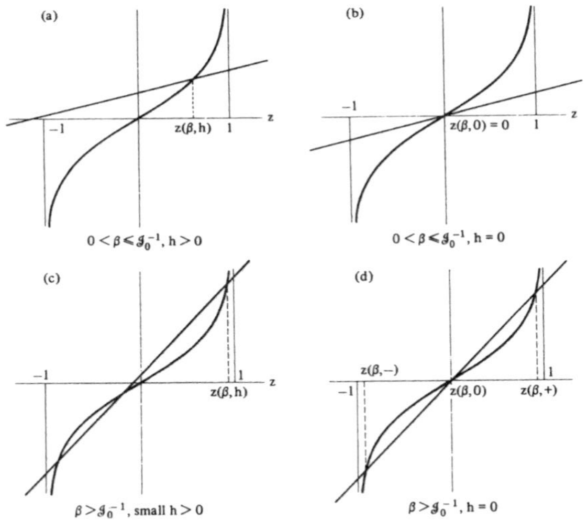

Figure IV.3. Solutions of the Curie–Weiss equation (4.14).

Values of β, h	Solutions of $\partial i_{\beta,h}(z)/\partial z = 0$	Minimum points of $i_{\beta,h}(z)$	Continuity properties
(a) $0 < \beta \le \mathscr{J}_0^{-1}$, $h \neq 0$	unique $z(\beta, h)$	$z(\beta, h)$	$z(\beta, h) \to z(\beta, 0) = 0$ as $h \to 0$
(b) $0 < \beta \le \mathscr{J}_0^{-1}$, $h = 0$	unique $z(\beta, 0) = 0$	$z(\beta, 0) = 0$	
(c) $\beta > \mathscr{J}_0^{-1}$, $h \neq 0$	$z(\beta, h)$ (same sign as h) plus 2 roots of opposite sign for small h	$z(\beta, h)$	$z(\beta, h) \to \begin{cases} z(\beta, +) > 0 & \text{as } h \to 0^+ \\ z(\beta, -) < 0 & \text{as } h \to 0^- \end{cases}$
(d) $\beta > \mathscr{J}_0^{-1}$, $h = 0$	$z(\beta, +) > z(\beta, 0) = 0 > z(\beta, -)$	$z(\beta, +)$, $z(\beta, -)$	

(c) $\dfrac{S_n}{n} \xrightarrow{\exp} m(\beta,h)$ *for* $\beta > 0, h \neq 0$ *and* $0 < \beta \leq \mathscr{J}_0^{-1}, h = 0$.

$\dfrac{S_n}{n} \xrightarrow{\mathscr{D}} \tfrac{1}{2}\delta_{m(\beta,+)}(dz) + \tfrac{1}{2}\delta_{m(\beta,-)}(dz)$ *for* $\beta > \mathscr{J}_0^{-1}, h = 0$.*

Proof. (a) Denote by $Q_{n,\beta,h}$ the $P_{n,\beta,h}$-distribution of S_n/n. According to (4.10)–(4.11), if A is a Borel subset of $\mathbb{R}$, then $Q_{n,\beta,h}\{A\}$ equals

$$\int_A \exp[n(\tfrac{1}{2}\beta\mathscr{J}_0 z^2 + \beta hz)]Q_n^{(1)}(dz) \cdot \frac{1}{\int_{\mathbb{R}} \exp[n(\tfrac{1}{2}\beta\mathscr{J}_0 z^2 + \beta hz)]Q_n^{(1)}(dz)}.$$

By Theorem II.7.2(a), $\{Q_{n,\beta,h}\}$ has a large deviation property with $a_n = n$ and entropy function

$$I_{\beta,h}(z) = i_{\beta,h}(z) - \inf_{z\in\mathbb{R}} i_{\beta,h}(z) \qquad \text{where } i_{\beta,h}(z) = I_\rho^{(1)}(z) - (\tfrac{1}{2}\beta\mathscr{J}_0 z^2 + \beta hz). \tag{4.17}$$

For $\beta > 0, h \neq 0$ and $0 < \beta \leq \mathscr{J}_0^{-1}, h = 0$, let K be any closed set which does not contain the unique minimum point $z(\beta,h)$ of $i_{\beta,h}$. By Theorem II.7.2(b), there exists a number $N = N(K) > 0$ such that

$$Q_{n,\beta,h}\{K\} \leq e^{-nN} \qquad \text{for all sufficiently large } n. \tag{4.18}$$

This yields the first line of (4.15). For $\beta > \mathscr{J}_0^{-1}$ and $h = 0$, if K is the closed set

$$\{z\in\mathbb{R}: |z - z(\beta,+)| \geq \varepsilon \text{ and } |z - z(\beta,-)| \geq \varepsilon\} \qquad \text{where } 0 < \varepsilon < z(\beta,+),$$

then for all sufficiently large n, $Q_{n,\beta,0}\{K\} \leq e^{-nN}$ for some $N = N(K) > 0$. This yields the second line of (4.15) since the measures $\{Q_{n,\beta,0}\}$ are symmetric.

(b) The range of S_n/n is contained in the interval $[-1, 1]$. If f is a function in $\mathscr{C}(\mathbb{R})$ such that $f(x) = x$ for $-1 \leq x \leq 1$, then part (b) follows from part (a) and the behavior of $z(\beta,h)$ as $h \to 0^+$ and $h \to 0^-$.

(c) For $\beta > 0, h \neq 0$ and $0 < \beta \leq \mathscr{J}_0^{-1}, h = 0$, $m(\beta,h)$ equals $z(\beta,h)$. The bound (4.18) implies that S_n/n converges exponentially to $m(\beta,h)$. The convergence in distribution for $\beta > \mathscr{J}_0^{-1}$ and $h = 0$ follows from the second line of (4.15) since f is an arbitrary function in $\mathscr{C}(\mathbb{R})$. □

Remark IV.4.2. By (4.11) and Theorem II.7.3(a)

$$\lim_{n\to\infty} \frac{1}{n} \log Z(n,\beta,h) = \sup_{z\in\mathbb{R}} \{\tfrac{1}{2}\beta\mathscr{J}_0 z^2 + \beta hz - I_\rho^{(1)}(z)\}. \tag{4.19}$$

The function $\psi(\beta,h) = -\beta^{-1}\lim_{n\to\infty} n^{-1}\log Z(n,\beta,h)$ is called the *specific Gibbs free energy* for the Curie–Weiss model. The limit (4.19) can be derived without using large deviations [see Problem IV.9.4]. In Ellis and Newman (1978b), the limit (4.15) is derived without using large deviations.

*The second line of part (c) means that S_n/n converges in distribution to a random variable distributed by $\tfrac{1}{2}\delta_{m(\beta,+)} + \tfrac{1}{2}\delta_{m(\beta,-)}$.

Let us consider the form of the entropy function $I_{\beta,h}(z)$ defined in (4.17). For $\beta > 0, h \neq 0$ and $0 < \beta \leq \mathscr{J}_0^{-1}, h = 0, I_{\beta,h}(z)$ is a convex function on $\mathbb{R}$ with a unique minimum point at $z(\beta, h)$ $(I_{\beta,h}(z(\beta,h)) = 0)$. However, for $\beta > \mathscr{J}_0^{-1}$ and $h = 0$, $I_{\beta,h}(z)$ is not a convex function. It has minimum points at $z(\beta, +)$ and $z(\beta, -)$ and is positive for all other values of z. Another example of a nonconvex entropy function was given in Example II.6.2.

This completes our discussion of spontaneous magnetization for the Curie–Weiss model. Other aspects of the model will be studied in Section V.9.

IV.5. Spontaneous Magnetization for General Ferromagnets on $\mathbb{Z}$

In the Curie–Weiss model the interaction $J(i-j)$ equals $\mathscr{J}_0/|\Lambda|$ for each i and j in Λ. Thus the interaction depends on the set Λ. We now consider other finite-volume Gibbs states $P_{\Lambda,\beta,h}$ on symmetric intervals Λ of $\mathbb{Z}$. The interactions J are assumed to satisfy the following hypotheses*: J is independent of Λ, J is nonnegative on $\mathbb{Z}$ (*ferromagnetic*) and is not identically zero, J is symmetric, and $\sum_{k\in\mathbb{Z}} J(k) < \infty$. This summability hypothesis restricts the interaction strength between distant spins. J is called a *summable ferromagnetic interaction* on $\mathbb{Z}$ and the corresponding spin models are called *general ferromagnets* on $\mathbb{Z}$. These models cannot be treated by the large deviation technique which was used in the Curie–Weiss case. Less direct methods are required. In this section, we illustrate one of the most powerful of these methods, which is convexity. The main fact about general ferromagnets on $\mathbb{Z}$ is that unless the interaction has infinite range, there is no spontaneous magnetization. The behavior of models on $\mathbb{Z}$ contrasts sharply with the behavior of models on $\mathbb{Z}^D$, $D \geq 2$. The latter exhibit spontaneous magnetization for all nontrivial interactions, regardless of whether they have finite range or infinite range [Theorem V.5.1].

A useful function for studying the magnetization is the *Gibbs free energy*. It is defined as

$$\Psi(\Lambda,\beta,h) = -\beta^{-1}\log Z(\Lambda,\beta,h), \tag{4.20}$$

where $Z(\Lambda,\beta,h) = \int_{\Omega_\Lambda} \exp[-\beta H_{\Lambda,h}(\omega)]\pi_\Lambda P_\rho(d\omega)$. $\Psi(\Lambda,\beta,h)$ has a simple relation to the magnetization:

$$\begin{aligned} -\frac{\partial\Psi(\Lambda,\beta,h)}{\partial h} &= -\beta^{-1}\frac{\partial Z(\Lambda,\beta,h)}{\partial h}\cdot\frac{1}{Z(\Lambda,\beta,h)} \\ &= \sum_{j\in\Lambda}\sum_{\omega\in\Omega_\Lambda}\omega_j \exp[-\beta H_{\Lambda,h}(\omega)]2^{-|\Lambda|}\frac{1}{Z(\Lambda,\beta,h)} = M(\Lambda,\beta,h). \end{aligned} \tag{4.21}$$

Our proofs of the existence and properties of the specific magnetization $m(\beta,h) = \lim_{\Lambda\uparrow\mathbb{Z}} |\Lambda|^{-1}M(\Lambda,\beta,h)$ will be based on the function

* More general interactions are discussed in Appendix C.3.

(4.22) $$\psi(\beta, h) = \lim_{\Lambda \uparrow \mathbb{Z}} \frac{1}{|\Lambda|} \Psi(\Lambda, \beta, h),$$

called the *specific Gibbs free energy*. The existence of the limit (4.22) for a summable interaction J is proved in Appendix D.1. Another proof based on level-3 large deviations is given in Theorem IV.7.3 below.

There is a degenerate case of the specific Gibbs free energy that is worth pointing out. If the interaction J is identically zero, then $Z(\Lambda, \beta, h) = \int_{\Omega_\Lambda} \exp(\beta h \sum_{j \in \Lambda} \omega_j) \pi_\Lambda P_\rho(d\omega)$ and $\psi(\beta, h) = -\beta^{-1} \log \int_{\{1,-1\}} e^{\beta h x} \rho(dx) = -\beta^{-1} \log \cosh \beta h$. Thus $\psi(\beta, h)$ equals $-\beta^{-1} c_\rho(\beta h)$, where $c_\rho(t)$ is the free energy function of the measure ρ. Properties of the specific Gibbs free energy are proved next.

Theorem IV.5.1. *Let J be a summable ferromagnetic interaction on $\mathbb{Z}$. Then the following conclusions hold.*

(a) *For each $\beta > 0$, the specific Gibbs free energy $\psi(\beta, h)$ is a concave, even function of h real and is a continuously differentiable function of $h \neq 0$.*

(b) *For $\beta > 0$ and h real, the specific magnetization $m(\beta, h) = \lim_{\Lambda \uparrow \mathbb{Z}} |\Lambda|^{-1} M(\Lambda, \beta, h)$ exists. For $\beta > 0$ and $h \neq 0$, $m(\beta, h) = -\partial \psi(\beta, h)/\partial h$.*

Proof. (a) Concavity is preserved under pointwise limits. Therefore the concavity of $\psi(\beta, h)$ will follow from the concavity of $\Psi(\Lambda, \beta, h)$. The latter is equivalent to the inequality

(4.23) $$Z(\Lambda, \beta, \lambda h_1 + (1 - \lambda) h_2) \leq Z(\Lambda, \beta, h_1)^\lambda \cdot Z(\Lambda, \beta, h_2)^{1-\lambda}$$

for any h_1 and h_2 real and $0 < \lambda < 1$. The left-hand side equals

(4.24) $$\int_{\Omega_\Lambda} \exp\left(\beta \lambda h_1 \sum_{j \in \Lambda} \omega_j\right) \exp\left(\beta(1 - \lambda) h_2 \sum_{j \in \Lambda} \omega_j\right) \cdot \exp\left(\frac{\beta}{2} \sum_{i,j \in \Lambda} J(i - j) \omega_i \omega_j\right) \pi_\Lambda P_\rho(d\omega),$$

and so (4.23) follows from Hölder's inequality* applied to the functions $\exp(\beta \lambda h_1 \sum_{j \in \Lambda} \omega_j)$ and $\exp(\beta(1 - \lambda) h_2 \sum_{j \in \Lambda} \omega_j)$. The evenness of $\psi(\beta, \cdot)$ follows from the fact that $H_{\Lambda, -h}(\omega) = H_{\Lambda, h}(-\omega)$.

(b) The following proof is due to Preston (1974a). The Gibbs free energy is related to the magnetization by the formula

(4.25) $$\frac{1}{|\Lambda|}(\Psi(\Lambda, \beta, h) - \Psi(\Lambda, \beta, 0)) = -\int_0^h \frac{1}{|\Lambda|} M(\Lambda, \beta, s) ds,$$

which is equivalent to (4.21). We would like to prove the existence of the specific magnetization by passing to the limit $\Lambda \uparrow \mathbb{Z}$ in (4.25). The left-hand side becomes $\psi(\beta, h) - \psi(\beta, 0)$, but care is needed in handling the integral on the right. Consider $h \geq 0$ ($h < 0$ is handled similarly). Since $0 \leq M(\Lambda, \beta, h)$

*Corollary VI.4.2 with $p = 1/\lambda$, $q = 1/(1 - \lambda)$.

$\leq |\Lambda|$, there exists an infinite subsequence $\{\Lambda'\}$ of $\{\Lambda\}$ such that $m(\beta, h) = \lim_{\Lambda' \uparrow \mathbb{Z}} |\Lambda'|^{-1} M(\Lambda', \beta, h)$ exists for every rational number $h \geq 0$. $M(\Lambda', \beta, 0)$ equals 0 and thus $m(\beta, 0)$ equals 0. Since $M(\Lambda', \beta, h)$ is a concave function of $h \geq 0$, a standard convexity result implies that $m(\beta, h)$ exists for all $h \geq 0$ [Theorem VI.3.3(a)]. The limit $m(\beta, h)$ is concave for $h \geq 0$ and hence is continuous for $h > 0$ [Theorem VI.3.1]. By the Lebesgue dominated convergence theorem, $m(\beta, h)$ satisfies

$$\psi(\beta, h) - \psi(\beta, 0) = -\int_0^h m(\beta, s)ds. \tag{4.26}$$

The preceding argument may be repeated for the functions $\{|\bar{\Lambda}|^{-1} M(\bar{\Lambda}, \beta, h)\}$, where $\{\bar{\Lambda}\}$ is an arbitrary infinite subsequence of $\{\Lambda\}$. This leads to a limit function, say $\bar{m}(\beta, h)$, which satisfies $\bar{m}(\beta, 0) = 0$ as well as equation (4.26). Therefore, $\bar{m}(\beta, h)$ equals $m(\beta, h)$. We conclude that the limit $\lim_{\Lambda \uparrow \mathbb{Z}} |\Lambda|^{-1} M(\Lambda, \beta, h)$ exists along the entire sequence $\{\Lambda\}$ and equals $m(\beta, h)$. Finally, (4.26) implies that $\psi(\beta, h)$ is a continuously differentiable function of $h \neq 0$ and $\partial\psi(\beta, h)/\partial h = -m(\beta, h)$. □

The specific Gibbs free energy $\psi(\beta, h)$ need not be differentiable at $h = 0$. The relationship between spontaneous magnetization and the existence of $\partial\psi(\beta, 0)/\partial h$ is made explicit in the following theorem. Recall that spontaneous magnetization is said to occur at inverse temperature β if $m(\beta, +) = \lim_{h \to 0^+} m(\beta, h)$ is positive.

Theorem IV.5.2.* *Let J be a summable ferromagnetic interaction on $\mathbb{Z}$. Then the following conclusions hold.*

(a) *For $\beta > 0$ and $h \neq 0$, $m(\beta, h) = -\partial\psi(\beta, h)/\partial h$.*

(b) *For each $\beta > 0$, $m(\beta, 0) = 0$; $m(\beta, h)$ is a non-negative, concave function of $h \geq 0$ and a nondecreasing function of h real. It satisfies $m(\beta, -h) = -m(\beta, h)$ and $|m(\beta, h)| \leq 1$.*

(c) *For each $h \geq 0$, $m(\beta, h)$ is a non-negative, nondecreasing function of $\beta > 0$.*

(d) *For each $\beta > 0$, the limits $m(\beta, +) = \lim_{h \to 0^+} m(\beta, h)$ and $m(\beta, -) = \lim_{h \to 0^-} m(\beta, h)$ exist and $m(\beta, -) = -m(\beta, +)$; $m(\beta, +)$ is a non-negative, nondecreasing function of $\beta > 0$.*

(e) *For each $\beta > 0$,*

$$m(\beta, +) = -\frac{\partial\psi(\beta, 0)}{\partial h^+} \geq m(\beta, 0) = 0 \geq m(\beta, -) = -\frac{\partial\psi(\beta, 0)}{\partial h^-}. \tag{4.27}$$

Thus spontaneous magnetization occurs at β if and only if $\psi(\beta, h)$ is not differentiable at $h = 0$.

These properties of $m(\beta, h)$ may be read from Figure IV.1. Part (a) of

*Other properties of $m(\beta, h)$ are derived in Problem V.13.1.

Theorem IV.5.2 was proved in Theorem IV.5.1. The properties listed in parts (b) and (c) were stated for $M(\Lambda, \beta, h)$ in Theorem IV.3.4 and are preserved under passage to the limit $\Lambda \uparrow \mathbb{Z}$. Since $\psi(\beta, h)$ is concave for h real and differentiable for $h \neq 0$, $\partial\psi/\partial h$ is nonincreasing for $h \neq 0$. Hence $m(\beta, +) = \lim_{h\to 0^+} m(\beta, h) = -\lim_{h\to 0^+} \partial\psi(\beta, h)/\partial h$ exists and equals the right-hand derivative of $-\psi$ at $h = 0$ ($-\partial\psi(\beta, 0)/\partial h^+$); similarly for $m(\beta, -)$. Since for each $h \geq 0$, $m(\beta, h)$ is a non-negative, nondecreasing function of $\beta > 0$, the same properties hold as for $m(\beta, +)$. This proves part (d) as well as (4.27). Since $\psi(\beta, h)$ is differentiable at $h = 0$ if and only if $\partial\psi(\beta, 0)/\partial h^+ = \partial\psi(\beta, 0)/\partial h^-$, we obtain the last assertion in part (e).

According to part (e), spontaneous magnetization corresponds to a discontinuity in the first-order derivative $\partial\psi/\partial h$. Hence it is known as a *first-order phase transition.*

Spontaneous magnetization indicates a strong cooperation among the individual spins, and it occurs only if the alignment effects built into the finite-volume Gibbs states persist in the limit $\Lambda \uparrow \mathbb{Z}$. Thus, one might expect spontaneous magnetization to occur only if distant spins have a suitably strong interaction and the temperature is sufficiently low. The next theorem justifies this intuition.

Theorem IV.5.3. *Let J be a summable ferromagnetic interaction on $\mathbb{Z}$ and define the positive number $\mathscr{J}_0 = \sum_{k\in\mathbb{Z}} J(k)$. Define the critical inverse temperature*

$$\beta_c = \sup\{\beta > 0: m(\beta, +) = 0\}.$$

Then the following conclusions hold.

(a) *β_c is well-defined and $\mathscr{J}_0^{-1} \leq \beta_c \leq \infty$; in particular $\beta_c > 0$.*

(b) *$\beta_c = \infty$ if $\sum_{k\in\mathbb{Z}} |k| J(k) < \infty$.*

(c) *$\beta_c < \infty$ if $J(k) = k^{-2}$ ($k \neq 0$) or if $J > 0$ and $J(k) \sim |k|^{-\alpha}$, some $1 < \alpha < 2$.**

(d) *For $0 < \beta < \beta_c$, $\psi(\beta, h)$ is differentiable at $h = 0$ and $m(\beta, +) = 0$. For $\beta > \beta_c$, the differentiability fails and spontaneous magnetization occurs:*

$$m(\beta, +) = -\frac{\partial\psi(\beta, 0)}{\partial h^+} > 0 > m(\beta, -) = -\frac{\partial\psi(\beta, 0)}{\partial h^-}. \tag{4.28}$$

Comments on proof. (a) Since $m(\beta, +)$ is a non-negative, nondecreasing function of $\beta > 0$, β_c is well-defined. Let $m^{\text{CW}}(\beta, h)$ denote the specific magnetization for the Curie–Weiss model with interaction $\mathscr{J}_0/|\Lambda|$, where $\mathscr{J}_0$ equals $\sum_{k\in\mathbb{Z}} J(k)$. Pearce (1981) shows that for $\beta > 0$ and $h \geq 0$, $m(\beta, h) \leq m^{\text{CW}}(\beta, h)$. Since $m^{\text{CW}}(\beta, +) = \lim_{h\to 0^+} m^{\text{CW}}(\beta, h)$ is positive for $\beta > \mathscr{J}_0^{-1}$ [Theorem IV.4.1(b)], it follows that $\beta_c \geq \mathscr{J}_0^{-1}$. Pearce's method is outlined in Problem IV.9.7.[8]

(b), (c) These parts are discussed below.

* $J(k) \sim |k|^{-\alpha}$ means $\lim_{|k|\to\infty} [\log J(k)/\log |k|] = -\alpha$.

(d) For $0 < \beta < \beta_c$, $m(\beta, +)$ equals 0 and so $\psi(\beta, h)$ is differentiable at $h = 0$. For $\beta > \beta_c$, $m(\beta, +)$ is positive and (4.28) follows from (4.27).[9]

We show a special case of part (b) by proving that β_c equals ∞ for the Ising model.

Example IV.5.4. For the Ising model,

$$Z(\Lambda, \beta, h) = \int_{\Omega_\Lambda} \exp\left(\beta \mathcal{J} \sum_{j=-N}^{N-1} \omega_j \omega_{j+1} + \beta h \sum_{j=-N}^{N} \omega_j\right) \pi_\Lambda P_\rho(d\omega). \tag{4.29}$$

$Z(\Lambda, \beta, h)$ can be expressed in terms of a matrix product. For $\alpha_1, \alpha_2 \in \{1, -1\}$, define $B(\alpha_1, \alpha_2) = \frac{1}{2}\exp[\beta \mathcal{J} \alpha_1 \alpha_2 + \beta h(\alpha_1 + \alpha_2)/2]$ and let $B = B_{\beta,h}$ be the 2×2 symmetric matrix

$$\begin{pmatrix} B(1,1) & B(1,-1) \\ B(-1,1) & B(-1,-1) \end{pmatrix}.$$

B is called the *transfer matrix* of the model. Λ equals $\{j \in \mathbb{Z} : |j| \le N\}$ and

$$\begin{aligned} Z(\Lambda, \beta, h) &= \frac{1}{2} \sum_{\{\omega_j = \pm 1;\, j \in \Lambda\}} a(\omega_{-N}) B(\omega_{-N}, \omega_{-N+1}) \cdots B(\omega_{N-1}, \omega_N) a(\omega_N) \\ &= \frac{1}{2} \sum_{\omega_N, \omega_{-N} = \pm 1} a(\omega_{-N}) B^{2N}(\omega_{-N}, \omega_N) a(\omega_N), \end{aligned} \tag{4.30}$$

where $a(x) = \exp(\frac{1}{2}\beta h x)$. The larger eigenvalue of $B_{\beta,h}$ is

$$\lambda(B_{\beta,h}) = \tfrac{1}{2} e^{\beta \mathcal{J}} [\cosh \beta h + (\sinh^2 \beta h + e^{-4\beta \mathcal{J}})^{1/2}].$$

By Lemma IX.4.1(d) (Perron–Frobenius)

$$-\beta \psi(\beta, h) = \lim_{\Lambda \uparrow \mathbb{Z}} \frac{1}{|\Lambda|} \log Z(\Lambda, \beta, h) = \log \lambda(B_{\beta,h}). \tag{4.31}$$

For each $\beta > 0$, $\psi(\beta, h)$ is a real analytic function of real h. Hence $\partial \psi(\beta, 0)/\partial h$ exists for each $\beta > 0$ and β_c equals ∞. This calculation can be generalized to any finite-range interaction, and as with the Ising model, one finds that β_c equals ∞ [Ruelle (1969, Section 5.6)]. See Appendix C.6 for details.

According to part (b) of Theorem IV.5.3, β_c equals ∞ not just for interactions of finite range but whenever $\sum_{k \in \mathbb{Z}} |k| J(k)$ is finite. A method of proof is outlined in Problem IV.9.8. This leaves the infinite-range case where $\sum_{k \in \mathbb{Z}} J(k)$ is finite but $\sum_{k \in \mathbb{Z}} |k| J(k)$ diverges. It follows from Dyson (1969a) that β_c is finite if J is positive and $J(k) \sim |k|^{-\alpha}$ for some $1 < \alpha < 2$. Fröhlich and Spencer (1982a) proved that β_c is finite also in the borderline case $J(k) = k^{-2}$ $(k \neq 0)$. These proofs of spontaneous magnetization are difficult and are omitted.[10]

Finally, we study convergence properties of the random variables $S_\Lambda/|\Lambda|$, which define the spin per site in Λ. By (4.7), the expectation of $S_\Lambda/|\Lambda|$ with respect to the finite-volume Gibbs state $P_{\Lambda,\beta,h}$ gives the magnetization per site, $M(\Lambda,\beta,h)/|\Lambda|$. By analogy with the Curie–Weiss model, we expect that as $\Lambda \uparrow \mathbb{Z}$, $S_\Lambda/|\Lambda|$ converges exponentially to the specific magnetization $m(\beta,h)$ for all $\beta > 0$, $h \neq 0$, and $0 < \beta < \beta_c$, $h = 0$.

Theorem IV.5.5. *Let J be a summable ferromagnetic interaction on $\mathbb{Z}$. Then there exists a constant m such that*

$$S_\Lambda/|\Lambda| \xrightarrow{\exp} m \qquad \textit{with respect to } \{P_{\Lambda,\beta,h}\} \tag{4.32}$$

if and only if $\psi(\beta,h)$ is differentiable with respect to h. In this case m equals $m(\beta,h) = -\partial\psi(\beta,h)/\partial h$. Thus (4.32) *holds for all $\beta > 0$, $h \neq 0$ and for all $0 < \beta < \beta_c$, $h = 0$;* (4.32) *fails for $\beta > \beta_c$ and $h = 0$ [Theorems* IV.5.1 *and* IV.5.3].

Proof. According to Section II.6, the exponential convergence can be proved by considering the free energy function $c_{\beta,h}(t)$ of the sequence $\{S_\Lambda\}$. For t real, $c_{\beta,h}(t) = \lim_{\Lambda\uparrow\mathbb{Z}} c_{\Lambda,\beta,h}(t)$, where

$$\begin{aligned} c_{\Lambda,\beta,h}(t) &= \frac{1}{|\Lambda|}\log\int_{\Omega_\Lambda}\exp[tS_\Lambda(\omega)]P_{\Lambda,\beta,h}(d\omega) \\ &= \frac{1}{|\Lambda|}\log\int_{\Omega_\Lambda}\exp[-\beta H_{\Lambda,h}(\omega) + tS_\Lambda(\omega)]\pi_\Lambda P_\rho(d\omega)\cdot\frac{1}{Z(\Lambda,\beta,h)}. \end{aligned}$$

For each $\omega\in\Omega_\Lambda$,

$$\begin{aligned} -\beta H_{\Lambda,h}(\omega) + tS_\Lambda(\omega) &= \frac{\beta}{2}\sum_{i,j\in\Lambda} J(i-j)\omega_i\omega_j + (\beta h + t)\sum_{j\in\Lambda}\omega_j \\ &= -\beta H_{\Lambda,h+t/\beta}(\omega). \end{aligned}$$

Hence

$$\begin{aligned} c_{\Lambda,\beta,h}(t) &= \frac{1}{|\Lambda|}\log\frac{Z(\Lambda,\beta,h+t/\beta)}{Z(\Lambda,\beta,h)} \\ &= -\beta\frac{1}{|\Lambda|}[\Psi(\Lambda,\beta,h+t/\beta) - \Psi(\Lambda,\beta,h)] \end{aligned}$$

and $c_{\beta,h}(t) = -\beta[\psi(\beta,h+t/\beta) - \psi(\beta,h)]$. By Theorem II.6.3, there exists a constant m such that $S_\Lambda/|\Lambda| \xrightarrow{\exp} m$ if and only if $c_{\beta,h}(t)$ is differentiable at $t = 0$. In this case, m equals $c'_{\beta,h}(0)$. But $c'_{\beta,h}(0)$ exists if and only if $\partial\psi(\beta,h)/\partial h$ exists, and then $c'_{\beta,h}(0) = -\partial\psi(\beta,h)/\partial h$. This completes the proof. □

Theorem IV.5.5 implies that spontaneous magnetization occurs at β if and only if $S_\Lambda/|\Lambda|$ fails to converge exponentially to $m(\beta,0) = 0$. For this reason, we are justified in calling spontaneous magnetization a *level*-1 *phase transition*. Let us interpret the level-1 phase transition in terms of entropy. According to the proof of Theorem IV.5.5, the free energy function of the sequence $\{S_\Lambda\}$ with respect to $\{P_{\Lambda,\beta,h}\}$ is

(4.33) $$c_{\beta,h}(t) = -\beta[\psi(\beta, h + t/\beta) - \psi(\beta, h)].$$

For $0 < \beta < \beta_c$ and any h real, $c_{\beta,h}(t)$ is differentiable for all t real. Hence by Theorem II.6.1, the $P_{\Lambda,\beta,h}$-distributions of $S_\Lambda/|\Lambda|$ have a large deviation property with $a_\Lambda = |\Lambda| = 2N + 1$ and entropy function

(4.34) $$\begin{aligned} I_{\beta,h}(z) &= \sup_{t \in \mathbb{R}} \{tz - c_{\beta,h}(t)\} = \sup_{t \in \mathbb{R}} \{tz + \beta\psi(\beta, h + t/\beta)\} - \psi(\beta, h) \\ &= \sup_{x \in \mathbb{R}} \{\beta xz + \beta\psi(\beta, x)\} - [\beta hz + \beta\psi(\beta, h)]. \end{aligned}$$

$I_{\beta,h}(z)$ is a convex function and it attains its infimum of 0 at the unique point $c'_{\beta,h}(0) = m(\beta, h)$ [Theorem II.6.3]. The situation is different for $\beta > \beta_c$. Let us focus on the case $h = 0$. The function $c_{\beta,0}(t)$ is not differentiable at $t = 0$, and $I_{\beta,0}(z)$ does not attain its infimum at a unique point. The theory of Legendre–Fenchel transforms shows that $I_{\beta,0}(z)$ attains its infimum on the whole interval

$$\begin{aligned} [(c_{\beta,0})'_-(0), (c_{\beta,0})'_+(0)] &= [-\partial\psi(\beta,0)/\partial h^-, -\partial\psi(\beta,0)/\partial h^+] \\ &= [m(\beta, -), m(\beta, +)], \end{aligned}$$

where $m(\beta, +)$ is the spontaneous magnetization [Theorem VII.2.1(g)]. According to part (b) of Theorem II.6.1,

$$\limsup_{\Lambda \uparrow \mathbb{Z}} \frac{1}{|\Lambda|} \log P_{\Lambda,\beta,0}\{S_\Lambda/|\Lambda| \in K\} \le -\inf_{z \in K} I_{\beta,0}(z) \quad \text{for each closed set } K \text{ in } \mathbb{R}.$$

Thus, if K is disjoint from the interval $[m(\beta, -), m(\beta, +)]$, then the probabilities $P_{\Lambda,\beta,0}\{S_\Lambda/|\Lambda| \in K\}$ decay exponentially as $\Lambda \uparrow \mathbb{Z}$. However, since $c_{\beta,0}(t)$ fails to be differentiable at $t = 0$, we are unable to apply part (c) of Theorem II.6.1 to conclude that the $P_{\Lambda,\beta,0}$-distributions of $S_\Lambda/|\Lambda|$ have a large deviation property. If the large deviation property does hold with some entropy function $I(z)$, then $I_{\beta,0}(z)$ equals the closed convex hull of $I(z)$ [see Problem VII.8.2].

This completes our discussion of finite-volume Gibbs states. In the next section, we consider probability measures which describe the infinite-volume ferromagnet.

IV.6. Infinite-Volume Gibbs States and Phase Transitions

The finite-volume Gibbs states studied in the previous three sections are probability measures on the finite sets $\Omega_\Lambda = \{1, -1\}^\Lambda$. In this section we study states of general ferromagnets on $\mathbb{Z}$. The states are probability measures on the infinite-volume configuration space $\Omega = \{1, -1\}^{\mathbb{Z}}$. These measures are obtained from the finite-volume Gibbs states by weak limits ($\Lambda \uparrow \mathbb{Z}$). Equivalent notions of infinite-volume measures are discussed in Appendix C.

The set $\{1, -1\}$ is topologized by the discrete topology and the set Ω by the product topology. According to Tychonoff's theorem, Ω is compact. The σ-field generated by the open sets of the product topology is called the Borel σ-field of Ω and is denoted by $\mathscr{B}(\Omega)$. $\mathscr{B}(\Omega)$ coincides with the σ-field generated by the cylinder sets of Ω [Propositions A.3.2 and A.3.5(b)]. Let $\mathscr{M}(\Omega)$ be the set of probability measures on $\mathscr{B}(\Omega)$ and $\mathscr{M}_s(\Omega)$ the subset of $\mathscr{M}(\Omega)$ consisting of strictly stationary probability measures. Strict stationarity is a natural condition since most of the infinite-volume Gibbs states which we obtain are strictly stationary with respect to spatial translations, or (as it is usually said) are translation invariant. A translation invariant, infinite-volume Gibbs state is called a *phase*.

Throughout this section, we fix a summable ferromagnetic interaction J on $\mathbb{Z}$. The finite-volume Gibbs state $P_{\Lambda,\beta,h}$ defined in Section IV.3 will be modified by means of *external conditions*. The measure $P_{\Lambda,\beta,h}$ models a ferromagnet on the set $\Lambda = \{j \in \mathbb{Z} : |j| \le N\}$, where N is a non-negative integer. External conditions correspond physically to the situation where an experimenter prepares the complement of Λ, $\Lambda^c = \mathbb{Z}\backslash\Lambda$, by fixing a configuration on that set. Let $\tilde{\omega}$ be a point in the set $\Omega_{\Lambda^c} = \{1, -1\}^{\Lambda^c}$. The coordinates $\tilde{\omega}_j$, $j \in \Lambda^c$, denote the values of the fixed external spins at the sites of Λ^c. When we want to indicate the dependence of $\tilde{\omega}$ upon Λ, we will write $\tilde{\omega}(\Lambda)$. We define the Hamiltonian of a configuration $\omega \in \Omega_\Lambda = \{1, -1\}^\Lambda$ to be*

$$H_{\Lambda,h,\tilde{\omega}}(\omega) = -\frac{1}{2} \sum_{i,j\in\Lambda} J(i-j)\omega_i\omega_j - \sum_{i\in\Lambda}\left(h + \sum_{j\in\Lambda^c} J(i-j)\tilde{\omega}_j\right)\omega_i. \tag{4.35}$$

Thus each $\tilde{\omega}_j$, $j \in \Lambda^c$, interacts with the spins in Λ through the given interaction. Compare $H_{\Lambda,h,\tilde{\omega}}$ with the Hamiltonian $H_{\Lambda,h}$ in (4.3). The external condition $\tilde{\omega}$ changes $H_{\Lambda,h}$ by altering the external field acting at each site $i \in \Lambda$ from the value h to the value $h_i = h + \sum_{j\in\Lambda^c} J(i-j)\tilde{\omega}_j$. Since J is summable, h_i is well-defined. The finite-volume Gibbs state on Λ with external condition $\tilde{\omega}$ is defined to be the probability measure $P_{\Lambda,\beta,h,\tilde{\omega}}$ on $\mathscr{B}(\Omega_\Lambda)$ which assigns to each $\{\omega\}$, $\omega \in \Omega_\Lambda$, the probability

$$P_{\Lambda,\beta,h,\tilde{\omega}}\{\omega\} = \exp[-\beta H_{\Lambda,h,\tilde{\omega}}(\omega)]\pi_\Lambda P_\rho\{\omega\} \cdot \frac{1}{Z(\Lambda,\beta,h,\tilde{\omega})}. \tag{4.36}$$

In this formula, β is positive and $Z(\Lambda,\beta,h,\tilde{\omega})$ is the normalization $\int_{\Omega_\Lambda} \exp[-\beta H_{\Lambda,h,\tilde{\omega}}(\omega)]\pi_\Lambda P_\rho(d\omega)$. There are two important choices of $\tilde{\omega}$. If each $\tilde{\omega}_j = 1$ (resp., -1), then the external condition is called *plus* (resp., *minus*) and the measure is written as $P_{\Lambda,\beta,h,+}$ (resp., $P_{\Lambda,\beta,h,-}$). Expectation with respect to $P_{\Lambda,\beta,h,\tilde{\omega}}$ will be denoted by $\langle - \rangle_{\Lambda,\beta,h,\tilde{\omega}}$.

We have defined external conditions by means of points $\tilde{\omega}$ in Ω_{Λ^c}. One can allow other external conditions such as *free* (each $\tilde{\omega}_j = 0$ in (4.35)) or *periodic*

*We omit from (4.35) the interaction between $\tilde{\omega}_i$ and $\tilde{\omega}_j$ and between h and $\tilde{\omega}_i$. These interactions are constants independent of ω. If included in (4.35), they would cancel out in (4.36).

(one modifies the definition of $J(i-j)$). These external conditions are useful in certain applications. However, restricting external conditions to be points in Ω_{Λ^c} allows for a cleaner formulation of infinite-volume Gibbs states and facilitates the proof of the equivalence between these states and other notions of infinite-volume measures [Appendix C].

Let $\mathscr{C}(\Omega)$ be the space of bounded, continuous, real-valued functions on Ω with the supremum norm. We say that a sequence $\{P_n; n = 1, 2, \ldots\}$ in $\mathscr{M}(\Omega)$ *converges weakly* to $P \in \mathscr{M}(\Omega)$, and write $P_n \Rightarrow P$ or $P = w\text{-}\lim_{n\to\infty} P_n$, if $\int_\Omega f dP_n \to \int_\Omega f dP$ for every $f \in \mathscr{C}(\Omega)$. If a weak limit P exists, then it is unique. With respect to weak convergence, $\mathscr{M}(\Omega)$ is a compact metric space [Theorem A.11.2].

Let $\{P_n; n = 1, 2, \ldots\}$ be an arbitrary sequence in $\mathscr{M}(\Omega)$. We call a subset $\mathscr{S}$ of $\mathscr{C}(\Omega)$ a *convergence-determining class* if the existence of the limit $\lim_{n\to\infty} \int_\Omega f dP_n$ for each $f \in \mathscr{S}$ implies that $\{P_n; n = 1, 2, \ldots\}$ converges weakly to some probability measure P. Since Ω is compact, $\mathscr{C}(\Omega)$ itself is a convergence-determining class because of the Riesz representation theorem. Other well-known examples are the subset consisting of all product functions $f(\omega) = \prod_{i\in B} \omega_i$ for B a finite subset of $\mathbb{Z}$ (define $f(\omega) = 1$ if B is empty) and the subset consisting of all functions $f(\omega) = \chi_\Sigma(\omega)$ for Σ a cylinder set in Ω. These examples are discussed in Theorem A.11.3. Another useful convergence-determining class is given in the next lemma.

Lemma IV.6.1. *For B a nonempty finite subset of $\mathbb{Z}$, define $f_B(\omega) = \prod_{i\in B}[\frac{1}{2}(1+\omega_i)]$. For B the empty set, define $f_B(\omega) = 1$. Then the subset of $\mathscr{C}(\Omega)$ consisting of all functions $f_B(\omega)$ is a convergence-determining class.*

Proof. The limit $\lim_{n\to\infty} \int_\Omega f_B dP_n$ exists for any finite set B if and only if the limit $\lim_{n\to\infty} \int_\Omega f dP_n$ exists for all functions $f(\omega) = \prod_{i\in B} \omega_i$, where $f(\omega) = 1$ if B is empty. Hence the lemma is a consequence of Theorem A.11.3(a). □

In order to define infinite-volume Gibbs states, we extend each finite-volume Gibbs state $P_{\Lambda,\beta,h,\tilde{\omega}}$ to a probability measure on $\mathscr{B}(\Omega)$. Let $B_{\Lambda,\tilde{\omega}}$ denote the set $\{\omega \in \Omega : \omega_j = \tilde{\omega}_j$ for each $j \in \Lambda^c\}$ and π_Λ the projection of Ω onto Ω_Λ defined by $(\pi_\Lambda \omega)_i = \omega_i$, $i \in \Lambda$. We define the extension $\bar{P}_{\Lambda,\beta,h,\tilde{\omega}}$ of the finite-volume Gibbs state by setting

$$\bar{P}_{\Lambda,\beta,h,\tilde{\omega}}\{A\} = P_{\Lambda,\beta,h,\tilde{\omega}}\{\pi_\Lambda(A \cap B_{\Lambda,\tilde{\omega}})\} = \sum_{\omega \in \pi_\Lambda(A\cap B_{\Lambda,\tilde{\omega}})} P_{\Lambda,\beta,h,\tilde{\omega}}\{\omega\} \tag{4.37}$$

for A a Borel subset of Ω. The right-hand side of (4.37) is given by (4.36). Since the support of $\bar{P}_{\Lambda,\beta,h,\tilde{\omega}}$ is the set $B_{\Lambda,\tilde{\omega}}$, the extension is compatible with the external condition. We note a useful property of $\bar{P}_{\Lambda,\beta,h,\tilde{\omega}}$. Let f be a function in $\mathscr{C}(\Omega)$ such that the value $f(\omega)$ depends on only finitely many coordinates of ω; let these coordinates be $\omega_{i_1}, \ldots, \omega_{i_r}$. If Λ is any symmetric interval containing the sites $i_1, \ldots, i_r$, then the restriction of f to the coordinates $\{\omega_i; i \in \Lambda\}$ defines a function in $\mathscr{C}(\Omega_\Lambda)$. We denote the restriction by the same symbol f; if $\omega = \pi_\Lambda \bar{\omega}$, $\omega \in \Omega_\Lambda$ and $\bar{\omega} \in \Omega$, then $f(\omega) = f(\bar{\omega})$. It is easy to check [Problem IV.9.9] that

$$\int_\Omega f(\omega) \bar{P}_{\Lambda,\beta,h,\tilde{\omega}}(d\omega) = \int_{\Omega_\Lambda} f(\omega) P_{\Lambda,\beta,h,\tilde{\omega}}(d\omega). \tag{4.38}$$

An example of such a function f is the function f_B in the previous lemma. From now on, we will denote the extension $\bar{P}_{\Lambda,\beta,h,\tilde{\omega}}$ by the same symbol $P_{\Lambda,\beta,h,\tilde{\omega}}$ used for the original measure. The extension will be called a finite-volume Gibbs state.

Since the space $\mathcal{M}(\Omega)$ is compact, any sequence of finite-volume Gibbs states $\{P_{\Lambda,\beta,h,\tilde{\omega}(\Lambda)}; \Lambda \uparrow \mathbb{Z}\}$ with arbitrary external conditions $\{\tilde{\omega}(\Lambda)\}$ has a convergent subsequence $\{P_{\Lambda',\beta,h,\tilde{\omega}(\Lambda')}\}$. Each of these weak limits is an infinite-volume state for the given interaction. It is natural to investigate the dependence of the limits upon the choice of external conditions. For different values of β and h several situations occur: there is a unique weak limit regardless of the choice of external conditions and the limit is translation invariant; the weak limit depends upon the choice of external conditions and is either translation invariant or not. In the second situation, a phase transition is said to occur. We shall call this a *level-3 phase transition* in order to distinguish it from the level-1 phase transition which is spontaneous magnetization.

Consider the set of limits

$$\mathcal{G}^0_{\beta,h} = \{P \in \mathcal{M}(\Omega) : P = \underset{\Lambda' \uparrow \mathbb{Z}}{w\text{-lim}}\, P_{\Lambda',\beta,h,\tilde{\omega}(\Lambda')}\}, \tag{4.39}$$

where $\{\Lambda'\}$ is any increasing sequence of symmetric intervals whose union is $\mathbb{Z}$ and $\tilde{\omega}(\Lambda')$ is any external condition for Λ'. Define $\mathcal{G}_{\beta,h}$ to be the closed convex hull of $\mathcal{G}^0_{\beta,h}$. This is the intersection of all closed convex subsets of $\mathcal{M}(\Omega)$ containing $\mathcal{G}^0_{\beta,h}$. Equivalently, $\mathcal{G}_{\beta,h}$ is the closure, with respect to weak convergence, of the set of convex combinations

$$\left\{P \in \mathcal{M}(\Omega) : P = \sum_{j=1}^{r} \lambda_j P_j, \lambda_j > 0, \sum_{j=1}^{r} \lambda_j = 1, P_j \in \mathcal{G}^0_{\beta,h}\right\}. \tag{4.40}$$

Each measure $P \in \mathcal{G}_{\beta,h}$ is called an *infinite-volume Gibbs state*. A level-3 phase transition is said to occur if $\mathcal{G}_{\beta,h}$ consists of more than one measure.

The passage from $\mathcal{G}^0_{\beta,h}$ to $\mathcal{G}_{\beta,h}$ has an interesting physical interpretation. We denote the limit P in (4.39) by $P_{\beta,h,\{\omega(\Lambda')\}}$. This measure corresponds physically to the situation where an experimenter is sure that the external condition for each Λ' is $\omega(\Lambda')$. The experimenter may also be uncertain about the external conditions. Such an uncertainty is represented by the convex combination P in (4.40), where $\lambda_1, \lambda_2, \ldots, \lambda_r$ are the probabilities of r different choices.

Before proving properties of infinite-volume Gibbs states, we extend the definition of Gibbs free energy to the state $P_{\Lambda,\beta,h,\tilde{\omega}}$ with external condition $\tilde{\omega}$. Recall that for the finite-volume Gibbs state $P_{\Lambda,\beta,h}$ without external condition, we defined $\Psi(\Lambda,\beta,h) = -\beta^{-1}\log Z(\Lambda,\beta,h)$, where $Z(\Lambda,\beta,h) = \int_{\Omega_\Lambda}\exp[-\beta H_{\beta,h}(\omega)]\pi_\Lambda P_\rho(d\omega)$. If $\sum_{k\in\mathbb{Z}} J(k) < \infty$, then the specific Gibbs free energy $\psi(\beta,h) = \lim_{\Lambda\uparrow\mathbb{Z}}|\Lambda|^{-1}\Psi(\Lambda,\beta,h)$ exists [Appendix D.1]. Given an external condition $\tilde{\omega} = \tilde{\omega}(\Lambda)$, we define the Gibbs free energy in the state $P_{\Lambda,\beta,h,\tilde{\omega}}$ by the formula

$$\begin{aligned}\Psi(\Lambda,\beta,h,\tilde{\omega}) &= -\beta^{-1}\log Z(\Lambda,\beta,h,\tilde{\omega})\\ &= -\beta^{-1}\log\int_{\Omega_\Lambda}\exp[-\beta H_{\Lambda,h,\tilde{\omega}}(\omega)]\pi_\Lambda P_\rho(d\omega).\end{aligned}$$

The next lemma shows that for any sequence of external conditions $\{\tilde{\omega}(\Lambda);\ \Lambda\uparrow\mathbb{Z}\}$ the limit $\lim_{\Lambda\uparrow\mathbb{Z}}|\Lambda|^{-1}\Psi(\Lambda,\beta,h,\tilde{\omega}(\Lambda))$ exists and is independent of the sequence chosen.

Lemma IV.6.2. *Let J be a summable ferromagnetic interaction on $\mathbb{Z}$. Then for $\beta > 0$ and h real, $\lim_{\Lambda\uparrow\mathbb{Z}}|\Lambda|^{-1}\Psi(\Lambda,\beta,h,\tilde{\omega}(\Lambda))$ exists and is independent of the choice of $\{\tilde{\omega}(\Lambda)\}$. The limit equals $\lim_{\Lambda\uparrow\mathbb{Z}}|\Lambda|^{-1}\Psi(\Lambda,\beta,h) = \psi(\beta,h)$.*

Proof. We use the comparison lemma. Lemma II.7.4. For any probability measure P on $\mathscr{B}(\Omega_\Lambda)$ and real-valued functions f and g on Ω_Λ

$$\left|\log\int_{\Omega_\Lambda} e^f dP - \log\int_{\Omega_\Lambda} e^g dP\right| \le \|f-g\|_\infty,$$

where $\|-\|_\infty$ denotes the maximum over Ω_Λ. It follows that

$$(4.41)\qquad \frac{1}{|\Lambda|}|\Psi(\Lambda,\beta,h,\tilde{\omega}) - \Psi(\Lambda,\beta,h)| \le \frac{1}{|\Lambda|}\|H_{\Lambda,h,\tilde{\omega}} - H_{\Lambda,h}\|_\infty.$$

Since $\psi(\beta,h) = \lim_{\Lambda\uparrow\mathbb{Z}}|\Lambda|^{-1}\Psi(\Lambda,\beta,h)$ exists, it suffices to prove that $|\Lambda|^{-1}\|H_{\Lambda,h,\tilde{\omega}} - H_{\Lambda,h}\|_\infty \to 0$ as $\Lambda\uparrow\mathbb{Z}$. For any $\varepsilon > 0$, there exists a positive integer N_ε so that $\sum_{|k|>N_\varepsilon} J(k) < \varepsilon$. For any $\omega\in\Omega_\Lambda$

$$\frac{1}{|\Lambda|}|H_{\Lambda,h,\tilde{\omega}}(\omega) - H_{\Lambda,h}(\omega)| \le \frac{1}{|\Lambda|}\sum_{i\in\Lambda}\sum_{j\in\Lambda^c} J(i-j) \le \frac{1}{|\Lambda|}\sum{}' J(i-j) + \varepsilon,$$

(4.42)

where $\sum'$ denotes the sum over all $i\in\Lambda$ and $j\in\Lambda^c$ such that $|i-j|\le N_\varepsilon$. Since $\sum' J(i-j) \le \max_{k\in\mathbb{Z}}(J(k))\cdot 2N_\varepsilon^2$, the proof is complete. □

We recall from Theorems IV.5.1 and IV.5.3(d) that for all $\beta > 0$, $h \ne 0$ and $0 < \beta < \beta_c$, $h = 0$, $\partial\psi(\beta,h)/\partial h$ exists and equals minus the specific magnetization $m(\beta,h) = \lim_{\Lambda\uparrow\mathbb{Z}}|\Lambda|^{-1}M(\Lambda,\beta,h)$. $M(\Lambda,\beta,h)$ is the magnetization in the finite-volume Gibbs state $P_{\Lambda,\beta,h}$ without external condition. This relation will be extended to the states $\{P_{\Lambda,\beta,h,\tilde{\omega}}\}$ by means of the following useful convexity result.

Lemma IV.6.3. *Let $\{f_n; n = 1, 2, \ldots\}$ be a sequence of convex functions on an open interval A of $\mathbb{R}$ such that $f(t) = \lim_{n\to\infty} f_n(t)$ exists for every $t \in A$. If each f_n and f are differentiable at some point $t_0 \in A$, then $\lim_{n\to\infty} f_n'(t_0)$ exists and equals $f'(t_0)$.*

Proof. Define $g_n(t) = (f_n(t) - f_n(t_0))/(t - t_0)$ and $g(t) = (f(t) - f(t_0))/(t - t_0)$ for $t \in A$, $t \neq t_0$. Then $g_n(t) \to g(t)$ since by hypothesis $f_n \to f$ on A. The convexity of f_n implies that $g_n(s) \leq f_n'(t_0) \leq g_n(t)$ for $s < t_0 < t$ $(s, t \in A)$.* Taking $n \to \infty$, we see that

$$\sup_{\{s\in A: s<t_0\}} g(s) \leq \liminf_{n\to\infty} f_n'(t_0) \leq \limsup_{n\to\infty} f_n'(t_0) \leq \inf_{\{t\in A: t>t_0\}} g(t). \tag{4.43}$$

As the pointwise limit of convex functions, f is convex, and so $f'(t_0) = \sup_{\{s\in A: s<t_0\}} g(s) = \inf_{\{t\in A: t>t_0\}} g(t)$. Inserting this in (4.43) completes the proof. □

Given an external condition $\tilde{\omega} = \tilde{\omega}(\Lambda)$, define $M(\Lambda, \beta, h, \tilde{\omega}) = \sum_{i\in\Lambda} \langle\omega_i\rangle_{\Lambda,\beta,h,\tilde{\omega}}$. This sum is the magnetization in the finite-volume Gibbs state $P_{\Lambda,\beta,h,\tilde{\omega}}$. By the same calculation as in (4.21), $\partial\Psi(\Lambda, \beta, h, \tilde{\omega})/\partial h$ equals $-M(\Lambda, \beta, h, \tilde{\omega})$, and as in the proof of Theorem IV.5.1, $\Psi(\Lambda, \beta, h, \tilde{\omega})$ is a concave function of h real. Since $|\Lambda|^{-1}\Psi(\Lambda, \beta, h, \tilde{\omega}(\Lambda)) \to \psi(\beta, h)$ and $\partial\psi(\beta, h)/\partial h$ exists for $\beta > 0$, $h \neq 0$ and $0 < \beta < \beta_c$, $h = 0$, Lemma IV.6.3 yields the following important fact.

Lemma IV.6.4. *Let J be be a summable ferromagnetic interaction on $\mathbb{Z}$. Then for $\beta > 0$, $h \neq 0$ and $0 < \beta < \beta_c$, $h = 0$, $\lim_{\Lambda\uparrow\mathbb{Z}} |\Lambda|^{-1} M(\Lambda, \beta, h, \tilde{\omega}(\Lambda))$ exists and is independent of the choice of $\{\tilde{\omega}(\Lambda)\}$. For these values of β and h,*

$$\lim_{\Lambda\uparrow\mathbb{Z}} \frac{1}{|\Lambda|} M(\Lambda, \beta, h, \tilde{\omega}(\Lambda)) = -\frac{\partial\psi(\beta, h)}{\partial h} = m(\beta, h). \tag{4.44}$$

Let T denote the shift mapping on Ω. A probability measure P on $\mathscr{B}(\Omega)$ is said to be *translation invariant* if for each Borel set A $P\{T^{-1}A\} = P\{A\}$. Let $\mathscr{M}_s(\Omega)$ denote the set of translation invariant probability measures on $\mathscr{B}(\Omega)$. A measure $P \in \mathscr{M}_s(\Omega)$ is called *ergodic* if $P\{A\}$ equals 0 or 1 for any Borel set A which satisfies $T^{-1}A = A$. We have denoted the set of infinite-volume Gibbs states by $\mathscr{G}_{\beta,h}$. A measure P in $\mathscr{G}_{\beta,h}$ is called a *phase* if P is translation invariant. Let $m(\beta, h)$ be the specific magnetization and let $m(\beta, +) = \lim_{h\to 0^+} m(\beta, h)$ and $m(\beta, -) = \lim_{h\to 0^-} m(\beta, h)$. According to Theorem IV.5.3 there exists a critical inverse temperature $\beta_c \in (0, \infty]$ such that spontaneous magnetization occurs at all $\beta > \beta_c$ but not at any $0 < \beta < \beta_c$; i.e.,

$$m(\beta, +) = 0 = m(\beta, -) \qquad \text{for } 0 < \beta < \beta_c,$$
$$m(\beta, +) > 0 > m(\beta, -) \qquad \text{for } \beta > \beta_c.$$

* See page 214.

The next theorem describes phases of the ferromagnet and relates the occurrence of a level-3 phase transition to that of spontaneous magnetization.

Theorem IV.6.5. *Let J be a summable ferromagnetic interaction on $\mathbb{Z}$. Then the following conclusions hold.*

(a) *For each $\beta > 0$ and h real, the weak limits*

$$P_{\beta,h,+} = \underset{\Lambda \uparrow \mathbb{Z}}{\text{w-lim}}\, P_{\Lambda,\beta,h,+}, \qquad P_{\beta,h,-} = \underset{\Lambda \uparrow \mathbb{Z}}{\text{w-lim}}\, P_{\Lambda,\beta,h,-} \tag{4.45}$$

exist and are translation invariant. Thus $\mathscr{G}_{\beta,h}$ and $\mathscr{G}_{\beta,h} \cap \mathscr{M}_s(\Omega)$ are nonempty. The measures $P_{\beta,h,+}$ and $P_{\beta,h,-}$ are ergodic.

(b) *$P_{\beta,h,+}$ equals $P_{\beta,h,-}$ if and only if $\partial\psi(\beta,h)/\partial h$ exists. Thus, for $\beta > 0$, $h \neq 0$ and for $0 < \beta < \beta_c$, $h = 0$, $P_{\beta,h,+}$ equals $P_{\beta,h,-}$. For these values of β and h, define $P_{\beta,h} = P_{\beta,h,+} = P_{\beta,h,-}$.*

(c) *If $\partial\psi(\beta,h)/\partial h$ exists, then $P_{\beta,h}$ is the unique measure in $\mathscr{G}_{\beta,h}$ and thus in $\mathscr{G}_{\beta,h} \cap \mathscr{M}_s(\Omega)$. No level-3 phase transition occurs. The mean $\int_\Omega \omega_0 P_{\beta,h}(d\omega)$ equals the specific magnetization $m(\beta,h)$.*

(d) *For $\beta > \beta_c$, $P_{\beta,0,+} \neq P_{\beta,0,-}$. In fact,*

$$\int_\Omega \omega_0 P_{\beta,0,+}(d\omega) = m(\beta,+) > 0 > \int_\Omega \omega_0 P_{\beta,0,-}(d\omega) = m(\beta,-). \tag{4.46}$$

Thus for $\beta > \beta_c$ and $h = 0$, a level-3 phase transition occurs.

(e) *For $\beta > \beta_c$, $\mathscr{G}_{\beta,0} \cap \mathscr{M}_s(\Omega)$ contains (at least) all the measures $P_{\beta,0}^{(\lambda)} = \lambda P_{\beta,0,+} + (1-\lambda) P_{\beta,0,-}$, $0 \leq \lambda \leq 1$.*

The proof of this theorem requires several new results which we will present as lemmas at the end of the section. First, we interpret the contents of the theorem. See Note 11 for further comments on the structure of $\mathscr{G}_{\beta,h}$ and $\mathscr{G}_{\beta,h} \cap \mathscr{M}_s(\Omega)$.

Part (a). Let $S_\Lambda(\omega) = \sum_{j\in\Lambda} \omega_j$ be the spin in a symmetric interval Λ. The ergodic theorem [Theorem A.11.5] implies that if $P \in \mathscr{M}_s(\Omega)$ is ergodic, then

$$\lim_{\Lambda \uparrow \mathbb{Z}} \frac{S_\Lambda(\omega)}{|\Lambda|} = \int_\Omega \omega_0 P(d\omega) \qquad P\text{-a.s.} \tag{4.47}$$

Let P be the ergodic infinite-volume Gibbs state $P_{\beta,h,+}$ or $P_{\beta,h,-}$. Then $\lim_{\Lambda\uparrow\mathbb{Z}} S_\Lambda(\omega)/|\Lambda|$, which is the limiting spin per site in a sample ω drawn from the magnet, is a constant independent of ω P-a.s. The measures $P_{\beta,h,+}$ and $P_{\beta,h,-}$ are called *pure phases* of the magnet. Stronger clustering properties of $P_{\beta,h,+}$ and $P_{\beta,h,-}$ are given in Corollary A.11.12.

Parts (b)–(c). According to Lemma IV.6.4, for $\beta > 0$, $h \neq 0$ and $0 < \beta < \beta_c$, $h = 0$, the limiting magnetization per site, $\lim_{\Lambda\uparrow\mathbb{Z}} |\Lambda|^{-1} M(\Lambda,\beta,h,\tilde{\omega}(\Lambda))$, exists and is independent of the choice of external conditions $\{\tilde{\omega}(\Lambda)\}$. It equals the specific magnetization $m(\beta, h)$. For these values of β and h, the independence of external conditions extends to the infinite-volume Gibbs states; that is, $P_{\beta,h} = P_{\beta,h,+} = P_{\beta,h,-}$ is the unique measure in $\mathscr{G}_{\beta,h}$ as well as in $\mathscr{G}_{\beta,h} \cap \mathscr{M}_s(\Omega)$. The ergodic limit (4.47) and the fact that*

$$\int_\Omega \omega_0 P_{\beta,h}(d\omega) = m(\beta,h) > 0 \qquad \text{for } h > 0 \tag{4.48}$$

imply that with respect to $P_{\beta,h}$, for $h > 0$ almost every $\omega \in \Omega$ has a majority of the spins $+1$. There is a similar statement for $h < 0$. This support property of $P_{\beta,h}$ is the infinite-volume analog of the alignment effect built into the finite-volume Gibbs states $\{P_{\Lambda,\beta,h}\}$.

Parts (d)–(e). For $\beta > \beta_c$, the distinct measures $P_{\beta,0,+}$ and $P_{\beta,0,-}$ are both ergodic, and so there exist disjoint subsets $A_{\beta,+}$ and $A_{\beta,-}$ of Ω such that $P_{\beta,0,+}\{A_{\beta,+}\} = 1$ and $P_{\beta,0,-}\{A_{\beta,-}\} = 1$ [Theorem A.11.7(b)]. The ergodic limit (4.47) and the fact that $\int_\Omega \omega_0 P_{\beta,0,+}(d\omega) = m(\beta,+) > 0$ imply that with respect to $P_{\beta,0,+}$, almost every $\omega \in A_{\beta,+}$ has a majority of the spins $+1$. Similarly, with respect to $P_{\beta,0,-}$, almost every $\omega \in A_{\beta,-}$ has a majority of the spins -1. The measures $P_{\beta,0,+}$ and $P_{\beta,0,-}$ are called a *pure plus phase* and a *pure minus phase*, respectively. These measures are simply related by the formula $P_{\beta,0,+}(d(-\omega)) = P_{\beta,0,-}(d\omega)$, which follows from the same relation for the finite-volume Gibbs states $P_{\Lambda,\beta,0,+}$ and $P_{\Lambda,\beta,0,-}$. For $0 < \lambda < 1$, the nonergodic measure $P_{\beta,0}^{(\lambda)} = \lambda P_{\beta,0,+} + (1-\lambda) P_{\beta,0,-}$ is called a *mixed phase*. With respect to this measure, the limiting spin per site, $\lim_{\Lambda \uparrow \mathbb{Z}} S_\Lambda(\omega)/|\Lambda|$, is not a constant but depends on the choice of ω [see Theorem IV.6.6(c)].

Level-3 phase transition. That $\mathscr{G}_{\beta,h}$ consists of a unique measure for $\beta > 0$, $h \neq 0$ and $0 < \beta < \beta_c$, $h = 0$, contrasts with the nonuniqueness of measures in $\mathscr{G}_{\beta,0}$ for $\beta > \beta_c$. We have called this nonuniqueness a level-3 phase transition. Formula (4.46) shows that spontaneous magnetization is the contraction of the level-3 phase transition onto level-1. The level-3 phase transition can also be looked at as a *symmetry-breaking transition*.[12] For $h = 0$, the microscopic interaction energy between each pair of spins ω_i, ω_j equals $-J(i-j)\omega_i\omega_j$. For all i and j, these energy terms are invariant with respect to sign changes $(\omega_i, \omega_j) \to (-\omega_i, -\omega_j)$ in the spins. However for $\beta > \beta_c$ neither of the states $P_{\beta,0,+}$, $P_{\beta,0,-}$ retains this invariance.

The next result refines the ergodic theorem by giving additional convergence properties of the microscopic sums $\{S_\Lambda/|\Lambda|\}$ as $\Lambda \uparrow \mathbb{Z}$. We see that exponential convergence to a constant distinguishes the values $\beta > 0$, $h \neq 0$ and $0 < \beta < \beta_c$, $h = 0$ from the values $\beta > \beta_c$. $h = 0$.

*The positivity of $m(\beta, h)$ for $h > 0$ is proved in a footnote on page 165. Also see Problem V.13.1(b).

Theorem IV.6.6. *Let J be a summable ferromagnetic interaction on $\mathbb{Z}$. Then the following conclusions hold.*

(a) *For $\beta > 0$, $h \neq 0$ and for $0 < \beta < \beta_c$, $h = 0$,*

$$S_\Lambda/|\Lambda| \xrightarrow{\text{a.s.}} m(\beta,h) \quad \text{and} \quad S_\Lambda/|\Lambda| \xrightarrow{\exp} m(\beta,h) \quad \text{w.r.t. } P_{\beta,h} \text{ as } \Lambda \uparrow \mathbb{Z}^D.$$

(b) *For $\beta > \beta_c$,*

$$\frac{S_\Lambda}{|\Lambda|} \xrightarrow{\text{a.s.}} m(\beta,+) \quad \text{w.r.t. } P_{\beta,0,+},$$

$$\frac{S_\Lambda}{|\Lambda|} \xrightarrow{\text{a.s.}} m(\beta,-) \quad \text{w.r.t. } P_{\beta,0,-} \quad \text{as } \Lambda \uparrow \mathbb{Z}.$$

In each case, exponential convergence fails.

(c) *For $\beta > \beta_c$ and each $0 < \lambda < 1$, there exists a random variable $Y_\beta^{(\lambda)}$ on Ω with distribution $\lambda\delta_{m(\beta,+)} + (1-\lambda)\delta_{m(\beta,-)}$ such that $S_\Lambda/|\Lambda| \xrightarrow{\text{a.s.}} Y_\beta^{(\lambda)}$ w.r.t. $P_{\beta,0}^{(\lambda)} = \lambda P_{\beta,0,+} + (1-\lambda)P_{\beta,0,-}$ as $\Lambda \uparrow \mathbb{Z}$.*

With respect to the various measures in this theorem, large deviation bounds for $S_\Lambda/|\Lambda|$ can be derived from Theorem II.6.1. We omit the formulas, which are easily worked out as in the discussion at the end of Section IV.5 (see Lemma IV.6.11 for the calculation of the free energy function).[13]

We now turn to the proofs of Theorems IV.6.5 and IV.6.6. The first step is to prove that the weak limits $P_{\beta,h,+} = \text{w-lim}_{\Lambda\uparrow\mathbb{Z}} P_{\Lambda,\beta,h,+}$ and $P_{\beta,h,-} = \text{w-lim}_{\Lambda\uparrow\mathbb{Z}} P_{\Lambda,\beta,h,-}$ exist. According to Lemma IV.6.1, it suffices to prove that for each finite set B the limits $\lim_{\Lambda\uparrow\mathbb{Z}} \int_\Omega f_B dP_{\Lambda,\beta,h,+}$ and $\lim_{\Lambda\uparrow\mathbb{Z}} \int_\Omega f_B dP_{\Lambda,\beta,h,-}$ exist, where $f_B(\omega) = \prod_{i\in B}[\frac{1}{2}(1+\omega_i)]$ for B nonempty. By the discussion after that lemma, each integral $\int_\Omega f_B dP_{\Lambda,\beta,h,+}$ or $\int_\Omega f_B dP_{\Lambda,\beta,h,-}$ equals $\int_{\Omega_\Lambda} f_B dP_{\Lambda,\beta,h,+}$ or $\int_{\Omega_\Lambda} f_B dP_{\Lambda,\beta,h,-}$, respectively, provided Λ contains B. The proof depends upon a powerful monotonicity result due to Fortuin, Kastelyn, and Ginibre (1971) and known as the FKG inequality.

The FKG inequality is valid for a more general measure than the finite-volume Gibbs state $P_{\Lambda,\beta,h,\tilde\omega}$. Let Λ be an arbitrary nonempty finite subset of $\mathbb{Z}$, $\{J_{ij}; i,j\in\Lambda\}$ a set of non-negative real numbers, and $\{h_i; i\in\Lambda\}$ a set of real numbers. Let P be the probability measure on $\mathscr{B}(\Omega_\Lambda)$ which assigns to each $\{\omega\}$, $\omega\in\Omega_\Lambda$, the probability

$$P\{\omega\} = \exp[-H(\omega)]\pi_\Lambda P_\rho\{\omega\}\cdot\frac{1}{Z}. \tag{4.49}$$

In this formula, $H(\omega)$ equals $-\frac{1}{2}\sum_{i,j\in\Lambda} J_{ij}\omega_i\omega_j - \sum_{i\in\Lambda} h_i\omega_i$ and Z equals $\int_{\Omega_\Lambda} \exp[-H(\omega)]\pi_\Lambda P_\rho(d\omega)$. Expectation with respect to P is denoted by $\langle - \rangle_{\Lambda,\{h_i\}}$ or by $\langle - \rangle$. The measure P reduces to $P_{\Lambda,\beta,h,\tilde\omega}$ if Λ is a symmetric interval, $J_{ij} = \beta J(i-j)$, and $h_i = h + \sum_{j\in\Lambda^c} J(i-j)\tilde\omega_j$.

Let ω and $\bar\omega$ be points in Ω_Λ. We write $\omega \le \bar\omega$ if $\omega_i \le \bar\omega_i$ for each $i\in\Lambda$. A real-valued function f on Ω_Λ is said to be *nondecreasing* if $f(\omega) \le f(\bar\omega)$ whenever $\omega \le \bar\omega$. For example, the function $Y(\omega) = \omega_i$ for $i\in\Lambda$ is nondecreasing as is $f_B(\omega)$ for any subset B of Λ. However, $f(\omega) = \omega_i\omega_j$ ($i \neq j$ in Λ) is not nondecreasing. The FKG inequality is stated in part (a) of the next theorem. A useful consequence of the inequality is stated in part (b).[14]

Theorem IV.6.7. *Let f and g be nondecreasing functions on Ω_Λ. Then the following conclusions hold.*

(a) *For any values of $\{h_i\}$, the covariance of f and g is non-negative:*

$$\langle fg\rangle_{\Lambda,\{h_i\}} - \langle f\rangle_{\Lambda,\{h_i\}}\langle g\rangle_{\Lambda,\{h_i\}} \geq 0.$$

(b) $\{h_i\} \leq \{\bar{h}_i\}$ *implies* $\langle f\rangle_{\Lambda,\{h_i\}} \leq \langle f\rangle_{\Lambda,\{\bar{h}_i\}}$.

Proof. (a) The following proof is due to Battle and Rosen (1980). The argument is by induction on the number of sites $|\Lambda|$ in Λ. We have

$$\langle fg\rangle - \langle f\rangle\langle g\rangle = \int_{\Omega_\Lambda}\int_{\Omega_\Lambda} [f(\omega) - f(\tilde{\omega})][g(\omega) - g(\tilde{\omega})]P(d\omega)P(d\tilde{\omega}). \tag{4.50}$$

If $|\Lambda| = 1$, then the integrand is non-negative since f and g are nondecreasing. Hence $\langle fg\rangle - \langle f\rangle\langle g\rangle \geq 0$. Assume now that the inequality has been proved for $|\Lambda| = 1, \ldots, n-1$, some $n \geq 2$, and consider the inequality for $|\Lambda| = n$. Fix any site α in Λ. We set $\omega = (\omega', \omega_\alpha)$, where ω' has the $n-1$ components $\{\omega_i; i \in \Lambda\backslash\{\alpha\}\}$, and rewrite $H(\omega)$ in the form

$$H(\omega', \omega_\alpha) = -\frac{1}{2}\sum_{i,j\in\Lambda\backslash\{\alpha\}} J_{ij}\omega_i\omega_j - \sum_{i\in\Lambda\backslash\{\alpha\}}\left(h_i + \frac{1}{2}(J_{i\alpha} + J_{\alpha i})\omega_\alpha\right)\omega_i$$
$$-\frac{1}{2}J_{\alpha\alpha} - h_\alpha\omega_\alpha.$$

For fixed $\omega_\alpha \in \{1, -1\}$, let $\nu_{\omega_\alpha}(d\omega')$ be the probability measure on $\Omega_{\Lambda\backslash\{\alpha\}}$ defined by

$$\nu_{\omega_\alpha}(d\omega') = \exp[-H(\omega', \omega_\alpha)]\pi_{\Lambda\backslash\{\alpha\}}P_\rho(d\omega')\cdot\frac{1}{Z(\omega_\alpha)},$$

where $Z(\omega_\alpha)$ equals $\int_{\Omega_{\Lambda\backslash\{\alpha\}}}\exp[-H(\omega', \omega_\alpha)]\pi_{\Lambda\backslash\{\alpha\}}P_\rho(d\omega')$. We now write

$$\langle fg\rangle = \int_{\Omega_\Lambda} f(\omega)g(\omega)P(d\omega)$$
$$= \int_{\{1,-1\}}\left[\int_{\Omega_{\Lambda\backslash\{\alpha\}}} f(\omega', \omega_\alpha)g(\omega', \omega_\alpha)\nu_{\omega_\alpha}(d\omega')\right] Z(\omega_\alpha)\rho(d\omega_\alpha)\cdot\frac{1}{Z}.$$

The inductive hypothesis clearly applies to the ω'-integral. It follows that

$$\langle fg\rangle \geq \int_{\{1,-1\}} \phi(\omega_\alpha)\gamma(\omega_\alpha)Z(\omega_\alpha)\rho(d\omega_\alpha)\cdot\frac{1}{Z}, \tag{4.51}$$

where $\phi(\omega_\alpha) = \int_{\Omega_{\Lambda\setminus\{\alpha\}}} f(\omega', \omega_\alpha)\nu_{\omega_\alpha}(d\omega')$ and $\gamma(\omega_\alpha) = \int_{\Omega_{\Lambda\setminus\{\alpha\}}} g(\omega', \omega_\alpha)\nu_{\omega_\alpha}(d\omega')$. We prove just below that ϕ and γ are nondecreasing. The result for $|\Lambda| = 1$ and (4.51) yield (4.50):

$$\begin{aligned}\langle fg\rangle &\geq \int_{\{1,-1\}} \phi(\omega_\alpha)Z(\omega_\alpha)\rho(d\omega_\alpha)\cdot\frac{1}{Z}\cdot\int_{\{1,-1\}} \gamma(\omega_\alpha)Z(\omega_\alpha)\rho(d\omega_\alpha)\cdot\frac{1}{Z}\\ &= \langle f\rangle\langle g\rangle.\end{aligned}$$

We prove that $\phi(-1) \leq \phi(1)$. The same proof shows that γ is nondecreasing. In the definition of $\nu_{\omega_\alpha}(d\omega')$, replace $\omega_\alpha \in \{1, -1\}$ by a real parameter t. We prove that the function $t \to \int_{\Omega_{\Lambda\setminus\{\alpha\}}} f(\omega', 1)\nu_t(d\omega')$ is nondecreasing. Since f is nondecreasing, it will then follow that

$$\begin{aligned}\phi(-1) &= \int_{\Omega_{\Lambda\setminus\{\alpha\}}} f(\omega', -1)\nu_{-1}(d\omega') \leq \int_{\Omega_{\Lambda\setminus\{\alpha\}}} f(\omega', 1)\nu_{-1}(d\omega')\\ &\leq \int_{\Omega_{\Lambda\setminus\{\alpha\}}} f(\omega', 1)\nu_1(d\omega') = \phi(1).\end{aligned}$$

We have

$$\begin{aligned}\frac{d}{dt}\int_{\Omega_{\Lambda\setminus\{\alpha\}}} &f(\omega', 1)\nu_t(d\omega')\\ &= \int_{\Omega_{\Lambda\setminus\{\alpha\}}} f(\omega', 1)\left[-\frac{\partial H(\omega', t)}{\partial t}\right]\nu_t(d\omega')\\ &\quad - \int_{\Omega_{\Lambda\setminus\{\alpha\}}} f(\omega', 1)\nu_t(d\omega')\cdot\int_{\Omega_{\Lambda\setminus\{\alpha\}}}\left[-\frac{\partial H(\omega', t)}{\partial t}\right]\nu_t(d\omega').\end{aligned}$$

The function $-\partial H(\omega', t)/\partial t$ equals $\sum_{i\in\Lambda\setminus\{\alpha\}} \frac{1}{2}(J_{i\alpha} + J_{\alpha i})\omega_i + h_\alpha$, and so it is a nondecreasing function on $\Omega_{\Lambda\setminus\{\alpha\}}$. By the inductive hypothesis, we conclude that $d\int_{\Omega_{\Lambda\setminus\{\alpha\}}} f(\omega', 1)\nu_t(d\omega')/dt \geq 0$. The FKG inequality is proved.

(b) Since $\partial H/\partial h_i = -\omega_i$ and

$$\frac{\partial Z}{\partial h}\cdot\frac{1}{Z} = \int_{\Omega_\Lambda} \omega_i \exp[-H(\omega)]\pi_\Lambda P_\rho(d\omega)\cdot\frac{1}{Z} = \langle\omega_i\rangle,$$

we have

$$\begin{aligned}\frac{\partial}{\partial h_i}\langle f\rangle &= \frac{\partial}{\partial h_i}\left\{\int_{\Omega_\Lambda} f(\omega)\exp[-H(\omega)]\pi_\Lambda P_\rho(d\omega)\cdot\frac{1}{Z}\right\}\\ &= \int_{\Omega_\Lambda} f(\omega)\cdot\omega_i\exp[-H(\omega)]\pi_\Lambda P_\rho(d\omega)\cdot\frac{1}{Z} - \langle f\rangle Z_{h_i}\cdot\frac{1}{Z} \qquad (4.52)\\ &= \langle f\cdot\omega_i\rangle - \langle f\rangle\langle\omega_i\rangle.\end{aligned}$$

This is non-negative as f and ω_i are nondecreasing. Thus $\langle f\rangle_{\Lambda,\{h_i\}}$ is a nondecreasing function of each h_i. Part (b) is proved. □

Let J be a non-negative, summable, symmetric function on $\mathbb{Z}$. Denote by $\langle - \rangle_{\Lambda,\beta,h,+}$ (resp., $\langle - \rangle_{\Lambda,\beta,h,-}$) expectation with respect to the measure P in (4.49) with $J_{ij} = \beta J(i-j)$ and $h_i = h + \sum_{j\in\Lambda^c} J(i-j)(+1)$ (resp., $h_i = h + \sum_{j\in\Lambda^c} J(i-j)(-1)$).

Corollary IV.6.8. *Let $\{\Lambda_n; n = 1, 2, \ldots\}$ be an increasing sequence of finite subsets of $\mathbb{Z}$ such that $\Lambda_n \uparrow \mathbb{Z}$ as $n \to \infty$. For B a nonempty finite set in $\mathbb{Z}$, pick an integer $n(B)$ so that B is contained in Λ_n for all $n \geq n(B)$; $f_B(\omega) = \prod_{i\in B} [(1 + \omega_i)/2]$ is a nondecreasing function on Ω_{Λ_n} whenever $n \geq n(B)$. For $\beta > 0$ and h real, the following conclusions hold.*

(a) *For $n \geq n(B)$, $\langle f_B \rangle_{\Lambda_n,\beta,h,+}$ is a nonincreasing sequence as $n \to \infty$; $\langle f_B \rangle_{\Lambda_n,\beta,h,-}$ is a nondecreasing sequence as $n \to \infty$.*

(b) *The limits $\langle f_B \rangle_{\beta,h,+} = \lim_{\Lambda_n\uparrow\mathbb{Z}} \langle f_B \rangle_{\Lambda_n,\beta,h,+}$ and $\langle f_B \rangle_{\beta,h,-} = \lim_{\Lambda_n\uparrow\mathbb{Z}} \langle f_B \rangle_{\Lambda_n,\beta,h,-}$ both exist.*

Proof. (a) The external field h can be chosen to be site-dependent, say $\tilde{h}_i$. If $\tilde{h}_i$ tends to ∞, then h_i tends to ∞. Choose $\bar{n} > n \geq n(B)$, so that $\Lambda_n \subset \Lambda_{\bar{n}}$. The FKG inequality implies that $\langle f_B \rangle_{\Lambda_{\bar{n}},\beta,h,+} \leq \langle f_B \rangle_{\Lambda_n,\beta,h,+}$ because the latter can be obtained from the former by taking $h_i \to \infty$ for each $i \in \Lambda_{\bar{n}} \backslash \Lambda_n$. A similar proof shows that if $\bar{n} > n \geq n(B)$, then $\langle f_B \rangle_{\Lambda_{\bar{n}},\beta,h,-} \geq \langle f_B \rangle_{\Lambda_n,\beta,h,+}$.

(b) This follows from part (a). □

We now prove two lemmas from which Theorem IV.6.5 will follow. Let Λ be a symmetric interval of $\mathbb{Z}$. For $\beta > 0$ and h real, let $P_{\Lambda,\beta,h,\tilde{\omega}}$ be the finite-volume Gibbs state on Λ with external condition $\tilde{\omega}$ corresponding to a summable ferromagnetic interaction J. We write $\langle - \rangle_{\Lambda,h,\tilde{\omega}}$ for expectation with respect to $P_{\Lambda,\beta,h,\tilde{\omega}}$.

Lemma IV.6.9. *Let $f_B(\omega) = \prod_{i\in B} [\frac{1}{2}(1 + \omega_i)]$ for B a nonempty finite subset of $\mathbb{Z}$. For $\beta > 0$ and h real, the following conclusions hold.*

(a) *The limits $\langle f_B \rangle_{h,+} = \lim_{\Lambda\uparrow\mathbb{Z}} \langle f_B \rangle_{\Lambda,h,+}$ and $\langle f_B \rangle_{h,-} = \lim_{\Lambda\uparrow\mathbb{Z}} \langle f_B \rangle_{\Lambda,h,-}$ exist and for any $k \in \mathbb{Z}$ $\langle f_{B+k} \rangle_{h,+} = \langle f_B \rangle_{h,+}$ and $\langle f_{B+k} \rangle_{h,-} = \langle f_B \rangle_{h,-}$.*

(b) *For any real number h_0, $\lim_{h\to h_0^+} \langle f_B \rangle_{h,+} = \langle f_B \rangle_{h_0,+}$ and $\lim_{h\to h_0^-} \langle f_B \rangle_{h,-} = \langle f_B \rangle_{h_0,-}$.*

(c) *For any symmetric interval Λ containing B and any external condition $\tilde{\omega}$, $\langle f_B \rangle_{\Lambda,h,-} \leq \langle f_B \rangle_{\Lambda,h,\tilde{\omega}} \leq \langle f_B \rangle_{\Lambda,h,+}$.*

(d) $0 \leq \langle f_B \rangle_{h,+} - \langle f_B \rangle_{h,-} \leq |B| (\langle \omega_0 \rangle_{h,+} - \langle \omega_0 \rangle_{h,-})$.

Proof. (a) By Corollary IV.6.8(b), the limits

$$\langle f_B \rangle_{h,+} = \lim_{\Lambda\uparrow\mathbb{Z}} \langle f_B \rangle_{\Lambda,h,+} \quad \text{and} \quad \langle f_{B+k} \rangle_{h,+} = \lim_{\Lambda\uparrow\mathbb{Z}} \langle f_{B+k} \rangle_{\Lambda,h,+}$$

both exist. The interaction strength $J(i-j)$ between each pair of sites i, j is translation invariant. Hence if Λ contains B, then $\langle f_{B+k} \rangle_{\Lambda+k,h,+} = \langle f_B \rangle_{\Lambda,h,+}$ and

$$\lim_{\Lambda \uparrow \mathbb{Z}} \langle f_{B+k} \rangle_{\Lambda+k,h,+} = \lim_{\Lambda \uparrow \mathbb{Z}} \langle f_B \rangle_{\Lambda,h,+} = \langle f_B \rangle_{h,+}.$$

We can find a sequence $\{\bar{\Lambda}_n; n = 1, 2, \ldots\}$ of symmetric intervals such that $\bar{\Lambda}_1 \subseteq \bar{\Lambda}_2 + k \subseteq \bar{\Lambda}_3 \subseteq \bar{\Lambda}_4 + k \subseteq \ldots$ and $\bigcup_{n=1}^{\infty} \bar{\Lambda}_n = \mathbb{Z}$. By Corollary IV.6.8(b), $\lim_{\Lambda_n \uparrow \mathbb{Z}} \langle f_{B+k} \rangle_{\Lambda_n,h,+}$ exists, where $\Lambda_n = \bar{\Lambda}_n$ for n odd, $\Lambda_n = \bar{\Lambda}_n + k$ for n even. The existence of this limit implies that

$$\lim_{\Lambda \uparrow \mathbb{Z}} \langle f_{B+k} \rangle_{\Lambda,h,+} = \lim_{\Lambda \uparrow \mathbb{Z}} \langle f_{B+k} \rangle_{\Lambda+k,h,+}.$$

Combining this with the previous two displays, we conclude that $\langle f_{B+k} \rangle_{h,+}$ equals $\langle f_B \rangle_{h,+}$. That $\langle f_{B+k} \rangle_{h,-}$ equals $\langle f_B \rangle_{h,-}$ is proved similarly.

(b) Let Λ contain B. Since $\langle f_B \rangle_{h,+} \le \langle f_B \rangle_{\Lambda,h,+}$,

$$\text{(4.53)} \qquad \limsup_{h \to h_0^+} \langle f_B \rangle_{h,+} \le \lim_{h \to h_0^+} \langle f_B \rangle_{\Lambda,h,+} = \langle f_B \rangle_{\Lambda,h_0,+}.$$

Hence $\limsup_{h \to h_0^+} \langle f_B \rangle_{h,+} \le \lim_{\Lambda \uparrow \mathbb{Z}} \langle f_B \rangle_{\Lambda,h_0,+} = \langle f_B \rangle_{h_0,+}$. By the FKG inequality, if $h > h_0$, then $\langle f_B \rangle_{h_0,+} \le \langle f_B \rangle_{h,+}$, and so $\langle f_B \rangle_{h_0,+} \le \liminf_{h \to h_0^+} \langle f_B \rangle_{h,+}$. It follows that $\lim_{h \to h_0^+} \langle f_B \rangle_{h,+} = \langle f_B \rangle_{h_0,+}$. The second half of part (b) is proved similarly.

(c) For any external condition $\tilde{\omega}$, the field $h_i = h + \sum_{j \in \Lambda^c} J(i-j)\tilde{\omega}_j$ acting at site $i \in \Lambda$ lies between the field corresponding to the plus external condition and the field corresponding to the minus external condition. Hence the FKG inequality yields part (c).

(d) The function $f(\omega) = \sum_{i \in B} \omega_i - f_B(\omega)$ is nondecreasing. Hence by the FKG inequality, if Λ contains B, then

$$\text{(4.54)} \qquad 0 \le \langle f_B \rangle_{\Lambda,h,+} - \langle f_B \rangle_{\Lambda,h,-} \le \sum_{i \in B} [\langle \omega_i \rangle_{\Lambda,h,+} - \langle \omega_i \rangle_{\Lambda \cdot h,-}].$$

Take $\Lambda \uparrow \mathbb{Z}$ and use the fact that $\langle \omega_i \rangle_{h,+} = \langle \omega_0 \rangle_{h,+}$ and $\langle \omega_i \rangle_{h,-} = \langle \omega_0 \rangle_{h,-}$ to complete the proof of part (d). □

Spontaneous magnetization occurs at β if and only if

$$m(\beta, +) = -\frac{\partial \psi(\beta, 0)}{\partial h^+} = -\lim_{h \to 0^+} \frac{\partial \psi(\beta, h)}{\partial h}$$

is positive. The next lemma relates $m(\beta, h)$, $m(\beta, +)$, and $m(\beta, -)$ to the quantities $\langle \omega_0 \rangle_{\beta,h,+}$ and $\langle \omega_0 \rangle_{\beta,h,-}$, defined as

$$\langle \omega_0 \rangle_{\beta,h,+} = \lim_{\Lambda \uparrow \mathbb{Z}} \int_{\Omega} \omega_0 P_{\Lambda,\beta,h,+}(d\omega),$$

$$\langle \omega_0 \rangle_{\beta,h,-} = \lim_{\Lambda \uparrow \mathbb{Z}} \int_{\Omega} \omega_0 P_{\Lambda,\beta,h,-}(d\omega).$$

According to Corollary IV.6.8(b), the limits exist. Later, we will identify $\langle \omega_0 \rangle_{\beta,h,+}$ and $\langle \omega_0 \rangle_{\beta,h,-}$ as the means of infinite-volume Gibbs states $P_{\beta,h,+}$ and $P_{\beta,h,-}$, respectively.

Lemma IV.6.10. *Let J be a summable ferromagnetic interaction on $\mathbb{Z}$. Then the following conclusions hold.*

(a) *For $\beta > 0$ and $h \neq 0$,*

$$\langle \omega_0 \rangle_{\beta,h,+} = m(\beta,h) = -\frac{\partial\psi(\beta,h)}{\partial h} = \langle \omega_0 \rangle_{\beta,h,-}.$$

(b) *For $\beta > 0$ and $h = 0$,*

$$(4.55) \qquad \begin{aligned} \langle \omega_0 \rangle_{\beta,0,+} &= m(\beta,+) = -\frac{\partial\psi(\beta,0)}{\partial h^+} \geq 0, \\ \langle \omega_0 \rangle_{\beta,0,-} &= m(\beta,-) = -\frac{\partial\psi(\beta,0)}{\partial h^-} \leq 0. \end{aligned}$$

Thus spontaneous magnetization occurs at β if and only if $\langle \omega_0 \rangle_{\beta,0,+} = m(\beta,+) > 0$.

Proof. (a) For any $\varepsilon > 0$, let Λ_ε be a symmetric interval such that $\langle \omega_0 \rangle_{\Lambda_\varepsilon,\beta,h,+} \leq \langle \omega_0 \rangle_{\beta,h,+} + \varepsilon$. Given another symmetric interval Λ which contains Λ_ε, define $B_\varepsilon(\Lambda) = \{i \in \Lambda : \Lambda_\varepsilon + i \subseteq \Lambda\}$. The sequence $\{\langle \omega_i \rangle_{\Lambda,\beta,h,+}\}$ is nonincreasing as $\Lambda \uparrow \mathbb{Z}$ [Corollary IV.6.8(a)], and so for any $i \in B_\varepsilon(\Lambda)$

$$(4.56) \qquad \begin{aligned} \langle \omega_0 \rangle_{\beta,h,+} &= \langle \omega_i \rangle_{\beta,h,+} \leq \langle \omega_i \rangle_{\Lambda,\beta,h,+} \leq \langle \omega_i \rangle_{\Lambda_\varepsilon+i,\beta,h,+} \\ &= \langle \omega_0 \rangle_{\Lambda_\varepsilon,\beta,h,+} \leq \langle \omega_0 \rangle_{\beta,h,+} + \varepsilon. \end{aligned}$$

Thus for all symmetric intervals Λ containing Λ_ε

$$(4.57) \qquad \langle \omega_0 \rangle_{\beta,h,+} \leq \frac{1}{|B_\varepsilon(\Lambda)|} \sum_{i \in B_\varepsilon(\Lambda)} \langle \omega_i \rangle_{\Lambda,\beta,h,+} \leq \langle \omega_0 \rangle_{\beta,h,+} + \varepsilon.$$

Recall the quantity $M(\Lambda,\beta,h,+) = \sum_{i\in\Lambda} \langle \omega_i \rangle_{\Lambda,\beta,h,+}$, which is the magnetization in the state $P_{\Lambda,\beta,h,+}$. Since $|\Lambda| \cdot |B_\varepsilon(\Lambda)|^{-1} \to 1$ as $\Lambda \uparrow \mathbb{Z}$, it follows from (4.57) that $\lim_{\Lambda\uparrow\mathbb{Z}} |\Lambda|^{-1} M(\Lambda,\beta,h,+) = \langle \omega_0 \rangle_{\beta,h,+}$ for any real h. But for $h \neq 0$ this limit also equals $m(\beta,h)$ [Lemma IV.6.4]. Thus for $h \neq 0$ $\langle \omega_0 \rangle_{\beta,h,+} = m(\beta,h) = -\partial\psi(\beta,h)/\partial h$. A similar proof shows that for $h \neq 0$ $\langle \omega_0 \rangle_{\beta,h,-} = m(\beta,h) = -\partial\psi(\beta,h)/\partial h$.

(b) In part (a) take $h \to 0^+$ and use Lemma IV.6.9(b) to obtain $\langle \omega_0 \rangle_{\beta,0,+} = m(\beta,+) = -\partial\psi(\beta,0)/\partial h^+$. This gives half of (4.55). The other half is proved similarly. □

Proof of Theorem IV.6.5. (a) The existence and translation invariance of the measures $P_{\beta,h,+} = w\text{-}\lim_{\Lambda\uparrow\mathbb{Z}} P_{\Lambda,\beta,h,+}$ and $P_{\beta,h,-} = w\text{-}\lim_{\Lambda\uparrow\mathbb{Z}} P_{\Lambda,\beta,h,-}$ follow from Lemmas IV.6.1 and IV.6.9(a). Thus, $P_{\beta,h,+}$ and $P_{\beta,h,-}$ belong to $\mathcal{G}_{\beta,h} \cap \mathcal{M}_s(\Omega)$. Let $\mathcal{D}$ be a nonempty subset of $\mathcal{M}(\Omega)$. A measure $P \in \mathcal{D}$ is called an *extremal point* of $\mathcal{D}$ if $P = \lambda P_1 + (1-\lambda)P_2$ for $0 < \lambda < 1$ and $P_1, P_2 \in \mathcal{D}$ implies $P_1 = P_2 = P$. The set of extremal points of $\mathcal{G}_{\beta,h} \cap \mathcal{M}_s(\Omega)$ is the set of ergodic measures in $\mathcal{G}_{\beta,h}$ [Theorem A.11.8(b)]. Hence in order to show that $P_{\beta,h,+}$ is ergodic, it suffices to prove that $P_{\beta,h,+}$ is extremal in

$\mathscr{G}_{\beta,h} \cap \mathscr{M}_s(\Omega)$.* We prove the stronger statement that $P_{\beta,h,+}$ is extremal in $\mathscr{G}_{\beta,h}$. Assume that $P_{\beta,h,+} = \lambda P_1 + (1-\lambda)P_2$ for $0 < \lambda < 1$ and $P_1, P_2 \in \mathscr{G}_{\beta,h}$. Lemma IV.6.9(c) implies that for any measure $P \in \mathscr{G}_{\beta,h}$, $\int_\Omega f_B dP \leq \int_\Omega f_B dP_{\beta,h,+}$ for all finite sets B in $\mathbb{Z}$ ($f_\phi(\omega) = 1$). The integrals $\{\int_\Omega f_B dP;\ B$ finite$\}$ determine P. Hence if either measure P_i in the decomposition of $P_{\beta,h,+}$ differs from $P_{\beta,h,+}$, then $\int_\Omega f_B dP_i < \int_\Omega f_B dP_{\beta,h,+}$ for some finite set B, and so

$$\int_\Omega f_B dP_{\beta,h,+} = \lambda \int_\Omega f_B dP_1 + (1-\lambda)\int_\Omega f_B dP_2 < \int_\Omega f_B dP_{\beta,h,+}. \tag{4.58}$$

This contradiction proves that $P_{\beta,h,+}$ is extremal in $\mathscr{G}_{\beta,h}$ and thus is ergodic. A similar proof shows that $P_{\beta,h,-}$ is extremal in $\mathscr{G}_{\beta,h}$ and thus is ergodic.

(b) If $\partial\psi(\beta,h)/\partial h$ exists, then by Lemma IV.6.10(b) $\langle\omega_0\rangle_{\beta,h,+} = \langle\omega_0\rangle_{\beta,h,-}$. Lemma IV.6.9(d) implies that for any finite subset B in $\mathbb{Z}$, $\langle f_B\rangle_{\beta,h,+} = \langle f_B\rangle_{\beta,h,-}$. Thus, $P_{\beta,h,+} = P_{\beta,h,-}$. Conversely, suppose that $P_{\beta,h,+} = P_{\beta,h,-}$. If $h \neq 0$, then $\partial\psi(\beta,h)/\partial h$ exists by Theorem IV.5.1(a). If $h = 0$, then

$$\begin{aligned}\langle\omega_0\rangle_{\beta,0,+} &= \lim_{\Lambda\uparrow\mathbb{Z}} \int_\Omega \omega_0 P_{\Lambda,\beta,0,+}(d\omega) = \int_\Omega \omega_0 P_{\beta,0,+}(d\omega) = \int_\Omega \omega_0 P_{\beta,0,-}(d\omega)\\ &= \lim_{\Lambda\uparrow\mathbb{Z}} \int_\Omega \omega_0 P_{\Lambda,\beta,0,-}(d\omega) = \langle\omega_0\rangle_{\beta,0,-},\end{aligned}$$

and $\partial\psi(\beta,0)/\partial h$ exists by Lemma IV.6.10(b).

(c) Lemma IV.6.9(c) implies that if $P_{\beta,h,+} = P_{\beta,h,-} = P_{\beta,h}$, then $\mathscr{G}_{\beta,h}$ and thus $\mathscr{G}_{\beta,h} \cap \mathscr{M}_s(\Omega)$ consist of the unique measure $P_{\beta,h}$. We have defined $\langle\omega_0\rangle_{\beta,h,+} = \lim_{\Lambda\uparrow\mathbb{Z}} \int_\Omega \omega_0 P_{\Lambda,\beta,h,+}(d\omega)$. But if $\partial\psi(\beta,h)/\partial h$ exists, then $P_{\Lambda,\beta,h,+} \Rightarrow P_{\beta,h}$, and so $\langle\omega_0\rangle_{\beta,h,+} = \int_\Omega \omega_0 P_{\beta,h}(d\omega)$. The equality $\int_\Omega \omega_0 P_{\beta,h}(d\omega) = m(\beta,h)$ follows from Lemma IV.6.10.

(d) For $\beta > \beta_c$, $\partial\psi(\beta,h)/\partial h$ does not exist at $h = 0$, and thus $\langle\omega_0\rangle_{\beta,h,+} \neq \langle\omega_0\rangle_{\beta,h,-}$. Therefore $P_{\beta,0,+} \neq P_{\beta,0,-}$. Formula (4.46) follows from Lemma IV.6.10(b) and the fact that $P_{\Lambda,\beta,0,+} \Rightarrow P_{\beta,0,+}$ and $P_{\Lambda,\beta,0,-} \Rightarrow P_{\beta,0,-}$.

(e) Since $\mathscr{G}_{\beta,0} \cap \mathscr{M}_s(\Omega)$ is convex, it must contain all the measures $\{P_{\beta,0}^{(\lambda)}; 0 \leq \lambda \leq 1\}$. This completes the proof of Theorem IV.6.5. □

The key step in proving Theorem IV.6.6 is to calculate the free energy functions of the sequence $\{S_\Lambda\}$ with respect to the infinite-volume Gibbs state $P_{\beta,h,+}$ and $P_{\beta,h,-}$, respectively.

Lemma IV.6.11. *Let J be a summable ferromagnetic interaction on $\mathbb{Z}$. For $\beta > 0$, h real, and t real, define*

$$c_{\beta,h,+}(t) = \lim_{\Lambda\uparrow\mathbb{Z}} \frac{1}{|\Lambda|} \log\langle\exp(tS_\Lambda)\rangle_{\beta,h,+},$$

$$c_{\beta,h,-}(t) = \lim_{\Lambda\uparrow\mathbb{Z}} \frac{1}{|\Lambda|} \log\langle\exp(tS_\Lambda)\rangle_{\beta,h,-}.$$

*The ergodicity of $P_{\beta,h,+}$ and $P_{\beta,h,-}$ also follows from Corollary A.11.12(a).

Then

(4.59) $$c_{\beta,h,+}(t) = c_{\beta,h,-}(t) = -\beta[\psi(\beta, h + t/\beta) - \psi(\beta, h)].$$

Proof. We first evaluate $c_{\beta,h,+}(t)$ for $t \geq 0$. Define

$$c_\Lambda(t) = \frac{1}{|\Lambda|}\log\langle\exp(tS_\Lambda)\rangle_{\beta,h,+}.$$

For Λ' a symmetric interval containing Λ, define $g_{\Lambda,\Lambda'}(t) = \langle\exp(tS_\Lambda)\rangle_{\Lambda',\beta,h,+}$. Since $P_{\Lambda',\beta,h,+} \Rightarrow P_{\beta,h,+}$ as $\Lambda' \uparrow \mathbb{Z}$, $|\Lambda|^{-1}\log g_{\Lambda,\Lambda'}(t) \to c_\Lambda(t)$ as $\Lambda' \uparrow \mathbb{Z}$ for each $t \geq 0$. We claim that

(4.60) $$\langle\exp(tS_\Lambda)\rangle_{\Lambda,\beta,h,\tilde{\omega}} \leq g_{\Lambda,\Lambda'}(t) \leq \langle\exp(tS_\Lambda)\rangle_{\Lambda,\beta,h,+} \qquad \text{for } t \geq 0,$$

where $\tilde{\omega}$ is the external condition $\tilde{\omega}_j = 1$ for $j \in (\Lambda')^c$, $\tilde{\omega}_j = -1$ for $j \in \Lambda' \backslash \Lambda$. Indeed for $t \geq 0$, the function $\exp(t\sum_{j\in\Lambda}\omega_j)$ is nondecreasing on Ω_Λ, and the right-hand (resp., left-hand) side of (4.60) can be obtained from the middle term by letting the external field $h_i \to \infty$ (resp., $h_i \to -\infty$) for each $i \in \Lambda' \backslash \Lambda$. Hence (4.60) follows from the FKG inequality. Another application of FKG gives $\langle\exp(tS_\Lambda)\rangle_{\Lambda,\beta,h,-} \leq \langle\exp(tS_\Lambda)\rangle_{\Lambda,\beta,h,\tilde{\omega}}$. As in the proof of Theorem IV.5.5,

$$\frac{1}{|\Lambda|}\log\langle\exp(tS_\Lambda)\rangle_{\Lambda,\beta,h,-} = -\beta\frac{1}{|\Lambda|}[\Psi(\Lambda,\beta,h + t/\beta, -) - \Psi(\Lambda,\beta,h,-)],$$

$$\frac{1}{|\Lambda|}\log\langle\exp(tS_\Lambda)\rangle_{\Lambda,\beta,h,+} = -\beta\frac{1}{|\Lambda|}[\Psi(\Lambda,\beta,h + t/\beta, +) - \Psi(\Lambda,\beta,h,+)].$$

(4.61)

Thus for $t \geq 0$

(4.62) $$-\beta\frac{1}{|\Lambda|}[\Psi(\Lambda,\beta,h + t/\beta, -) - \Psi(\Lambda,\beta,h,-)] \leq c_\Lambda(t) \leq -\beta\frac{1}{|\Lambda|}[\Psi(\Lambda,\beta,h + t/\beta, +) - \Psi(\Lambda,\beta,h,+)].$$

By Lemma IV.6.2 (independence of the specific Gibbs free energy of the choice of external conditions), we conclude that

$$c_{\beta,h,+}(t) = \lim_{\Lambda\uparrow\mathbb{Z}} c_\Lambda(t) = -\beta[\psi(\beta, h + t/\beta) - \psi(\beta,h)].$$

A similar proof yields (4.59) for $c_{\beta,h,-}(t)$, $t \geq 0$. For $t < 0$, the function $-\exp(t\sum_{j\in\Lambda}\omega_j)$ is nondecreasing on Ω_Λ, and so (4.60) holds with the senses of the inequalities reversed. As above, we obtain (4.59) for $c_{\beta,h,+}(t)$, $t < 0$. A similar proof yields (4.59) for $c_{\beta,h,-}(t)$, $t < 0$.

Proof of Theorem IV.6.6. (a) For $\beta > 0$, $h \neq 0$ and $0 < \beta < \beta_c$, $h = 0$, $\partial\psi(\beta,h)/\partial h$ exists. Hence by the previous lemma and Theorem II.6.3,

$S_\Lambda/|\Lambda| \xrightarrow{\exp} -\partial\psi(\beta,h)/\partial h = m(\beta,h)$. The almost sure convergence follows from the ergodic theorem or Theorem II.6.4.

(b) By the ergodic theorem, $S_\Lambda/|\Lambda| \to \langle\omega_0\rangle_{\beta,0,+} = m(\beta,+)$ $P_{\beta,0,+}$-a.s. and $S_\Lambda/|\Lambda| \to \langle\omega_0\rangle_{\beta,0,-} = m(\beta,-)$ $P_{\beta,0,-}$-a.s. In each case, the almost sure convergence cannot be strengthened to exponential convergence since by Theorem II.6.3 exponential convergence is equivalent to the existence of $(c_{\beta,0,+})'(0)$ or $(c_{\beta,0,-})'(0)$, respectively. But $\partial\psi(\beta,0)/\partial h$ does not exist for $\beta > \beta_c$, and so by the previous lemma $(c_{\beta,0,+})'(0)$ and $(c_{\beta,0,-})'(0)$ do not exist for $\beta > \beta_c$.

(c) This follows from the ergodic theorem and Theorem A.11.7(b). □

We have now completed the proof of the existence of infinite-volume Gibbs states and studied convergence properties of the spin per site $S_\Lambda/|\Lambda|$ with respect to these measures. In the next section, we show how to characterize the set of translation invariant infinite-volume Gibbs states in terms of a variational principle.

IV.7. The Gibbs Variational Formula and Principle

Let J be a summable ferromagnetic interaction on $\mathbb{Z}$. The set $\mathscr{G}_{\beta,h}$ of infinite-volume Gibbs states was defined as the closed convex hull of the set of weak limits of finite-volume Gibbs states. The Gibbs variational principle is another approach to studying the infinite-volume ferromagnet. It characterizes the set of translation invariant infinite-volume Gibbs states directly, eliminating the need to consider weak limits at all. The Gibbs variational principle expresses the specific Gibbs free energy $\psi(\beta,h)$ as the supremum of an energy functional minus an entropy functional over $\mathscr{M}_s(\Omega)$. The set of measures at which the supremum is attained is exactly the set $\mathscr{G}_{\beta,h} \cap \mathscr{M}_s(\Omega)$ [Theorem IV.7.3]. We recall that in the Gibbs variational principle for the discrete ideal gas [Theorem III.8.2], there was a unique solution for each value of β. By contrast, for $\beta > \beta_c$, $h = 0$, the Gibbs variational principle for ferromagnets has nonunique solutions since the set $\mathscr{G}_{\beta,0} \cap \mathscr{M}_s(\Omega)$ contains the distinct measures $P_{\beta,0,+}$, $P_{\beta,0,-}$, and all convex combinations.

Before stating the Gibbs variational principle, we consider an analog which characterizes the finite-volume Gibbs state $P_{\Lambda,\beta,h,\tilde\omega}$. Let Λ be a symmetric interval of $\mathbb{Z}$ and $\tilde\omega$ an external condition. Given a probability measure P on $\mathscr{B}(\Omega_\Lambda)$, we define the energy in P to be

$$U(\Lambda,h,\tilde\omega;P) = \int_{\Omega_\Lambda} H_{\Lambda,h,\tilde\omega}(\omega)P(d\omega),$$

where $H_{\Lambda,h,\tilde\omega}$ is the Hamiltonian defined in (4.35). $I^{(2)}_{\pi_\Lambda P_\rho}(P)$ denotes the relative entropy of P with respect to $\pi_\Lambda P_\rho$,

$$\int_{\Omega_\Lambda} \log \frac{dP}{d(\pi_\Lambda P_\rho)}(\omega)\, P(d\omega) = \sum_{\omega \in \Omega_\Lambda} \log \frac{P\{\omega\}}{\pi_\Lambda P_\rho\{\omega\}} \cdot P\{\omega\}.$$

$Z(\Lambda, \beta, h, \tilde{\omega})$ is the partition function $\int_{\Omega_\Lambda} \exp[-\beta H_{\Lambda,h,\tilde{\omega}}(\omega)] \pi_\Lambda P_\rho(d\omega)$.

Proposition IV.7.1. *For any $\beta > 0$, h real, and external condition $\tilde{\omega}$,*

$$\log Z(\Lambda, \beta, h, \tilde{\omega}) = \sup_{P \in \mathcal{M}(\Omega_\Lambda)} \{-\beta U(\Lambda, h, \tilde{\omega}; P) - I^{(2)}_{\pi_\Lambda P_\rho}(P)\},$$

and the supremum is attained at the unique measure $P = P_{\Lambda,\beta,h,\tilde{\omega}}$.

Proof. For any probability measure P on Ω_Λ,

$$\beta U(\Lambda, h, \tilde{\omega}; P) + I^{(2)}_{\pi_\Lambda P_\rho}(P) + \log Z(\Lambda, \beta, h, \tilde{\omega})$$

$$= \sum_{\omega \in \Omega_\Lambda} \log\left(\frac{P\{\omega\}}{\exp[-\beta H_{\Lambda,h,\tilde{\omega}}(\omega)] \cdot \pi_\Lambda P_\rho\{\omega\}/Z(\Lambda, \beta, h, \tilde{\omega})}\right) \cdot P\{\omega\}.$$

The sum equals the relative entropy of P with respect to $P_{\Lambda,\beta,h,\tilde{\omega}}$. Hence

$$\beta U(\Lambda, h, \tilde{\omega}; P) + I^{(2)}_{\pi_\Lambda P_\rho}(P) + \log Z(\Lambda, \beta, h, \tilde{\omega}) \geq 0$$

and equality holds if and only if $P = P_{\Lambda,\beta,h,\tilde{\omega}}$ [Proposition I.4.1(b)]. □

We now introduce the functionals which appear in the Gibbs variational principle. Let P be a translation invariant probability measure on $\mathscr{B}(\Omega)$, Λ a symmetric interval, and π_Λ the projection of Ω onto Ω_Λ defined by $(\pi_\Lambda \omega)_i = \omega_i$, $i \in \Lambda$. Define a probability measure $\pi_\Lambda P$ on $\mathscr{B}(\Omega_\Lambda)$ by requiring $\pi_\Lambda P\{F\} = P\{\pi_\Lambda^{-1} F\}$ for subsets F of Ω_Λ and consider the functional

$$U(\Lambda, h, \tilde{\omega}(\Lambda); \pi_\Lambda P) = \int_{\Omega_\Lambda} H_{\Lambda,h,\tilde{\omega}(\Lambda)}(\omega) \pi_\Lambda P(d\omega),$$

where $\tilde{\omega}(\Lambda)$ is an external condition. $I^{(2)}_{\pi_\Lambda P_\rho}(\pi_\Lambda P)$ denotes the relative entropy of $\pi_\Lambda P$ with respect to $\pi_\Lambda P_\rho$.

Lemma IV.7.2. *Let J be a summable ferromagnetic interaction on $\mathbb{Z}$. Then for $\beta > 0$ and h real, the following conclusions hold.*

(a) $\lim_{\Lambda \uparrow \mathbb{Z}} |\Lambda|^{-1} \log Z(\Lambda, \beta, h, \tilde{\omega}(\Lambda))$ *exists and is independent of the choice of $\{\tilde{\omega}(\Lambda)\}$. The limit equals $-\beta\psi(\beta, h)$, where $\psi(\beta, h)$ is the specific Gibbs free energy.*

(b) *For any $P \in \mathcal{M}_s(\Omega)$, $u(h, P) = \lim_{\Lambda \uparrow \mathbb{Z}} |\Lambda|^{-1} U(\Lambda, h, \tilde{\omega}(\Lambda); \pi_\Lambda P)$ exists and is independent of the choice of $\{\tilde{\omega}(\Lambda)\}$. The limit is given by*

$$u(h; P) = -\frac{1}{2} \sum_{k \in \mathbb{Z}} J(k) \int_\Omega \omega_0 \omega_k P(d\omega) - h \int_\Omega \omega_0 P(d\omega) \tag{4.63}$$

and is a bounded, affine, continuous functional of $P \in \mathcal{M}_s(\Omega)$. The functional $u(h; P)$ is called the specific energy in P.

(c) *For any* $P \in \mathcal{M}_s(\Omega)$, $\lim_{\Lambda \uparrow \mathbb{Z}} |\Lambda|^{-1} I^{(2)}_{\pi_\Lambda P_\rho}(\pi_\Lambda P)$ *exists and equals* $I_\rho^{(3)}(P)$, *the mean relative entropy of* P *with respect to* P_ρ. $I_\rho^{(3)}(P)$ *is an affine, lower semicontinuous function of* $P \in \mathcal{M}_s(\Omega)$.

Proof. (a) Lemma IV.6.2.

(b) Since P is translation invariant, $U(\Lambda, h, \tilde{\omega}(\Lambda); \pi_\Lambda P)$ equals

$$-\frac{1}{2}\sum_{k\in\mathbb{Z}} J(k)N(\Lambda,k)\int_\Omega \omega_0\omega_k P(d\omega) - \sum_{i\in\Lambda}\sum_{j\in\Lambda^c} J(i-j)\tilde{\omega}_j \int_\Omega \omega_0 P(d\omega) \tag{4.64}$$
$$- h|\Lambda| \int_\Omega \omega_0 P(d\omega),$$

where $N(\Lambda, k)$ is the number of ordered pairs i, j in Λ for which $i - j = k$. As in the proof of Lemma IV.6.2, the $\tilde{\omega}$-term is $o(|\Lambda|)$ as $\Lambda \uparrow \mathbb{Z}$ [see (4.42)]. Since for each k $N(\Lambda, k) \le |\Lambda|$ and $|\Lambda|^{-1} N(\Lambda, k) \to 1$ as $\Lambda \uparrow \mathbb{Z}$, (4.63) follows. The continuous functionals $u_n(h, P) = -\frac{1}{2}\sum_{|k|\le n} J(k)\int_\Omega \omega_0\omega_k P(d\omega) - h\int_\Omega \omega_0 P(d\omega)$ converge uniformly over $\mathcal{M}_s(\Omega)$ to $u(h; P)$ as $n \to \infty$. Hence $u(h; P)$ is continuous. The boundedness and affineness of $u(h; P)$ are obvious.

(c) This is proved in Section IX.2. □

The next theorem is due to Ruelle (1967) and Lanford and Ruelle (1969). Part (a) is called the *Gibbs variational formula*. Part (b), which characterizes the translation invariant infinite-volume Gibbs states as solutions of this formula, is called the *Gibbs variational principle*. Heuristically, the theorem follows from Proposition IV.7.1 by dividing each term in the latter by $|\Lambda|$ and taking $\Lambda \uparrow \mathbb{Z}$.

Theorem IV.7.3. *Let* J *be a summable ferromagnetic interaction on* $\mathbb{Z}$. *Then for* $\beta > 0$ *and* h *real, the following conclusions hold.*

(a) $-\beta\psi(\beta, h) = \sup_{P\in\mathcal{M}_s(\Omega)}\{-\beta u(h; P) - I_\rho^{(3)}(P)\}$.

(b) *The set of* $P \in \mathcal{M}_s(\Omega)$ *at which the supremum in part* (a) *is attained equals* $\mathcal{G}_{\beta,h} \cap \mathcal{M}_s(\Omega)$, *the set of translation invariant infinite-volume Gibbs states.*

First, we will check the consistency of the theorem with Theorem IV.6.5, which analyzed the structure of the set $\mathcal{G}_{\beta,h} \cap \mathcal{M}_s(\Omega)$. Then, we will prove the Gibbs variational formula using level-3 large deviations. In Appendix C.5 we sketch a proof of the Gibbs variational formula and principle for a much larger class of models than we are now considering. The proof is due to Föllmer (1973) and Preston (1976). In Appendix C.6 we solve the Gibbs variational formula for finite-range interactions, using techniques to be developed in Chapter IX.

The Gibbs variational principle makes explicit an energy-entropy competition which underlies the ferromagnetic phase transition. First consider $\beta = 0$. Then in the Gibbs variational formula the energy term is absent, and $\sup_{P\in\mathcal{M}_s(\Omega)}\{-I_\rho^{(3)}(P)\} = -\inf_{P\in\mathcal{M}_s(\Omega)} I_\rho^{(3)}(P)$ is attained at the unique measure P_ρ. This is consistent with Theorem IV.6.5 since for small β

$\mathscr{G}_{\beta,h} \cap \mathscr{M}_s(\Omega)$ consists of the unique measure $P_{\beta,h}$ and $P_{\beta,h} \Rightarrow P_\rho$ as $\beta \to 0^+$ [Problem V.13.1(d)]. Now consider large β. Then in the Gibbs variational formula the energy term dominates. For $h > 0$

$$\sup_{P \in \mathscr{M}_s(\Omega)} \{-u(h; P)\} = \sup_{P \in \mathscr{M}_s(\Omega)} \left\{\frac{1}{2} \sum_{k \in \mathbb{Z}} J(k) \int_\Omega \omega_0 \omega_k dP + h \int_\Omega \omega_0 dP\right\}$$

is attained at the unique measure P_{δ_1}, which is the infinite product measure with identical one-dimensional marginals δ_1. This is consistent with Theorem IV.6.5 since for all $h > 0$ $\mathscr{G}_{\beta,h} \cap \mathscr{M}_s(\Omega)$ consists of the unique measure $P_{\beta,h}$ and $P_{\beta,h} \Rightarrow P_{\delta_1}$ as $\beta \to \infty$ [Problem V.13.1(d)]. P_{δ_1} is supported on the totally aligned plus-ground state $\bar{\omega}_+$ ($\bar{\omega}_{+,j} = 1$ for all $j \in \mathbb{Z}$). On the other hand, if $h = 0$, then $\sup_{P \in \mathscr{M}_s(\Omega)} \{-u(0; P)\}$ is attained at all the measures $P^{(\lambda)} = \lambda P_{\delta_1} + (1 - \lambda) P_{\delta_{-1}}$, $0 \le \lambda \le 1$.* This would be consistent with Theorem IV.6.5(d)–(e) if whenever β_c is finite, $P_{\beta,0,+}$ (resp., $P_{\beta,0,-}$) converged weakly to P_{δ_1} (resp., $P_{\delta_{-1}}$) as $\beta \to \infty$. A proof of this statement for models on $\mathbb{Z}^D$, $D \ge 2$, is sketched in Problem V.13.1(d).

We now derive the Gibbs variational formula using level-3 large deviations. The interval Λ consists of the $2N + 1$ integers j with $|j| \le N$. In order to ease the notation, we consider $h = 0$ and write $Z(\Lambda, \beta, 0)$ as

$$Z(n, \beta) = \int_\Omega \exp\left[\frac{\beta}{2} \sum_{i,j=1}^{n} J(i-j)\omega_i \omega_j\right] P_\rho(d\omega),$$

where $n = 2N + 1$. A similar proof is valid for $h \neq 0$. A relatively easy case is the Ising model on $\mathbb{Z}$, for which

$$Z(n, \beta) = \int_\Omega \exp\left[\beta \mathscr{J} \sum_{j=1}^{n-1} \omega_j \omega_{j+1}\right] P_\rho(d\omega), \qquad \mathscr{J} > 0.$$

This equals the quantity $Z_n^{(3)}$ in Theorem II.7.3(c) with $G(\omega_j, \omega_{j+1}) = \beta \mathscr{J} \omega_j \omega_{j+1}$. Hence by (2.42)

$$-\beta\psi(\beta, 0) = \lim_{n\to\infty} \frac{1}{n} \log Z(n, \beta) = \sup_{P \in \mathscr{M}_s(\Omega)} \left\{\beta \mathscr{J} \int_\Omega \omega_1 \omega_2 P(d\omega) - I_\rho^{(3)}(P)\right\}.$$

This gives the Gibbs variational formula since $\int_\Omega \omega_1 \omega_2 dP = \int_\Omega \omega_0 \omega_1 \, dP$.[15] We now prove the Gibbs variational formula for any finite-range interaction J. We write

$$\begin{aligned} \frac{1}{2} \sum_{i,j=1}^{n} J(i-j)\omega_i\omega_j &= \frac{n}{2} J(0) + \sum_{1 \le j < i \le n} J(i-j)\omega_i\omega_j \\ &= \frac{n}{2} J(0) + \sum_{k=1}^{n-1} J(k) \sum_{j=1}^{n-k} \omega_j \omega_{k+j}. \end{aligned}$$

*The set of maximizing measures for $\sup_{P \in \mathscr{M}_s(\Omega)} \{-u(0; P)\}$ may contain other measures besides $\{P^{(\lambda)}, 0 \le \lambda \le 1\}$ (depending on J).

Thus if J has range L and $n > L$ then

$$Z(n,\beta) = \int_\Omega \exp\left\{\beta\left[\frac{n}{2}J(0) + \sum_{k=1}^{L} J(k)\sum_{j=1}^{n-k}\omega_j\omega_{k+j}\right]\right\} P_\rho(d\omega).$$

Let $R_n(\omega,\cdot)$ denote the empirical process $n^{-1}\sum_{k=0}^{n-1}\delta_{T^kY(n,\omega)}(\cdot)$, where T is the shift mapping on Ω and $Y(n,\omega)$ is the periodic point in Ω obtained by repeating $(Y_1(\omega), Y_2(\omega), \ldots, Y_n(\omega))$ periodically $(Y_j(\omega) = \omega_j)$. We have

$$\int_\Omega \bar{\omega}_1\bar{\omega}_{k+1}R_n(\omega,d\bar{\omega}) = \frac{1}{n}\sum_{j=1}^{n-k}\omega_j\omega_{k+j} + \frac{1}{n}\times k \text{ cyclic terms},$$

where the k cyclic terms are $\omega_j\omega_{k+j-n}$, $j = n-k+1, \ldots, n$. Thus uniformly in ω

$$\left| n\sum_{k=1}^{L} J(k)\int_\Omega \bar{\omega}_1\bar{\omega}_{k+1}R_n(\omega,d\bar{\omega}) - \sum_{k=1}^{L}J(k)\sum_{j=1}^{n-k}\omega_j\omega_{k+j}\right| = O(L).$$

By the comparison lemma, Lemma II.7.4, $Z(n,\beta)$ has the same leading order asymptotic behavior as

$$\bar{Z}(n,\beta) = \int_\Omega \exp\left\{n\beta\left[\frac{1}{2}J(0) + \sum_{k=1}^{L}J(k)\int_\Omega \bar{\omega}_1\bar{\omega}_{k+1}R_n(\omega,d\bar{\omega})\right]\right\} P_\rho(d\omega)$$

$$= \int_{\mathscr{M}_s(\Omega)} \exp[-n\beta u(0;P)]Q_n^{(3)}(dP),$$

where $Q_n^{(3)}$ is the distribution of $R_n(\omega,\cdot)$ on $\mathscr{M}_s(\Omega)$. By Theorem IX.1.1, $\{Q_n^{(3)}\}$ has a large deviation property with $a_n = n$ and entropy function $I_\rho^{(3)}$. Varadhan's theorem, Theorem II.7.1, yields the Gibbs variational formula:

$$-\beta\psi(\beta,0) = \lim_{n\to\infty}\frac{1}{n}\log\bar{Z}(n,\beta) = \sup_{P\in\mathscr{M}_s(\Omega)}\{-\beta u(0;P) - I_\rho^{(3)}(P)\}.$$

We finally prove the Gibbs variational formula for an infinite-range, summable interaction J. Take $L > 0$ and define $J_L(k) = J(k)$ for $|k| \le L$ and $J_L(k) = 0$ for $|k| > L$. We use the result just proved for the finite-range interaction J_L. By taking L sufficiently large, we can make the quantities $-\beta\psi(\beta,0)$ and $\sup_{P\in\mathscr{M}_s(\Omega)}\{-\beta u(0,P) - I_\rho^{(3)}(P)\}$ corresponding to J_L arbitrarily close to the respective quantities corresponding to J. This completes the proof of Theorem IV.7.3(a).

For the general ferromagnet, the translation invariant infinite-volume Gibbs states are also characterized by an entropy principle which is equivalent to the Gibbs variational principle. The entropy principle involves an energy constraint, expressed in terms of the specific energy $u(h;P)$ in the state $P\in\mathscr{M}_s(\Omega)$. For each $\beta > 0$ and h real, define the set

$$A_{\beta,h} = \{u\in\mathbb{R}: u = u(h;P) \text{ for some } P\in\mathscr{G}_{\beta,h}\cap\mathscr{M}_s(\Omega)\}.$$

For $\beta > 0$, $h \neq 0$ and $0 < \beta < \beta_c$, $h = 0$, $\mathscr{G}_{\beta,h}\cap\mathscr{M}_s(\Omega)$ consists of the unique measure $P_{\beta,h}$, and so $A_{\beta,h}$ is the single point $\{u(h; P_{\beta,h})\}$. The situation for $\beta > \beta_c$, $h = 0$ can be different since it is possible for $\mathcal{G}_{\beta,0} \cap \mathcal{M}_s(\Omega)$ to contain two measures P_1 and P_2 for which $u(0; P_1) \neq u(0, P_2)$. Since $u(h; P)$ is a continuous functional of P, $A_{\beta,0}$ is a compact interval of $\mathbb{R}$. The next theorem generalizes the entropy principle in Theorem IV.3.3, which characterized the finite-volume Gibbs state $P_{\Lambda,\beta,h}$.[16]

Theorem IV.7.4. *Let J be a summable ferromagnetic interaction on $\mathbb{Z}$. For $\beta > 0$, h real, and $u \in A_{\beta,h}$ let $\mathcal{I}_{\beta,h,u}$ denote the subset of $\{P \in \mathcal{M}_s(\Omega): u(h; P) = u\}$ at which $\inf\{I_\rho^{(3)}(P): P \in \mathcal{M}_s(\Omega), u(h; P) = u\}$ is attained. Then the following conclusions hold.*

(a) $\bigcup_{u \in A_{\beta,h}} \mathcal{I}_{\beta,h,u} = \mathcal{G}_{\beta,h} \cap \mathcal{M}_s(\Omega)$.

(b) *For $\beta > \beta_c$, $h = 0$, and $u = u(0; P_{\beta,0,+})$, the set $\mathcal{I}_{\beta,0,u}$ contains (at least) all the infinite-volume Gibbs states $P_{\beta,0}^{(\lambda)} = \lambda P_{\beta,0,+} + (1 - \lambda) P_{\beta,0,-}$, $0 \le \lambda \le 1$. Thus $\inf\{I_\rho^{(3)}(P): P \in \mathcal{M}_s(\Omega), u(0; P) = u(0; P_{\beta,0,+})\}$ is not attained at a unique measure.*

Proof. (a) Given $\bar{P} \in \mathcal{G}_{\beta,h} \cap \mathcal{M}_s(\Omega)$, set $u = u(h; \bar{P})$. Then by Theorem IV.7.3

$$-\beta\psi(\beta, h) = -\beta u - I_\rho^{(3)}(\bar{P}) \le -\beta u - \inf\{I_\rho^{(3)}(P): P \in \mathcal{M}_s(\Omega), u(h; P) = u\}$$

$$\le \sup_{P \in \mathcal{M}_s(\Omega)} \{-\beta u(h; P) - I_\rho^{(3)}(P)\} = -\beta\psi(\beta, h). \tag{4.65}$$

This implies that $\bar{P} \in \mathcal{I}_{\beta,h,u}$. Now assume that $\bar{P} \in \mathcal{I}_{\beta,h,u}$, some $u \in A_{\beta,h,}$ does not belong to $\mathcal{G}_{\beta,h} \cap \mathcal{M}_s(\Omega)$. Then for any $P_0 \in \mathcal{G}_{\beta,h} \cap \mathcal{M}_s(\Omega)$ with $u(h; P_0) = u$

$$-\beta u - I_\rho^{(3)}(\bar{P}) = -\beta u(h; \bar{P}) - I_\rho^{(3)}(\bar{P})$$
$$< -\beta\psi(\beta, h) = -\beta u(h; P_0) - I_\rho^{(3)}(P_0) = -\beta u - I_\rho^{(3)}(P_0).$$

Thus $I_\rho^{(3)}(P_0) < I_\rho^{(3)}(\bar{P})$. This contradicts the hypothesis that $\bar{P}$ belongs to $\mathcal{I}_{\beta,h,u}$. Part (a) is proved.

(b) By symmetry $u(0; P_{\beta,0,+})$ equals $u(0; P_{\beta,0,-})$. Hence for all $0 \le \lambda \le 1$, $u(0; P_{\beta,0}^{(\lambda)})$ equals $u(0; P_{\beta,0,+})$. Since $P_{\beta,0}^{(\lambda)}$ belongs to $\mathcal{G}_{\beta,h} \cap \mathcal{M}_s(\Omega)$, the proof of part (a) shows that $P_{\beta,0}^{(\lambda)}$ belongs to $\mathcal{I}_{\beta,h,u(0;P_{\beta,0,+})}$. □

The nonuniqueness property of $I_\rho^{(3)}$ expressed in part (b) of Theorem IV.7.4 is a level-3 analog of a nonuniqueness property of the function $I_{\beta,0}(z)$, $\beta > \beta_c$, discussed at the end of Section IV.5. The set of means $\{\int_\Omega \omega_0 P_{\beta,0}^{(\lambda)}(d\omega); 0 \le \lambda \le 1\}$ of the measures $\{P_{\beta,0}^{(\lambda)}; 0 \le \lambda \le 1\}$ is exactly the interval $[m(\beta, -), m(\beta, +)]$ on which $I_{\beta,0}(z)$ attains its infimum of 0.

One of the main results in this chapter is that if β_c if finite, then for $\beta > \beta_c$ and $h = 0$ there exist nonunique, translation invariant infinite-volume Gibbs states $P_{\beta,0,+}$ and $P_{\beta,0,+}$. In the next chapter we will study ferromagnetic models on the lattices $\mathbb{Z}^D$, $D \in \{2, 3, \dots\}$. These models share many features with the models on $\mathbb{Z}$. An important contrast is the fact that β_c is finite for any nontrivial interaction.

IV.8. Notes

Many of these notes apply with little or no change to ferromagnetic models on $\mathbb{Z}^D$, $D \in \{1, 2, \ldots\}$. These models will be treated in the next chapter. The notes which do apply are so indicated.

1 (page 88) ($\mathbb{Z}^D, D \geq 1$). The similarities between the phase transitions for liquid–gas systems and ferromagnetic systems are discussed by Stanley (1971, Chapters 1 and 2). A model that is based on analogies between the two kinds of systems is the lattice–gas model [Lee and Yang (1952), Stanley (1971, Appendix A)].

2 (page 89) ($\mathbb{Z}^D, D \geq 1$) (a) Introductions to ferromagnetic models and related lattice systems can be found in Griffiths (1971, 1972), Spitzer (1971), Georgii (1972), Thompson (1972), Kindermann and Snell (1980), and Gross (1982). Ruelle (1969, 1978), Lanford (1973), Preston (1974b, 1976), Israel (1979), and Simon (1985) are advanced references. Wightman (1979) is a beautiful overview of the thermodynamics of phase transitions (based on Gibbs's geometric approach through convexity) and the mathematics of lattice systems. Lebowitz (1975) is a useful review of properties of ferromagnetic models. Also see Gallavotti (1972a).

(b) A rough sketch of the Ising model first appeared in a 1920 paper of Lenz, but the model was named after his student, E. Ising. Ising (1925) concluded that there is no phase transition for $D = 1$ but erroneously tried to generalize his argument to $D = 2$. Brush (1967) discusses the history of the model.

3 (page 93) ($\mathbb{Z}^D, D \geq 1$) The discussion of correlations in Section IV.2 is based upon unpublished lecture notes of A. Sokal and upon Wilson (1979).

4 (page 94) ($\mathbb{Z}^D, D \geq 1$) The divergence of the specific magnetic susceptibility $\chi(\beta, 0) = \partial m(\beta, 0)/\partial h$ at $\beta = \beta_c$ is related, in a liquid-gas system, to the phenomenon of critical opalescence, which is the strong scattering of light by the system at the critical point [Stanley (1971, Chapter 1)]. The strong scattering is caused by abnormally large density fluctuations in the system.

5 (page 95) ($\mathbb{Z}^D, D \geq 1$) (a) A useful generalization of the finite-volume Gibbs state $P_{\Lambda,\beta,h}$ in (4.4) is to allow many-body interactions [see Appendix C.3]. A second generalization is to replace the measure $\rho = \frac{1}{2}\delta_1 + \frac{1}{2}\delta_{-1}$ by a nondegenerate symmetric probability measure ρ on $\mathbb{R}$ for which $\int_{\mathbb{R}} e^{\sigma x^2} \rho(dx)$ is finite for all $\sigma > 0$. The measure ρ is called a *single-site distribution*. For example, see Lebowitz and Presutti (1976), Newman (1976a, 1976c), Ruelle (1976), Sylvester (1976b), Cassandro et al. (1978), and van Beijeren and Sylvester (1978). A third generalization is to allow vector-valued spins. For example, the d-vector spin model corresponds to spins taking values in the surface of the unit sphere in $\mathbb{R}^d$, $d \in \{2, 3, \ldots\}$; the single-site distribution is $\rho(dx) = \delta_1(\|x\|)$, $x \in \mathbb{R}^d$. The Heisenberg model is the case $d = 3$.

(b) Ferromagnetic models have been applied in a number of different areas. A useful technique for studying quantum fields is first to study the fields on a lattice, then to let the lattice spacing shrink to zero. The lattice approximations are ferromagnetic models whose single-site distributions have the form const $\cdot \exp(-P(x))dx$, $P(x)$ an even polynomial [Simon (1974), Guerra, Rosen, and Simon (1975), Rosen (1977), Glimm and Jaffe (1981)]. Stochastic models closely related to spin systems are studied in percolation theory [Kesten (1982), Aizenman and Newman (1984), Durrett (1984a, 1984b)] and in the theory of interacting particle systems [Spitzer (1970), Holley and Stroock (1976a, 1976b, 1977), Liggett (1977, 1985), Griffeath (1978), Durrett (1981)].

6 (page 97) $(\mathbb{Z}^D, D \geq 1)$ In order to avoid the nonphysical values $\beta \leq 0$ in Theorem IV.3.3, restrict U to the interval $(U_{\min}, U_0)$, where $U_0 = \lim_{\beta \to 0^+} U(\Lambda, h; P_{\Lambda,\beta,h}) = \int_{\Omega_\Lambda} H_{\Lambda,h}(\omega)\pi_\Lambda P_\rho(d\omega) = -\frac{1}{2}J(0)|\Lambda|$.

7 (page 98) $(\mathbb{Z}^D, D \geq 1)$ The Curie–Weiss model is discussed in Kac (1968) and in Thompson (1972). It is also known as the Husimi–Temperley model [Husimi (1953), Temperley (1954)].

8 (page 106) $(\mathbb{Z}^D, D \geq 1)$ The first Curie–Weiss bounds on β_c and $m(\beta, h)$ were found by Fisher (1967a), Griffiths (1967c), and Thompson (1971). For subsequent work on such bounds, see Cassandro et al. (1978), Simon (1980b), Pearce (1981), Sokal (1982a), Slawny (1983), and the references listed in these papers. Newman (1981b) proves the bound $m(\beta, h) \leq m^{\mathrm{CW}}(\beta, h)$ for $\beta > 0$ and $h \geq 0$ by using a connection between the Curie–Weiss model and Burger's equation.

9 (page 107) $(\mathbb{Z}^D, D \geq 1)$ Whether $m(\beta_c, +)$ equals 0 or is positive depends on the model [Lebowitz and Martin-Löf (1972, page 282)]. The first holds for the Ising model on $\mathbb{Z}^2$ [see (5.18)] while the second is believed to hold for the model on $\mathbb{Z}$ with $J(k) = k^{-2}, k \neq 0$ [see Note 10c]. In general, the value of $m(\beta_c, +)$ is not known.

10 (page 107) ($\mathbb{Z}$ only) (a) Simon and Sokal (1981) have an entropy-energy proof of the fact that $\sum_{k \in \mathbb{Z}} |k| J(k) < \infty$ implies that β_c is infinite. Dobrushin (1968c), Ruelle (1968), and Bricmont, Lebowitz, and Pfister (1979) show by different methods that if $\sum_{k \in \mathbb{Z}} |k| J(k)$ is finite, then there exists a unique infinite-volume Gibbs state for all $\beta > 0$ and h real. This implies that β_c is infinite [Theorem IV.6.5]. The latter three papers prove analogous results for interactions J of arbitrary sign.

(b) We show that β_c is finite if $J > 0$ and $J(k) \sim |k|^{-\alpha}$, some $1 < \alpha < 2$ [Theorem IV.5.3(c)]. There exist $b > 0$ and $1 < \gamma < 2$ such that $J(k) \geq b|k|^{-\gamma}$ for all $k \neq 0$. By Theorem V.4.3(f), β_c corresponding to J is less than or equal to β_c corresponding to the interaction $J(k) = b|k|^{-\gamma}$, $k \neq 0$. For the latter interaction, Dyson (1969a) proves that

$$\liminf_{|i-j| \to \infty} \lim_{\Lambda \uparrow \mathbb{Z}} \int_{\Omega_\Lambda} \omega_i \omega_j P_{\Lambda,\beta,0}(d\omega) > 0$$

for all sufficiently large β (long-range order).

Proposition IV.8.1. *If long-range order holds, then β_c is finite.*

Proof. By the GKS-2 inequality [see Remark V.4.1(a)], $\int_{\Omega_\Lambda} \omega_i \omega_j P_{\Lambda,\beta,0}(d\omega) \le \int_{\Omega_\Lambda} \omega_i \omega_j P_{\Lambda,\beta,0,+}(d\omega)$, where $P_{\Lambda,\beta,0,+}$ is the finite-volume Gibbs state with the plus external condition. By Theorem IV.6.5(a), Lemma IV.6.10(b), and Corollary A.11.12(b),

$$\lim_{|i-j|\to\infty} \lim_{\Lambda\uparrow\mathbb{Z}} \int_{\Omega_\Lambda} \omega_i \omega_j P_{\Lambda,\beta,0,+}(d\omega) = [m(\beta,+)]^2.$$

Thus $0 < \liminf_{|i-j|\to\infty} \lim_{\Lambda\uparrow\mathbb{Z}} \int_{\Omega_\Lambda} \omega_i \omega_j P_{\Lambda,\beta,0}(d\omega) \le [m(\beta,+)]^2$ for all sufficiently large β. It follows that β_c is finite. □

(c) The phase transition for the interaction $J(k) = k^{-2}$ $(k \ne 0)$ [Fröhlich and Spencer (1982a)] is believed to be an unusual kind. Namely, $m(\beta,+)$ is a discontinuous function of β at $\beta = \beta_c$: $m(\beta,+) \ge \text{const} > 0$ for $\beta \ge \beta_c$ and $m(\beta,+) = 0$ for $0 < \beta < \beta_c$. This was first discussed by Thouless (1969) and proved in a related hierarchical model by Dyson (1971). Also see Simon and Sokal (1981) and Sokal (1982b).

(d) There is a large literature concerning models on $\mathbb{Z}$. For example, see Dyson (1969b, 1972), Dobrushin (1973), Kolomytsev and Rokhlenko (1978, 1979), Cassandro and Olivieri (1981), Rogers and Thompson (1981), Simon (1981), and Imbrie (1982).

11 (page 115) (a) $(\mathbb{Z}^D, D \ge 1)$. The proof that the measures $P_{\beta,h,+}$ and $P_{\beta,h,-}$ are ergodic follows Slawny (1974, page 300). The latter paper uses the GKS-2 inequality instead of the FKG inequality. The rest of Theorem IV.6.5 and Lemmas IV.6.9–IV.6.10 are due to Lebowitz and Martin-Löf (1972).

(b) ($\mathbb{Z}$ only). Fannes, Vanheuverzwijn, and Verbeure (1982) prove that if $J(k)$ is monotonically decreasing for k sufficiently large (e.g., $J(k) = |k|^{-\alpha}$, $k \ne 0$, for $1 < \alpha \le 2$), then every infinite-volume Gibbs state is translation invariant. The proof is based on energy-entropy estimates.

(c) $(\mathbb{Z}^D, D \ge 1)$. The fact that $\mathscr{G}_{\beta,h}$ consists of a unique measure for all sufficiently small β [Theorem IV.6.5(c)] also follows from a general uniqueness theorem of Dobrushin (1968a). See Lanford (1973, Section C2) and Simon (1979b).

(d) $(\mathbb{Z}^D, D \ge 1)$. Part (e) of Theorem IV.6.5 can be strengthened. The following theorem is due to Lebowitz (1977).

Theorem IV.8.2. *Let J be a summable ferromagnetic interaction which is irreducible on $\mathbb{Z}$* [see page 96]. *Pick $\beta \ge \beta_c$ such that $\partial(\beta\psi(\beta,0))/\partial\beta$ exists. Then $\mathscr{G}_{\beta,0} \cap \mathscr{M}_s(\Omega)$ consists precisely of all the measures $P_{\beta,0}^{(\lambda)} = \lambda P_{\beta,0,+} + (1-\lambda)P_{\beta,0,-}$, $0 \le \lambda \le 1$; $\partial(\beta\psi(\beta,0))/\partial\beta$ exists for all but at most countably many values of $\beta \ge \beta_c$.*

The quantity $\partial(\beta\psi(\beta,0))/\partial\beta$ is called the *specific energy* [Problem IV.9.11].

12 (page 116) $(\mathbb{Z}^D, D \ge 1)$ Symmetry breaking is discussed further in Glimm and Jaffe (1981, Section 5.3).

13 (page 117) Here is an interesting open problem. According to Theorem II.6.1(b), for $\beta > \beta_c$ and any $\varepsilon > 0$, $P_{\beta,0,+}\{S_\Lambda/|\Lambda| \geq m(\beta,+) + \varepsilon\}$ converges to 0 exponentially fast as $\Lambda \uparrow \mathbb{Z}$. By Theorem IV.6.6(b), for $\beta > \beta_c$ and $0 < \varepsilon < 2\mathrm{m}(\beta,+)$, $P_{\beta,0,+}\{S_\Lambda/|\Lambda| \leq m(\beta,+) - \varepsilon\}$ converges to 0 but not exponentially fast. What is the decay rate of these probabilities?

14 (page 118) The FKG inequality originated in work on percolation models [Harris (1960)] and has been generalized and applied in many ways. See Battle and Rosen (1980), Newman (1980, 1984), Eaton (1982), Graham (1983), and the references listed in these papers.

15 (page 128) Let $\Gamma = \{1, -1\}$ and define $\mathcal{M}_s(\Gamma^2)$ to be the subset of $\mathcal{M}(\Gamma^2)$ consisting of probability measures τ with equal one-dimensional marginals. Clearly, if P belongs to $\mathcal{M}_s(\Omega)$, then $\tau = \pi_2 P$ belongs to $\mathcal{M}_s(\Gamma^2)$. For the Ising model on $\mathbb{Z}$, we have

$$\begin{aligned} -\beta\psi(\beta,0) &= \sup_{P\in\mathcal{M}_s(\Omega)} \left\{\beta\mathcal{J}\int_\Omega \omega_1\omega_2 P(d\omega) - I_\rho^{(3)}(P)\right\} \\ &= \sup_{\tau\in\mathcal{M}_s(\Gamma^2)} \left\{\beta\mathcal{J}\int_{\Gamma^2} \omega_1\omega_2 \tau(d\omega) \right. \\ &\qquad \left. - \inf\{I_\rho^{(3)}(P): P\in\mathcal{M}_s(\Omega), \pi_2 P = \tau\}\right\}. \end{aligned}$$

By Theorem IX.3.3, $\inf\{I_\rho^{(3)}(P): P\in\mathcal{M}_s(\Omega), \pi_2 P = \tau\}$ equals the function $I_{\rho,2}^{(3)}(\tau)$ defined in (9.6). Hence,

$$(4.66)\qquad -\beta\psi(\beta,0) = \sup_{\tau\in\mathcal{M}_s(\Gamma^2)} \left\{\beta\mathcal{J}\int_{\Gamma^2} \omega_1\omega_2 \tau(d\omega) - I_{\rho,2}^{(3)}(\tau)\right\}.$$

The latter can be derived directly if one expresses the partition function in terms of the empirical pair measure and uses Theorem IX.4.3 for $\alpha = 2$. Equation (4.31) in Section IV.5 gives another formula for $-\beta\psi(\beta,0)$ in terms of the larger eigenvalue of a 2×2 positive matrix. Theorem IX.4.4 shows the equality of the expressions for $-\beta\psi(\beta,0)$ given in (4.31) and (4.66). Any finite-range interaction on $\mathbb{Z}$ can be handled like the Ising model on $\mathbb{Z}$ [Appendix C.6].

16 (page 130) ($\mathbb{Z}^D, D \geq 1$) There is a contraction principle related to the entropy principle in Theorem IV.7.4. The function

$$\bar{I}_\rho^{(1)}(h;u) = \inf\{I_\rho^{(3)}(P): P\in\mathcal{M}_s(\Omega), u(h;P) = u\}$$

is the entropy function of the $\pi_\Lambda P_\rho$-distributions of $\{H_{\Lambda,h}/|\Lambda|\}$. The function $-\bar{I}_\rho^{(1)}(h;u)$ is called the *specific microcanonical entropy*. See Lanford (1973, Chapter B), Aizenman and Lieb (1981), and Simon (1985).

IV.9. Problems

Many of these problems extend with little or no change to ferromagnetic models on $\mathbb{Z}^D$, $D\in\{1,2,\ldots\}$. These models will be treated in the next chapter. The problems which do extend are so indicated.

IV.9.1. ($\mathbb{Z}^D, D \geq 1$). Prove Proposition IV.3.2.

IV.9.2. ($\mathbb{Z}^D, D \geq 1$). Prove that if $h > 0$, then as $\beta \to \infty$ the finite-volume Gibbs states $\{P_{\Lambda,\beta,0}; \beta > 0\}$ converge weakly to the unit point measure $\delta_{\bar{\omega}_+}$; $\bar{\omega}_+$ is the configuration in Ω_Λ defined by $\bar{\omega}_{+,j} = 1$ for each $j \in \Lambda$.

The next four problems concern the Curie–Weiss model [Section IV.4].

IV.9.3. Verify formula (4.12) and the table accompanying Figure IV.3.

IV.9.4 [Kac (1968, page 247)]. This problem shows how to derive the limit (4.19) without using large deviations.

(a) By substituting in (4.11) the identity

$$\exp(\tfrac{1}{2}y^2) = \frac{1}{\sqrt{2\pi}} \int_{\mathbb{R}} \exp(ty - \tfrac{1}{2}t^2)dt, \qquad y = \sqrt{n\beta \mathscr{J}_0}\, z, \tag{4.66}$$

and carrying out the $Q_n^{(1)}(dz)$ integration, prove that

$$\lim_{n\to\infty} \frac{1}{n} \log Z(n, \beta, h) = \sup_{t \in \mathbb{R}} \left\{ \log \cosh t - \frac{(t - \beta h)^2}{2\beta \mathscr{J}_0} \right\}.$$

(b) Using Problem VI.7.14, prove that

$$\sup_{t \in \mathbb{R}} \{\log \cosh t - \tfrac{1}{2}(\beta \mathscr{J}_0)^{-1}(t - \beta h)^2\} = \sup_{z \in \mathbb{R}} \{\tfrac{1}{2}\beta \mathscr{J}_0 z^2 + \beta h z - I_\rho^{(1)}(z)\}.$$

IV.9.5. Equation (4.19) shows that for the Curie–Weiss model, the specific Gibbs free energy $\psi(\beta, h)$ equals $-\beta^{-1} \sup_{z \in \mathbb{R}} \{\frac{1}{2}\beta \mathscr{J}_0 z^2 + \beta h z - I_\rho^{(1)}(z)\}$.

(a) For $\beta > 0$ and $h \neq 0$, prove that $\partial\psi(\beta, h)/\partial h$ exists and $-\partial\psi(\beta, h)/\partial h = z(\beta, h)$, where $z(\beta, h)$ is specified in the table accompanying Figure IV.3.

(b) For $\beta > 0$ and $h > 0$, prove that $(I_\rho^{(1)})''(z(\beta, h)) > \beta \mathscr{J}_0$.

(c) For $\beta > 0$ and $h > 0$, prove that $\partial z(\beta, h)/\partial h > 0$, $\partial z(\beta, h)/\partial \beta > 0$, and $\partial^2 z(\beta, h)/\partial h^2 < 0$.

(d) Verify the conclusions of Theorems IV.5.1 and IV.5.2 for $\psi(\beta, h)$ and for the Curie–Weiss specific magnetization.

IV.9.6. [Ellis (1981)]. Let $c_{\beta,h}(t) = \lim_{n\to\infty} n^{-1} \log \int_{\Omega_n} \exp[tS_n(\omega)] P_{n,\beta,h}(d\omega)$, where $P_{n,\beta,h}$ is defined in (4.9). The function $c_{\beta,h}$ is the free energy function of the sequence $\{S_n; n = 1, 2, \ldots\}$ for the Curie–Weiss model.

(a) Prove that $c_{\beta,h}(t) = \sup_{z \in \mathbb{R}} \{tz - i_{\beta,h}(z)\} + \inf_{z \in \mathbb{R}} i_{\beta,h}(z)$, where $i_{\beta,h}(z) = I_\rho^{(1)}(z) - (\frac{1}{2}\beta \mathscr{J}_0 z^2 + \beta h z)$.

(b) $c'_{\beta,h}(0)$ exists for $\beta > 0$, $h \neq 0$ and $0 < \beta \leq \mathscr{J}_0^{-1}$, $h = 0$, but $c'_{\beta,h}(0)$ does not exist for $\beta > \mathscr{J}_0^{-1}$, $h = 0$. Prove this statement first by explicit calculation and then by applying Theorems II.6.3 and IV.4.1.

(c) Evaluate the Legendre–Fenchel transform of $c_{\beta,h}$. What is the relationship between the Legendre–Fenchel transform and the function $I_{\beta,h}$ in (4.17)? [*Hint:* Theorem VI.5.8.]

The remaining problems concern general ferromagnets on $\mathbb{Z}$.

IV.9.7 ($\mathbb{Z}^D, D \geq 1$) [Pearce (1981)]. Let J be a summable ferromagnetic interaction on $\mathbb{Z}$ and set $\mathscr{J}_0 = \sum_{k \in \mathbb{Z}} J(k)$. Let $m(\beta, h)$ be the specific magnetization corresponding to J and $m^{\mathrm{CW}}(\beta, h)$ the specific magnetization for the Curie–Weiss model with interaction $\mathscr{J}_0/|\Lambda|$. For $h = 0$, $m^{\mathrm{CW}}(\beta, h)$ equals 0 and for $h > 0$, $m^{\mathrm{CW}}(\beta, h)$ equals the unique positive root $z(\beta, h)$ of (4.14). The present problem shows that for $\beta > 0$ and $h \geq 0$, $m^{\mathrm{CW}}(\beta, h) \geq m(\beta, h)$ and thus that $\beta_c \geq \mathscr{J}_0^{-1}$ [Theorem IV.5.3(a)].

Let Λ be a symmetric interval. For each $i, j \in \Lambda$, define $J_\Lambda(i-j) = \sum_{k \in \mathbb{Z}} J(i - j - k|\Lambda|)$. Fix $\beta > 0$ and $h \geq 0$. Let $P_{\Lambda,\beta,h}$ be the finite-volume Gibbs state corresponding to J and $P_{\Lambda,\beta,h,p}$ the finite-volume Gibbs state corresponding to J_Λ (p stands for the periodic boundary condition). For $i \in \Lambda$, define

$$\langle \omega_i \rangle_\Lambda = \int_{\Omega_\Lambda} \omega_i P_{\Lambda,\beta,h}(d\omega) \text{ and } \langle \omega_i \rangle_{\Lambda,p} = \int_{\Omega_\Lambda} \omega_i P_{\Lambda,\beta,h,p}(d\omega).$$

Since $h \geq 0$ and $J_{ij}^{(\Lambda)} \geq J(i-j)$, Table V.1(c) implies $\langle \omega_i \rangle_{\Lambda,p} \geq \langle \omega_i \rangle_\Lambda$ [page 147].

(a) Show that the sum $\sum_{j \in \Lambda} J_\Lambda(i-j)$ is independent of i and Λ and equals $\mathscr{J}_0 = \sum_{k \in \mathbb{Z}} J(k)$.

(b) Denote by $\langle - \rangle_0$ expectation with respect to $\exp(\beta(\mathscr{J}_0 z + h)\sum_{j \in \Lambda} \omega_j) \cdot \pi_\Lambda P_\rho(d\omega) \cdot (1/Z_0)$, where Z_0 is a normalization. Prove that

$$z - \langle \omega_i \rangle_{\Lambda,p} = \left\langle (z - \omega_i) \exp\left\{\frac{\beta}{2} \sum_{i,j \in \Lambda} J_\Lambda(i-j)(z-\omega_i)(z-\omega_j)\right\}\right\rangle_0 \cdot \frac{1}{Z},$$

where $z = z(\beta, h)$ and $Z = \langle \exp\{\frac{\beta}{2}\sum_{i,j \in \Lambda} J_\Lambda(i-j)(z-\omega_i)(z-\omega_j)\}\rangle_0$.

(c) Prove that for each positive integer r and site $j \in \Lambda$

$$\int_{\{1,-1\}} (z - \omega_j)^r \exp[\beta(\mathscr{J}_0 z + h)\omega_j]\rho(d\omega_j)$$
$$= \tfrac{1}{2}(z+1)^r \exp[\beta(\mathscr{J}_0 z + h)]\{a(\beta, h) + (-1)^r a(\beta, h)^r\}$$

is positive, where $z = z(\beta, h)$ and $a(\beta, h) = \exp[-2\beta(\mathscr{J}_0 z + h)]$. [*Hint:* Use (4.14).]

(d) By expanding the exponential in part (b) and using part (c), prove that $z(\beta, h) > \langle \omega_i \rangle_{\Lambda,p} \geq \langle \omega_i \rangle_\Lambda$. Deduce that

$$m^{\mathrm{CW}}(\beta, h) \geq m(\beta, h) = \lim_{\Lambda \uparrow \mathbb{Z}} \frac{1}{|\Lambda|} \sum_{i \in \Lambda} \langle \omega_i \rangle_\Lambda \quad \text{and} \quad \beta_c \geq \mathscr{J}_0^{-1}.$$

IV.9.8 ($\mathbb{Z}$ only) [Sakai (1976), Bricmont, Lebowitz, and Pfister (1979)]. Let J be a summable ferromagnetic interaction on $\mathbb{Z}$ for which $\sum_{k \in \mathbb{Z}} |k| J(k)$ is finite. This problem shows that for all $\beta > 0$ and $h = 0$ there exists a unique translation invariant infinite-volume Gibbs state $P_{\beta,0}$. Theorem IV.6.5 implies that β_c equals ∞, thus proving Theorem IV.5.3(b).

By Lemma IV.6.9(c), it suffices to prove that $P_{\beta,0,+}$ equals $P_{\beta,0,-}$ for all $\beta > 0$.

(a) Let Σ be any cylinder set in Ω. Prove that for all sufficiently large symmetric intervals Λ

$$P_{\Lambda,\beta,0,-}\{\Sigma\} \leq \exp[4\beta \sum_{k\in\mathbb{Z}} |k|J(k)] \cdot P_{\Lambda,\beta,0,+}\{\Sigma\}.$$

[*Hint:* First prove the bound for cylinder sets of the form $\{\omega\in\Omega: \omega_i = \bar{\omega}_i$ for each $i\in\bar{\Lambda}\}$, where $\bar{\Lambda}$ is a symmetric interval and $\bar{\omega}$ is a configuration in $\Omega_{\bar{\Lambda}}$.]

(b) By Theorem IV.6.5(a), $P_{\Lambda,\beta,0,+} \Rightarrow P_{\beta,0,+}$ and $P_{\Lambda,\beta,0,-} \Rightarrow P_{\beta,0,-}$ as $\Lambda \uparrow \mathbb{Z}$. Deduce a contradiction to part (a) if $P_{\beta,0,+} \neq P_{\beta,0,-}$ [*Hint:* Theorem A.11.7(b)].

IV.9.9. ($\mathbb{Z}^D, D \geq 1$) Prove (4.38). [*Hint:* Suppose that $f\in\mathscr{C}(\Omega)$ depends only on the coordinates $\omega_{i_1}, \ldots, \omega_{i_r}$. If Λ contains the sites $i_1, \ldots, i_r$, then f can be written in the form $\sum_{i=1}^{s} a_i\chi_{\Sigma_i}$, where s is a positive integer, $a_1, \ldots, a_s$ are real numbers, and $\Sigma_1, \ldots, \Sigma_s$ are cylinder sets such that $\pi_\Lambda(\Sigma_i \cap B_{\Lambda,\tilde{\omega}}) = \pi_\Lambda\Sigma_i$.]

IV.9.10. ($\mathbb{Z}$ only). Fill in the details in the proof of the Gibbs variational formula for an infinite-range, summable interaction on $\mathbb{Z}$ and for an arbitrary value of h.

IV.9.11. ($\mathbb{Z}^D, D \geq 1$). Let J be a summable ferromagnetic interaction on $\mathbb{Z}$. $H_{\Lambda,h}$ denotes the Hamiltonian defined in (4.3); $P_{\Lambda,\beta,h}$ the corresponding finite-volume Gibbs state defined in (4.4); $Z(\Lambda,\beta,h)$ the partition function $\int_{\Omega_\Lambda} \exp[-\beta H_{\Lambda,h}(\omega)]\pi_\Lambda P_\rho(d\omega)$; $\Psi(\Lambda,\beta,h)$ the Gibbs free energy $-\beta^{-1}\log Z(\Lambda,\beta,h)$, and $\psi(\beta,h)$ the specific Gibbs free energy $\lim_{\Lambda\uparrow\mathbb{Z}} |\Lambda|^{-1}\Psi(\Lambda,\beta,h)$.

(a) Prove that $\beta\Psi(\Lambda,\beta,h)$ is a concave function of $\beta > 0$ and that

$$\frac{\partial(\beta\Psi(\Lambda,\beta,h))}{\partial\beta} = \int_{\Omega_\Lambda} H_{\Lambda,h}(\omega)P_{\Lambda,\beta,h}(d\omega).$$

(b) Prove that if $\partial(\beta\psi(\beta,h))/\partial\beta$ exists, then

$$\lim_{\Lambda\uparrow\mathbb{Z}} \frac{1}{|\Lambda|}\int_{\Omega_\Lambda} H_{\Lambda,h}(\omega)P_{\Lambda,\beta,h}(d\omega) = \frac{\partial(\beta\psi(\beta,h))}{\partial\beta}.$$

The limit is called the *specific energy*.

Chapter V

Magnetic Models on $\mathbb{Z}^D$ and on the Circle

V.1. Introduction

This chapter consists of two parts. Part 1 extends the results of Chapter IV to ferromagnetic models on the integer lattices $\mathbb{Z}^D$, $D \in \{1, 2, \ldots\}$. These results included properties of the specific Gibbs free energy, the specific magnetization, and infinite-volume Gibbs states. The dimension of the lattice enters in a dramatic way in the proof of spontaneous magnetization. For $D = 1$, the critical inverse temperature β_c is finite only for interactions of infinite range; e.g., $J(k) = |k|^{-\alpha}$, $k \neq 0$, for some $1 < \alpha \leq 2$. By contrast, for $D \geq 2$, β_c is finite for any nontrivial interaction J [Theorem V.5.1]. We give a complete proof of spontaneous magnetization for $D \geq 2$. First, a set of powerful moment inequalities is proved which allow us to reduce from a general model to the Ising model on $\mathbb{Z}^2$. Then spontaneous magnetization is shown for the latter by means of a combinatorial argument due to Peierls. The moment inequalities also yield monotonicity and concavity properties of the specific magnetization which were stated in Chapter IV without proof.

Part 2 of Chapter V presents new material, which includes a central limit theorem for the total spin and limit theorems for a magnetic model on the circle which is related to the Curie–Weiss model. Let $\mathscr{U}$ denote the set of points (β, h) with $\beta > 0, h \neq 0$ and $0 < \beta < \beta_c, h = 0$. We recall from Theorem IV.6.5 that for (β, h) in $\mathscr{U}$, there exists a unique infinite-volume Gibbs state $P_{\beta,h}$. The proof extends to models on $\mathbb{Z}^D$, $D \geq 2$, without change. For ω a point in the infinite-volume configuration space $\Omega = \{1, -1\}^{\mathbb{Z}^D}$, let $Y_k(\omega) = \omega_k$ be the spin random variable at site $k \in \mathbb{Z}^D$. We denote by S_Λ the total spin $\sum_{k \in \Lambda} Y_k$ in a symmetric hypercube Λ of $\mathbb{Z}^D$. Define the pair correlation

$$\langle Y_0 ; Y_k \rangle_{\beta,h} = \langle Y_0 Y_k \rangle_{\beta,h} - \langle Y_0 \rangle_{\beta,h} \langle Y_k \rangle_{\beta,h},$$

where $\langle - \rangle_{\beta,h}$ denote expectation with respect to $P_{\beta,h}$. By one of the moment inequalities proved in Part 1, $\langle Y_0 ; Y_k \rangle_{\beta,h}$ is positive. Under the assumption that the sum $\sigma^2(\beta, h) = \sum_{k \in \mathbb{Z}^D} \langle Y_0 ; Y_k \rangle_{\beta,h}$ is finite, S_Λ is shown to satisfy a central limit theorem: for each $f \in \mathscr{C}(\mathbb{R})$

$$\text{(5.1)} \qquad \lim_{\Lambda \uparrow \mathbb{Z}^D} \left\langle f\left(\frac{S_\Lambda - |\Lambda| m(\beta, h)}{\sqrt{|\Lambda|}}\right)\right\rangle_{\beta, h} = \int_{\mathbb{R}} f(x) N(0, \sigma^2(\beta, h))(dx),$$

where $N(0, \sigma^2)(dx)$ is the Gaussian measure $(2\pi\sigma^2)^{-1/2} \exp(-x^2/2\sigma^2)dx$. The finiteness of $\sigma^2(\beta, h)$ is a reasonable assumption since it means that the correlations among distant spins are weak.

The parameter $\sigma^2(\beta, h)$ has two interpretations which connect the microcopics and the macroscopics of the models. On the one hand, $\sigma^2(\beta, h)$ equals the limiting variance of $S_\Lambda/\sqrt{|\Lambda|}$:

$$\lim_{\Lambda \uparrow \mathbb{Z}^D} \frac{1}{|\Lambda|}\{\langle S_\Lambda^2 \rangle_{\beta, h} - \langle S_\Lambda \rangle_{\beta, h}^2\} = \sigma^2(\beta, h),$$

This follows formally from (5.1) by setting $f(x) = x^2$ (this function is not in $\mathscr{C}(\mathbb{R})$). On the other hand, $\sigma^2(\beta, h)$ is closely related to the geometry of the isotherms $m(\beta, h)$ for β fixed, h real. Figure IV.1 depicts the isotherms for the Ising model on $\mathbb{Z}^2$. The isotherm $m(\beta, h)$ for $0 < \beta < \beta_c$ passes smoothly through 0, while the critical isotherm $m(\beta_c, h)$ has an infinite slope at $h = 0$. The infinite slope signals the onset of spontaneous magnetization for $\beta > \beta_c$. The isotherm for $\beta > \beta_c$ has a jump discontinuity at $h = 0$ of magnitude $2m(\beta, +)$. The relation to be proved in Section V.7 is that $\sigma^2(\beta, h)$ equals β^{-1} times the slope $\partial m(\beta, h)/\partial h$. This slope is called the specific magnetic susceptibility and is denoted by $\chi(\beta, h)$.

What is the status of the central limit theorem at the critical point $(\beta_c, 0)$? Our results are much less complete, but physically the situation seems clear. The susceptibility $\chi(\beta_c, 0)$ and the limiting variance $\sigma^2(\beta_c, 0) = \beta_c^{-1}\chi(\beta_c, 0)$ are both infinite. Abnormally large fluctuations of S_Λ cause the central limit theorem to break down. An interesting question, discussed in Section V.8, is how to rescale S_Λ by a nonclassical scaling $|\Lambda|^\lambda$, $\lambda > \frac{1}{2}$, such that at the critical point $S_\Lambda/|\Lambda|^\lambda$ has a nondegenerate limit X in distribution. The value of λ and the form of the distribution of X (e.g., non-Gaussian or Gaussian) depend on the dimension D of the lattice. Theorem V.8.6 gives information on X under suitable hypotheses.

Section V.9 returns to the Curie–Weiss model, which was first studied in the previous chapter. One of the results shows that at the critical point $S_\Lambda/|\Lambda|^{3/4}$ converges in distribution to an explicitly determined, non-Gaussian random variable X. Section V.10 studies a magnetic model on the circle which generalizes the Curie–Weiss model and which is interesting because it exhibits a new kind of phase transition. A number of theorems in these last two sections are derived by large deviation techniques.

PART 1

V.2. Finite-Volume Gibbs States on $\mathbb{Z}^D$, $D \geq 1$

The definition of finite-volume Gibbs states on $\mathbb{Z}$ carries over to $\mathbb{Z}^D$ essentially without change. But it will ease the exposition if from the start we include external conditions in the definitions of these states. Points in the integer lattice $\mathbb{Z}^D$, $D \in \{1, 2, \ldots\}$, are denoted by $j = (j_1, \ldots, j_D)$. Given a non-negative integer N, let Λ be the symmetric subset of $\mathbb{Z}^D$ consisting of all points j satisfying $|j_\alpha| \leq N$ for each $\alpha = 1, \ldots, D$. Thus for $D = 2$, Λ is a square while for $D = 3$, Λ is a cube. For general D, we will refer to these sets as *symmetric hypercubes*. The number of points in Λ, denoted by $|\Lambda|$, is $(2N+1)^D$. To each site j there is assigned a spin which takes the value $\omega_j \in \{1, -1\}$. Configuration space is the set Ω_Λ of all sequences $\omega = \{\omega_j; j \in \Lambda\}$; thus $\Omega_\Lambda = \{1, -1\}^\Lambda$. We define the *Hamiltonian* or *interaction energy* of a spin configuration $\omega \in \Omega_\Lambda$ to be

$$(5.2) \qquad H_{\Lambda,h,\tilde{\omega}}(\omega) = -\tfrac{1}{2} \sum_{i,j \in \Lambda} J(i-j)\omega_i\omega_j - \sum_{i \in \Lambda} \left(h + \sum_{j \in \Lambda^c} J(i-j)\tilde{\omega}_j \right) \omega_i,$$

where $\Lambda^c = \mathbb{Z}^D \backslash \Lambda$. The interaction J is assumed to satisfy the following hypotheses: J is independent of Λ, J is non-negative (*ferromagnetic*) and is not identically zero, J is symmetric ($J(i-j) = J(j-i)$ for each i and j), and $\sum_{k \in \mathbb{Z}^D} J(k) < \infty$. J is called a *summable ferromagnetic interaction* on $\mathbb{Z}^D$.* For each i and j, the interaction strength $J(i-j)$ is translation invariant; i.e., $J(i-j) = J((i+k)-(j+k))$ for each $k \in \mathbb{Z}^D$. The real number h gives the strength of an external magnetic field acting at each site in Λ. The quantity $\tilde{\omega} = \{\tilde{\omega}_j; j \in \Lambda^c\}$ defines the external condition. We recall that in our definition of infinite-volume Gibbs states on $\mathbb{Z}$, $\tilde{\omega}_j$ took the value 1 or -1 [Section IV.6]. In Sections V.2–V.5, where only finite-volume Gibbs states are studied, we will allow more general external conditions in order to increase the applicability of the results. In the Hamiltonian $H_{\Lambda,h,\tilde{\omega}}$ in (5.2), the quantity $\tilde{\omega}_j$, $j \in \Lambda^c$, takes the value 1 (spin-up), -1 (spin-down), or 0 (site j unoccupied). If $\tilde{\omega}_j$ equals 0 for each $j \in \Lambda^c$, then the Hamiltonian $H_{\Lambda,h,\tilde{\omega}}$ reduces to the Hamiltonian $H_{\Lambda,h}$ defined in (4.3). This choice of $\tilde{\omega}$ is called

*More general interactions are discussed in Appendix C.3.

the *free external condition*. When we want to indicate the dependence of an $\tilde{\omega}$ upon Λ, we will write $\tilde{\omega}(\Lambda)$. Since J is summable, the quantity $h_i = h + \sum_{j \in \Lambda^c} J(i-j)\tilde{\omega}_j$ in (5.2) is well-defined. Let $\|-\|$ denote the Euclidean norm. An interaction J is said to have *finite range* if $J(k)$ equals 0 for all k with $\|k\|$ sufficiently large. The *range* is the smallest number L such that $J(k) = 0$ whenever $\|k\| > L$.

Let $\mathscr{B}(\Omega_\Lambda)$ denote the set of all subsets of Ω_Λ. We define the *finite-volume Gibbs state* on Λ with external condition $\tilde{\omega}$ to be the probability measure $P_{\Lambda,\beta,h,\tilde{\omega}}$ on $\mathscr{B}(\Omega_\Lambda)$ which assigns to each $\{\omega\}$, $\omega \in \Omega_\Lambda$, the probability

$$(5.3) \qquad P_{\Lambda,\beta,h,\tilde{\omega}}\{\omega\} = \exp[-\beta H_{\Lambda,h,\tilde{\omega}}(\omega)]\pi_\Lambda P_\rho\{\omega\} \cdot \frac{1}{Z(\Lambda,\beta,h,\tilde{\omega})}.$$

The parameter $\beta > 0$ is the inverse absolute temperature, $\pi_\Lambda P_\rho$ is the product measure on $\mathscr{B}(\Omega_\Lambda)$ with identical one-dimensional marginals $\rho = \frac{1}{2}\delta_1 + \frac{1}{2}\delta_{-1}$, and $Z(\Lambda,\beta,h,\tilde{\omega})$ equals $\int_{\Omega_\Lambda} \exp[-\beta H_{\Lambda,h,\tilde{\omega}}(\omega)]\pi_\Lambda P_\rho(d\omega)$. $Z(\Lambda,\beta,h,\tilde{\omega})$ is called the *partition function*. For A a subset of Ω_Λ, we have

$$P_{\Lambda,\beta,h,\tilde{\omega}}\{A\} = \sum_{\omega \in A} P_{\Lambda,\beta,h,\tilde{\omega}}\{\omega\}.$$

If for each site $j \in \Lambda^c$, $\tilde{\omega}_j$ equals 1 (resp., -1), then the external condition $\tilde{\omega}$ is called *plus* (resp., *minus*) and the measure is written as $P_{\Lambda,\beta,h,+}$ (resp., $P_{\Lambda,\beta,h,-}$). For the free external condition (each $\tilde{\omega}_j = 0$), we write $P_{\Lambda,\beta,h,f}$. This is the same as the measure $P_{\Lambda,\beta,h}$ defined in (4.4).

The ferromagnetic system which has been studied the most extensively is the Ising model on $\mathbb{Z}^D$ (especially $D = 2$). As we will see, much more is known about this model than about other ferromagnetic systems.

Example V.2.1. *The Ising model on* $\mathbb{Z}^D$. Fix a number $\mathscr{J} > 0$. We define $J(i-j)$ to be $\mathscr{J}$ if $\|i-j\| = 1$ and to be 0 if $\|i-j\| \neq 1$. This interaction, which couples only nearest neighbors, has range 1.

The definition of spontaneous magnetization for models on $\mathbb{Z}^D$ coincides with the definition given in the previous chapter. Let $\{Y_j; j \in \Lambda\}$ be the coordinate functions on Ω_Λ defined by $Y_j(\omega) = \omega_j$ and let $S_\Lambda(\omega) = \sum_{j \in \Lambda} Y_j(\omega)$. The magnetization in the state $P_{\Lambda,\beta,h,\tilde{\omega}}$ is defined to be

$$M(\Lambda,\beta,h,\tilde{\omega}) = \int_{\Omega_\Lambda} S_\Lambda(\omega) P_{\Lambda,\beta,h,\tilde{\omega}}(d\omega) = \sum_{j \in \Lambda} \langle Y_j \rangle_{\Lambda,\beta,h,\tilde{\omega}},$$

where $\langle - \rangle_{\Lambda,\beta,h,\tilde{\omega}}$ denotes expectation with respect to $P_{\Lambda,\beta,h,\tilde{\omega}}$. We write $M(\Lambda,\beta,h,\tilde{\omega})$ as $M(\Lambda,\beta,h,+)$, $M(\Lambda,\beta,h,-)$, or $M(\Lambda,\beta,h,f)$ if $\tilde{\omega}$ is plus, minus, or free, respectively. $M(\Lambda,\beta,h,f)$ is the same as the quantity $M(\Lambda,\beta,h)$ defined in (4.7). We will study the magnetization per site in the limit as the symmetric hypercubes Λ expand to fill $\mathbb{Z}^D$. This limit, called the *infinite-volume limit* or the *thermodynamic limit*, is denoted by $\Lambda \uparrow \mathbb{Z}^D$. The key fact is that for any $h \neq 0$ and any sequence of external conditions $\{\tilde{\omega}(\Lambda); \Lambda \uparrow \mathbb{Z}^D\}$

the specific magnetization $m(\beta, h) = \lim_{\Lambda \uparrow \mathbb{Z}^D} |\Lambda|^{-1} M(\Lambda, \beta, h, \tilde{\omega}(\Lambda))$ exists and is independent of the sequence chosen. This was proved in Lemma IV.6.4 for models on $\mathbb{Z}$, and the proof extends verbatim to models on $\mathbb{Z}^D$. In Section V.4 we show that the limit $m(\beta, +) = \lim_{h \to 0^+} m(\beta, h)$ exists and is non-negative. Spontaneous magnetization is said to occur at inverse temperature β if $m(\beta, +)$ is positive. Section V.5 shows that for any model on $\mathbb{Z}^D$, $D \geq 2$, with a nontrivial interaction, spontaneous magnetization occurs at all sufficiently large β. The proofs in Sections V.4 and V.5 are based on a set of powerful moment inequalities which are presented next.

V.3. Moment Inequalities*, [1]

The moment inequalities are valid for a more general measure than the finite-volume Gibbs state $P_{\Lambda, \beta, h, \tilde{\omega}}$. Let Λ be an arbitrary nonempty finite subset of $\mathbb{Z}^D$, $\{J_{ij}; i, j \in \Lambda\}$ a set of non-negative real numbers, and $\{h_i; i \in \Lambda\}$ a set of real numbers. Let P be the probability measure on $\mathscr{B}(\Omega_\Lambda)$ which assigns to each $\{\omega\}$, $\omega \in \Omega_\Lambda$, the probability

$$P\{\omega\} = \exp[-H(\omega)] \pi_\Lambda P_\rho\{\omega\} \cdot \frac{1}{Z}. \tag{5.4}$$

In this formula, $H(\omega) = -\frac{1}{2} \sum_{i,j \in \Lambda} J_{ij} \omega_i \omega_j - \sum_{i \in \Lambda} h_i \omega_i$ and

$$Z = \int_{\Omega_\Lambda} \exp[-H(\omega)] \pi_\Lambda P_\rho(d\omega).$$

Expectation with respect to P is denoted by $\langle - \rangle$ or by $\langle - \rangle_{\{h_i\}}$. The measure P reduces to $P_{\Lambda, \beta, h, \tilde{\omega}}$ if Λ is a symmetric hypercube, $J_{ij} = \beta J(i-j)$, and $h_i = h + \sum_{j \in \Lambda^c} J(i-j) \tilde{\omega}_j$. For B a nonempty subset of Λ, we define $\omega_B = \prod_{i \in B} \omega_i$.

Properties of the magnetization will be derived from the following four inequalities. Three of these inequalities require that each h_i be non-negative. Analogous results hold when each h_i is non-positive. The second and third inequalities involve covariances of the $\{\omega_i\}$ while the last one involves a triple moment.

GKS-1 Inequality [Griffiths (1967a), (1967b), Kelly and Sherman (1968)]. *Assume that each $h_i \geq 0$. Then for any nonempty subset B of Λ, $\langle \omega_B \rangle \geq 0$.*

GKS-2 Inequality [Griffiths (1967a), (1967b), Kelly and Sherman (1968)]. *Assume that each $h_i \geq 0$. Then for any nonempty subsets A and B of Λ, $\langle \omega_A \omega_B \rangle - \langle \omega_A \rangle \langle \omega_B \rangle \geq 0$.*

Percus Inequality [Percus (1975)]. *For arbitrary $\{h_i\}$ and sites $i, j \in \Lambda$, $\langle \omega_i \omega_j \rangle - \langle \omega_i \rangle \langle \omega_j \rangle \geq 0$.*†

* These are also called correlation inequalities.

† This is a special case of the FKG inequality as well as of the more general Percus inequality in Lemma V.3.1(b).

GHS Inequality [Griffiths, Hurst, and Sherman (1970)]. *Assume that each* $h_i \geq 0$. *Then for arbitrary sites* $i, j, k \in \Lambda$

$$\begin{aligned}\langle(\omega_i - \langle\omega_i\rangle)(\omega_j - \langle\omega_j\rangle)(\omega_k - \langle\omega_k\rangle)\rangle \\ = \langle\omega_i\omega_j\omega_k\rangle - \langle\omega_i\rangle\langle\omega_j\omega_k\rangle - \langle\omega_j\rangle\langle\omega_i\omega_k\rangle - \langle\omega_k\rangle\langle\omega_i\omega_j\rangle \\ + 2\langle\omega_i\rangle\langle\omega_j\rangle\langle\omega_k\rangle \leq 0.\end{aligned}$$

The FKG inequality is not needed now in order to prove properties of the magnetization, but we state it here for later use. Its importance has been demonstrated in Section IV.6.

FKG Inequality [Fortuin, Kastelyn, and Ginibre (1971)].* *Let f and g be real-valued, nondecreasing functions on* Ω_Λ. *Then for arbitrary* $\{h_i\}$, $\langle fg\rangle - \langle f\rangle\langle g\rangle \geq 0$. *In particular,* $\{h_i\} \leq \{\bar{h}_i\}$ *implies* $\langle f\rangle_{\{h_i\}} \leq \langle f\rangle_{\{\bar{h}_i\}}$.

The proofs of the moment inequalities follow Sylvester (1976a).

Proof of the GKS-1 *inequality*. By definition

$$\langle\omega_B\rangle = \int_{\Omega_\Lambda} \omega_B \exp\left(\frac{1}{2}\sum_{i,j\in\Lambda} J_{ij}\omega_i\omega_j + \sum_{i\in\Lambda} h_i\omega_i\right)\pi_\Lambda P_\rho(d\omega)\cdot\frac{1}{Z}.$$

We expand the exponential in its Taylor series and factor the resulting integrals over the sites $i \in \Lambda$. The proof is reduced to showing that $\int_{\{1,-1\}} \omega_i^m \rho(d\omega_i) \geq 0$ for each site i and all non-negative integers m. But the integral is 0 for m odd and 1 for m even. □

Proof of the GKS-2 *inequality and the Percus inequality*. The proof employs a method of duplicate variables. We construct a doubled spin system with configuration space $\Omega_\Lambda \times \Omega_\Lambda$, spins $\{\omega_i; i\in\Lambda\}$, $\{\sigma_i; i\in\Lambda\}$, Hamiltonian

$$\begin{aligned}H(\omega,\sigma) &= H(\omega) + H(\sigma) \\ &= -\frac{1}{2}\sum_{i,j\in\Lambda} J_{ij}\omega_i\omega_j - \sum_{i\in\Lambda} h_i\omega_i - \frac{1}{2}\sum_{i,j\in\Lambda} J_{ij}\sigma_i\sigma_j - \sum_{i\in\Lambda} h_i\sigma_i,\end{aligned}$$

and finite-volume Gibbs state

$$P(d\omega d\sigma) = \exp[-H(\omega,\sigma)]\pi_\Lambda P_\rho(d\omega)\pi_\Lambda P_\rho(d\sigma)\cdot\frac{1}{Z},$$

where Z is a normalization. Thus the doubled system consists of two copies of the original system which do not interact with each other. We denote expectation with respect to $P(d\omega d\sigma)$ by $\langle - \rangle^{(2)}$. The inequalities will follow from the next lemma. Part (a) is due to Ginibre (1970) and part (b) to Percus (1975).

Lemma V.3.1. *Define* $t_i = (\omega_i + \sigma_i)/\sqrt{2}$ *and* $q_i = (\omega_i - \sigma_i)/\sqrt{2}$.

(a) *If each* $h_i \geq 0$, *then for any nonempty subsets A and B of* Λ, $\langle q_A t_B\rangle^{(2)} \geq 0$.†

* Nondecreasing means $\{\omega_i\} \leq \{\bar{\omega}_i\} \Rightarrow f(\{\omega_i\}) \leq f(\{\bar{\omega}_i\})$. The proof of the FKG inequality given in Theorem IV.6.7 applies to $D \geq 1$ without change.

† $q_A = \prod_{i\in A} q_i$, $t_B = \prod_{i\in B} t_i$.

(b) *For arbitrary* $\{h_i\}$ *and any nonempty subset* A *of* Λ, $\langle q_A \rangle^{(2)} \geq 0$.

Proof. (a) In terms of t_i and q_i,

$$H(\omega, \sigma) = -\frac{1}{2} \sum_{i,j \in \Lambda} J_{ij} \left[\left[\frac{t_i + q_i}{\sqrt{2}} \right] \left[\frac{t_j + q_j}{\sqrt{2}} \right] + \left[\frac{t_i - q_i}{\sqrt{2}} \right] \left[\frac{t_j - q_j}{\sqrt{2}} \right] \right]$$

$$(5.5) \qquad - \sum_{i \in \Lambda} h_i \left[\frac{t_i + q_i}{\sqrt{2}} + \frac{t_i - q_i}{\sqrt{2}} \right]$$

$$= -\frac{1}{2} \sum_{i,j \in \Lambda} J_{ij}(t_i t_j + q_i q_j) - \sqrt{2} \sum_{i \in \Lambda} h_i t_i.$$

We substitute this expression into the formula

$$\langle q_A t_B \rangle^{(2)} = \int_{\Omega_\Lambda \times \Omega_\Lambda} q_A t_B \exp[-H(\omega, \sigma)] \pi_\Lambda P_\rho(d\omega) \pi_\Lambda P_\rho(d\sigma) \cdot \frac{1}{Z},$$

expand the exponential in its Taylor series, and factor the resulting integrals over the sites $i \in \Lambda$. The proof is reduced to showing that for each site i and all non-negative integers m and n

$$\int_{\{1,-1\} \times \{1,-1\}} q_i^m t_i^n \rho(d\omega_i) \rho(d\sigma_i)$$

$$= \int_{\{1,-1\} \times \{1,-1\}} \left(\frac{\omega_i - \sigma_i}{\sqrt{2}} \right)^m \left(\frac{\omega_i + \sigma_i}{\sqrt{2}} \right)^n \rho(d\omega_i) \rho(d\sigma_i) \geq 0.$$

Since the spins take values 1 or -1, the integrand is 0 if both m and n are positive. If either m or n equals 0, then we are left with moments of t_i alone or of q_i alone, and these are either 0 (odd moments) or positive (even moments).

(b) By (5.5), $\langle q_A \rangle^{(2)}$ equals

$$\int_{\Omega_\Lambda \times \Omega_\Lambda} q_A \exp\left(\frac{1}{2} \sum_{i,j \in \Lambda} J_{ij} q_i q_j \right) \exp\left(\frac{1}{2} \sum_{i,j \in \Lambda} J_{ij} t_i t_j + \sqrt{2} \sum_{i \in \Lambda} h_i t_i \right)$$

$$\cdot \prod_{i \in \Lambda} \rho(d\omega_i) \rho(d\sigma_i) \cdot \frac{1}{Z}.$$

We expand the first exponential in its Taylor series. It suffices to show that for all non-negative integers m_i

$$(5.6) \qquad \int_{\Omega_\Lambda \times \Omega_\Lambda} \prod_{i \in \Lambda} q_i^{m_i} \exp\left(\frac{1}{2} \sum_{i,j \in \Lambda} J_{ij} t_i t_j + \sqrt{2} \sum_{i \in \Lambda} h_i t_i \right) \prod_{i \in \Lambda} \rho(d\omega_i) \rho(d\sigma_i) \geq 0.$$

By symmetry, the integral is 0 unless all the m_i are even [Problem V.13.2]. If all the m_i are even, then the integral is positive. □

In order to prove the GKS-2 inequality, we express $\langle \omega_A \omega_B \rangle - \langle \omega_A \rangle \langle \omega_B \rangle$ in terms of the doubled system:

$$\langle \omega_A \omega_B \rangle - \langle \omega_A \rangle \langle \omega_B \rangle = \langle \omega_A \omega_B - \omega_A \sigma_B \rangle^{(2)}$$

$$(5.7) \qquad = \left\langle \prod_{i \in A} \left(\frac{t_i + q_i}{\sqrt{2}} \right) \left[\prod_{i \in B} \left(\frac{t_i + q_i}{\sqrt{2}} \right) - \prod_{i \in B} \left(\frac{t_i - q_i}{\sqrt{2}} \right) \right] \right\rangle^{(2)}.$$

When the product $\prod_{i \in B}(t_i - q_i)/\sqrt{2}$ is expanded, any term with a negative coefficient is cancelled by a corresponding term in the expansion of $\prod_{i \in B}(t_i + q_i)/\sqrt{2}$. Hence the right-hand side of (5.7) is the expectation of a polynomial in $\{q_i\}$ and $\{t_i\}$ which has non-negative coefficients. Lemma V.3.1(a) shows that the expectation is non-negative. This proves the GKS-2 inequality.

The Percus inequality follows from Lemma V.3.1(b):

$$\langle \omega_i \omega_j \rangle - \langle \omega_i \rangle \langle \omega_j \rangle = \tfrac{1}{2} \langle \omega_i \omega_j + \sigma_i \sigma_j - \omega_i \sigma_j - \omega_j \sigma_i \rangle^{(2)} = \langle q_i q_j \rangle^{(2)} \geq 0. \quad \square$$

Proof of the GHS *inequality*. We introduce a quadrupled spin system with configuration space $\Omega_\Lambda \times \Omega_\Lambda \times \Omega_\Lambda \times \Omega_\Lambda$; spins $\{\omega_i; i \in \Lambda\}$, $\{\sigma_i; i \in \Lambda\}$, $\{\omega_i'; i \in \Lambda\}$, $\{\sigma_i'; i \in \Lambda\}$; Hamiltonian $H(\omega, \sigma, \omega', \sigma') = H(\omega) + H(\sigma) + H(\omega') + H(\sigma')$; and finite-volume Gibbs state

$$P(d\omega d\sigma d\omega' d\sigma')$$
$$= \exp[-H(\omega, \sigma, \omega', \sigma')] \pi_\Lambda P_\rho(d\omega) \pi_\Lambda P_\rho(d\sigma) \pi_\Lambda P_\rho(d\omega') \pi_\Lambda P_\rho(d\sigma') \cdot \frac{1}{Z},$$

where Z is a normalization. Expectation with respect to this measure is denoted by $\langle - \rangle^{(4)}$. The following lemma is due to Ellis and Monroe (1975).

Lemma V.3.2. *Define*

$$(5.8) \qquad \begin{aligned} \alpha_i &= \frac{\omega_i + \sigma_i + \omega_i' + \sigma_i'}{2}, & \beta_i &= \frac{\omega_i + \sigma_i - \omega_i' - \sigma_i'}{2}, \\ \gamma_i &= \frac{\omega_i - \sigma_i + \omega_i' - \sigma_i'}{2}, & \delta_i &= \frac{-\omega_i + \sigma_i + \omega_i' - \sigma_i'}{2}. \end{aligned}$$

If each $h_i \geq 0$, *then for any nonempty subsets* A, B, C, *and* D *in* F, $\langle \alpha_A \beta_B \gamma_C \delta_D \rangle^{(4)} \geq 0$.*

Proof. Since the transformation (5.8) is orthogonal,

$$-H(\omega) - H(\sigma) - H(\omega') - H(\sigma')$$
$$= \frac{1}{2} \sum_{i, j \in \Lambda} J_{ij}(\alpha_i \alpha_j + \beta_i \beta_j + \gamma_i \gamma_j + \delta_i \delta_j) + 2 \sum_{i \in \Lambda} h_i \alpha_i.$$

This is a polynomial with non-negative coefficients. By expanding the exponential in $\langle \alpha_A \beta_B \gamma_C \delta_D \rangle^{(4)}$ and factoring the resulting integrals over the sites $i \in F$, we reduce the proof to showing that for each site i and all non-negative integers k, l, m and n,

* $\alpha_A = \prod_{i \in A} \alpha_i$, etc.

(5.9) $$\int_{\{1,-1\}^4} \alpha_i^k \beta_i^l \gamma_i^m \delta_i^n \rho(d\omega_i)\rho(d\sigma_i)\rho(d\omega_i')\rho(d\sigma_i') \geq 0.$$

By symmetry, the integral is 0 unless k, l, m, and n all have the same parity [Problem V.13.3]. When the parity is even, the integrand is non-negative and (5.9) holds. When the parity is odd, since $\omega_i^2 = \sigma_i^2 = (\omega_i')^2 = (\sigma_i')^2 = 1$, we find that

$$\alpha_i^k \beta_i^l \gamma_i^m \delta_i^n = \alpha_i^{k-1}\beta_i^{l-1}\gamma_i^{m-1}\delta_i^{n-1}\alpha_i\beta_i\gamma_i\delta_i \quad \text{and} \quad \alpha_i\beta_i\gamma_i\delta_i = \tfrac{1}{4}(\omega_i\sigma_i - \omega_i'\sigma_i')^2.$$

Hence the integrand is again non-negative, and (5.9) holds. □

We now prove the GHS inequality. Define

$$t_i = \frac{\omega_i + \sigma_i}{\sqrt{2}}, \qquad q_i = \frac{\omega_i - \sigma_i}{\sqrt{2}}, \qquad t_i' = \frac{\omega_i' + \sigma_i'}{\sqrt{2}}, \qquad q_i' = \frac{\omega_i' - \sigma_i'}{\sqrt{2}}.$$

As noted by Lebowitz (1974),

$$\langle \omega_i\omega_j\omega_k\rangle - \langle\omega_i\rangle\langle\omega_j\omega_k\rangle - \langle\omega_j\rangle\langle\omega_i\omega_k\rangle - \langle\omega_k\rangle\langle\omega_i\omega_j\rangle + 2\langle\omega_i\rangle\langle\omega_j\rangle\langle\omega_k\rangle$$
$$= -\sqrt{2}\langle t_k q_i' q_j' - t_k q_i q_j\rangle^{(4)}.$$

The latter term can be written as

$$-\sqrt{2}\left\langle \frac{\alpha_k + \beta_k}{\sqrt{2}}\left[\left(\frac{\gamma_i + \delta_i}{\sqrt{2}}\right)\left(\frac{\gamma_j + \delta_j}{\sqrt{2}}\right) - \left(\frac{\gamma_i - \delta_i}{\sqrt{2}}\right)\left(\frac{\gamma_j - \delta_j}{\sqrt{2}}\right)\right]\right\rangle^{(4)}$$
$$= -\langle(\alpha_k + \beta_k)(\gamma_i\delta_j + \gamma_j\delta_i)\rangle^{(4)}.$$

This is non-positive by Lemma V.3.2. The proofs of the moment inequalities are complete.

V.4. Properties of the Magnetization and the Gibbs Free Energy

Spontaneous magnetization for models on $\mathbb{Z}^D$, $D \geq 2$, will be proved in the next section. The proof depends upon properties of the magnetization $M(\Lambda, \beta, h, f)$ as well as on connections between the magnetization and the Gibbs free energy.

Let Λ be a symmetric hypercube in $\mathbb{Z}^D$. For $\beta > 0$ and h real, let $P_{\Lambda,\beta,h,\bar{\omega}}$ be the finite-volume Gibbs state on Λ with external condition $\bar{\omega}$ corresponding to a summable ferromagnetic interaction J. The quantity

$$M(\Lambda, \beta, h, \bar{\omega}) = \sum_{l\in\Lambda} \langle Y_l\rangle_{\Lambda,\beta,h,\bar{\omega}}$$

denotes the magnetization in the state $P_{\Lambda,\beta,h,\bar{\omega}}$. The parameters h and $\bar{\omega}$ appear in $P_{\Lambda,\beta,h,\bar{\omega}}$ in the combination $h_i = h + \sum_{j\in\Lambda^c} J(i-j)\bar{\omega}_j$, $i \in \Lambda$. Properties of $\langle Y_l\rangle_{\Lambda,\beta,h,\bar{\omega}}$ as functions of β, of the interaction strengths

Table V.1. Properties of $\langle Y_l\rangle_{\Lambda,\beta,h,\tilde{\omega}}$

Property of $\langle Y_l\rangle$	Moment Inequality Used in Proof	Conditions on $h_i = h + \sum_{j\in\Lambda^c} J(i-j)\tilde{\omega}_j$, $i\in\Lambda$
(a) $\langle Y_l\rangle \geq 0$	GKS-1	Each $h_i \geq 0$
(b) $\dfrac{\partial}{\partial\beta}\langle Y_l\rangle \geq 0$	GKS-2	Each $h_i \geq 0$
(c) $\dfrac{\partial}{\partial J(k)}\langle Y_l\rangle \geq 0$	GKS-2	Each $h_i \geq 0$, each $\tilde{\omega}_j \geq 0$ ($j\in\Lambda^c$)
(d) $\dfrac{\partial}{\partial h_i}\langle Y_l\rangle \geq 0$	Percus	Arbitrary h_i
(e) $\dfrac{\partial^2}{\partial h_i \partial h_j}\langle Y_l\rangle \leq 0$	GHS	Each $h_i \geq 0$

i, j, l are sites in Λ (not necessarily distinct); $\beta > 0$; $J(k) \geq 0$, $k\in\mathbb{Z}^D$. We have $h_i \geq 0$ if, for example, $h \geq 0$ and $\tilde{\omega}$ is plus or free.

$\{J(k); k\in\mathbb{Z}^D\}$, and of $\{h_i; i\in\Lambda\}$ are listed in Table V.1. $M(\Lambda,\beta,h,\tilde{\omega})$ satisfies the same properties (just sum over $l\in\Lambda$). All but one of these properties require that each h_i be non-negative. This non-negativity restricts the allowed values of h and $\tilde{\omega}$. Analogous results hold when each h_i is nonpositive.

Proof of Properties of $\langle Y_l\rangle_{\Lambda,\beta,h,\tilde{\omega}}$. We write H for

$$H_{\Lambda,h,\tilde{\omega}} = -\tfrac{1}{2}\sum_{i,j\in\Lambda} J(i-j)\omega_i\omega_j - \sum_{i\in\Lambda} h_i\omega_i,$$

Z for $Z(\Lambda,\beta,h,\tilde{\omega}) = \int_{\Omega_\Lambda} \exp[-\beta H(\omega)]\pi_\Lambda P_\rho(d\omega)$, and $\langle - \rangle$ for $\langle - \rangle_{\Lambda,\beta,h,\tilde{\omega}}$. The following easily verified calculations are needed:

$$(5.10)\quad \frac{\partial Z}{\partial\beta}\frac{1}{Z} = -\int_{\Omega_\Lambda} H(\omega)\exp[-\beta H(\omega)]\pi_\Lambda P_\rho(d\omega)\cdot\frac{1}{Z} = -\langle H\rangle,$$

$$(5.11)\quad \frac{\partial H(\omega)}{\partial J(k)} = -\frac{1}{2}\sum_{\substack{i,j\in\Lambda\\ i-j=k}} \omega_i\omega_j - \sum_{\substack{i\in\Lambda, j\in\Lambda^c\\ i-j=k}} \tilde{\omega}_j\omega_i, \qquad \frac{\partial H(\omega)}{\partial h_i} = -\omega_i,$$

$$(5.12)\quad \frac{\partial Z}{\partial J(k)}\frac{1}{Z} = -\beta\cdot\frac{1}{2}\sum_{\substack{i,j\in\Lambda\\ i-j=k}} \langle\omega_i\omega_j\rangle - \beta\sum_{\substack{i\in\Lambda, j\in\Lambda^c\\ i-j=k}} \tilde{\omega}_j\langle\omega_i\rangle,$$

$$(5.13)\quad \frac{\partial Z}{\partial h_i}\frac{1}{Z} = \beta\int_{\Omega_\Lambda} \omega_i\exp[-\beta H(\omega)]\pi_\Lambda P_\rho(d\omega)\cdot\frac{1}{Z} = \beta\langle\omega_i\rangle = \beta\langle Y_i\rangle.$$

The first sums in (5.11) and (5.12) are absent if there are no sites $i, j\in\Lambda$ which satisfy $i - j = k$; similarly for the second sums in (5.11) and (5.12).

(a) If each $h_i \geq 0$, then $\langle Y_l\rangle \geq 0$ by the GKS-1 inequality.

$$\text{(b)}\quad \frac{\partial}{\partial\beta}\langle Y_l\rangle = \frac{\partial}{\partial\beta}\left[\int_{\Omega_\Lambda} \omega_l\exp[-\beta H(\omega)]\pi_\Lambda P_\rho(d\omega)\cdot\frac{1}{Z}\right]$$

$$= -\int_{\Omega_\Lambda} \omega_l H(\omega)\exp[-\beta H(\omega)]\pi_\Lambda P_\rho(d\omega)\cdot\frac{1}{Z}$$
$$-\int \omega_l \exp[-\beta H(\omega)]\pi_\Lambda P_\rho(d\omega)\cdot\frac{\partial Z}{\partial\beta}\frac{1}{Z^2}$$
$$= -\langle\omega_l H\rangle + \langle\omega_l\rangle\langle H\rangle$$
$$= \frac{1}{2}\sum_{i,j\in\Lambda} J(i-j)[\langle\omega_l\omega_i\omega_j\rangle - \langle\omega_l\rangle\langle\omega_i\omega_j\rangle]$$
$$+ \sum_{i\in\Lambda} h_i(\langle\omega_l\omega_i\rangle - \langle\omega_l\rangle\langle\omega_i\rangle).$$

If each $h_i \geq 0$, then $\partial\langle Y_l\rangle/\partial\beta \geq 0$ by the GKS-2 inequality.

(c) We have

$$\frac{\partial}{\partial J(k)}\langle Y_l\rangle =$$
$$\frac{\beta}{2}\sum_{\substack{i,j\in\Lambda\\ i-j=k}}[\langle\omega_l\omega_i\omega_j\rangle - \langle\omega_l\rangle\langle\omega_i\omega_j\rangle] + \beta\sum_{\substack{i\in\Lambda, j\in\Lambda^c\\ i-j=k}}\tilde{\omega}_j[\langle\omega_l\omega_i\rangle - \langle\omega_l\rangle\langle\omega_i\rangle].$$

If each $h_i \geq 0$ and each $\tilde{\omega}_j \geq 0$, then $\partial\langle Y_l\rangle/\partial J(k) \geq 0$ by the GKS-2 inequality.

(d) $\dfrac{\partial}{\partial h_i}\langle Y_l\rangle = \beta[\langle\omega_l\omega_i\rangle - \langle\omega_l\rangle\langle\omega_i\rangle] \geq 0$ by the Percus inequality.

(e)
$$\frac{\partial^2}{\partial h_i\partial h_j}\langle Y_l\rangle = \beta\frac{\partial}{\partial h_i}(\langle\omega_l\omega_j\rangle - \langle\omega_l\rangle\langle\omega_j\rangle)$$
$$= \beta^2[\langle\omega_l\omega_j\omega_i\rangle - \langle\omega_l\omega_j\rangle\langle\omega_i\rangle$$
$$- (\langle\omega_l\omega_i\rangle - \langle\omega_l\rangle\langle\omega_i\rangle)\langle\omega_j\rangle$$
$$- \langle\omega_l\rangle(\langle\omega_j\omega_i\rangle - \langle\omega_j\rangle\langle\omega_i\rangle)]$$
$$= \beta^2[\langle\omega_l\omega_j\omega_i\rangle - \langle\omega_l\rangle\langle\omega_j\omega_i\rangle - \langle\omega_j\rangle\langle\omega_l\omega_i\rangle$$
$$- \langle\omega_i\rangle\langle\omega_l\omega_j\rangle + 2\langle\omega_l\rangle\langle\omega_j\rangle\langle\omega_i\rangle].$$

If each $h_i \geq 0$, then $\partial^2\langle Y_l\rangle/\partial h_i\partial h_j \leq 0$ by the GHS inequality.

Remark V.4.1. (a) Let A be a nonempty subset of Λ. The GKS-2 inequality implies that

$$\frac{\partial\langle\omega_A\rangle_{\Lambda,\beta,h,\tilde{\omega}}}{\partial\beta} \geq 0 \quad\text{and}\quad \frac{\partial\langle\omega_A\rangle_{\Lambda,\beta,h,\tilde{\omega}}}{\partial h_i} \geq 0 \quad\text{provided each } h_i \geq 0,$$
$$\frac{\partial\langle\omega_A\rangle_{\Lambda,\beta,h,\tilde{\omega}}}{\partial J(k)} \geq 0 \quad\text{provided } h_i \geq 0 \text{ and each } \tilde{\omega}_j \geq 0.$$

The proof is the same as that of properties (b) and (c) in Table V.1. For example, $\partial\langle\omega_A\rangle/\partial h_i = \beta[\langle\omega_A\omega_i\rangle - \langle\omega_A\rangle\langle\omega_i\rangle] \geq 0$ by the GKS-2 inequality.

(b) By (5.13), $\langle\omega_i\rangle = \beta^{-1}\partial\log Z/\partial h_i$. Let $i_1, \ldots, i_r$ be sites in Λ. The derivatives

$$u_r(i_1, \ldots, i_r) = \beta^{-r} \frac{\partial^r}{\partial h_{i_1} \cdots \partial h_{i_r}} \log Z$$

are called the *Ursell functions* or *cluster functions*. We have

$$u_1(i) = \langle \omega_i \rangle, \qquad u_2(i,j) = \langle \omega_i \omega_j \rangle - \langle \omega_i \rangle \langle \omega_j \rangle,$$

$$u_3(i,j,k) = \langle \omega_i \omega_j \omega_k \rangle - \langle \omega_i \rangle \langle \omega_j \omega_k \rangle - \langle \omega_j \rangle \langle \omega_i \omega_k \rangle - \langle \omega_k \rangle \langle \omega_i \omega_j \rangle + 2\langle \omega_i \rangle \langle \omega_j \rangle \langle \omega_k \rangle.$$

An immediate consequence of Table V.1 is the following theorem, which states properties of $M(\Lambda, \beta, h, f)$.

Theorem V.4.2.* *The magnetization $M(\Lambda, \beta, h, f)$ corresponding to the free external condition $\tilde{\omega} = 0$ has the following properties.*

(a) *For each $\beta > 0$, $M(\Lambda, \beta, 0, f) = 0$; $M(\Lambda, \beta, h, f)$ is a non-negative concave function of $h \geq 0$ and a nondecreasing function of h real. It satisfies $M(\Lambda, \beta, -h, f) = -M(\Lambda, \beta, h, f)$ and $|M(\Lambda, \beta, h, f)| \leq |\Lambda|$.*

(b) *For each $h \geq 0$, $M(\Lambda, \beta, h, f)$ is a non-negative, non-decreasing function of $\beta > 0$ and of each interaction strength $J(k) \geq 0$, $k \in \mathbb{Z}^D$.*

Proof. (a) The relation $M(\Lambda, \beta, -h, f) = -M(\Lambda, \beta, h, f)$ follows from the fact that $P_{\Lambda,\beta,-h,f}\{\omega\} = P_{\Lambda,\beta,h,f}\{-\omega\}$ for each $\omega \in \Omega_\Lambda$. It implies $M(\Lambda, \beta, 0, f) = 0$. For the free external condition, h_i equals h for each site $i \in \Lambda$. Hence if h is non-negative, then Table V.1 applies. By Table V.1(a), $M(\Lambda, \beta, h, f) = \sum_{l \in \Lambda} \langle Y_l \rangle_{\Lambda,\beta,h,f} \geq 0$ for $h \geq 0$. The concavity of M is equivalent to $\partial^2 M/\partial h^2 \leq 0$, and by Table V.1(e)

$$\frac{\partial^2 M}{\partial h^2} = \sum_{l \in \Lambda} \frac{\partial^2 \langle Y_l \rangle}{\partial h^2} = \sum_{i,j,l \in \Lambda} \frac{\partial^2 \langle Y_l \rangle}{\partial h_i \partial h_j} \leq 0 \qquad \text{for } h \geq 0.$$

For h real, $\partial M/\partial h \geq 0$ [Table V.1(d)]. We have

$$|M(\Lambda, \beta, h, f)| \leq \sum_{l \in \Lambda} |\langle Y_l \rangle_{\Lambda,\beta,h,f}| \leq |\Lambda|.$$

(b) For $h \geq 0$, $\partial M/\partial \beta \geq 0$ and $\partial M/\partial J(k) \geq 0$ by Table V.1(b) and (c), respectively. □

Connections between the magnetization and the Gibbs free energy were explored in Sections IV.5 and IV.6 for models on $\mathbb{Z}$. We now extend these results to the D-dimensional models. Fix a summable ferromagnetic interaction J on $\mathbb{Z}^D$. Given a symmetric hypercube Λ and an external condition $\tilde{\omega} = \tilde{\omega}(\Omega)$, we define the *Gibbs free energy*

$$\Psi(\Lambda, \beta, h, \tilde{\omega}) = -\beta^{-1} \log Z(\Lambda, \beta, h, \tilde{\omega}),$$

* $M(\Lambda, \beta, h, f)$ equals the function $M(\Lambda, \beta, h)$ in Theorem IV.3.4.

where $Z(\Lambda, \beta, h, \tilde{\omega}) = \int_{\Omega_\Lambda} \exp[-\beta H_{\Lambda,h,\tilde{\omega}}(\omega)] \pi_\Lambda P_\rho(d\omega)$. If $\tilde{\omega}$ is free, then we write $\Psi(\Lambda, \beta, h, f)$. The latter is the same as the function $\Psi(\Lambda, \beta, h)$ defined in (4.20). An important property of the Gibbs free energy is that for any sequence of external conditions $\{\tilde{\omega}(\Lambda), \Lambda \uparrow \mathbb{Z}^D\}$, the *specific Gibbs free energy* $\psi(\beta, h) = \lim_{\Lambda \uparrow \mathbb{Z}^D} |\Lambda|^{-1} \Psi(\Lambda, \beta, h, \tilde{\omega}(\Lambda))$ exists and is independent of the sequence chosen. The existence of the limit for the free external condition is proved in Appendix D.1. To prove that the limit exists for any sequence of external conditions and is independent of the sequence chosen, let us compare $\Psi(\Lambda, \beta, h, \tilde{\omega}(\Lambda))$ with $\Psi(\Lambda, \beta, h, f)$. We use inequality (4.41), which holds for arbitrary dimension D:

$$(5.14) \qquad \frac{1}{|\Lambda|} |\Psi(\Lambda, \beta, h, \tilde{\omega}) - \Psi(\Lambda, \beta, h, f)| \le \frac{1}{|\Lambda|} \|H_{\Lambda,h,\tilde{\omega}} - H_{\Lambda,h,f}\|_\infty .$$

$H_{\Lambda,h,f}$ is the Hamiltonian with the free external condition and $\|-\|_\infty$ denotes the maximum over Ω_Λ. Since $\psi(\beta, h) = \lim_{\Lambda \uparrow \mathbb{Z}^D} |\Lambda|^{-1} \Psi(\Lambda, \beta, h, f)$ exists, it suffices to prove that $|\Lambda|^{-1} \|H_{\Lambda,h,\tilde{\omega}} - H_{\Lambda,h,f}\|_\infty \to 0$ as $\Lambda \uparrow \mathbb{Z}^D$. For any $\varepsilon > 0$, there exists a positive integer N_ε so that $\sum_{\|k\| > N_\varepsilon} J(k) < \varepsilon$. For any $\omega \in \Omega_\Lambda$

$$\begin{aligned} \frac{1}{|\Lambda|} |H_{\Lambda,h,\tilde{\omega}}(\omega) - H_{\Lambda,h,f}(\omega)| &\le \frac{1}{|\Lambda|} \sum_{i \in \Lambda} \sum_{j \in \Lambda^c} J(i-j) \\ &\le \frac{1}{|\Lambda|} \sum{}' J(i-j) + \varepsilon, \end{aligned}$$

where $\sum'$ denotes the sum over all $i \in \Lambda$ and $j \in \Lambda^c$ such that $\|i - j\| \le N_\varepsilon$. We have $\sum' J(i-j) \le (\max_{k \in \mathbb{Z}^D} J(k)) \text{ const } |\Lambda|^{(D-1)/D} N_\varepsilon^{D+1}$, where the constant depends only on D. It follows that $\lim_{\Lambda \uparrow \mathbb{Z}^D} |\Lambda|^{-1} \Psi(\Lambda, \beta, h, \tilde{\omega}(\Lambda))$ exists and is independent of the choice of $\{\tilde{\omega}(\Lambda)\}$.

The next topic is smoothness properties of $\psi(\beta, h)$. We first note that for any external condition $\tilde{\omega}$, $\Psi(\Lambda, \beta, h, \tilde{\omega})$ is a concave function of h real. Indeed, the concavity is equivalent of the inequality

$$(5.15) \qquad Z(\Lambda, \beta, \lambda h_1 + (1-\lambda) h_2, \tilde{\omega}) \le Z(\Lambda, \beta, h_1, \tilde{\omega})^\lambda \cdot Z(\Lambda, \beta, h_2, \tilde{\omega})^{1-\lambda}$$

for any $h_1 \neq h_2$ real and $0 < \lambda < 1$. As in the proof of Theorem IV.5.1(a), (5.15) follows from Hölder's inequality. Since concavity is preserved under pointwise limits, we conclude that $\psi(\beta, h)$ is a concave function of h real. In addition, $\psi(\beta, h)$ is an even function of h since $\Psi(\Lambda, \beta, h, f)$ is an even function of h. We next show that $\psi(\beta, h)$ is a continuously differentiable function of $h \neq 0$. The first step is to relate $\Psi(\Lambda, \beta, h, \tilde{\omega})$ to the magnetization $M(\Lambda, \beta, h, \tilde{\omega})$ in the finite-volume Gibbs state $P_{\Lambda,\beta,h,\tilde{\omega}}$:

$$\begin{aligned} -\frac{\partial \Psi(\Lambda, \beta, h, \tilde{\omega})}{\partial h} &= -\beta^{-1} \frac{\partial Z(\Lambda, \beta, h, \tilde{\omega})}{\partial h} \cdot \frac{1}{Z(\Lambda, \beta, h, \tilde{\omega})} \\ &= \sum_{j \in \Lambda} \sum_{\omega \in \Omega_\Lambda} \omega_j \exp(-\beta H_{\Lambda,h,\tilde{\omega}}(\omega)) 2^{-|\Lambda|} \frac{1}{Z(\Lambda, \beta, h, \tilde{\omega})} \\ &= M(\Lambda, \beta, h, \tilde{\omega}). \end{aligned}$$

In particular, for $h \ge 0$ and the free external condition

$$\frac{1}{|\Lambda|}(\Psi(\Lambda,\beta,h,f)-\Psi(\Lambda,\beta,0,f)) = -\int_0^h \frac{1}{|\Lambda|}M(\Lambda,\beta,s,f)ds.$$

Since $0 \le M(\Lambda,\beta,h,f) \le |\Lambda|$, there exists an infinite subsequence $\{\Lambda'\}$ of $\{\Lambda\}$ such that $m(\beta,h) = \lim_{\Lambda'\uparrow\mathbb{Z}^D}|\Lambda'|^{-1}M(\Lambda',\beta,h,f)$ exists for every rational number $h \ge 0$. $M(\Lambda',\beta,0,f)$ equals 0 and thus $m(\beta,0)$ equals 0. Since $M(\Lambda',\beta,h,f)$ is a concave function of $h \ge 0$, a standard convexity result implies that $m(\beta,h)$ exists for all $h \ge 0$ [Theorem VI.3.3(a)]. The limit $m(\beta,h)$ is concave for $h \ge 0$ and hence is continuous for $h > 0$ [Theorem VI.3.1]. By the Lebesgue dominated convergence theorem, $m(\beta,h)$ satisfies

$$\psi(\beta,h)-\psi(\beta,0) = -\int_0^h m(\beta,s)ds.$$

This equation shows that $\psi(\beta,h)$ is a continuously differentiable function of $h>0$ and that $\partial\psi(\beta,h)/\partial h = -m(\beta,h)$. A similar proof works for $h<0$. Since $\Psi(\Lambda,\beta,h,\tilde{\omega}(\Lambda))$ is a concave function of h real and for $h \ne 0$ $\partial\psi(\beta,h)/\partial h = -m(\beta,h)$ exists, Lemma IV.6.3 implies that for $h \ne 0$, $\lim_{\Lambda\uparrow\mathbb{Z}^D}|\Lambda|^{-1}M(\Lambda,\beta,h,\tilde{\omega}(\Lambda))$ exists along the entire sequence $\{\Lambda\}$ and

$$\begin{aligned}\lim_{\Lambda\uparrow\mathbb{Z}^D}\frac{1}{|\Lambda|}M(\Lambda,\beta,h,\tilde{\omega}(\Lambda)) &= -\lim_{\Lambda\uparrow\mathbb{Z}^D}\frac{1}{|\Lambda|}\frac{\partial\Psi(\Lambda,\beta,h,\tilde{\omega}(\Lambda))}{\partial h}\\ &= -\frac{\partial\psi(\beta,h)}{\partial h} = m(\beta,h).\end{aligned} \tag{5.16}$$

Thus the limit is independent of the choice of external conditions $\{\tilde{\omega}(\Lambda)\}$.

The next theorem summarizes the facts about $\psi(\beta,h)$ and $m(\beta,h)$ which have just been proved. It also lists properties of $m(\beta,h)$ which follow from properties of $M(\Lambda,\beta,h,f)$ in Theorem V.4.2.

Theorem V.4.3.* *Let J be a summable feromagnetic interaction on $\mathbb{Z}^D$. Then the following conclusions hold.*

(a) *For $\beta>0$ and h real, the specific Gibbs free energy $\psi(\beta,h) = \lim_{\Lambda\uparrow\mathbb{Z}^D}|\Lambda|^{-1}\Psi(\Lambda,\beta,h,\tilde{\omega}(\Lambda))$ exists and is independent of the choice of $\{\tilde{\omega}(\Lambda)\}$.*

(b) *For each $\beta>0$ $\psi(\beta,h)$ is a concave, even function of h real and is a continuously differentiable function of $h \ne 0$.*

(c) *For $\beta>0$ and $h\ne 0$, the specific magnetization $m(\beta,h) = \lim_{\Lambda\uparrow\mathbb{Z}^D}|\Lambda|^{-1}M(\Lambda,\beta,h,\tilde{\omega}(\Lambda))$ exists and is independent of the choice of $\{\tilde{\omega}(\Lambda)\}$. Moreover $m(\beta,h) = -\partial\psi(\beta,h)/\partial h$.*

(d) *Define $m(\beta,0)=0$. For each $\beta>0$, $m(\beta,h)$ is a non-negative concave function of $h \ge 0$ and a nondecreasing function of h real. It satisfies $m(\beta,-h) = -m(\beta,h)$ and $|m(\beta,h)| \le 1$.*

(e) *For each $h \ge 0$, $m(\beta,h)$ is a non-negative, nondecreasing function of $\beta>0$ and of each interaction strength $J(k)\ge 0$, $k\in\mathbb{Z}^D$.*

(f) *For each $\beta>0$, the limits $m(\beta,+) = \lim_{h\to 0^+}m(\beta,h)$ and $m(\beta,-) =$*

*Other properties of $m(\beta,h)$ are derived in Problem V.13.1.

$\lim_{h\to 0^-} m(\beta,h)$ *exist and* $m(\beta,-) = -m(\beta,+)$; $m(\beta,+)$ *is a non-negative, nondecreasing function of* $\beta > 0$ *and of each interaction strength* $J(k) \geq 0$, $k \in \mathbb{Z}^D$.

Part (b) of the theorem states that for each $\beta > 0$ $\psi(\beta,h)$ is a continuously differentiable function of $h \neq 0$. We have the following important extension, which is a consequence of the Lee–Yang theorem [Lee and Yang (1952)].[2] The proof is omitted.

Theorem V.4.4. *Let J be a summable ferromagnetic interaction on $\mathbb{Z}^D$. Then for each $\beta > 0$ $\psi(\beta,h)$ is a real analytic function of $h \neq 0$.*

Spontaneous magnetization occurs at β if and only if

$$m(\beta,+) = -\frac{\partial\psi(\beta,0)}{\partial h^+} = -\lim_{h\to 0^+}\frac{\partial\psi(\beta,h)}{\partial h}$$

is positive. We next relate $m(\beta,h)$, $m(\beta,+)$, and $m(\beta,-)$ to the quantities $\langle\omega_0\rangle_{\beta,h,+}$ and $\langle\omega_0\rangle_{\beta,h,-}$, defined as

$$\langle\omega_0\rangle_{\beta,h,+} = \lim_{\Lambda\uparrow\mathbb{Z}^D}\int_\Omega \omega_0 P_{\Lambda,\beta,h,+}(d\omega),$$

$$\langle\omega_0\rangle_{\beta,h,-} = \lim_{\Lambda\uparrow\mathbb{Z}^D}\int_\Omega \omega_0 P_{\Lambda,\beta,h,-}(d\omega);$$

ω_0 denotes the spin at the origin of $\mathbb{Z}^D$. The existence of the limits follows from the FKG inequality [Corollary IV.6.8(b)]. Later, we will identity $\langle\omega_0\rangle_{\beta,h,+}$ and $\langle\omega_0\rangle_{\beta,h,-}$ as the means of infinite-volume Gibbs states $P_{\beta,h,+}$ and $P_{\beta,h,-}$, respectively. The following lemma is proved exactly like Lemma IV.6.10, and so the proof is omitted.

Lemma V.4.5. *Let J be a summable ferromagnetic interaction on $\mathbb{Z}^D$. Then the following conclusions hold.*

(a) *For $\beta > 0$ and $h \neq 0$,*

$$\langle\omega_0\rangle_{\beta,h,+} = m(\beta,h) = -\frac{\partial\psi(\beta,h)}{\partial h} = \langle\omega_0\rangle_{\beta,h,-}.$$

(b) *For $\beta > 0$ and $h = 0$,*

$$\langle\omega_0\rangle_{\beta,0,+} = m(\beta,+) = -\frac{\partial\psi(\beta,0)}{\partial h^+} \geq 0,$$

(5.17)

$$\langle\omega_0\rangle_{\beta,0,-} = m(\beta,-) = -\frac{\partial\psi(\beta,0)}{\partial h^-} \leq 0.$$

Thus spontaneous magnetization occurs at β if and only if $\langle\omega_0\rangle_{\beta,0,+} = m(\beta,+) > 0$.

Lemma V.4.5 will be used in the next section in order to prove spontaneous magnetization for models on $\mathbb{Z}^D$, $D \geq 2$.

V.5. Spontaneous Magnetization on $\mathbb{Z}^D$, $D \geq 2$, Via the Peierls Argument

In the previous chapter, we saw that for models on $\mathbb{Z}$ spontaneous magnetization does not occur unless the interaction has infinite range. It occurs, for example, if $J(k)$ equals $|k|^{-\alpha}$ $(k \neq 0)$ for some $1 < \alpha \leq 2$, but not if $\sum_{k \in \mathbb{Z}} |k| J(k)$ converges. Let us contrast this with the behavior of models on $\mathbb{Z}^D$, $D \geq 2$. Spontaneous magnetization occurs for the Ising model, whose interaction has range 1. More generally it occurs for any model for which there exist two linearly independent vectors i and j such that $J(i)$ and $J(j)$ are positive. The range of the interaction is not a factor. The next theorem proves spontaneous magnetization under the extra assumption that i and j are unit vectors. The general case is Problem V.13.8(b).

Theorem V.5.1. *Let J be a summable ferromagnetic interaction on $\mathbb{Z}^D$, $D \geq 2$, and define the positive number $\mathscr{J}_0 = \sum_{k \in \mathbb{Z}^D} J(k)$. Define the critical inverse temperature*

$$\beta_c = \sup\{\beta > 0 : m(\beta, +) = 0\}.$$

(a) *Then β_c is well-defined and $\mathscr{J}_0^{-1} \leq \beta_c$; in particular, $\beta_c > 0$.*

(b) *Assume that there exist two linearly independent unit vectors i and j in $\mathbb{Z}^D$ such that $J(i)$ and $J(j)$ are positive. Then β_c is finite. Spontaneous magnetization occurs at all $\beta > \beta_c$; i.e., $\langle \omega_0 \rangle_{\beta, 0, +} = m(\beta, +) > 0$ for $\beta > \beta_c$.*

For the Ising model on $\mathbb{Z}^2$ with zero external field, Onsager (1944) found the exact value of the specific Gibbs free energy $\psi(\beta, 0)$. This calculation, one of the most famous in mathematical physics, involved the transfer matrix formalism [see Example IV.5.4]. Five years later, he announced without proof the exact values of β_c and $m(\beta, +)$ for this model [Onsager (1949)]:

$$(5.18) \qquad \sinh 2\beta_c \mathscr{J} = 1, \quad m(\beta, +) = [1 - (\sinh(2\beta\mathscr{J}))^{-4}]^{1/8} \quad \text{for} \quad \beta \geq \beta_c.$$

$\mathscr{J} > 0$ is the nearest-neighbor interaction strength. These values were later confirmed to be correct.[3] The lower bound on β_c given in Theorem V.5.1(a) is not a bad estimate of the true value. By Theorem V.5.1(a) $\beta_c \mathscr{J} \geq 0.25$ (since $\mathscr{J}_0 = 4\mathscr{J}$) while according to (5.18) $\beta_c \mathscr{J} = \frac{1}{2} \sinh^{-1} 1 = 0.44+$. Notice that for the Ising model on $\mathbb{Z}^2$, $m(\beta_c, +) = 0$; i.e., there is no spontaneous magnetization at the critical inverse temperature. For other ferromagnetic models, the value of $m(\beta_c, +)$ is not known.

The proof of Theorem V.5.1 uses the moment inequalities developed in the previous section. These inequalities reduce the proof of spontaneous magnetization for an arbitrary model to showing spontaneous magnetization for the Ising model on $\mathbb{Z}^2$. Spontaneous magnetization for the Ising model is implied, of course, by the Onsager values in (5.18). We shall present another proof which is based on a combinatorial argument due to Peierls (1936). The Peierls argument has been generalized and applied to many different models in mathematical physics and is one of the most useful tools for understanding phase transitions.[4]

Proof of Theorem V.5.1 (a) The number β_c is well-defined since $m(\beta, +)$ is a non-negative, nondecreasing function of $\beta > 0$. The bound $\beta_c \geq \mathscr{J}_0^{-1}$, which is a Curie–Weiss lower bound on β_c, follows from Pearce (1981) [see the comments after Theorem IV.5.3 and Note 8 in Chapter IV].

(b) Pick $\mathscr{J} > 0$ such that $J(i) \geq \mathscr{J}$, $J(j) \geq \mathscr{J}$, and compare the specific magnetization $m(\beta, h)$ with the specific magnetization $m_2(\beta, h)$ of the Ising model on $\mathbb{Z}^2$ with interaction strength $J((1,0)) = J((0,1)) = \mathscr{J}$. Let $\tilde{J}$ be the interaction obtained from the original J by reducing to 0 all the interaction strengths $J(k)$, $k \neq i, j$, and by reducing to $\mathscr{J}$ the interaction strengths $J(i)$ and $J(j)$. For $h \geq 0$, $m(\beta, h)$ is a nondecreasing function of each $J(k)$ [Theorem V.4.3(e)]. Since the specific magnetization of the model on $\mathbb{Z}^D$ with interaction $\tilde{J}$ equals $m_2(\beta, h)$, we see that if $h \geq 0$, then $m(\beta, h) \geq m_2(\beta, h)$. This implies that $m(\beta, +) \geq m_2(\beta, +)$. Hence spontaneous magnetization for the original model will follow once we prove that $m_2(\beta, +)$ is positive for sufficiently large β.

Let $P_{\Lambda,\beta,h,+}$ be the finite-volume Gibbs state with the plus external condition for the Ising model on a symmetric square Λ in $\mathbb{Z}^2$. We prove that $m_2(\beta, +)$ is positive for sufficiently large β by proving that $\langle \omega_0 \rangle_{\Lambda,\beta,h,+} \geq$ const > 0 for all $\beta \geq \bar{\beta} > 0$ and all Λ [see Lemma V.4.5(b)]. Since

$$\begin{aligned} \langle \omega_0 \rangle_{\Lambda,\beta,0,+} &= P_{\Lambda,\beta,0,+}\{\omega \in \Omega_\Lambda : \omega_0 = 1\} - P_{\Lambda,\beta,0,+}\{\omega \in \Omega_\Lambda : \omega_0 = -1\} \\ &= 1 - 2P_{\Lambda,\beta,0,+}\{\omega \in \Omega_\Lambda : \omega_0 = -1\}, \end{aligned} \tag{5.19}$$

it suffices to prove a uniform bound

$$P_{\Lambda,\beta,0,+}\{\omega \in \Omega_\Lambda : \omega_0 = -1\} \leq \tfrac{1}{2} - \delta < \tfrac{1}{2} \text{ for all } \beta \geq \bar{\beta} > 0 \text{ and all } \Lambda. \tag{5.20}$$

Peierls' idea was to associate a set of closed curves with each configuration $\omega \in \Omega_\Lambda$. Define the boundary of Λ, bd Λ, to be the set of all sites $j \in \Lambda^c$ which are a distance 1 from Λ. Since the interaction has range 1, bd Λ contains all the sites in Λ^c which interact with sites in Λ. The external condition fixes each spin on bd Λ to have the value $+1$. Given $\omega \in \Omega_\Lambda$, draw a vertical or horizontal segment of unit length halfway between every adjacent pair of plus spin and minus spin in $\Lambda \cup$ bd Λ. Every simple closed curve consisting of such segments is called a border contained in the configuration. For example, Figure V.1 shows five borders, three of length 4, one of length 8, and one of length 18. The ambiguity that arises when two borders touch at a corner is resolved by cutting the corners adjacent to the sites with minus spins.

The key observation is that if the spin at the origin is -1, then because of the plus external condition, the configuration must contain a border which surrounds the origin. Probabilities associated with such borders are not hard to estimate. For γ a positive integer, let Θ_γ be the set of all borders of length γ which surround the origin. We have

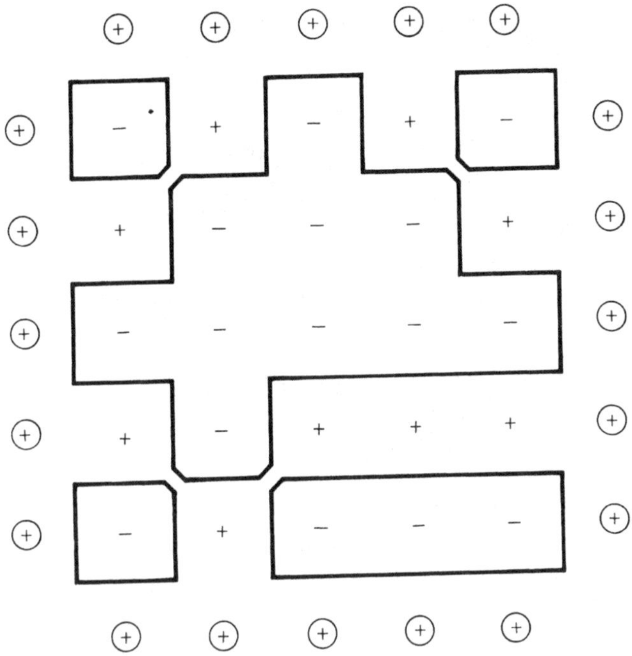

Figure V.1. A particular configuration on a 5 × 5 square to illustrate the Peierls argument. The origin is at the center. The boundary spins (inside circles) are +1 for all configurations.*

$$P_{\Lambda,\beta,0,+}\{\omega\in\Omega_\Lambda : \omega_0 = -1\} \leq \sum_{\gamma\geq 1}\sum_{\theta\in\Theta_\gamma} P_{\Lambda,\beta,0,+}\{\omega\in\Omega_\Lambda : \omega \text{ contains } \theta\}.$$

(5.21)

The next lemma gives an entropy-energy estimate on the cardinality of Θ_γ and on the probability that $\omega\in\Omega_\Lambda$ contains a border in Θ_γ. Since long borders mean many minus spins, it is not surprising that the $P_{\Lambda,\beta,0,+}$-probability of a long border is small. The proof of the next lemma follows the presentation in Spitzer (1971, Chapter 7).

Lemma V.5.2. *The following bounds are valid for any symmetric square* Λ *in* $\mathbb{Z}^2$.

(a) $|\Theta_\gamma| \leq \gamma^2 3^\gamma$.

(b) *For each border* $\theta\in\Theta_\gamma$ *of length* γ

(5.22) $$P_{\Lambda,\beta,0,+}\{\omega\in\Omega_\Lambda : \omega \text{ contains } \theta\} \leq \exp(-\gamma\beta\mathcal{J}).$$

Proof. (a) This proof uses a counting argument (entropy estimate). A border of length γ surrounding the origin either passes just to the right of the origin or it can be obtained by translation of a border which does. The

* Figure V.1 is adapted from Figure 8 in Griffiths (1972, page 61).

number of borders of length γ which pass just to the right of the origin is at most 3^γ, since each unit segment of γ can point in at most three directions. The number of possible translations is at most γ^2 (γ possibilities vertically and γ horizontally). Hence $|\Theta_\gamma| \le \gamma^2 3^\gamma$.

(b) This is proved by studying symmetries of the Hamiltonian (energy estimate). Given a fixed border $\theta \in \Theta_\gamma$, let $\Omega_{\Lambda,\theta}$ be the set of configurations $\omega \in \Omega_\Lambda$ which contain θ. We map $\omega \in \Omega_{\Lambda,\theta}$ into a new configuration $\bar{\omega} \in \Omega_\Lambda$ defined by

$$\bar{\omega}_j = \begin{cases} \omega_j & \text{if } j \text{ is outside the border } \theta, \\ -\omega_j & \text{if } j \text{ is inside the border } \theta. \end{cases}$$

This map is well-defined by the elementary Jordan curve theorem for simple closed polygons. Consider the Hamiltonian

$$H_{\Lambda,0,+}(\omega) = -\frac{1}{2}\sum_{i,j \in \Lambda} J(i-j)\omega_i\omega_j - \sum_{i \in \Lambda, j \in \Lambda^c} J(i-j)\tilde{\omega}_j\omega_i,$$

where $\tilde{\omega}_j$ equals $+1$ and $J(i-j)$ equals $\mathscr{J} > 0$ if $\|i-j\| = 1$ and otherwise equals 0. For every segment of unit length of the border θ, the map $\omega \to \bar{\omega}$ changes a term of the form $\omega_i\omega_j = -1$ or $\tilde{\omega}_j\omega_i = -1$ (i and j on opposite sides of this segment) from -1 to 1. No other changes take place. Hence $H_{\Lambda,0,+}(\omega)$ decreases by $\mathscr{J}$ for every unit segment of θ and so $H_{\Lambda,0,+}(\bar{\omega}) = H_{\Lambda,0,+}(\omega) - \gamma\mathscr{J}$ if θ has length γ. Substituting this into the definition

$$P_{\Lambda,\beta,0,+}\{\omega \in \Omega_\Lambda : \omega \text{ contains } \theta\} = \frac{\sum_{\omega \in \Omega_{\Lambda,\theta}} \exp[-\beta H_{\Lambda,0,+}(\omega)]}{\sum_{\omega \in \Omega_\Lambda} \exp[-\beta H_{\Lambda,0,+}(\omega)]},$$

we see that

$$P_{\Lambda,\beta,0,+}\{\omega \in \Omega_\Lambda : \omega \text{ contains } \theta\} = \exp(-\gamma\beta\mathscr{J}) \frac{\sum_{\omega \in \Omega_{\Lambda,\theta}} \exp[-\beta H_{\Lambda,0,+}(\bar{\omega})]}{\sum_{\omega \in \Omega_\Lambda} \exp[-\beta H_{\Lambda,0,+}(\omega)]}.$$

Each configuration $\bar{\omega}$ which occurs in the sum over $\Omega_{\Lambda,\theta}$ occurs at most once, since two different configurations containing θ cannot be mapped into the same $\bar{\omega}$. Hence the sum over $\Omega_{\Lambda,\theta}$ is a positive number which is less than the sum over Ω_Λ in the numerator. This proves (5.22). □

We now complete the proof of Theorem V.5.1 by showing spontaneous magnetization for the Ising model on $\mathbb{Z}^2$ at all sufficiently large β. By (5.21) and the lemma,

$$(5.23) \qquad P_{\Lambda,\beta,0,+}\{\omega \in \Omega_\Lambda : \omega_0 = -1\} \le f(\beta) = \sum_{\gamma \ge 1} \gamma^2 3^\gamma \exp(-\gamma\beta\mathscr{J}).$$

If $\beta\mathscr{J} > \log 3$, then $f(\beta)$ is finite, and $f(\beta) \to 0$ as $\beta \to \infty$. Hence given $\delta > 0$, we can pick a number $\bar{\beta}$ such that $f(\beta) \le \frac{1}{2} - \delta < \frac{1}{2}$ for all $\beta \ge \bar{\beta}$. As $\bar{\beta}$ is independent of Λ, we have proved (5.20), and we are done.

V.6. Infinite-Volume Gibbs States and Phase Transitions

In Section IV.6, we introduced the set $\mathscr{G}_{\beta,h}$ of infinite-volume Gibbs states for ferromagnetic models on $\mathbb{Z}$. For different values of β and h, we analyzed the structure of $\mathscr{G}_{\beta,h}$ [Theorem IV.6.5] as well as the asymptotic behavior of the spin per site $S_\Lambda/|\Lambda|$ with respect to certain translation invariant measures in $\mathscr{G}_{\beta,h}$ [Theorem IV.6.6]. These theorems will be generalized in this section to ferromagnetic models on $\mathbb{Z}^D$.

The finite-volume Gibbs state $P_{\Lambda,\beta,h,\tilde{\omega}}$ defined in (5.3) is a probability measure on $\Omega_\Lambda = \{1, -1\}^\Lambda$. In Sections V.2–V.5, the external condition was defined by the quantity $\tilde{\omega} = \{\tilde{\omega}_j; j \in \Lambda^c\}$, where each $\tilde{\omega}_j$ takes the value 1, -1, or 0. In order to stay consistent with the definitions in Section IV.6, we now restrict the allowed external conditions by requiring each $\tilde{\omega}_j$, $j \in \Lambda^c$, to take the value 1 or -1. Thus $\tilde{\omega} = \{\tilde{\omega}_j; j \in \Lambda^c\}$ is a point in $\Omega_{\Lambda^c} = \{1, -1\}^{\Lambda^c}$. The probability measures studied in the present section are probability measures on the infinite-volume configuration space $\Omega = \{1, -1\}^{\mathbb{Z}^D}$. These measures are obtained from the finite-volume Gibbs states by weak limits ($\Lambda \uparrow \mathbb{Z}^D$).

The set $\{1, -1\}$ is topologized by the discrete topology and the set Ω by the product topology. According to Tychonoff's theorem, Ω is compact. The σ-field generated by the open sets of the product topology is called the Borel σ-field of Ω and is denoted by $\mathscr{B}(\Omega)$. $\mathscr{B}(\Omega)$ coincides with the σ-field generated by the cylinder sets of Ω [Propositions A.3.2 and A.3.5(b)]. We denote by $\mathscr{M}(\Omega)$ the set of probability measures on $\mathscr{B}(\Omega)$. Let Λ be a symmetric hypercube in $\mathbb{Z}^D$, $\tilde{\omega} = \tilde{\omega}(\Lambda)$ an external condition, $B_{\Lambda,\tilde{\omega}}$ the set $\{\omega \in \Omega : \omega_j = \tilde{\omega}_j$, each $j \in \Lambda^c\}$, and π_Λ the projection of Ω onto Ω_Λ defined by $(\pi_\Lambda \omega)_i = \omega_i$, $i \in \Lambda$. We extend $P_{\Lambda,\beta,h,\tilde{\omega}}$ to a probability measure $\bar{P}_{\Lambda,\beta,h,\tilde{\omega}}$ on $\mathscr{B}(\Omega)$ by setting

$$\text{(5.24)} \qquad \bar{P}_{\Lambda,\beta,h,\tilde{\omega}}\{A\} = P_{\Lambda,\beta,h,\tilde{\omega}}\{\pi_\Lambda(A \cap B_{\Lambda,\tilde{\omega}})\} = \sum_{\omega \in \pi_\Lambda(A \cap B_{\Lambda,\tilde{\omega}})} P_{\Lambda,\beta,h,\tilde{\omega}}\{\omega\}$$

for A a Borel subset of Ω. The right-hand side of (5.24) is given by (5.3). As in Section IV.6, the extension will be denoted by the same symbol $P_{\Lambda,\beta,h,\tilde{\omega}}$ used for the original measure. We say that a sequence $\{P_n; n = 1, 2, \ldots\}$ in $\mathscr{M}(\Omega)$ *converges weakly* to $P \in \mathscr{M}(\Omega)$, and write $P_n \Rightarrow P$ or $P = w\text{-}\lim_{n\to\infty} P_n$, if $\int_\Omega f dP_n \to \int_\Omega f dP$ for every $f \in \mathscr{C}(\Omega)$. With respect to weak convergence $\mathscr{M}(\Omega)$ is a compact metric space [Theorem A.11.2].

Consider the set of limits

$$\mathscr{G}^0_{\beta,h} = \{P \in \mathscr{M}(\Omega) : P = w\text{-}\lim_{\Lambda' \uparrow \mathbb{Z}^D} P_{\Lambda',\beta,h,\tilde{\omega}(\Lambda')}\},$$

where $\{\Lambda'\}$ is any increasing sequence of symmetric hypercubes whose union is $\mathbb{Z}^D$ and $\tilde{\omega}(\Lambda')$ is any external condition for Λ'. Define $\mathscr{G}_{\beta,h}$ to be the closed convex hull of $\mathscr{G}^0_{\beta,h}$. Each measure $P \in \mathscr{G}_{\beta,h}$ is called an *infinite-volume Gibbs state*.* The compactness of $\mathscr{M}(\Omega)$ assures that $\mathscr{G}^0_{\beta,h}$ and thus $\mathscr{G}_{\beta,h}$ are non-

*Equivalent notions of infinite-volume measures are discussed in Appendix C.

empty. A level-3 phase transition is said to occur if $\mathscr{G}_{\beta,h}$ consists of more than one measure.

For each $\alpha \in \{1, \ldots, D\}$ let T_α denote the shift mapping on Ω defined by $(T_\alpha \omega)_j = \omega_{j+u_\alpha}$, where u_α is the unit coordinate vector in the αth direction. The shifts T_α commute. A probability measure P on Ω is said to be *translation invariant* if for each Borel set A and index α $P\{T_\alpha^{-1}A\} = P\{A\}$. Let $\mathscr{M}_s(\Omega)$ denote the set of translation invariant probability measures on $\mathscr{B}(\Omega)$. A measure $P \in \mathscr{M}_s(\Omega)$ is called *ergodic* if $P\{A\}$ equals 0 or 1 for any Borel set A which satisfies $T_\alpha^{-1}A = A$ for all indices α. By definition, a *phase* is a translation invariant infinite-volume Gibbs state. A phase is called *pure* if it is ergodic.

The next theorem generalizes Theorems IV.6.5 and IV.6.6. The proofs of these theorems used the FKG inequality and properties of the specific Gibbs free energy, but since the dimension of the lattice was not a factor, the proofs extend with essentially no change to models on $\mathbb{Z}^D$. Define $S_\Lambda(\omega) = \sum_{j \in \Lambda} Y_j(\omega)$, where Λ is a symmetric hypercube in $\mathbb{Z}^D$ and $Y_j(\omega) = \omega_j$ are the coordinate functions on Ω.

Theorem V.6.1.* *Let J be a summable ferromagnetic interaction on $\mathbb{Z}^D$. Then the following conclusions hold.*

(a) *For each $\beta > 0$ and h real, the weak limits*

$$P_{\beta,h,+} = \underset{\Lambda \uparrow \mathbb{Z}^D}{w\text{-lim}}\, P_{\Lambda,\beta,h,+} \quad \text{and} \quad P_{\beta,h,-} = \underset{\Lambda \uparrow \mathbb{Z}^D}{w\text{-lim}}\, P_{\Lambda,\beta,h,-}$$

exist and are translation invariant. Thus $\mathscr{G}_{\beta,h}$ and $\mathscr{G}_{\beta,h} \cap \mathscr{M}_s(\Omega)$ are nonempty. The measures $P_{\beta,h,+}$ and $P_{\beta,h,-}$ are ergodic.

(b) *$P_{\beta,h,+}$ equals $P_{\beta,h,-}$ if and only if $\partial\psi(\beta,h)/\partial h$ exists. Thus, for $\beta > 0$, $h \neq 0$ and for $0 < \beta < \beta_c$, $h = 0$, $P_{\beta,h,+}$ equals $P_{\beta,h,-}$. For these values of β and h, define $P_{\beta,h} = P_{\beta,h,+} = P_{\beta,h,-}$.*

(c) *If $\partial\psi(\beta,h)/\partial h$ exists, then $P_{\beta,h}$ is the unique measure in $\mathscr{G}_{\beta,h}$ and thus in $\mathscr{G}_{\beta,h} \cap \mathscr{M}_s(\Omega)$. No level-3 phase transition occurs. The mean $\int_\Omega \omega_0 P_{\beta,h}(d\omega)$ equals the specific magnetization $m(\beta,h)$. With respect to $P_{\beta,h}$,*

$$S_\Lambda/|\Lambda| \xrightarrow{\text{a.s.}} m(\beta,h) \quad \text{and} \quad S_\Lambda/|\Lambda| \xrightarrow{\text{exp}} m(\beta,h) \quad \text{as } \Lambda \uparrow \mathbb{Z}^D.$$

(d) *For $\beta > \beta_c$, $P_{\beta,0,+} \neq P_{\beta,0,-}$. In fact,*

$$\int_\Omega \omega_0 P_{\beta,0,+}(d\omega) = m(\beta,+) > 0 > \int_\Omega \omega_0 P_{\beta,0,-}(d\omega) = m(\beta,-).$$

Thus for $\beta > \beta_c$ and $h = 0$, a level-3 phase transition occurs. We have

$$S_\Lambda/|\Lambda| \xrightarrow{\text{a.s.}} m(\beta,+) \quad \text{w.r.t. } P_{\beta,0,+},$$

$$S_\Lambda/|\Lambda| \xrightarrow{\text{a.s.}} m(\beta,-) \quad \text{w.r.t. } P_{\beta,0,-} \quad \text{as } \Lambda \uparrow \mathbb{Z}^D.$$

In each case, exponential convergence fails.

* See Note 11 in Chapter IV for further comments on the structure of $\mathscr{G}_{\beta,h}$ and $\mathscr{G}_{\beta,h} \cap \mathscr{M}_s(\Omega)$.

(e) *For* $\beta > \beta_c$, $\mathscr{G}_{\beta,0} \cap \mathscr{M}_s(\Omega)$ *contains (at least) all the measures* $P_{\beta,0}^{(\lambda)} = \lambda P_{\beta,0,+} + (1-\lambda)P_{\beta,0,-}$, $0 \le \lambda \le 1$. *For each* $0 < \lambda < 1$, *there exists a random variable* $Y_\beta^{(\lambda)}$ *on* Ω *with distribution* $\lambda\delta_{m(\beta,+)} + (1-\delta)\delta_{m(\beta,-)}$ *such that* $S_\Lambda/|\Lambda| \overset{\text{a.s.}}{\to} Y_\beta^{(\lambda)}$ w.r.t. $P_{\beta,0}^{(\lambda)}$ *as* $\Lambda \uparrow \mathbb{Z}^D$.

We emphasize that as in Theorem IV.6.5 the translation invariance of the measures $P_{\beta,h,+}$ and $P_{\beta,h,-}$ follows from the translation invariance of the interaction strength $J(i-j)$ between each pair of sites i, j. The almost sure convergence in part (c) follows from the ergodic theorem [Theorem A.11.5] or from the exponential convergence in part (c) and Theorem II.6.4. The ergodic theorem also yields the almost sure convergence in parts (d) and (e).

With respect to the various measures in Theorem V.6.1, large deviation bounds for $S_\Lambda/|\Lambda|$ can be derived from Theorem II.6.1. We omit the formulas, which are easily worked out as in the discussion at the end of Section IV.5 (see Lemma IV.6.11 for the calculation of the free energy function).

Some refinements can be made in Theorem V.6.1 for the Ising model. The theorem states that for any ferromagnetic model on $\mathbb{Z}^D$, $\mathscr{G}_{\beta,h}$ consists of a unique measure $P_{\beta,h}$ for $\beta > 0$, $h \neq 0$ and for $0 < \beta < \beta_c$, $h = 0$. This fact follows from the existence of $\partial\psi(\beta,h)/\partial h$ for $\beta > 0$, $h \neq 0$ and for $0 < \beta < \beta_c$, $h = 0$. What happens at the critical point $(\beta_c, 0)$? For the Ising model on $\mathbb{Z}^2$, (5.18) shows that $m(\beta_c, +) = 0$. By Lemma V.4.5(b), $\partial\psi(\beta_c, 0)/\partial h$ exists, and so we conclude that $G_{\beta_c,0}$ consists of a unique measure $P_{\beta_c,0}$. For any ferromagnetic model, a natural question is whether for $\beta > \beta_c$ $\mathscr{G}_{\beta,0}$ contains measures which are not translation invariant. For the Ising model on $\mathbb{Z}^D$ with $D \ge 3$, there exist nontranslation invariant infinite-volume Gibbs states at all sufficiently large $\beta > \beta_c$. This was proved by Dobrushin (1972) by analyzing the finite-volume Gibbs states $\{P_{\Lambda,\beta,0\tilde{\omega}}\}$ with the nontranslation invariant external condition $\tilde{\omega}_{(j_1,\cdots,j_D)} = \text{signum}(j_D + \frac{1}{2})$. This gives a plus-external condition on the upper half of the hypercube Λ and a minus-external condition on the bottom half of Λ. Dobrushin's result has been generalized and the proof simplified by van Beijeren (1975). The same construction does not produce nontranslation invariant states for $D = 2$, as was shown by Gallavotti (1972b). In fact, the infinite-volume Gibbs state obtained with this external condition is the average $\frac{1}{2}P_{\beta,0,+} + \frac{1}{2}P_{\beta,0,-}$. Aizenman (1979, 1980) and independently Higuchi (1982) proved the important extension that for $D = 2$ every measure $P \in \mathscr{G}_{\beta,0}$ has the form $P_{\beta,0}^{(\lambda)} = \lambda P_{\beta,0,+} + (1-\lambda)P_{\beta,0,-}$ for some $0 \le \lambda \le 1$. In particular, for the Ising model on $\mathbb{Z}^2$, any infinite-volume Gibbs state must be translation invariant. For each $0 < \lambda < 1$, Abraham and Reed (1976) have produced a sequence of external conditions $\{\tilde{\omega}_\lambda(\Lambda)\}$ such that $P_{\Lambda,\beta,0,\tilde{\omega}_\lambda(\Lambda)} \Rightarrow P_{\beta,0}^{(\lambda)}$ as $\Lambda \uparrow \mathbb{Z}^2$. See the review by Pfister (1983).

We recall that for models on $\mathbb{Z}$ the translation invariant infinite-volume Gibbs states are characterized by the Gibbs variational principle [Theorem IV.7.3(b)]. This principle involves the functionals $u(h; P)$ and $I_\rho^{(3)}(P)$, which are, respectively, the specific energy in P and the mean relative entropy of P

with respect to P_ρ. We now state a generalization of Theorem IV.7.3 for models on $\mathbb{Z}^D$. Let P be a measure in $\mathcal{M}_s(\Omega)$; Λ a symmetric hypercube; π_Λ the projection of Ω onto Ω_Λ defined by $(\pi_\Lambda\omega)_i = \omega_i$, $i \in \Lambda$; and $\pi_\Lambda P$ the probability measure on $\mathscr{B}(\Omega_\Lambda)$ defined by $\pi_\Lambda P\{F\} = P\{\pi_\Lambda^{-1}F\}$ for subsets F of Ω_Λ. Let P_ρ equal the infinite product measure in $\mathcal{M}_s(\Omega)$ with identical one-dimensional marginals $\rho = \frac{1}{2}\delta_1 + \frac{1}{2}\delta_{-1}$. Given an external condition $\tilde{\omega}(\Lambda)$, define the functional

$$U(\Lambda, h, \tilde{\omega}(\Lambda); \pi_\Lambda P) = \int_{\Omega_\Lambda} H_{\Lambda,h,\tilde{\omega}(\Lambda)}(\omega)\pi_\Lambda P(d\omega),$$

where $H_{\Lambda,h,\tilde{\omega}(\Lambda)}$ is the Hamiltonian defined in (5.2). $I^{(2)}_{\pi_\Lambda P_\rho}(\pi_\Lambda P)$ denotes the relative entropy of $\pi_\Lambda P$ with respect to $\pi_\Lambda P_\rho$,

$$\int_{\Omega_\Lambda} \log \frac{d(\pi_\Lambda P)}{d(\pi_\Lambda P_\rho)}(\omega)\pi_\Lambda P(d\omega).$$

Lemma V.6.2. *Let J be a summable ferromagnetic interaction on $\mathbb{Z}^D$. Then for any h real, the following conclusions hold.*

(a) *For any $P \in \mathcal{M}_s(\Omega)$, $u(h; P) = \lim_{\Lambda\uparrow\mathbb{Z}^D}|\Lambda|^{-1}U(\Lambda, h, \tilde{\omega}(\Lambda); P)$ exists and is independent of the choice of external conditions $\{\tilde{\omega}(\Lambda)\}$. The limit is given by*

$$u(h; P) = -\frac{1}{2}\sum_{k\in\mathbb{Z}^D} J(k)\int_\Omega \omega_0\omega_k P(d\omega) - h\int_\Omega \omega_0 P(d\omega)$$

and is a bounded, affine, continuous functional of $P \in \mathcal{M}_s(\Omega)$.

(b) *For any $P \in \mathcal{M}_s(\Omega)$, $\lim_{\Lambda\uparrow\mathbb{Z}^D}|\Lambda|^{-1}I^{(2)}_{\pi_\Lambda P_\rho}(\pi_\Lambda P)$ exists and defines an affine, lower semicontinuous functional of $P \in \mathcal{M}_s(\Omega)$. This functional is denoted by $I^{(3)}_\rho(P)$.*

Part (a) of the lemma is proved exactly like Lemma IV.7.2(b). Part (b) is due to Robinson and Ruelle (1967). Also see Israel (1979, Section II.2) for a proof.

The next theorem states the *Gibbs variational formula* and the *Gibbs variational principle* (parts (a) and (b), respectively). The theorem is due to Ruelle (1967) and Lanford and Ruelle (1969).

Theorem V.6.3. *Let J be a summable ferromagnetic interaction on $\mathbb{Z}^D$. Then for $\beta > 0$ and h real, the following conclusions hold.*

(a) $-\beta\psi(\beta, h) = \sup_{P\in\mathcal{M}_s(\Omega)}\{-\beta u(h; P) - I^{(3)}_\rho(P)\}$, *where $\psi(\beta, h)$ is the specific Gibbs free energy.*

(b) *The set of $P \in \mathcal{M}_s(\Omega)$ at which the supremum in part* (a) *is attained equals $\mathscr{G}_{\beta,h} \cap \mathcal{M}_s(\Omega)$, the set of translation invariant infinite-volume Gibbs states.*

In Chapter IV, where we studied models on $\mathbb{Z}$, the Gibbs variational formula in part (a) was proved by means of level-3 large deviations. For models on $\mathbb{Z}^D$, $D \geq 2$, a large deviations proof of the formula is at present not available. In Appendix C.5, we sketch a proof of the Gibbs variational formula and principle for a much larger class of models than we are now considering. The proof is due to Föllmer (1973) and Preston (1976). Formula (3.3) of Föllmer (1973) generalizes to $\mathbb{Z}^D$ formula (2.20) in Chapter II which expresses the mean relative entropy in terms of a regular conditional distribution. The measures $P \in \mathscr{G}_{\beta,h} \cap \mathscr{M}_s(\Omega)$ are also characterized by an entropy principle as in Theorem IV.7.4. The proof extends to models on $\mathbb{Z}^D$ without change.

This completes Part 1 of Chapter V. We have succeeded in generalizing to models on $\mathbb{Z}^D$ all of the results presented in the previous chapter for models on $\mathbb{Z}$. Part 2 is devoted to new material.

PART 2

V.7. Infinite-Volume Gibbs States and the Central Limit Theorem

Let J be a summable ferromagnetic interaction on $\mathbb{Z}^D$ and $m(\beta, h)$ the corresponding specific magnetization. The critical inverse temperature β_c is defined as $\sup\{\beta > 0: m(\beta, +) = 0\}$, where $m(\beta, +) = \lim_{h \to 0^+} m(\beta, h)$. We denote by $\mathcal{U}$ the set of points (β, h) with $\beta > 0$, $h \neq 0$ and $0 < \beta < \beta_c$, $h = 0$. For each $(\beta, h) \in \mathcal{U}$ there exists a unique infinite-volume Gibbs state $P_{\beta,h}$. With respect to $P_{\beta,h}$, the spin per site $S_\Lambda/|\Lambda| = \sum_{j \in \Lambda} Y_j/|\Lambda|$ satisfies a law of large numbers with limiting mean $m(\beta, h)$ [Theorem V.6.1(c)]. The main result in this section, Theorem V.7.2, proves a central limit theorem for the spin $S_\Lambda = \sum_{j \in \Lambda} Y_j$ under the reasonable assumption that the correlations among the $\{Y_j\}$ are weak. A fascinating problem occurs at the critical point $(\beta_c, 0)$, where the correlations are abnormally large and the central limit theorem is thought to fail. This problem will be discussed in Section V.8.

For $(\beta, h) \in \mathcal{U}$ and $k \in \mathbb{Z}^D$, define the quantity

$$\langle Y_0 ; Y_k \rangle_{\beta,h} = \langle Y_0 Y_k \rangle_{\beta,h} - \langle Y_0 \rangle_{\beta,h} \langle Y_k \rangle_{\beta,h},$$

where Y_0 denotes the spin at the origin of $\mathbb{Z}^D$ and $\langle - \rangle_{\beta,h}$ denotes expectation with respect to $P_{\beta,h}$. $\langle Y_0 ; Y_k \rangle_{\beta,h}$ is called a *pair correlation* and is non-negative by the Percus inequality. We say that S_Λ satisfies a central limit theorem with respect to $P_{\beta,h}$ if $V_\Lambda = [S_\Lambda - |\Lambda| m(\beta, h)]/\sqrt{|\Lambda|}$ converges in distribution to a Gaussian random variable with mean 0 and some variance $\sigma^2(\beta, h) > 0$; i.e., for any $f \in \mathscr{C}(\mathbb{R})$

$$\lim_{\Lambda \uparrow \mathbb{Z}^D} \langle f(V_\Lambda) \rangle_{\beta,h} = \int_{\mathbb{R}} f(x) N(0, \sigma^2(\beta, h))(dx),$$

where $N(0, \sigma^2)(dx) = (2\pi\sigma^2)^{-1/2} \exp(-x^2/2\sigma^2) dx$. This limit is summarized by writing $V_\Lambda \xrightarrow{\mathscr{D}} N(0, \sigma^2(\beta, h))$. If S_Λ satisfies a central limit theorem, then a reasonable guess for the variance of the limiting Gaussian random variable is $\sigma^2(\beta, h) = \lim_{\Lambda \uparrow \mathbb{Z}^D} |\Lambda|^{-1} \Sigma_\Lambda^2(\beta, h)$, where $\Sigma_\Lambda^2(\beta, h)$ is the variance of S_Λ. Substituting $S_\Lambda = \sum_{j \in \Lambda} Y_j$, we find that

$$\Sigma_\Lambda^2(\beta, h) = \langle S_\Lambda^2 \rangle_{\beta,h} - \langle S_\Lambda \rangle_{\beta,h}^2 = \sum_{i \in \Lambda} \sum_{j \in \Lambda} \langle Y_i ; Y_j \rangle_{\beta,h}.$$

The following lemma establishes the existence of $\sigma^2(\beta,h)$ and gives an explicit formula for it. The proof is due to Newman (1980).

Lemma V.7.1. *For $(\beta,h)\in\mathscr{U}$, define $\Sigma_\Lambda^2(\beta,h)=\langle S_\Lambda^2\rangle_{\beta,h}-\langle S_\Lambda\rangle_{\beta,h}^2$. Then the limit $\sigma^2(\beta,h)=\lim_{\Lambda\uparrow\mathbb{Z}^D}|\Lambda|^{-1}\Sigma_\Lambda^2(\beta,h)$ exists as an extended real-valued number and $\sigma^2(\beta,h)=\sum_{k\in\mathbb{Z}^D}\langle Y_0\,;Y_k\rangle_{\beta,h}$.*

Proof. By translation invariance, $\langle Y_i;Y_j\rangle_{\beta,h}$ is a function of $i-j$ only, say $C(i-j)$. Since $C(i-j)\geq 0$,

$$\frac{1}{|\Lambda|}\Sigma_\Lambda^2(\beta,h)\leq\frac{1}{|\Lambda|}\sum_{i\in\Lambda}\sum_{j\in\mathbb{Z}^D}C(i-j)=\sum_{k\in\mathbb{Z}^D}C(k),$$

and so $\limsup_{\Lambda\uparrow\mathbb{Z}^D}|\Lambda|^{-1}\Sigma_\Lambda^2(\beta,h)\leq\sum_{k\in\mathbb{Z}^D}C(k)$. If $\Lambda=\{j\in\mathbb{Z}^D:|j_\alpha|\leq N,\ \alpha=1,\ldots,D\}$, then for any $0<\varepsilon<1$ define $\Lambda(\varepsilon)=\{j\in\mathbb{Z}^D:|j_\alpha|\leq(1-\varepsilon)N,\ \alpha=1,\ldots,D\}$. Note that if $i\in\Lambda(\varepsilon)$ and $\|i-j\|\leq\varepsilon N$, then $j\in\Lambda$. Hence

$$\begin{aligned}\frac{1}{|\Lambda|}\Sigma_\Lambda^2(\beta,h)&=\frac{1}{|\Lambda|}\sum_{i\in\Lambda}\sum_{j\in\Lambda}C(i-j)\geq\frac{1}{|\Lambda|}\sum_{i\in\Lambda(\varepsilon)}\sum_{j\in\Lambda}C(i-j)\\&\geq\frac{1}{|\Lambda|}\sum_{i\in\Lambda(\varepsilon)}\sum_{\|i-j\|\leq\varepsilon N}C(i-j)=\frac{|\Lambda(\varepsilon)|}{|\Lambda|}\sum_{\|k\|\leq\varepsilon N}C(k).\end{aligned}$$

Choosing $\varepsilon=\varepsilon_N$ such that $\varepsilon_N\to 0$ and $\varepsilon_N N\to\infty$ as $N\to\infty$, we see that $\liminf_{\Lambda\uparrow\mathbb{Z}^D}|\Lambda|^{-1}\Sigma_\Lambda^2(\beta,h)\geq\sum_{k\in\mathbb{Z}^D}C(k)$. It follows that

$$\sigma^2(\beta,h)=\lim_{\Lambda\uparrow\mathbb{Z}^D}\frac{1}{|\Lambda|}\Sigma_\Lambda^2(\beta,h)=\sum_{k\in\mathbb{Z}^D}\langle Y_0\,;Y_k\rangle_{\beta,h}.\qquad\square$$

The central limit theorem for S_Λ is stated next, under the assumption that $\sigma^2(\beta,h)$ is finite.[5] We prove that $\sigma^2(\beta,h)$ is finite for all $\beta>0$ and $h\neq 0$, using the fact that for each $\beta>0$ the specific Gibbs free energy $\psi(\beta,h)$ is a real analytic function of $h\neq 0$ [Theorem V.4.4]. The finiteness of $\sigma^2(\beta,h)$ for $0<\beta<\beta_c$ and $h=0$ will be discussed in Theorems V.7.6 and V.7.7, which appear after the proof of the central limit theorem. For the Ising model on $\mathbb{Z}^D$, $D\geq 2$, we will see that $\sigma^2(\beta,h)$ is finite for all $(\beta,h)\in\mathscr{U}$ and thus that the central limit theorem holds for all $(\beta,h)\in\mathscr{U}$.

Theorem V.7.2. *Let J be a summable ferromagnetic interaction on $\mathbb{Z}^D$ and (β,h) a point in $\mathscr{U}$.*

(a) *If $\sigma^2(\beta,h)=\sum_{k\in\mathbb{Z}^D}\langle Y_0\,;Y_k\rangle_{\beta,h}$ is finite,* then*

$$\frac{S_\Lambda-|\Lambda|m(\beta,h)}{\sqrt{|\Lambda|}}\xrightarrow{\mathscr{D}}N(0,\sigma^2(\beta,h))\qquad\text{with respect to }P_{\beta,h}\qquad\text{as }\Lambda\uparrow\mathbb{Z}^D.$$

(b) *$\sigma^2(\beta,h)$ is finite for all $\beta>0$ and $h\neq 0$.*

The proof of this theorem is lengthy and depends on several auxiliary results. Lemma V.7.4 will be referred to in the next section. The rest of the proof can be skipped with no loss of continuity.

Lemma V.7.4 yields part (b) of the theorem by relating $\sigma^2(\beta,h)$ to the

* $\sigma^2(\beta,h)$ is positive by Lemma V.7.3(a) below.

quantity $\chi(\beta, h) = \partial m(\beta, h)/\partial h$. This function, which measures the response of the specific magnetization to changes in h, is called the *specific magnetic susceptibility*. The lemma shows that for $\beta > 0$ and $h \neq 0$, $\sigma^2(\beta, h)$ is finite, $\chi(\beta, h)$ exists, and

$$(5.25) \qquad \beta \cdot \sigma^2(\beta, h) = \beta \cdot \lim_{\Lambda \uparrow \mathbb{Z}^D} \frac{1}{|\Lambda|} \{\langle S_\Lambda^2 \rangle_{\beta,h} - \langle S_\Lambda \rangle_{\beta,h}^2\} = \chi(\beta, h).$$

To see that (5.25) is reasonable, consider the free energy function $c_{\beta,h}(t)$ of the sequence $\{S_\Lambda\}$ with respect to the infinite-volume Gibbs state $P_{\beta,h}$. In Lemma IV.6.11, the proof of which applies verbatim to models on $\mathbb{Z}^D$, we showed that for any real t

$$\begin{aligned} c_{\beta,h}(t) &= \lim_{\Lambda \uparrow \mathbb{Z}^D} c_\Lambda(t) \\ (5.26) \qquad &= \lim_{\Lambda \uparrow \mathbb{Z}^D} \frac{1}{|\Lambda|} \log \langle \exp(t S_\Lambda) \rangle_{\beta,h} \\ &= -\beta[\psi(\beta, h + t/\beta) - \psi(\beta, h)]. \end{aligned}$$

The function $c_{\beta,h}(t)$ is differentiable at $t = 0$, and this gave the law of large numbers for S_Λ with limiting mean

$$\lim_{\Lambda \uparrow \mathbb{Z}^D} \langle S_\Lambda/|\Lambda| \rangle_{\beta,h} = \lim_{\Lambda \uparrow \mathbb{Z}^D} c'_\Lambda(0) = c'_{\beta,h}(0) = -\frac{\partial \psi(\beta, h)}{\partial h} = m(\beta, h).$$

An easy calculation yields $c''_\Lambda(0) = |\Lambda|^{-1}\{\langle S_\Lambda^2 \rangle_{\beta,h} - \langle S_\Lambda \rangle_{\beta,h}^2\}$, and since this converges to $\sigma^2(\beta, h)$, it is tempting to conclude that

$$\begin{aligned} \sigma^2(\beta, h) &= \lim_{\Lambda \uparrow \mathbb{Z}^D} c''_\Lambda(0) = c''_{\beta,h}(0) = -\beta^{-1} \frac{\partial^2 \psi(\beta, h)}{\partial h^2} \\ &= \beta^{-1} \frac{\partial m(\beta, h)}{\partial h} = \beta^{-1} \chi(\beta, h). \end{aligned}$$

Lemma V.7.3. *Let (β, h) be a point in $\mathscr{U}$. Then the following conclusions hold.*

(a) *For $h \geq 0$, $0 < \sum_{k \in \Lambda} \langle Y_0; Y_k \rangle_{\Lambda,\beta,h,+} \uparrow \sum_{k \in \mathbb{Z}^D} \langle Y_0; Y_k \rangle_{\beta,h} \leq \infty$ as $\Lambda \uparrow \mathbb{Z}^D$.*

(b) *For $k \in \mathbb{Z}^D$ and $h \geq 0$, $0 \leq \langle Y_0; Y_k \rangle_{\beta,t} \leq \langle Y_0; Y_k \rangle_{\beta,h}$ for all $t \geq h$.*

(c) *For $k \in \mathbb{Z}^D$ and $h \geq 0$, $\langle Y_0; Y_k \rangle_{\beta,t} \uparrow \langle Y_0; Y_k \rangle_{\beta,h}$ as $t \to h^+$.*

Proof. (a) Given symmetric hypercubes $\Lambda \subseteq \Lambda'$ and a site $k \in \Lambda$, $\langle Y_0; Y_k \rangle_{\Lambda,\beta,h,+}$ can be obtained from $\langle Y_0; Y_k \rangle_{\Lambda',\beta,h,+}$ by taking $h_i \to \infty$ for each site $i \in \Lambda' \backslash \Lambda$. We have [see the calculation on page 148]

$$\begin{aligned} \frac{\partial}{\partial h_i} \langle Y_0; Y_k \rangle_{\Lambda',\beta,h,+} &= \frac{\partial}{\partial h_i} [\langle Y_0 Y_k \rangle - \langle Y_0 \rangle \langle Y_k \rangle] \\ &= \langle Y_0 Y_k Y_i \rangle - \langle Y_0 \rangle \langle Y_k Y_i \rangle - \langle Y_k \rangle \langle Y_0 Y_i \rangle \\ &\quad - \langle Y_i \rangle \langle Y_0 Y_k \rangle + 2 \langle Y_0 \rangle \langle Y_k \rangle \langle Y_i \rangle. \end{aligned}$$

For $h \geq 0$, this sum is nonpositive by the GHS inequality. Thus for $h \geq 0$

$\langle Y_0; Y_k\rangle_{\Lambda,\beta,h,+} \leq \langle Y_0; Y_k\rangle_{\Lambda',\beta,h,+}$, and as $\Lambda \uparrow \mathbb{Z}^D$ $\langle Y_0; Y_k\rangle_{\Lambda,\beta,h,+}$ increases to $\langle Y_0; Y_k\rangle_{\beta,h}$. Each term $\langle Y_0; Y_k\rangle_{\Lambda,\beta,h,+}$ is non-negative (Percus inequality) and $\sum_{k\in\Lambda}\langle Y_0; Y_k\rangle_{\Lambda,\beta,h,+}$ increases to $\sum_{k\in\mathbb{Z}^D}\langle Y_0; Y_k\rangle_{\beta,h}$ (monotone convergence theorem). By explicit calculation, $\langle Y_0; Y_0\rangle_{\{0\},\beta,h,+}$ is positive, and so for any symmetric hypercube Λ

$$0 < \langle Y_0; Y_0\rangle_{\{0\},\beta,h,+} \leq \langle Y_0; Y_0\rangle_{\Lambda,\beta,h,+}.$$

It follows that $\sum_{k\in\Lambda}\langle Y_0; Y_k\rangle_{\Lambda,\beta,h,+}$ is positive.

(b) The Percus inequality implies that $\langle Y_0; Y_k\rangle_{\beta,t} \geq 0$. The GHS inequality implies that

$$\langle Y_0; Y_k\rangle_{\beta,t} \leq \langle Y_0; Y_k\rangle_{\beta,h} \qquad \text{for all } t \geq h \geq 0 \text{ and sites } k \in \mathbb{Z}^D. \tag{5.27}$$

(c) This is a consequence of (5.27) and Lemma IV.6.9(b). The proof of the latter uses the FKG inequality and extends verbatim to models on $\mathbb{Z}^D$. □

Our proof of the next lemma follows Sokal (1981, Appendix).

Lemma V.7.4. *Let J be a summable ferromagnetic interaction on $\mathbb{Z}^D$. Then the following conclusions hold.*

(a) *For $\beta > 0$ and $h \neq 0$, the specific magnetic susceptibility $\chi(\beta,h) = \partial m(\beta,h)/\partial h = -\partial^2\psi(\beta,h)/\partial h^2$ exists, the sum $\sigma^2(\beta,h) = \sum_{k\in\mathbb{Z}^D}\langle Y_0; Y_k\rangle_{\beta,h}$ is finite, and**

$$\chi(\beta,h) = -\frac{\partial^2\psi(\beta,h)}{\partial h^2} = \beta\cdot\sigma^2(\beta,h) > 0. \tag{5.28}$$

(b) *For $0 < \beta < \beta_c$ and $h = 0$, $\chi(\beta,0)$ exists and is finite if and only if $\sigma^2(\beta,0)$ is finite. Formula (5.28) holds in the sense that the three terms are all finite or all infinite together.*

Proof. (a) For $(\beta,h_0)\in\mathscr{U}$, $\partial\psi(\beta,h)/\partial h$ exists for all h in some neighborhood of h_0 and $-\partial\psi(\beta,h)/\partial h$ equals $m(\beta,h) = \langle Y_0\rangle_{\beta,h} = \lim_{\Lambda\uparrow\mathbb{Z}^D}\langle Y_0\rangle_{\Lambda,\beta,h,+}$ [Theorem V.4.3 and Lemma V.4.5]. Fix $h_0 \geq 0$ and denote by $D_+^2\psi(\beta,h_0)$ the right-hand h-derivative of $\partial\psi/\partial h$ at (β,h_0). Since $\partial\langle Y_0\rangle_{\Lambda,\beta,h,+}/\partial h = \beta\cdot\sum_{k\in\Lambda}\langle Y_0; Y_k\rangle_{\Lambda,\beta,h,+}$, we have

$$\begin{aligned}
-D_+^2\psi(\beta,h_0) &= -\lim_{h\to 0^+}\frac{1}{h}\left(\frac{\partial\psi(\beta,h_0+h)}{\partial h} - \frac{\partial\psi(\beta,h_0)}{\partial h}\right)\\
&= \lim_{h\to 0^+}\lim_{\Lambda\uparrow\mathbb{Z}^D}\frac{1}{h}(\langle Y_0\rangle_{\Lambda,\beta,h_0+h,+} - \langle Y_0\rangle_{\Lambda,\beta,h_0,+})\\
&= \lim_{h\to 0^+}\lim_{\Lambda\uparrow\mathbb{Z}^D}\frac{1}{h}\int_{h_0}^{h_0+h}\frac{\partial\langle Y_0\rangle_{\Lambda,\beta,t,+}}{\partial t}\,dt\\
&= \beta\cdot\lim_{h\to 0^+}\lim_{\Lambda\uparrow\mathbb{Z}^D}\frac{1}{h}\int_{h_0}^{h_0+h}\sum_{k\in\Lambda}\langle Y_0; Y_k\rangle_{\Lambda,\beta,t,+}\,dt.
\end{aligned} \tag{5.29}$$

* For any $\beta > 0$, $m(\beta,+)$ is non-negative. For $h > 0$ (5.28) implies $\partial m(\beta,h)/\partial h > 0$. Hence for $h > 0$ $m(\beta,h)$ is positive. This fact was needed on page 116.

Because of Lemma V.7.3(a), we may interchange the limit $\Lambda \uparrow \mathbb{Z}^D$ with the integral in the last term in (5.29) to obtain

$$(5.30) \qquad -D_+^2\psi(\beta, h_0) = \beta \cdot \lim_{h \to 0^+} \frac{1}{h} \int_{h_0}^{h_0+h} \sum_{k \in \mathbb{Z}^D} \langle Y_0; Y_k \rangle_{\beta,t} dt.$$

Assume that $\sigma^2(\beta, h_0)$ is finite. Lemma V.7.3(b) justifies bringing the sum over k in (5.30) outside the limit $h \to 0^+$. Lemma V.7.3(c) implies

$$\begin{aligned} -D_+^2\psi(\beta, h_0) &= \beta \sum_{k \in \mathbb{Z}^D} \lim_{h \to 0^+} \frac{1}{h} \int_{h_0}^{h_0+h} \langle Y_0; Y_k \rangle_{\beta,t} dt \\ &= \beta \sum_{k \in \mathbb{Z}^D} \langle Y_0; Y_k \rangle_{\beta,h_0} = \beta \cdot \sigma^2(\beta, h_0). \end{aligned}$$

Now assume that $\sigma^2(\beta, h_0)$ is infinite. Then by Lemma V.7.3(c), given arbitrary $L > 0$ there exist positive numbers R and h such that $\sum_{\|k\| \le R} \langle Y_0; Y_k \rangle_{\beta,t} \ge L$ for all $t \in [h_0, h_0 + h]$. Hence

$$-D_+^2\psi(\beta, h_0) \ge \beta \cdot \lim_{h \to 0^+} \frac{1}{h} \int_{h_0}^{h_0+h} \sum_{\|k\| \le R} \langle Y_0; Y_k \rangle_{\beta,t} dt \ge \beta L,$$

and since L is arbitrary, $-D_+^2\psi(\beta, h_0) = \infty$. The conclusion is that for $h_0 \ge 0$, $\sigma^2(\beta, h_0)$ is finite if and only if $D_+^2\psi(\beta, h_0)$ exists and is finite, and then the latter equals $-\sigma^2(\beta, h_0)$.

According to Theorem V.4.4, for each $\beta > 0$ $\psi(\beta, h)$ is a real analytic function of $h \neq 0$. This implies that for $h > 0$, $D_+^2\psi(\beta, h)$ exists and equals $\partial^2\psi(\beta, h)/\partial h^2$. By our work in the previous paragraph $\sigma^2(\beta, h)$ is then finite and equals $-\partial^2\psi(\beta, h)/\partial h^2$; $\sigma^2(\beta, h)$ is positive by Lemma V.7.3(a). This proves part (a) of Lemma V.7.4 for $h > 0$. Since $\psi(\beta, h)$ is an even function of h, we obtain part (a) for $h < 0$ by symmetry.

(b) $D_-^2\psi(\beta, 0)$ equals $D_+^2\psi(\beta, 0)$ (if either exists) since $\partial\psi(\beta, h)/\partial h$ is an odd function of h and $\partial\psi(\beta, 0)/\partial h = 0$. According to the proof of part (a), $\partial^2\psi(\beta, 0)/\partial h^2$ exists and is finite if and only if $\sigma^2(\beta, 0)$ is finite, and then (5.28) holds. This completes the proof of Lemma V.7.4. □

We are now ready to prove the central limit theorem.

Proof of Theorem V.7.2. Part (b) is implied by Lemma V.7.4(a). The proof of part (a) is based on convexity. We give the proof for $h \ge 0$; $h < 0$ is handled similarly. Define $V_\Lambda = [S_\Lambda - |\Lambda| m(\beta, h)]/\sqrt{|\Lambda|}$. It suffices to prove that

$$(5.31) \qquad \lim_{\Lambda \uparrow \mathbb{Z}^D} \langle \exp(tV_\Lambda) \rangle_{\beta,h} = \exp(\tfrac{1}{2}\sigma^2(\beta, h)t^2) \qquad \text{for all } t \in [0, \alpha),$$

some $\alpha > 0$ [Theorem A.8.7(a)]. We express $\langle \exp(tV_\Lambda) \rangle_{\beta,h}$ in terms of the function $c_\Lambda(t) = |\Lambda|^{-1} \log\langle \exp(tS_\Lambda) \rangle_{\beta,h}$. Note that $c_\Lambda(0) = 0$, $c'_\Lambda(0) = |\Lambda|^{-1}\langle S_\Lambda \rangle_{\beta,h} = m(\beta, h)$, and $c''_\Lambda(0) = |\Lambda|^{-1}\{\langle S_\Lambda^2 \rangle_{\beta,h} - \langle S_\Lambda \rangle_{\beta,h}^2\}$. By Lemma V.7.1

$$(5.32) \qquad c''_\Lambda(0) = \frac{1}{|\Lambda|} \Sigma_\Lambda^2(\beta, h) \to \sigma^2(\beta, h) \qquad \text{as } \Lambda \uparrow \mathbb{Z}^D.$$

Take $t > 0$ and set $t_\Lambda = t/\sqrt{|\Lambda|}$. We have by Taylor's theorem with remainder

$$\begin{aligned}\log\langle\exp(tV_\Lambda)\rangle_{\beta,h} &= |\Lambda|c_\Lambda(t_\Lambda) - t\sqrt{|\Lambda|}m(\beta,h)\\ &= t^2[t_\Lambda^{-2}c_\Lambda(t_\Lambda) - t_\Lambda^{-1}c'_\Lambda(0)]\\ &= t^2\int_0^1 (1-s)c''_\Lambda(st_\Lambda)ds.\end{aligned} \tag{5.33}$$

The following properties of c_Λ are proved below.

(5.34) For $h \geq 0$, c'_Λ is concave on $[0, \infty)$.

(5.35) For $h \geq 0$, $\lim_{\Lambda\uparrow\mathbb{Z}^D} c''_\Lambda(t_\Lambda) = \sigma^2(\beta,h)$ for all $t\in[0,\alpha)$, some $\alpha > 0$.

The concavity of c'_Λ on $[0,\infty)$ implies that c''_Λ is nonincreasing on $[0,\infty)$. Hence for each $0 \leq s \leq 1$ $c''_\Lambda(t_\Lambda) \leq c''_\Lambda(st_\Lambda) \leq c''_\Lambda(0)$. By (5.33), for each $t > 0$

$$\tfrac{1}{2}c''_\Lambda(t_\Lambda)t^2 \leq \log\langle\exp(tV_\Lambda)\rangle_{\beta,h} \leq \tfrac{1}{2}c''_\Lambda(0)t^2.$$

By (5.32) and (5.35), the two extreme terms in the last display both converge to $\frac{1}{2}\sigma^2(\beta,h)t^2$ for all $0 < t < \alpha$. This gives (5.31) for all $0 < t < \alpha$, and (5.31) is obvious for $t = 0$. Hence the central limit theorem will be proved once we show (5.34) and (5.35).

Proof of property (5.34). For $h \geq 0$, the concavity of c'_Λ on $[0,\infty)$ is a consequence of the GHS inequality. Consider the sequence of finite-volume Gibbs states on symmetric hypercubes $\{\bar\Lambda\}$ in $\mathbb{Z}^D$:

$$P_{\bar\Lambda,\beta,h,+}(d\omega) = \exp[-\beta H_{\bar\Lambda,h,+}(\omega)]\pi_{\bar\Lambda}P_\rho(d\omega)\cdot\frac{1}{Z(\bar\Lambda,\beta,h,+)},$$

where

$$Z(\bar\Lambda,\beta,h,+) = \int_{\Omega_{\bar\Lambda}} \exp[-\beta H_{\bar\Lambda,h,+}(\omega)]\pi_{\bar\Lambda}P_\rho(d\omega).$$

Since $P_{\bar\Lambda,\beta,h,+} \Rightarrow P_{\beta,h}$ as $\bar\Lambda\uparrow\mathbb{Z}^D$, we have

$$\begin{aligned}c'_\Lambda(t) &= \frac{1}{|\Lambda|}\langle S_\Lambda\exp(tS_\Lambda)\rangle_{\beta,h}\cdot\frac{1}{\langle\exp(tS_\Lambda)\rangle_{\beta,h}}\\ &= \frac{1}{|\Lambda|}\lim_{\bar\Lambda\uparrow\mathbb{Z}^D}\langle S_\Lambda\exp(tS_\Lambda)\rangle_{\bar\Lambda,\beta,h,+}\cdot\frac{1}{\langle\exp(tS_\Lambda)\rangle_{\bar\Lambda,\beta,h,+}}\\ &= \frac{1}{|\Lambda|}\lim_{\bar\Lambda\uparrow\mathbb{Z}^D}\sum_{k\in\Lambda}\langle Y_k\exp(tS_\Lambda)\rangle_{\bar\Lambda,\beta,h,+}\cdot\frac{1}{\langle\exp(tS_\Lambda)\rangle_{\bar\Lambda,\beta,h,+}}.\end{aligned} \tag{5.36}$$

We rewrite the expectations in the last display. Take $\bar\Lambda \supseteq \Lambda$ and let $h_i = h + \sum_{j\in\bar\Lambda^c} J(i-j)(+1)$ for $i\in\bar\Lambda$. Define new external fields $h_i(t) = h_i + t/\beta$ for $i\in\Lambda$ and $h_i(t) = h_i$ for $i\in\bar\Lambda\backslash\Lambda$. Then for each $\omega\in\Omega_{\bar\Lambda}$

$$\begin{aligned}tS_\Lambda(\omega) - \beta H_{\bar\Lambda,h,+}(\omega) &= \frac{\beta}{2}\sum_{i,j\in\bar\Lambda} J(i-j)\omega_i\omega_j + \beta\sum_{i\in\bar\Lambda} h_i\omega_i + t\sum_{i\in\Lambda}\omega_i\\ &= -\beta H_{\bar\Lambda,\{h_i(t)\},+}.\end{aligned}$$

Thus, (5.36) can be rewritten as $c'_\Lambda(t) = |\Lambda|^{-1} \lim_{\bar{\Lambda}\uparrow\mathbb{Z}^D} \sum_{k\in\Lambda} \langle Y_k\rangle_{\bar{\Lambda},\beta,\{h_i(t)\},+}$. Since each $h_i(t)$ is non-negative for $t \geq -\beta h$, we have by Table V.1(e)

$$\frac{\partial^2}{\partial t^2}\langle Y_k\rangle_{\bar{\Lambda},\beta,\{h_i(t)\},+} = \beta^{-2}\sum_{i\in\Lambda}\frac{\partial^2}{\partial h_i^2}\langle Y_k\rangle_{\bar{\Lambda},\beta,\{h_i(t)\},+} \leq 0 \qquad \text{for } t \geq -\beta h.$$

Thus $\langle Y_k\rangle_{\bar{\Lambda},\beta,\{h_i(t)\},+}$ is a concave function of t for $t \geq -\beta h$. We conclude that

$$(5.37) \qquad c'_\Lambda \text{ is concave on } [-\beta h, \infty) \qquad \text{for } h \geq 0.$$

This verifies property (5.34).

The proof of property (5.35) depends on the next lemma.

Lemma V.7.5. *Let $\{f_n; n = 1, 2, \ldots\}$ be a sequence of convex functions on an open interval A of $\mathbb{R}$ such that $f(t) = \lim_{n\to\infty} f_n(t)$ exists for every $t\in A$. Let $\{t_n; n = 1, 2, \ldots\}$ be a sequence in A which converges to a point $t_0\in A$. If $f'_n(t_n)$ and $f'(t_0)$ exist, then $\lim_{n\to\infty} f'_n(t_n)$ exists and equals $f'(t_0)$.*

Proof. If each point t_n equals t_0, then $f'_n(t_n) \to f'(t_0)$ by Lemma IV.6.3. We modify the proof of this lemma to handle a general sequence $t_n \to t_0$. Define $h_n(t) = (f_n(t) - f_n(t_n))/(t - t_n)$ for $t\in A$, $t \neq t_n$, and $g(t) = (f(t) - f(t_0))/(t - t_0)$ for $t\in A$, $t\neq t_0$. The sequence $\{f_n\}$ converges to f uniformly on each compact subset of A [Theorem VI.3.3(b)]. Since f is continuous relative to A [Theorem VI.3.1],

$$|f_n(t_n) - f(t_0)| \leq |f_n(t_n) - f(t_n)| + |f(t_n) - f(t_0)| \to 0 \qquad \text{as } n\to\infty.$$

Therefore, for each $t\in A$, $t\neq t_0$, $h_n(t)\to g(t)$. The convexity of f_n implies that $h_n(s) \leq f'_n(t_n) \leq h_n(t)$ for $s < t_n < t \quad (s, t\in A)$. Taking $n\to\infty$, we see that

$$(5.38) \qquad \sup_{\{s\in A: s<t_0\}} g(s) \leq \liminf_{n\to\infty} f'_n(t_n) \leq \limsup_{n\to\infty} f'_n(t_n) \leq \sup_{\{t\in A: t>t_0\}} g(t).$$

As the extreme terms both equal $f'(t_0)$, the proof is done. □

Proof of property (5.35). We first consider $h > 0$. According to (5.37), each function $-c'_\Lambda$ is convex on $[-\beta h, \infty)$. Since $h > 0$, this interval contains the origin in its interior. Suppose we show that $f(t) = \lim_{\Lambda\uparrow\mathbb{Z}^D}(-c'_\Lambda(t))$ exists for all $t\in(-\alpha, \alpha)$, some $\alpha > 0$, and that $f'(0)$ exists and equals $-\sigma^2(\beta, h)$. Lemma V.7.5 then implies that for any $0 \leq t < \alpha$

$$-c''_\Lambda(t/\sqrt{|\Lambda|}) \to f'(0) = -\sigma^2(\beta, h) \qquad \text{as } \Lambda\uparrow\mathbb{Z}^D.$$

This is property (5.35) for $h > 0$. We first prove that $f(t)$ exists. By Hölder's inequality $c_\Lambda(t)$ is a convex function of t real, and as noted in (5.26), $c_\Lambda(t)$ converges to the free energy function $c_{\beta,h}(t) = -\beta[\psi(\beta, h + t/\beta) - \psi(\beta, h)]$. For $h > 0$, $\partial\psi(\beta, h_0)/\partial h$ exists for all h_0 in some neighborhood of h. Hence there exists $\alpha > 0$ such that $c'_{\beta,h}(t) = -\partial\psi(\beta, h + t/\beta)/\partial h$ exists for all

$-\alpha < t < \alpha$. Now Lemma IV.6.3 implies that $c'_\Lambda(t) \to c'_{\beta,h}(t)$ for all $-\alpha < t < \alpha$. We conclude that for $-\alpha < t < \alpha$ $f(t) = \lim_{\Lambda \uparrow \mathbb{Z}^D}(-c'_\Lambda(t))$ exists and $f(t)$ equals $-c'_{\beta,h}(t)$. Since $\sigma^2(\beta,h)$ is finite, Lemma V.7.4(a) shows that $\partial^2\psi(\beta,h)/\partial h^2$ exists and equals $-\beta\cdot\sigma^2(\beta,h)$. Hence $f'(0) = -c''_{\beta,h}(0) = \beta^{-1}\partial^2\psi(\beta,h)/\partial h^2$ exists and equals $-\sigma^2(\beta,h)$. This completes the proof of property (5.35) for $h > 0$.

The proof of (5.35) for $h > 0$ used the fact that all the functions $\{-c'_\Lambda\}$ are convex on an interval which contains the origin in its interior. But for $h = 0$, $-c'_\Lambda$ is convex on $[0,\infty)$ and by symmetry is concave on $(-\infty, 0]$. Hence the above proof must be modified. We fix $(\beta, 0) \in \mathscr{U}$ (i.e., $0 < \beta < \beta_c$), and assume that $\sigma^2(\beta,0)$ is finite. Since for $0 < \beta < \beta_c$ $\partial\psi(\beta,h)/\partial h$ exists for all h real, $c_{\beta,0}(t) = -\beta[\psi(\beta,t/\beta) - \psi(\beta,0)]$ is differentiable for all t real. Define new functions

$$g_\Lambda(t) = \begin{cases} -c'_\Lambda(t) & \text{for } t \geq 0, \\ -c''_\Lambda(0)\cdot t & \text{for } t < 0, \end{cases} \qquad g(t) = \begin{cases} -c'_{\beta,0}(t) & \text{for } t \geq 0, \\ -\sigma^2(\beta,0)\cdot t & \text{for } t < 0; \end{cases}$$

g_Λ is continuous at 0 ($c'_\Lambda(0) = m(\beta,0) = 0$) and is convex on $\mathbb{R}$. By (5.32), $c''_\Lambda(0) \to \sigma^2(\beta,h)$, and so $g_\Lambda(t) \to g(t)$ for all t real. Since $\sigma^2(\beta,0)$ is assumed to be finite, $g'(0) = -c''_{\beta,0}(0) = \beta^{-1}\partial^2\psi(\beta,0)/\partial h^2$ exists and equals $-\beta\sigma^2(\beta,0)$. Lemma V.7.5 implies that for any $t > 0$

$$g'_\Lambda(t/\sqrt{|\Lambda|}) = -c''_\Lambda(t/\sqrt{|\Lambda|}) \to g'(0) = -\sigma^2(\beta,0) \qquad \text{as } \Lambda \uparrow \mathbb{Z}^D.$$

This proves property (5.35) for $h = 0$ and completes the proof of the central limit theorem. □

Let (β,h) be a point in $\mathscr{U}$ ($\beta > 0$, $h \neq 0$ or $0 < \beta < \beta_c$, $h = 0$). According to Lemma V.7.4(a), for any summable ferromagnetic interaction J on $\mathbb{Z}^D$, the variance $\sigma^2(\beta,h) = \sum_{k\in\mathbb{Z}^D}\langle Y_0; Y_k\rangle_{\beta,h}$ is finite for $\beta > 0$ and $h \neq 0$. Set $\mathscr{J}_0 = \sum_{k\in\mathbb{Z}^D} J(k)$. By Theorem V.5.1(a), $\mathscr{J}_0^{-1} \leq \beta_c$. We claim that if the interaction decays exponentially (thus, in particular, if the interaction has finite range), then for $h = 0$ $\sigma^2(\beta,0)$ is finite at least for all $0 < \beta < \mathscr{J}_0^{-1}$. This fact is a consequence of the next result on the exponential decay of the correlations $\langle Y_0; Y_k\rangle_{\beta,0}$.[6]

Theorem V.7.6. *Assume that there exist positive constants c_1 and c_2 such that $0 \leq J(k) \leq c_1\exp(-c_2\|k\|)$ for all $k\in\mathbb{Z}^D$. Set $\mathscr{J}_0 = \sum_{k\in\mathbb{Z}^D} J(k)$. Then the following conclusions hold.*

(a) *$\langle Y_0; Y_k\rangle_{\beta,0}$ decays exponentially at least for all $0 < \beta < \mathscr{J}_0^{-1}$. That is, for each $0 < \beta < \mathscr{J}_0^{-1}$ and $h = 0$, there exist positive constants a and x such that $0 \leq \langle Y_0; Y_k\rangle_{\beta,0} \leq a\exp(-x\|k\|)$ for all $k\in\mathbb{Z}^D$. In particular, $\sigma^2(\beta,0) = \sum_{k\in\mathbb{Z}^D}\langle Y_0; Y_k\rangle_{\beta,0}$ is finite.*

(b) *For each $\beta > 0$ and $h \neq 0$, $\langle Y_0; Y_k\rangle_{\beta,h}$ decays exponentially.*

Part (a) is proved in Sokal (1982a). Also see the references listed in that paper. The proof of part (a) is outlined in Problem V.13.12. Part (b) is proved in Lebowitz and Penrose (1968) for finite-range interactions and in Duneau, Souillard, and Iagolnitzer (1975) for exponentially decaying interactions. A reasonable conjecture is that for any exponentially decaying interaction on $\mathbb{Z}^D$, $\langle Y_0; Y_k\rangle_{\beta,0}$ decays exponentially for all $0 < \beta < \beta_c$. This is in fact known to hold for the Ising model on $\mathbb{Z}^D$, $D \geq 2$.

Theorem V.7.7. *For the Ising model on $\mathbb{Z}^D$, $D \geq 2$, $\langle Y_0 ; Y_k\rangle_{\beta,h}$ decays exponentially for $0 < \beta < \beta_c$, $h = 0$ and $\beta > 0$, $h \neq 0$; i.e., for all $(\beta, h) \in \mathscr{U}$. In particular, $\sigma^2(\beta, h) = \sum_{k\in\mathbb{Z}^D} \langle Y_0 ; Y_k\rangle_{\beta,h}$ is finite for all $(\beta, h) \in \mathscr{U}$.*

The exponential decay for $\beta > 0$, $h \neq 0$ is a consequence of part (b) of Theorem V.7.6. As Lebowitz (1972, Section III) points out, the exponential decay for $D = 2$ and $0 < \beta < \beta_c$, $h = 0$ follows from Onsager (1944). Aizenman (1984, 1985) proves the exponential decay for $D > 2$ and $0 < \beta < \beta_c$, $h = 0$.

The main result in this section is the central limit theorem, Theorem V.7.2. The theorem is valid for (β, h) in $\mathscr{U}$ provided $\sigma^2(\beta, h)$ is finite. For $\beta > 0$, $h \neq 0$, and any summable interaction, the finiteness was proved in Lemma V.7.4(a). For $0 < \beta < \mathscr{J}_0^{-1}$, $h = 0$, and any exponentially decaying interaction, the finiteness follows from Theorem V.7.6(a). In the special case of the Ising model on $\mathbb{Z}^2$, $\sigma^2(\beta, h)$ is finite for all $(\beta, h) \in \mathscr{U}$. The next section discusses the contrasting situation at the critical point $(\beta_c, 0)$, where the central limit theorem is expected to break down.

V.8. Critical Phenomena and the Breakdown of the Central Limit Theorem

The term critical phenomena refers to the behavior of a ferromagnetic system for (β, h) near or equal to the critical point $(\beta_c, 0)$. These phenomena include the behavior of functions such as the specific magnetization $m(\beta, h)$ and the specific magnetic susceptibility $\chi(\beta, h) = \partial m(\beta, h)/\partial h$ and the asymptotic behavior of random quantities such as the spin $S_\Lambda = \sum_{j\in\Lambda} Y_j$. In the previous section, we proved a central limit theorem for S_Λ under the assumption that $\sigma^2(\beta, h)$ or equivalently $\chi(\beta, h)$ is finite. For reasons to be explained below, one expects that at the critical point $\chi(\beta_c, 0)$ diverges and that the central limit theorem for S_Λ breaks down. An interesting question, discussed in the present section, is how to rescale S_Λ by a nonclassical scaling $|\Lambda|^\lambda$, $\lambda > \frac{1}{2}$, so that $S_\Lambda/|\Lambda|^\lambda$ has a nondegenerate limit in distribution. In this context, the spin S_Λ is often called a *block spin variable* and the limit in distribution of $S_\Lambda/|\Lambda|^\lambda$ a *block spin scaling limit*.[7] As we discuss at the end of the section, the form of this limit (e.g., non-Gaussian or Gaussian) is believed to depend upon the dimension D of the lattice. Theorem V.8.6 gives information on the limit under suitable hypotheses.

Before studying block spin scaling limits, we must fill in background material concerning critical phenomena. It will be useful first to review the discussion in Section IV.2 for the Ising model on $\mathbb{Z}^2$. For $h = 0$, the model exhibits dramatic new behavior as β increases through the inverse critical temperature β_c. For $0 < \beta < \beta_c$ the spontaneous magnetization $m(\beta, +) = \lim_{h \to 0^+} m(\beta, h)$ is 0 and there exists a unique phase $P_{\beta,0}$, which is pure. Pair correlations decay exponentially:

$$\langle Y_i Y_j \rangle_{\beta,0} \sim \exp[-\|i-j\|/\xi(\beta,0)] \qquad \text{as } \|i-j\| \to \infty.$$

The positive number $\xi(\beta, 0)$ is called the *correlation length* and is a rough measure of the distance over which correlations between spins are significant. By contrast, for $\beta > \beta_c$, $m(\beta, +)$ is positive and there are two distinct pure phases. As β approaches β_c from below, the model anticipates its new behavior by making adjustments on a microscopic scale. These adjustments appear in the form of spin fluctuations, or islands of correlated spins, which grow in size as the critical inverse temperature is approached. At $\beta = \beta_c$, the correlation length is infinite, correlations decay by a power law, and the spin fluctuations are extremely sensitive to small perturbations in h. This is reflected macroscopically in the divergence of the specific magnetic susceptibility $\chi(\beta, h) = \partial m(\beta, h)/\partial h$ at $(\beta, h) = (\beta_c, 0)$. This divergence can be seen in Figure IV.1.

The behavior of the Ising model on $\mathbb{Z}^2$ suggests how one might expect other models to behave. We restrict ourselves to finite-range ferromagnetic interactions on $\mathbb{Z}^D$, $D \geq 2$.* Define $\mathscr{U}$ to be the set of points (β, h) with $\beta > 0$, $h \neq 0$ and $0 < \beta < \beta_c$, $h = 0$; β_c is the critical inverse temperature corresponding to the interaction.

Definition V.8.1. The point $(\beta, h) = (\beta_c, 0)$ is called a *normal critical point* if the following properties hold.

(a) $0 < \beta_c < \infty$, where $\beta_c = \sup\{\beta > 0 : m(\beta, +) = \lim_{h \to 0^+} m(\beta, h) = 0\}$.

(b) $m(\beta_c, +) = 0$.

(c) The specific magnetic susceptibility $\chi(\beta, h) = \partial m(\beta, h)/\partial h$ is finite for all $(\beta, h) \in \mathscr{U}$ and $\chi(\beta_c, 0) = \infty$.[8]

The definition of β_c implies that $m(\beta, +) = 0$ for $0 < \beta < \beta_c$ and $m(\beta, +) > 0$ for $\beta > \beta_c$. Property V.8.1(b) requires that $m(\beta, +)$ vanish also at β_c.

The only model which is known rigorously to have a normal critical point is the Ising model on $\mathbb{Z}^2$. This model will be discussed again at the end of the section. Nevertheless, it is not unreasonable to expect that whenever β_c is finite, the critical point is normal (at least for finite-range interactions).

The next lemma extends to a normal critical point some results proved earlier for points $(\beta, h) \in \mathscr{U}$.

*In other words, the translation invariant interaction strengths $J(i-j)$ are assumed to be independent of Λ, non-negative, not all zero, symmetric and to vanish for all i and j with $\|i-j\|$ sufficiently large (finite range). The assumption of finite range is needed for Lemma V.8.3(a) ($\eta \leq 1$).

Lemma V.8.2. *Assume that* $(\beta_c, 0)$ *is a normal critical point. Then the following conclusions hold.*

(a) $\mathscr{G}_{\beta_c,0} \cap \mathscr{M}_s(\Omega)$ *consists of a unique measure* $P_{\beta_c,0}$, *and for any site* $k \in \mathbb{Z}^D$, $\langle Y_k \rangle_{\beta_c,0} = m(\beta_c, +) = 0$.

(b) *With respect to* $P_{\beta_c,0}$, $S_\Lambda/|\Lambda| \stackrel{\text{a.s.}}{\to} m(\beta_c, +) = 0$ *and* $S_\Lambda/|\Lambda| \stackrel{\text{exp}}{\to} m(\beta_c, +) = 0$ *as* $\Lambda \uparrow \mathbb{Z}^D$.

(c) $\langle Y_0 ; Y_k \rangle_{\beta_c,0} \to 0$ *as* $\|k\| \to \infty$.

(d) $\partial\psi(\beta_c, 0)/\partial h = 0$ *but*

$$(5.39) \qquad -\frac{\partial^2\psi(\beta_c, 0)}{\partial h^2} = \chi(\beta_c, 0) = \beta_c \cdot \sigma^2(\beta_c, 0) = \beta_c \cdot \sum_{k \in \mathbb{Z}^D} \langle Y_0 ; Y_k \rangle_{\beta_c,0} = \infty.$$

(e) *Let* $\Sigma_\Lambda^2(\beta_c, 0)$ *be the variance* $\langle S_\Lambda^2 \rangle_{\beta_c,0}$ ($\langle S_\Lambda \rangle_{\beta_c,0} = 0$ *by part* (a)). *Then* $|\Lambda|^{-1}\Sigma_\Lambda^2(\beta_c, 0) \to \sigma^2(\beta_c, 0) = \infty$ *as* $\Lambda \uparrow \mathbb{Z}^D$.

(f) *For each t real,*

$$\begin{aligned} c_{\beta_c,0}(t) &= \lim_{\Lambda \uparrow \mathbb{Z}^D} c_\Lambda(t) \\ &= \lim_{\Lambda \uparrow \mathbb{Z}^D} \frac{1}{|\Lambda|} \log\langle \exp(tS_\Lambda) \rangle_{\beta_c,0} \\ &= -\beta_c[\psi(\beta_c, t/\beta_c) - \psi(\beta_c, 0)]. \end{aligned}$$

Proof. (a) Since $m(\beta_c, +) = 0$, Lemma V.4.5(b) implies that $\partial\psi(\beta_c, 0)/\partial h$ exists and equals $-m(\beta_c, +) = 0$; also $\langle Y_k \rangle_{\beta_c,0} = \langle Y_0 \rangle_{\beta_c,0} = 0$. Theorem V.6.1(c) implies $\mathscr{G}_{\beta_c,0} = \{P_{\beta_c,0}\}$.

(b) This is proved like Theorem IV.6.6(a).

(c) Corollary A.11.12(b).

(d) By the proof of part (a) $\partial\psi(\beta_c, 0)/\partial h = 0$. The proof of Lemma V.7.4(b), given for $0 < \beta < \beta_c$, $h = 0$, applies verbatim to $\beta = \beta_c$, $h = 0$. Since $\chi(\beta_c, 0) = \infty$, (5.39) follows.

(e) The proof of Lemma V.7.1, given for $(\beta, h) \in \mathscr{U}$, applies verbatim to $\beta = \beta_c$, $h = 0$. Hence by (5.39) $|\Lambda|^{-1}\,\Sigma_\Lambda^2(\beta_c, 0) \to \sigma^2(\beta_c, 0) = \infty$ as $\Lambda \uparrow \mathbb{Z}^D$.

(f) This is proved like Lemma IV.6.11. □

In order to discuss block spin scaling limits for ferromagnetic models, it is useful to introduce a set of indices called *critical exponents*. These indices describe the behavior, at or near the critical point, of various functions associated with the models. Fix a finite-range ferromagnetic interaction J on $\mathbb{Z}^D, D \geq 2$, and assume that the critical point $(\beta_c, 0)$ is normal. We consider two functions, $\langle Y_0 ; Y_k \rangle_{\beta_c,0}$ as $\|k\| \to \infty$ and $m(\beta_c, h)$ as $h \to 0^+$. Lemma V.8.2(c) shows that as $\|k\| \to \infty$, $\langle Y_0 ; Y_k \rangle_{\beta_c,0} \to 0$. What is a reasonable guess for the rate at which $\langle Y_0 ; Y_k \rangle_{\beta_c,0}$ tends to 0? We appeal to an important theoretical tool known as the renormalization group, which is the basis for most of our information about critical phenomena in ferromagnetic models.[9] According to Theorem V.7.6, for exponentially decaying in-

Table V.2. Critical Exponents η and δ for Models on $\mathbb{Z}^D$

Function	Power law*	Critical exponent
$\langle Y_0; Y_k\rangle_{\beta_c,0}$	$\sim \|k\|^{-(D-2+\eta)}$	η
$m(\beta_c,h)$	$\sim h^{1/\delta}$	δ

*For $D=4$ the power laws must be modified by logarithmic corrections. See page 179.

teractions, $\langle Y_0; Y_k\rangle_{\beta,h}$ decays exponentially for (most) $(\beta,h)\in\mathscr{U}$. But if $\langle Y_0; Y_k\rangle_{\beta_c,0}$ also decayed exponentially, then $\sum_{k\in\mathbb{Z}^D}\langle Y_0; Y_k\rangle_{\beta_c,0}$ would converge, violating (5.39). Renormalization group calculations suggest that to leading order in $\|k\|$, $\langle Y_0; Y_k\rangle_{\beta_c,0}$ is described by a power law function. We write $\langle Y_0; Y_k\rangle_{\beta_c,0} \sim \|k\|^x$, $x<0$, if

$$x = \lim_{\|k\|\to\infty} \frac{1}{\log\|k\|}\cdot\log\langle Y_0; Y_k\rangle_{\beta_c,0} \tag{5.40}$$

exists. The number x is customarily denoted by $-(D-2+\eta)$, where D is the dimension of the lattice; η is known as the critical exponent corresponding to $\langle Y_0; Y_k\rangle_{\beta_c,0}$. If $\langle Y_0; Y_k\rangle_{\beta_c,0} \sim \|k\|^x = \|k\|^{-(D-2+\eta)}$ holds, then given any $\varepsilon>0$ there exists a number $R>1$ such that $\|k\|^{x-\varepsilon} \le \langle Y_0; Y_k\rangle_{\beta_c,0} \le \|k\|^{x+\varepsilon}$ for all $k\in\mathbb{Z}^D$ with $\|k\|\ge R$.

The second function whose critical point behavior we discuss is $m(\beta_c,h)$. As $h\to 0^+$, $m(\beta_c,h)$ converges to 0 because of property V.8.1(b) of a normal critical point ($m(\beta_c,+)=0$). Renormalization group calculations suggest that to leading order in h, $m(\beta_c,h)$ is also described by a power law function. We write $m(\beta_c,h)\sim h^y$, $y>0$, if

$$y = \lim_{h\to 0^+} \frac{1}{\log h}\cdot\log m(\beta_c,h) \tag{5.41}$$

exists. The number y is customarily denoted by $1/\delta$; δ is known as the critical exponent corresponding to $m(\beta_c,h)$. If $m(\beta_c,h)\sim h^y = h^{1/\delta}$ holds, then given any $\varepsilon>0$ there exists a number $h_0\in(0,1)$ such that $h^{y+\varepsilon}\le m(\beta_c,h)\le h^{y-\varepsilon}$ for all $0\le h\le h_0$. The definitions of η and δ are summarized in Table V.2.

Other critical exponents are defined for functions such as the specific magnetic susceptibility $\chi(\beta,0)$ as $\beta\to\beta_c^-$; the spontaneous magnetization $m(\beta,+)$ as $\beta\to\beta_c^+$; and the specific heat $\partial u(\beta,0)/\partial T = -\beta^2\partial u(\beta,0)/\partial\beta$ as $\beta\to\beta_c^-$ or $\beta\to\beta_c^+$, where $u(\beta,0)$ is the specific energy $\partial(\beta\psi(\beta,0))/\partial\beta$ [see Problem IV.9.11]. Critical exponents are discussed in detail in Sokal (1981) and in the references listed in Note 9. It is widely believed that each critical exponent depends only on the dimension D of the lattice and not on the details of the particular interaction. This belief, which is based on renormalization group arguments, is called *universality*. It will be discussed further at the end of the section.

The main results in this section are an inequality involving the critical exponents η and δ [Theorem V.8.4] and information on block spin scaling limits of S_Λ [Theorem V.8.6]. The next lemma gives bounds on η and δ and relates η to other quantities of interest. The hypercube Λ has been defined as the set $\{j \in \mathbb{Z}^D : |j_\alpha| \le N \text{ for } \alpha = 1, \ldots, D\}$, where N is a non-negative integer. If $n = 2N + 1$, then the number of sites in Λ is $|\Lambda| = n^D$. We shall denote the variance $\Sigma_\Lambda^2(\beta_c, 0)$ of S_Λ by Σ_n^2 and write Σ_n for $(\Sigma_\Lambda^2(\beta_c, 0))^{1/2}$.

Lemma V.8.3. *Let J be a finite-range ferromagnetic interaction on $\mathbb{Z}^D$, $D \ge 2$. Assume that $(\beta_c, 0)$ is a normal critical point, $\langle Y_0 ; Y_k \rangle_{\beta_c, 0} \sim \|k\|^{-(D-2+\eta)}$, and $m(\beta_c, h) \sim h^{1/\delta}$. Define $G(R) = \sum_{\|k\| \le R} \langle Y_0; Y_k \rangle_{\beta_c, 0}$, $R \ge 1$. Then the following conclusions hold.*

(a) $\eta \le 1$.
(b) $\delta \ge 1$.
(c) $G(R) \sim R^{2-\eta}$ *and* $\Sigma_n^2 \sim n^{D+2-\eta}$.*

Comments on proof. (a) Simon (1980a) proves that $\eta \le 1$. The proof, given for nearest-neighbor interactions, extends readily to finite-range interactions.

(b) If $\delta < 1$, then $\lim_{h \to 0^+} m(\beta_c, h)/h = 0$. But this limit equals

$$\lim_{h \to 0^+} \frac{-\partial \psi(\beta_c, h)/\partial h}{h} = -\frac{\partial^2 \psi(\beta_c, 0)}{\partial h^2},$$

which diverges by Lemma V.8.2(d). The contradiction proves $\delta \ge 1$.

(c) Heuristically, $G(R) \sim R^{2-\eta}$ holds since for large R, $G(R)$ should behave like the sum $\sum_{1 \le \|k\| \le R} \|k\|^{-(D-2+\eta)}$, and this in turn should behave like the integral $\int_{1 \le \|x\| \le R} \|x\|^{-(D-2+\eta)} \, d^D x$ over $\mathbb{R}^D$. Since $\eta \le 1$, the integral equals $\text{const}(R^{2-\eta} - 1)$.† The proof that $G(R) \sim R^{2-\eta}$ is left as Problem V.13.13.

The proof that $\Sigma_n^2 \sim n^{D+2-\eta}$ relies on the following inequalities. For any $i, j \in \Lambda$, we have $\|i - j\| \le \sqrt{D}n$, and so

$$\Sigma_n^2 = \sum_{i \in \Lambda} \sum_{j \in \Lambda} C(i-j) \le \sum_{i \in \Lambda} \sum_{\{j : \|i-j\| \le \sqrt{D}n\}} C(i-j) = n^D G(\sqrt{D}n),$$

where $C(i-j) = \langle Y_0 ; Y_{i-j} \rangle_{\beta_c, 0} = \langle Y_i ; Y_j \rangle_{\beta_c, 0}$. For fixed $0 < \varepsilon < 1$, let $\Lambda_n(\varepsilon) = \{j \in \mathbb{Z}^D : |j_\alpha| \le (1-\varepsilon)N \text{ for } \alpha = 1, \ldots, D\}$, where $n = 2N + 1$. As in the proof of Lemma V.7.1, for any $n \ge 3$

$$\Sigma_n^2 \ge |\Lambda_n(\varepsilon)| \sum_{\|k\| \le \varepsilon N} C(k) \ge |\Lambda_n(\varepsilon)| \sum_{\|k\| \le \varepsilon n/4} C(k) = (1-\varepsilon)^D n^D G(\tfrac{1}{4}\varepsilon n).$$

The last two displays imply that

$$\limsup_{n \to \infty} \frac{\log \Sigma_n^2}{\log(\sqrt{D}n)} \le D + \lim_{n \to \infty} \frac{\log G(\sqrt{D}n)}{\log(\sqrt{D}n)},$$

$$D + \lim_{n \to \infty} \frac{\log G(\frac{1}{4}\varepsilon n)}{\log(\frac{1}{4}\varepsilon n)} \le \liminf_{n \to \infty} \frac{\log \Sigma_n^2}{\log(\frac{1}{4}\varepsilon n)}.$$

Since $G(R) \sim R^{2-\eta}$, we conclude that $\Sigma_n^2 \sim n^{D+2-\eta}$. □

* $2 - \eta = \lim_{R \to \infty} \log G(R)/\log R$ and $D + 2 - \eta = \lim_{n \to \infty} \log \Sigma_n^2/\log n$.

† If η equaled 2, then the integral would equal $\text{const} \cdot \log R$.

We now turn to block spin scaling limits of S_Λ. First recall our results for $(\beta, h) \in \mathscr{U}$. By Lemma V.7.1, $|\Lambda|^{-1}\Sigma_\Lambda^2(\beta, h) \to \sigma^2(\beta, h) = \sum_{k \in \mathbb{Z}^D} \langle Y_0 ; Y_k \rangle_{\beta, h}$ as $\Lambda \uparrow \mathbb{Z}^D$, where $\Sigma_\Lambda^2(\beta, h)$ is the variance of S_Λ. Consequently, $\Sigma_\Lambda(\beta, h) \sim |\Lambda|^{1/2}$ whenever $\sigma^2(\beta, h)$ is finite (e.g., $\beta > 0$, $h \neq 0$). The central limit theorem states that whenever $\sigma^2(\beta, h)$ is finite, then $[S_\Lambda - |\Lambda| m(\beta, h)]/\sqrt{|\Lambda|}$ converges in distribution to $N(0, \sigma^2(\beta, h))$ [Theorem V.7.2(a)]. By the same proof, $[S_\Lambda - |\Lambda| m(\beta, h)]/\Sigma_\Lambda(\beta, h)$ converges in distribution to $N(0, 1)$. Contrast this with the situation at a normal critical point. Since $\sigma^2(\beta_c, 0) = \beta_c^{-1}\chi(\beta_c, 0) = \infty$ and $\langle (S_\Lambda/\sqrt{|\Lambda|})^2 \rangle_{\beta_c, 0} = |\Lambda|^{-1}\Sigma_\Lambda^2(\beta_c, 0) \to \infty$ as $\Lambda \uparrow \mathbb{Z}^D$ [Lemma V.8.2(e)], the central limit theorem is expected to fail. We now consider other scalings for S_Λ which can lead to non-Gaussian block spin scaling limits. If $\langle Y_0 ; Y_k \rangle_{\beta_c, 0} \sim \|k\|^{-(D-2+\eta)}$, then $\Sigma_n = (\Sigma_\Lambda^2(\beta, 0))^{1/2} \sim n^{(D+2-\eta)/2}$, and one possibility is to study the limit in distribution of $S_\Lambda/n^{(D+2-\eta)/2} = S_\Lambda/|\Lambda|^\lambda$, where $\lambda = (D + 2 - \eta)/(2D) > \frac{1}{2}$. The second possibility is to study the limit in distribution of S_Λ/Σ_n directly. This is technically simpler since for each n, S_Λ/Σ_n has fixed variance 1. From now on, we write S_n for S_Λ.

The nature of the limit of S_n/Σ_n is closely related to an inequality involving the critical exponents η and δ. This inequality, given in (5.42), was discovered by Buckingham and Gunton (1969); also see Fisher (1969). The proof given here is due to Newman (1979).

Theorem V.8.4. *Let J be a finite-range ferromagnetic interaction on $\mathbb{Z}^D$, $D \geq 2$. Assume that $(\beta_c, 0)$ is a normal critical point, $\langle Y_0 ; Y_k \rangle_{\beta_c, 0} \sim \|k\|^{-(D-2+\eta)}$, and $m(\beta_c, h) \sim h^{1/\delta}$. Then*

$$2 - \eta \leq D\frac{\delta - 1}{\delta + 1}. \tag{5.42}$$

Proof. For $t \in \mathbb{R}$, let $c_n(t) = n^{-D}\log\langle \exp(tS_n) \rangle_{\beta_c, 0}$. By Lemma V.8.2(f) ($|\Lambda| = n^D$),

$$c_{\beta_c, 0}(t) = \lim_{n \to \infty} c_n(t) = -\beta_c[\psi(\beta_c, t/\beta_c) - \psi(\beta_c, 0)]. \tag{5.43}$$

We write $c(t)$ for $c_{\beta_c, 0}(t)$. By means of the FKG inequality, Newman (1979, page 186) shows that $c_n(t) \leq c(t)$ for all $t \in \mathbb{R}$. To prove (5.42), we first show that

$$\liminf_{n \to \infty} n^D c(t/\Sigma_n) > 0 \qquad \text{for any } t \neq 0. \tag{5.44}$$

The $P_{\beta_c, 0}$ distribution of S_n/Σ_n is symmetric. Hence the inequality $c_n \leq c$ implies

$$\begin{aligned} \exp[n^D c(t/\Sigma_n)] \geq \exp[n^D c_n(t/\Sigma_n)] &= \langle \exp(tS_n/\Sigma_n) \rangle_{\beta_c, 0} \\ &\geq 1 + \tfrac{1}{2}t^2\langle (S_n/\Sigma_n)^2 \rangle_{\beta_c, 0} \geq 1. \end{aligned}$$

Suppose that (5.44) were not valid. Then for some infinite subsequence $\{n'\}$ and some $t \neq 0$, $(n')^D c(t/\Sigma_{n'})$ would converge to 0, and so by the last display $\langle (S_{n'}/\Sigma_{n'})^2 \rangle_{\beta_c, 0}$ would also converge to 0. But $\langle (S_n/\Sigma_n)^2 \rangle_{\beta_c, 0}$ equals 1 for all

n. This contradiction proves (5.44). We now prove (5.42). By Lemma V.8.3, $\Sigma_n \sim n^{(D+2-\eta)/2}$ and $\eta \le 1$. Also the relation $m(\beta_c, h) \sim h^{1/\delta}$ implies that

$$c(t) = c_{\beta_c,0}(t) = -\beta_c[\psi(\beta_c, t/\beta_c) - \psi(\beta_c, 0)] \sim t^{1+1/\delta} \qquad \text{as } t \to 0^+. \tag{5.45}$$

Hence for any $r < \frac{1}{2}(D + 2 - \eta)$ and all sufficiently large n, $\Sigma_n > n^r$. Also for any $p < 1 + 1/\delta$ and all sufficiently small $t > 0$, $c(t) < t^p$. Hence for all sufficiently large n and any $t > 0$

$$n^D c(t/\Sigma_n) < n^D (t/\Sigma_n)^p < t^p n^{D-pr}.$$

Inequality (5.44) implies that $D - pr \ge 0$, which in turn implies that $D - \frac{1}{2}(1 + 1/\delta)(D + 2 - \eta) \ge 0$. This is equivalent to (5.42) since $\delta \ge 1$. □

Inequality (5.42) allows us to improve the bound $\delta \ge 1$ [Lemma V.8.3(b)].

Corollary V.8.5. *Under the same hypotheses as the previous theorem,*

$$\delta \ge (D + 1)/(D - 1) > 1.$$

Proof. $\eta \le 1$ and so $1 \le D(\delta - 1)/(\delta + 1)$; hence $\delta \ge (D + 1)/(D - 1)$. □

It is strongly believed that if equality holds in the Buckingham–Gunton inequality (5.42) and if S_n/Σ_n has a limit in distribution, then the limit is non-Gaussian. Equality in (5.42) is designated by the term *hyperscaling*.* We first motivate a relationship between (5.42) and a scaling limit of $\{S_n/\Sigma_n\}$, then state a rigorous theorem. Let us suppose that for all real t $g(t) = \lim_{n\to\infty} \langle \exp(tS_n/\Sigma_n)\rangle_{\beta_c,0}$ exists. If it exists, then $g(t)$ is the moment generating function of the presumed scaling limit of S_n/Σ_n. We relate $g(t)$ heuristically to the free energy function $c_{\beta_c,0}(t) = \lim_{n\to\infty} c_n(t)$, where $c_n(t) = n^{-D} \log\langle \exp(tS_n)\rangle_{\beta_c,0}$. By Lemma V.8.3(c) $\Sigma_n \sim n^{(D+2-\eta)/2}$ and by (5.45) $c_{\beta_c,0}(t) \sim t^{1+1/\delta}$ as $t \to 0^+$. For large n it is tempting to write

$$\begin{aligned}\langle \exp(tS_n/\Sigma_n)\rangle_{\beta_c,0} &= \exp[n^D c_n(t/\Sigma_n)] \approx \exp[n^D c_{\beta_c,0}(t/\Sigma_n)] \\ &\approx \exp[n^D \text{const} \cdot |t/\Sigma_n|^{1+1/\delta}] \\ &\approx \exp[\text{const} \cdot |t|^{1+1/\delta} n^A],\end{aligned}$$

where $A = D - \frac{1}{2}(1 + 1/\delta)(D + 2 - \eta)$. If there is equality in (5.42), then $A = 0$. These calculations suggest that

$$g(t) = \lim_{n\to\infty} \langle \exp(tS_n/\Sigma_n)\rangle_{\beta_c,0} \approx \exp(\text{const} \cdot |t|^{1+1/\delta}). \tag{5.46}$$

Since $\delta > 1$, $1 + 1/\delta$ does not exceed 2, and (5.46) is consistent with the scaling limit of S_n/Σ_n being non-Gaussian (if it exists). One cannot take (5.46) literally since $g(t) = \exp(\text{const} \cdot |t|^{1+1/\delta})$ cannot be the moment generating function of a random variable. Indeed, since $\delta > 1$, $g''(0)$ equals ∞.

*For discussions of hyperscaling, see Fisher (1973, 1983).

The following theorem due to Newman (1979) partially justifies these computations. It discusses the nature of the block spin scaling limit under the assumption that it exists. The existence of the limit is an unsolved problem [cf., Note 7].* Hypotheses (5.47) and (5.48) in the theorem are related to the statements that $G(R) \sim R^{2-\eta}$ [Lemma V.8.3(c)] and $c_{\beta_c,0}(t) \sim t^{1+1/\delta}$ as $t \to 0^+$ [see (5.45)], but the hypotheses are not implied by these statements. The proof of the theorem is omitted.

Theorem V.8.6. *Let J be a finite-range ferromagnetic interaction on $\mathbb{Z}^D$, $D \geq 2$. Assume that $(\beta_c, 0)$ is a normal critical point, $\langle Y_0; Y_k\rangle \sim \|k\|^{-(D-2+\eta)}$ and $m(\beta_c, h) \sim h^{1/\delta}$. Suppose that for some $\varepsilon > 0$*

$$G(R) = \sum_{\|k\| \leq R} \langle Y_0; Y_k\rangle_{\beta_c,0} \geq \varepsilon R^{2-\eta} \qquad \text{for all sufficiently large } R > 1, \tag{5.47}$$

that $2 - \eta = D(\delta - 1)/(\delta + 1)$ [equality in (5.42)], and that for some $K < \infty$

$$c_{\beta_c,0}(t) = -\beta_c[\psi(\beta_c, t/\beta_c) - \psi(\beta_c, 0)] \leq Kt^{1+1/\delta} \qquad \text{for all sufficiently small } t > 0. \tag{5.48}$$

If S_n/Σ_n converges in distribution to some random variable X, then X satisfies, for some $0 < a < \infty$,

$$\begin{aligned} E\{\exp(tX)\} &\leq \exp\left(\frac{|at|^{1+1/\delta}}{1+1/\delta}\right) && \text{for all } t \text{ real}, \\ P\{X \geq x\} &\leq \exp\left(-\frac{(x/a)^{1+\delta}}{1+\delta}\right) && \text{for all } x \geq 0. \end{aligned} \tag{5.49}$$

Since $\delta > 1$ [Corollary V.8.5], X is non-Gaussian.

A main hypothesis in the theorems in this section is that $(\beta_c, 0)$ be a normal critical point [Definition V.8.1]. While the validity of this hypothesis is not known for the general ferromagnetic model, it holds for the Ising model on $\mathbb{Z}^2$, as we now discuss. By (5.18), β_c is positive and finite [property V.8.1(a)] and $m(\beta_c, +)$ equals 0 [property V.8.1(b)]. Property V.8.1(c) requires that $\chi(\beta, h)$ be finite for all $(\beta, h) \in \mathscr{U}$ and that $\chi(\beta_c, 0)$ equal ∞. The finiteness for $(\beta, h) \in \mathscr{U}$ is a consequence of Lemma V.7.4(a) and Theorem V.7.7. McCoy and Wu (1973, Section IX.4) calculate the asymptotic behavior of $\langle Y_0; Y_k\rangle_{\beta_c,0}$ as k tends to infinity along the diagonal in $\mathbb{Z}^2$ ($k = (n, n) \in \mathbb{Z}^2$, $n \to \infty$). They show that

$$\langle Y_0; Y_{(n,n)}\rangle_{\beta_c,0} = \frac{A}{n^{1/4}}\left(1 + O\left(\frac{1}{n^2}\right)\right) \qquad \text{as } n \to \infty, \tag{5.50}$$

* Since $\langle (S_n/\Sigma_n)^2\rangle_{\beta_c,0} = 1$ for all n, every subsequence of $\{S_n/\Sigma_n\}$ has a subsubsequence that that converges in distribution to some random variable X. Any such X is described by (5.49).

Table V.3. Limit Theorems for the Ising Model on $\mathbb{Z}^2$

Limit theorem	Value(s) of (β, h)	Statement of limit	Where proved
Law of large numbers	$(\beta, h) \in \mathscr{U} \cup (\beta_c, 0)$	$\frac{S_\Lambda}{\|\Lambda\|} \to m(\beta, h)$ a.s. and exponentially	Theorem V.6.1(c) Lemma V.8.2(b)
Central limit theorem	$(\beta, h) \in \mathscr{U}$	$\frac{S_\Lambda - \|\Lambda\| m(\beta, h)}{\sqrt{\|\Lambda\|}} \xrightarrow{\mathscr{D}} N(0, \sigma^2(\beta, h))$	Theorem V.7.2 Theorem V.7.7
Non-central limit theorem (conjectured)	$(\beta_c, 0)$	$S_\Lambda/\Sigma_\Lambda \xrightarrow{\mathscr{D}} X$, where $\Sigma_\Lambda \sim \|\Lambda\|^{15/16}$; $P\{X \geq x\} \leq \exp(-cx^{16})$, $x \geq 0$	Theorem V.8.6

where A is a positive constant. Since for any $k \in \mathbb{Z}^2$ $\langle Y_0; Y_k \rangle_{\beta_c, 0} \geq 0$, it follows that

$$\chi(\beta_c, 0) = \sum_{k \in \mathbb{Z}^2} \langle Y_0; Y_k \rangle_{\beta_c, 0} \geq \sum_{n \in \mathbb{Z}} \langle Y_0; Y_{(n,n)} \rangle_{\beta_c, 0} = \infty.$$

We discuss the applicability of Theorem V.8.6 to the Ising model on $\mathbb{Z}^2$. Inequalities due to Messager and Miracle-Sole (1977) allow one to compare correlations $\langle Y_0; Y_k \rangle_{\beta_c, 0}$ for any $k \in \mathbb{Z}^2$ with correlations along the diagonal in $\mathbb{Z}^2$. These inequalities combined with (5.50) yield the value $\eta = 1/4$.[10] One can show that if the relation $m(\beta_c, h) \sim h^{1/\delta}$ is valid and certain other assumptions hold, then the critical exponent δ equals 15.[11] With the values $\eta = 1/4$ and $\delta = 15$, equality holds in the Buckingham–Gunton inequality (5.42): $2 - 1/4 = 2(15 - 1)/(15 + 1)$. It is not known whether the other hypotheses in Theorem V.8.6 are valid. If they are, then the theorem implies that the block spin scaling limit X of S_n/Σ_n is non-Gaussian and

$$P\{X \geq x\} \leq \exp(-\text{const} \cdot x^{16}) \qquad \text{for all } x \geq 0.$$

By Lemma V.8.3(c) $\Sigma_n^2 \sim n^{D+2-\eta} = n^{15/4} = |\Lambda|^{15/8}$. Table V.3 summarizes this conjectured limit together with other limit theorems proved earlier in the chapter.

We end this section by summarizing current beliefs about critical exponents and the nature of block spin scaling limits at the critical point. These beliefs are largely based on renormalization group arguments. Rigorous results confirming some of these beliefs have been obtained by Sokal (1979), Aizenman (1981, 1982), Brydges (1982), Fröhlich (1982), Gawędzki and Kupiainen (1982, 1983), Aizenman and Graham (1983), and Aragão de Carvalho, Caracciolo, and Fröhlich (1983). For each $D \geq 2$, it is strongly believed that all finite-range ferromagnetic models on $\mathbb{Z}^D$ have the same critical exponents (universality).* As we discuss below, these critical expo-

*As in the rest of this chapter, all interactions are assumed to be translation invariant. Ferromagnetic models with sufficiently rapidly decaying infinite-range interactions are believed to lie in the same *universality class* as finite-range ferromagnetic models; i.e., the former are believed to have the same critical exponents as the latter for each $D \geq 2$.

nents depend on D. In particular, for $D = 2$, η and δ are believed to have the Ising values $\frac{1}{4}$ and 15, respectively, so that (5.42) is an equality (hyperscaling). It is also strongly believed that for $D = 2$, S_n/Σ_n has a non-Gaussian scaling limit. Most workers now believe that for $D = 3$ hyperscaling holds and S_n/Σ_n has a non-Gaussian scaling limit, but this view has been challenged by some; compare Nickel and Sharpe (1979), Nickel (1981, 1982) versus Baker (1977), Baker and Kincaid (1981). We discuss $D = 4$ below. For all finite-range ferromagnetic models on $\mathbb{Z}^D$ with $D > 4$, it is widely believed that each critical exponent has a fixed value independent of D; namely, the value predicted by a simplified theory of ferromagnetism called mean field theory. This theory approximates the given interaction strengths $\{J(i-j)\}$ among the spins by an average or mean interaction. It will be discussed further in the next section. The mean field values of η and δ are 0 and 3, respectively, so that in particular hyperscaling fails; i.e., for $D > 4$, (5.42) is a strict inequality $(2 < D/2)$. With the value $\eta = 0$, we have $\Sigma_n \sim n^{(D+2)/2} > n^{D/2} = \sqrt{|\Lambda|}$. For $D > 4$, it is also believed that S_n/Σ_n has a Gaussian scaling limit. Many aspects of this picture for $D > 4$ have recently been proved rigorously (see the references listed at the beginning of this paragraph). For $D = 4$, it is believed that critical exponents take on their mean field values modulo logarithmic correction factors [Larkin and Khmel'nitskii (1969), Brezin et al. (1973), Wegner and Riedel (1973)] and that block spin scaling limits are Gaussian. However, few rigorous results for $D = 4$ are known. The dimension $D = 4$ is called the *upper critical dimension* for finite-range ferromagnetic models. $D = 1$ is called the *lower critical dimension.*

The basic hypothesis of mean field theory is that each spin feels an average interaction caused by all the other spins. This kind of average interaction is built into the Curie–Weiss model, which was studied in Section IV.4. We return to this model in the next section in order to study its critical phenomena and other aspects.

V.9. Three Faces of the Curie–Weiss Model

The most obvious aspect of the Curie–Weiss model is that it lends itself readily to exact calculations. As we will see in the first part of this section, certain features of the model coincide with those of a simple phenomenological theory of ferromagnetism called mean field theory. Because of this, the Curie–Weiss model and mean field theory are often considered to be synonymous. But the model also has two other faces. First, it approximates general ferromagnetic models on $\mathbb{Z}^D$. For example, the specific magnetization and the critical inverse temperature for such a model are bounded above and below, respectively, by the corresponding Curie–Weiss quantities. Second, for the Curie–Weiss model most of the beliefs expressed in the previous section concerning critical phenomena can be explicitly verified. In particular, in Theorem V.9.5, we will find the exact form of the block spin scaling limit at the critical point and see that it agrees with the conclusion of Theorem V.8.6. We will consider these three faces of Curie–Weiss after summarizing the results in Section IV.4, where the model was first studied. The notation will be changed slightly in order to stay consistent with the notation being used in Chapter V.

Let Λ be a symmetric hypercube in $\mathbb{Z}^D$ and $\mathscr{J}_0$ a positive number. We define the Hamiltonian

$$H_{\Lambda,h}^{\mathrm{CW}}(\omega) = -\frac{1}{2}\frac{\mathscr{J}_0}{|\Lambda|}\sum_{i,j\in\Lambda}\omega_i\omega_j - h\sum_{j\in\Lambda}\omega_j,$$

where h is real and $\omega = \{\omega_j; j\in\Lambda\}$ is a point in the finite configuration space $\Omega_\Lambda = \{1,-1\}^\Lambda$. The Curie–Weiss model is defined by the probability measure $P_{\Lambda,\beta,h}^{\mathrm{CW}}$ on $\mathscr{B}(\Omega_\Lambda)$ which assigns to each $\{\omega\}$, $\omega\in\Omega_\Lambda$, the probability

$$P_{\Lambda,\beta,h}^{\mathrm{CW}}\{\omega\} = \exp[-\beta H_{\Lambda,h}^{\mathrm{CW}}(\omega)]\pi_\Lambda P_\rho\{\omega\}\cdot\frac{1}{Z^{\mathrm{CW}}(\Lambda,\beta,h)}.$$

In this formula, β is positive, $\pi_\Lambda P_\rho$ is the product measure on $\mathscr{B}(\Omega_\Lambda)$ with identical one-dimensional marginals $\rho = \frac{1}{2}\delta_1 + \frac{1}{2}\delta_{-1}$, and $Z^{\mathrm{CW}}(\Lambda,\beta,h)$ is the nornalization $\int_{\Omega_\Lambda}\exp[-\beta H_{\Lambda,h}^{\mathrm{CW}}(\omega)]\pi_\Lambda P_\rho(d\omega)$. Expectation with respect to $P_{\Lambda,\beta,h}^{\mathrm{CW}}$ is denoted by $\langle - \rangle_{\Lambda,\beta,h}^{\mathrm{CW}}$. The form of the Curie–Weiss interaction has two consequences. First, since the interaction $\mathscr{J}_0/|\Lambda|$ depends on the hypercube Λ, the model is not covered by the results in this chapter. Second, since each pair of spins in Λ is coupled with equal interaction strength $\mathscr{J}_0/|\Lambda|$, the interaction completely ignores the lattice structure of $\mathbb{Z}^D$ and thus removes all geometry from the model. Because of this, we may replace the hypercube Λ by the subset of the integers consisting of the points $\{1,2,\ldots,|\Lambda|\}$. This allows us to use the results of Section IV.4, where spontaneous magnetization was proved by a large deviation analysis.

Let $\psi^{\mathrm{CW}}(\beta,h)$ be the specific Gibbs free energy for the model. According to Remark IV.4.2, $\psi^{\mathrm{CW}}(\beta,h)$ is given by the variational formula

$$(5.51)\qquad \begin{aligned} -\beta\psi^{\mathrm{CW}}(\beta,h) &= \lim_{|\Lambda|\to\infty}\frac{1}{|\Lambda|}\log Z^{\mathrm{CW}}(\Lambda,\beta,h) \\ &= \sup_{z\in\mathbb{R}}\{\tfrac{1}{2}\beta\mathscr{J}_0 z^2 + \beta h z - I_\rho^{(1)}(z)\}, \end{aligned}$$

where $I_\rho^{(1)}$ is the level-1 entropy function of the measure ρ:

$$I_\rho^{(1)}(z) = \begin{cases} \dfrac{1-z}{2}\log(1-z) + \dfrac{1+z}{2}\log(1+z) & \text{for } |z|\le 1, \\ \infty & \text{for } |z|>1. \end{cases}$$

If a point z gives the supremum in (5.51), then z satisfies the equation $\beta\mathscr{J}_0 z + \beta h = (I_\rho^{(1)})'(z) = \frac{1}{2}\log[(1+z)/(1-z)]$. By inversion, solutions of the latter coincide with solutions of

$$(5.52)\qquad z = \tanh[\beta(\mathscr{J}_0 z + h)].$$

According to the table accompanying Figure IV.3, for $0 < \beta \leq \mathscr{J}_0^{-1}$ and h real, this equation has a unique solution $z(\beta, h)$ which is 0 for $h = 0$. For $\beta > \mathscr{J}_0^{-1}$ and $h \neq 0$, it has a unique solution $z(\beta, h)$ with the same sign as h. For $\beta > \mathscr{J}_0^{-1}$ and $h = 0$, it has three solutions $z(\beta, +) > z(\beta, 0) = 0 > z(\beta, -)$. Of these three solutions, only $z(\beta, +)$ and $z(\beta, -)$ give the supremum in the formula $-\beta\psi^{\text{CW}}(\beta, 0) = \sup_{z \in \mathbb{R}}\{\frac{1}{2}\beta\mathscr{J}_0 z^2 - I_\rho^{(1)}(z)\}$. Let $Y_j(\omega) = \omega_j$ be the coordinate functions on Ω_Λ and define $S_\Lambda = \sum_{j \in \Lambda} Y_j$ and $m^{\text{CW}}(\beta, h) = \lim_{\Lambda \uparrow \mathbb{Z}^D} \langle S_\Lambda/|\Lambda| \rangle_{\Lambda,\beta,h}^{\text{CW}}$. The following facts were found.

(a) For $\beta > 0$ and h real, $m^{\text{CW}}(\beta, h)$ equals $z(\beta, h)$.

(b) Define $m^{\text{CW}}(\beta, +) = \lim_{h \to 0^+} m^{\text{CW}}(\beta, h)$. Spontaneous magnetization occurs if and only if $\beta > \mathscr{J}_0^{-1}$; in fact,

$$m^{\text{CW}}(\beta, +) = \begin{cases} 0 & \text{for } 0 < \beta \leq \mathscr{J}_0^{-1}, \\ z(\beta, +) > 0 & \text{for } \beta > \mathscr{J}_0^{-1}. \end{cases}$$

(c) With respect to $\{P_{\Lambda,\beta,0}^{\text{CW}}\}$,

$$S_\Lambda/|\Lambda| \overset{\exp}{\to} m^{\text{CW}}(\beta, 0) = 0 \qquad \text{for } 0 < \beta \leq \mathscr{J}_0^{-1},$$

$$S_\Lambda/|\Lambda| \overset{\mathscr{D}}{\to} \tfrac{1}{2}\delta_{z(\beta,+)} + \tfrac{1}{2}\delta_{z(\beta,-)} \qquad \text{for } \beta > \mathscr{J}_0^{-1}.$$

The value $\beta_c^{\text{CW}} = \mathscr{J}_0^{-1}$ is the critical inverse temperature for the Curie–Weiss model.

Face A. The Curie–Weiss Model and Mean Field Theory

Let J be a summable ferromagnetic interaction on $\mathbb{Z}^D$ and Λ a symmetric hypercube. Define the Hamiltonian

$$\text{(5.53)} \qquad H_{\Lambda,h}(\omega) = -\frac{1}{2}\sum_{i,j \in \Lambda} J(i-j)\omega_i\omega_j - h\sum_{j \in \Lambda}\omega_j, \qquad \omega \in \Omega_\Lambda$$

and let $P_{\Lambda,\beta,h}$ be the corresponding finite-volume Gibbs state.* The measures $\{P_{\Lambda,\beta,h}; \Lambda \text{ symmetric hypercubes}\}$ define a ferromagnetic model on $\mathbb{Z}^D$ which we call a *general ferromagnet* in order to distinguish it from the Curie–Weiss model. Mean field theory is an approximation which replaces $H_{\Lambda,h}$ by the Hamiltonian

$$\text{(5.54)} \qquad H_{\Lambda,h}^{\text{MF}}(\omega) = -\mathscr{J}_0 m \sum_{j \in \Lambda}\omega_j - h\sum_{j \in \Lambda}\omega_j,$$

where $\mathscr{J}_0 = \sum_{k \in \mathbb{Z}^D} J(k)$. The parameter m represents an average field which is caused by all the spins and with which each spin interacts. It satisfies a self-consistency condition to be specified below. The finite-volume Gibbs state $P_{\Lambda,\beta,h}$ is replaced by the probability measure

$$P_{\Lambda,\beta,h}^{\text{MF}}(d\omega) = \exp[-\beta H_{\Lambda,h}^{\text{MF}}(\omega)]\pi_\Lambda P_\rho(d\omega)\cdot\frac{1}{Z^{\text{MF}}(\Lambda,\beta,h)}.$$

*In this section, we work only with the free external condition. The external condition is not indicated in the notation for $H_{\Lambda,h}$ and $P_{\Lambda,\beta,h}$.

Using (5.54), we calculate

$$Z^{\text{MF}}(\Lambda, \beta, h) = \int_{\Omega_\Lambda} \exp[-\beta H_{\Lambda,h}^{\text{MF}}(\omega)]\pi_\Lambda P_\rho(d\omega) = (\cosh[\beta(\mathcal{J}_0 m + h)])^{|\Lambda|}.$$

Thus, $P_{\Lambda,\beta,h}^{\text{MF}}$ is just the product measure

$$(5.55) \quad P_{\Lambda,\beta,h}^{\text{MF}}(d\omega) = \prod_{j\in\Lambda} \left\{ \exp(\beta(\mathcal{J}_0 m + h)\omega_j)\rho(d\omega_j) \cdot \frac{1}{\cosh[\beta(\mathcal{J}_0 m + h)]} \right\}.$$

Since m represents an average field caused by all the spins, a reasonable self-consistency condition is to require for each j that $m = \int_{\Omega_\Lambda} \omega_j P_{\Lambda,\beta,h}^{\text{MF}}(d\omega)$. Equation (5.55) implies that

$$(5.56) \qquad m = \tanh[\beta(\mathcal{J}_0 m + h)].$$

This is the same as equation (5.52) which gives the specific magnetization in the Curie–Weiss model. In order to avoid ambiguity, we must specify m in those cases where (5.56) has multiple solutions. For $\beta > \mathcal{J}_0^{-1}$ and $h \neq 0$, m is set equal to the unique solution which has the same sign as h. For $\beta > \mathcal{J}_0^{-1}$ and $h = 0$, m is set equal to $z(\beta, +)$. We see that for $\beta > 0$, $h \neq 0$, and for $0 < \beta \leq \mathcal{J}_0^{-1}$, $h = 0$, m equals the Curie–Weiss specific magnetization $m^{\text{CW}}(\beta, h)$ and for $\beta > \mathcal{J}_0^{-1}$, $h = 0$, m equals the spontaneous magnetization $m^{\text{CW}}(\beta, +)$. In the next section, we will see that m approximates the specific magnetization in the general ferromagnet defined by the Hamiltonians $\{H_{\Lambda,h}\}$ in (5.53).

In mean field theory, certain critical exponents can be easily calculated. For example, the critical exponent δ is defined by the relation $m(\beta_c, h) \sim h^{1/\delta}$, $h \to 0^+$. The mean field value of δ is found by substituting $m = m^{\text{CW}}(\beta, h)$, $\beta = \beta_c^{\text{CW}} = \mathcal{J}_0^{-1}$, and $h > 0$ in (5.56). As $x \to 0$, $\tanh x = x - x^3/3 + 0(x^5)$. Since $m^{\text{CW}}(\beta_c^{\text{CW}}, h) \to 0$ as $h \to 0$, (5.56) with $\beta\mathcal{J}_0 = 1$ implies

$$m = m + \beta h - \tfrac{1}{3}(m + \beta h)^3 + O([m + \beta h]^5) \qquad \text{as } h \to 0^+.$$

This yields $m \sim h^{1/3}$, which gives the mean field value $\delta = 3$. Since for $h > 0$ m equals $m^{\text{CW}}(\beta_c^{\text{CW}}, h)$, δ also equals 3 for the Curie–Weiss model. Now consider the critical exponent η. For a general ferromagnet, η was defined by the relation $\langle Y_0; Y_k \rangle_{\beta_c, 0} \sim \|k\|^{-(D-2+\eta)}$, $\|k\| \to \infty$. However, with respect to the mean field product measure (5.55), the individual spins are independent, and so in mean field theory this definition of η makes no sense. In order to find an alternate definition, we note that mean field theory can be regarded as the lowest order of a well-defined, successive approximation scheme to general ferromagnets. The next order past mean field theory is called the Gaussian approximation [Brezin et al. (1976), Ma (1976)]. This approximation yields the asymptotic relation $\langle Y_0; Y_k \rangle_{\beta_c, 0} \sim \|k\|^{-(D-2)}$, which in turn gives the value $\eta = 0$. The latter is usually taken as the mean field/Curie–Weiss value of the critical exponent η.

We now turn to Face B of the Curie–Weiss model as we study how it approximates a general ferromagnet.

Face B. Approximating General Ferromagnets

Consider again the general ferromagnet on $\mathbb{Z}^D$ defined by the Hamiltonians $\{H_{\Lambda,h}\}$ in (5.53). We replace the spatially varying interaction strengths $\{J(i-j)\}$ between each pair of sites $i, j \in \Lambda$ by an average interaction strength $\mathscr{J}_0/|\Lambda| = \sum_{k \in \mathbb{Z}^D} J(k)/|\Lambda|$. Then $H_{\Lambda,h}$ is replaced by the Curie–Weiss Hamiltonian $H_{\Lambda,h}^{\text{CW}}$ and the finite-volume Gibbs state $P_{\Lambda,\beta,h}$ corresponding to $H_{\Lambda,h}$ is replaced by the Curie–Weiss measure $P_{\Lambda,\beta,h}^{\text{CW}}$. How accurately does $P_{\Lambda,\beta,h}^{\text{CW}}$ approximate $P_{\Lambda,\beta,h}$?

Let $m(\beta, h)$ denote the specific magnetization corresponding to $\{P_{\Lambda,\beta,h}\}$ and let β_c be the critical inverse temperature. Our first theorem compares $m(\beta, h)$ and β_c with the Curie–Weiss values $m^{\text{CW}}(\beta, h)$ and $\beta_c^{\text{CW}} = \mathscr{J}_0^{-1}$, respectively. It also compares the critical inverse temperature for the Ising model on $\mathbb{Z}^D$ (with nearest-neighbor interaction strength $\mathscr{J} = 1$) with the corresponding Curie–Weiss value $\beta_c^{\text{CW}}(D)$. Since each site in $\mathbb{Z}^D$ has $2D$ nearest neighbors, $\beta_c^{\text{CW}}(D)$ equals $(2D)^{-1}$.

Theorem V.9.1. (a) *Let J be a summable ferromagnetic interaction on $\mathbb{Z}^D$ and set $\mathscr{J}_0 = \sum_{k \in \mathbb{Z}^D} J(k)$. Then for $\beta > 0$ and $h \geq 0$,*

$$m^{\text{CW}}(\beta, h) \geq m(\beta, h) \quad \text{and} \quad \beta_c \geq \beta_c^{\text{CW}} = \mathscr{J}_0^{-1}. \tag{5.57}$$

(b) *Let $\beta_c(D)$ denote the critical inverse temperature for the Ising model on $\mathbb{Z}^D$ with $\mathscr{J} = 1$. Then*

$$\beta_c(D) \geq \beta_c^{\text{CW}}(D) \quad \text{and} \quad \beta_c(D)/\beta_c^{\text{CW}}(D) = 1 + (2D)^{-1} + o(D^{-1}) \quad \text{as } D \to \infty.$$

The bounds (5.57) follow from Pearce (1981) [see the comments after Theorem IV.5.3]. Note 8 in Chapter IV lists other references. For part (b), see Bricmont and Fontaine (1982).

Part (b) is particularly interesting. The Curie–Weiss model not only gives a rigorous lower bound on $\beta_c(D)$ valid for all D, but it gives a correct asymptotic limit as $D \to \infty$. This asymptotic validity of Curie–Weiss is related to the behavior of critical exponents as functions of dimension. In the previous section, we mentioned the strongly held belief that critical exponents for finite-range ferromagnetic models on $\mathbb{Z}^D$ equal their mean field/Curie–Weiss values whenever $D > 4$. This means, for example, that in calculating critical exponents for the Ising model on $\mathbb{Z}^D$, $D > 4$, one should get identical results whether one uses the given nearest-neighbor interaction or the long-range, averaged Curie–Weiss interaction. By contrast, when $1 < D < 4$, the critical exponents are sensitive to the range of the interaction, and a discrepancy between the Ising and mean field/Curie–Weiss values occurs. In fact, for $D = 2$, (η, δ) equals $(\frac{1}{4}, 15)$ versus $(0, 3)$, respectively. For $D = 3$, the mean field/Curie–Weiss values $(\eta, \delta) = (0, 3)$ cannot equal the Ising values since the former violate the Buckingham–Gunton inequality (5.42).

Part (b) of Theorem V.9.1 shows that for the Ising model on $\mathbb{Z}^D$, the Curie–Weiss critical inverse temperature is asymptotically correct in the limit of large dimension. Theorem V.9.3 below considers another limit (with D fixed) in which the specific Gibbs free energy and the specific magnetization for a general ferromagnet converge to the corresponding Curie–Weiss quantities. If $\mathscr{J}_0$ is a positive number, then $\psi^{\text{CW}}(\mathscr{J}_0;\beta,h)$ will denote the specific Gibbs free energy for the Curie–Weiss model with interaction strength $\mathscr{J}_0/|\Lambda|$:

$$(5.58) \qquad -\beta\psi^{\text{CW}}(\mathscr{J}_0;\beta,h) = \sup_{z\in\mathbb{R}}\{\tfrac{1}{2}\beta\mathscr{J}_0 z^2 + \beta hz - I_\rho^{(1)}(z)\}.$$

The corresponding specific magnetization is denoted by $m^{\text{CW}}(\mathscr{J}_0;\beta,h)$.

We first motivate Theorem V.9.3 by means of the Gibbs variational formula. We restrict to $D=1$ since we have proved the Gibbs variational formula only in this case. Let J be a summable ferromagnetic interaction on $\mathbb{Z}$ and Λ a symmetric interval. Define the Hamiltonian

$$H_{\Lambda,h}(\omega) = -\tfrac{1}{2}\sum_{i,j\in\Lambda} J(i-j)\omega_i\omega_j - h\sum_{j\in\Lambda}\omega_j, \qquad \omega\in\Omega_\Lambda,$$

and the specific Gibbs free energy

$$(5.59) \qquad -\beta\psi(\beta,h) = \lim_{\Lambda\uparrow\mathbb{Z}}\frac{1}{|\Lambda|}\log\int_{\Omega_\Lambda}\exp[-\beta H_{\Lambda,h}(\omega)]\pi_\Lambda P_\rho(d\omega).$$

The Gibbs variational formula, derived in Theorem IV.7.3(a) via level-3 large deviations, states that

$$(5.60) \qquad -\beta\psi(\beta,h) = \sup_{P\in\mathscr{M}_s(\Omega)}\{-\beta u(h;P) - I_\rho^{(3)}(P)\}.$$

$\mathscr{M}_s(\Omega)$ is the set of translation invariant probability measures on $\Omega = \{1,-1\}^{\mathbb{Z}}$,

$$u(h;P) = -\frac{1}{2}\sum_{k\in\mathbb{Z}} J(k)\int_\Omega \omega_0\omega_k P(d\omega) - h\int_\Omega \omega_0 P(d\omega)$$

is the specific energy in P, and $I_\rho^{(3)}(P)$ is the mean relative entropy of P with respect to the infinite product measure P_ρ.

The idea underlying mean field theory is to approximate the finite-volume Gibbs state of a general ferromagnet by means of a product measure. We apply a similar idea in order to bound $-\beta\psi(\beta,h)$. Let $\mathscr{M}_{\text{pr}}(\Omega)$ be the subset of $\mathscr{M}_s(\Omega)$ consisting of product measures. For $P\in\mathscr{M}_{\text{pr}}(\Omega)$, $\int_\Omega\omega_0\omega_k P(d\omega) = (\int_\Omega\omega_0 P(d\omega))^2$ for $k\neq 0$, and so

$$u(h;P) = -\frac{1}{2}J(0) - \frac{1}{2}\sum_{k\neq 0} J(k)\left(\int_\Omega \omega_0 P(d\omega)\right)^2 - h\int_\Omega \omega_0 P(d\omega).$$

The mean $\int_\Omega\omega_0 P(d\omega)$ is a number $z\in[-1,1]$. Equation (5.60) implies that

$$-\beta\psi(\beta,h) \geq \sup_{P\in\mathscr{M}_{\text{pr}}(\Omega)}\{-\beta u(h;P) - I_\rho^{(3)}(P)\}$$

$$= \sup_{P \in \mathscr{M}_{pr}(\Omega)} \left\{ \tfrac{1}{2}\beta J(0) + \tfrac{1}{2}\beta \bar{\mathscr{J}}_0 \left(\int_\Omega \omega_0 P(d\omega) \right)^2 + \beta h \int_\Omega \omega_0 P(d\omega) - I_\rho^{(3)}(P) \right\}$$

$$= \sup_{z \in [-1,1]} \left\{ \tfrac{1}{2}\beta J(0) + \tfrac{1}{2}\beta \bar{\mathscr{J}}_0 z^2 + \beta h z - \inf\{ I_\rho^{(3)}(P) : P \in \mathscr{M}_{pr}(\Omega), \int_\Omega \omega_0 P(d\omega) = z \} \right\},$$

where $\bar{\mathscr{J}}_0 = \sum_{k \neq 0} J(k)$. We have

$$\inf \left\{ I_\rho^{(3)}(P) : P \in \mathscr{M}_{pr}(\Omega), \int_\Omega \omega_0 P(d\omega) = z \right\} = \inf \left\{ I_\rho^{(3)}(P) : P \in \mathscr{M}_s(\Omega), \int_\Omega \omega_0 P(d\omega) = z \right\},$$

and the latter equals $I_\rho^{(1)}(z)$ by the contraction principle in Theorem II.5.4. Since $I_\rho^{(1)}(z)$ equals ∞ for $|z| > 1$, we have the following result. It is analogous to the lower bound $m(\beta, h) \geq m^{\mathrm{CW}}(\beta, h)$ in Theorem V.9.1(a).

Theorem V.9.2. *Let J be a summable ferromagnetic interaction on $\mathbb{Z}$ and $\psi(\beta, h)$ the corresponding specific Gibbs free energy. Set $\bar{\mathscr{J}}_0 = \sum_{k \neq 0} J(k)$. Then for $\beta > 0$ and h real,*

$$-\beta\psi(\beta, h) \geq \tfrac{1}{2}\beta J(0) - \beta\psi^{\mathrm{CW}}(\bar{\mathscr{J}}_0; \beta, h) = \tfrac{1}{2}\beta J(0) + \sup_{z \in \mathbb{R}} \{ \tfrac{1}{2}\beta \bar{\mathscr{J}}_0 z^2 + \beta h z - I_\rho^{(1)}(z) \}.$$

The next theorem determines a sequence of interactions $\{J^{(\gamma)}\}$ on $\mathbb{Z}^D$, indexed by $0 < \gamma \leq 1$, such that the corresponding specific Gibbs free energies $\{\psi^{(\gamma)}\}$ converge in the limit $\gamma \to 0^+$ to the Curie–Weiss specific free energy $\psi^{\mathrm{CW}}(\mathscr{J}_0; \beta, h)$ for some $\mathscr{J}_0 > 0$. This will lead to a limit involving the specific magnetizations $m^{(\gamma)}(\beta, h) = -\partial\psi^{(\gamma)}(\beta, h)/\partial h$ ($\beta > 0, h \neq 0$). Let $J \not\equiv 0$ be a non-negative, continuous, symmetric function on $\mathbb{R}^D$ with bounded support. For $0 < \gamma \leq 1$ and $k \in \mathbb{Z}^D$, define $J^{(\gamma)}(k) = \gamma^D J(\gamma k)$. Since J is bounded, $J^{(\gamma)}(k)$ converges to 0 as $\gamma \to 0^+$, and the interaction $J^{(\gamma)}$ has range of order $1/\gamma$. Thus as $\gamma \to 0^+$, $J^{(\gamma)}$ defines a weak, long-range interaction on $\mathbb{Z}^D$, and the total interaction strength $\sum_{k \in \mathbb{Z}^D} J^{(\gamma)}(k)$ converges to the constant $\mathscr{J}_0 = \int_{\mathbb{R}^D} J(x)dx$. As a weak, long-range interaction is reminiscent of the Curie–Weiss model, it is plausible that $-\beta\psi^{(\gamma)}(\beta, h)$ converges to $-\beta\psi^{\mathrm{CW}}(\mathscr{J}_0; \beta, h)$ as $\gamma \to 0^+$. This limit was first proved in a special case by Kac (1959a) and is now known as a Kac limit.[12] For $D = 1$, the lower bound

(5.61) $$\liminf_{\gamma \to 0^+} (-\beta\psi^{(\gamma)}(\beta, h)) \geq -\beta\psi^{\mathrm{CW}}(\mathscr{J}_0; \beta, h)$$

is immediate from Theorem V.9.2 and the fact that $I_\rho^{(1)}(z)$ equals ∞ for $|z| > 1$.

Theorem V.9.3. *Let $J \not\equiv 0$ be a non-negative, continuous, symmetric function on $\mathbb{R}^d$ with bounded support and set $\mathscr{J}_0 = \int_{\mathbb{R}^D} J(x)dx$. For $\beta > 0$, h real, and $0 < \gamma \leq 1$, let $\psi^{(\gamma)}(\beta, h)$ be the specific Gibbs free energy corresponding to the scaled interaction $J^{(\gamma)}(k) = \gamma^D J(\gamma k)$, $k \in \mathbb{Z}$. Then the following conclusions hold.*

(a) *For $\beta > 0$ and h real,*

$$\lim_{\gamma \to 0^+} (-\beta\psi^{(\gamma)}(\beta, h)) = -\beta\psi^{\mathrm{CW}}(\mathscr{J}_0; \beta, h). \tag{5.62}$$

(b) *For $\beta > 0$ and $h \neq 0$,*

$$\lim_{\gamma \to 0^+} m^{(\gamma)}(\beta, h) = \lim_{\gamma \to 0^+} \left(-\frac{\partial \psi^{(\gamma)}(\beta, h)}{\partial h} \right) = -\frac{\partial \psi^{\mathrm{CW}}(\mathscr{J}_0; \beta, h)}{\partial h} = m^{\mathrm{CW}}(\mathscr{J}_0; \beta, h), \tag{5.63}$$

where $m^{\mathrm{CW}}(\mathscr{J}_0; \beta, h)$ is the Curie–Weiss specific magnetization. For each $\beta > 0$, the convergence is uniform on compact subsets of $h > 0$ and of $h < 0$.

Comments of proof. (a) This is proved in Thompson (1972, Appendix C) for $D = 1$ and in Thompson and Silver (1973) for $D \geq 1$ and other models. Simon (1985) has further generalizations. The order of the limits in part (a) is important. The function $\psi^{(\gamma)}$ is defined by the infinite-volume limit $\Lambda \uparrow \mathbb{Z}^D$ as in (5.59). The limit $\gamma \to 0^+$ involves the scaled interactions $\{J^{(\gamma)}\}$. The function $\psi^{\mathrm{CW}}(\mathscr{J}_0; \beta, h)$ is also defined by the infinite-volume limit $|\Lambda| \to \infty$ [see (5.51)], but in this limit the Curie–Weiss interaction $\mathscr{J}_0/|\Lambda|$ is scaled simultaneously.

(b) Problem IV.9.5 shows that for $h \neq 0$ $\partial\psi^{\mathrm{CW}}(\mathscr{J}_0; \beta, h)/\partial h$ exists and equals $-m^{\mathrm{CW}}(\mathscr{J}_0; \beta, h)$. The convexity of $-\psi^{(\gamma)}(\beta, h)$ for h real allows us to interchange the limit $\gamma \to 0^+$ and the differentiation with respect to h [Lemma IV.6.3], yielding (5.63). Since $m^{(\gamma)}(\beta, h)$ is concave for $h \geq 0$ and convex for $h \leq 0$, the uniformity of the limit in (5.63) is a consequence of Theorem VI.3.3(b).

Theorems V.9.1–V.9.3 have shown that the Curie–Weiss model approximates general ferromagnets. We now study its block spin scaling limits.

Face C. Block Spin Scaling Limits for the Curie–Weiss Model

Scaling limits for the Curie–Weiss model will be derived for β in the range $0 < \beta < \beta_c^{\mathrm{CW}}$ and for $h = 0$. To ease the notation, the symmetric hypercube Λ in $\mathbb{Z}^D$ is replaced by the set $\Lambda = \{1, 2, \ldots, n\}$, where n is a positive integer. All quantities are indexed by n instead of by Λ. Also, $\mathscr{J}_0$ is set equal to 1, so that the critical inverse temperature β_c^{CW} equals 1. Thus, we work with the probability measure

(5.64) $$P_{n,\beta,0}^{\text{CW}}(d\omega) = \exp\left[\frac{\beta}{2n}\sum_{i,j=1}^{n}\omega_i\omega_j\right]\pi_n P_\rho(d\omega)\cdot\frac{1}{Z^{\text{CW}}(n,\beta,0)}$$

on the set Ω_n of sequences $\omega = (\omega_1, \omega_2, \ldots, \omega_n)$ with each $\omega_j \in \{1, -1\}$. $Z^{\text{CW}}(n,\beta,0)$ equals $\int_{\Omega_n}\exp[(\beta/2n)\sum_{i,j=1}^{n}\omega_i\omega_j]\pi_n P_\rho(d\omega)$. For $\omega \in \Omega_n$, define $S_n(\omega) = \sum_{j=1}^{n}\omega_j$. In our study of scaling limits for general ferromagnets in Sections V.7 and V.8, we worked with the infinite-volume Gibbs states for the models. However, for the Curie–Weiss model, infinite-volume Gibbs states have not been defined. In this section, we will study scaling limits of $\{S_n\}$ with respect to the finite-volume Gibbs states $\{P_{n,\beta,0}^{\text{CW}}\}$. Expectation with respect to $P_{n,\beta,0}^{\text{CW}}$ will be denoted by $\langle - \rangle_{n,\beta,0}^{\text{CW}}$.

The first result is a central limit theorem for $0 < \beta < \beta^{\text{CW}}$ and $h = 0$. A central limit theorem also holds for any $\beta > 0$ and $h \neq 0$ [Ellis and Newman (1978b)].

Theorem V.9.4. *Fix $0 < \beta < \beta_c^{\text{CW}} = 1$ and define $\sigma^2(\beta) = (1-\beta)^{-1}$. Then as $n \to \infty$,*

(5.65) $$\frac{S_n}{\sqrt{n}} \xrightarrow{\mathscr{D}} N(0, \sigma^2(\beta)) \qquad \text{w.r.t. the finite-volume Gibbs states } \{P_{n,\beta,0}^{\text{CW}}\}.$$

Proof. For any $f \in \mathscr{C}(\mathbb{R})$, $\langle f(S_n/\sqrt{n})\rangle_{n,\beta,0}$ equals

(5.66) $$\frac{\displaystyle\int_\Omega f\left(\frac{\sum_{j=1}^{n}\omega_j}{\sqrt{n}}\right)\exp\left[\frac{\beta}{2}\left(\frac{\sum_{j=1}^{n}\omega_j}{\sqrt{n}}\right)^2\right]P_\rho(d\omega)}{\displaystyle\int_\Omega \exp\left[\frac{\beta}{2}\left(\frac{\sum_{j=1}^{n}\omega_j}{\sqrt{n}}\right)\right]P_\rho(d\omega)},$$

where P_ρ is the infinite product measure on $\Omega = \{1, -1\}^{\mathbb{Z}}$ with identical one-dimensional marginals $\rho = \frac{1}{2}\delta_1 + \frac{1}{2}\delta_{-1}$. With respect to P_ρ, the coordinates $\{\omega_j\}$ are i.i.d. The "obvious" step is to apply the classical central limit theorem* and to conclude that the ratio in (5.66) converges to

$$\frac{\int_{\mathbb{R}} f(x)\exp(\frac{1}{2}\beta x^2)N(0,1)(dx)}{\int_{\mathbb{R}}\exp(\frac{1}{2}\beta x^2)N(0,1)(dx)} = \frac{\int_{\mathbb{R}} f(x)\exp[-\frac{1}{2}(1-\beta)x^2]dx}{\int_{\mathbb{R}}\exp[-\frac{1}{2}(1-\beta)x^2]dx}.$$

As this equals $\int_{\mathbb{R}} f(x)N(0,\sigma^2(\beta))(dx)$, (5.65) will be proved. However, these steps must be justified since the function $\exp(\frac{1}{2}\beta z^2)$ is unbounded. We will show that the non-negative random variables $W_n(\omega) = \exp[\frac{1}{2}\beta(\sum_{j=1}^{n}\omega_j/\sqrt{n})^2]$ satisfy $\sup_{n\geq 1}\int_{\{W_n \geq \alpha\}} W_n\, dP_\rho \to 0$ as $\alpha \to \infty$ (uniform integrability). Then (5.65) follows from Theorem A.8.6.

We will prove that

(5.67) $$\sup_{n\geq 1} P_\rho\{W_n \geq \alpha\} \leq 2\alpha^{-1/\beta} \qquad \text{for } \alpha > 1.$$

Integration by parts and (5.67) show that uniformly in n

$*\sum_{j=1}^{n}\omega_j/\sqrt{n} \xrightarrow{\mathscr{D}} N(0,1)$ w.r.t. P_ρ.

$$\int_{\{W_n \geq \alpha\}} W_n dP_\rho = \alpha P_\rho\{W_n \geq \alpha\} + \int_\alpha^\infty P_\rho\{W_n \geq t\} dt = O(\alpha^{1-1/\beta}).$$

This converges to 0 as $\alpha \to \infty$ since $0 < \beta < 1$, and so the uniform integrability will be proved. Note first that for any $x > 0$ and integer $j \geq 1$

$$\int_\Omega e^{x\omega_j} P_\rho(d\omega) = \cosh x = \sum_{m\geq 0} x^{2m}/(2m)! \leq \sum_{m\geq 0} x^{2m}/(2^m m!) = \exp(\tfrac{1}{2}x^2).$$

(5.68)

Given $\alpha > 1$, let $x = (2(\log\alpha)/\beta)^{1/2} > 0$. Then by Chebyshev's inequality and (5.68)

$$\begin{aligned} P_\rho\{W_n \geq \alpha\} &= P_\rho\left\{\left|\frac{\sum_{j=1}^n \omega_j}{\sqrt{n}}\right| \geq x\right\} \\ &= 2P_\rho\left\{\frac{\sum_{j=1}^n \omega_j}{\sqrt{n}} \geq x\right\} \leq 2\exp(-x^2)\cdot \int_\Omega \exp\left(x\frac{\sum_{j=1}^n \omega_j}{\sqrt{n}}\right) P_\rho(d\omega) \\ &= 2\exp(-x^2)\prod_{j=1}^n \int_\Omega \exp\left(x\frac{\omega_j}{\sqrt{n}}\right) P_\rho(d\omega) \leq 2\exp\left(-\frac{x^2}{2}\right). \end{aligned}$$

This bound, which is independent of n, yields (5.67). □

Since for $\beta = \beta_c^{\mathrm{CW}} = 1$ $\sigma^2(\beta)$ diverges, we expect the central limit theorem to fail at the critical point $(\beta_c^{\mathrm{CW}}, 0) = (1, 0)$. The block spin scaling limit at the critical point is considered in Theorem V.9.5 below, which shows that $S_n/n^{3/4}$ has a non-Gaussian limit in distribution as $n \to \infty$. The theorem will first be motivated by a heuristic argument based on large deviations.

Let Q_n denote the P_ρ-distribution of $\sum_{j=1}^n \omega_j/n$. Then for any real number λ, $P_\rho\{\sum_{j=1}^n \omega_j/n^\lambda \in dz\} = Q_n(d(z/n^{1-\lambda}))$, and so for any $f \in \mathscr{C}(\mathbb{R})$

(5.69) $$\left\langle f\left(\frac{S_n}{n^\lambda}\right)\right\rangle_{n,1,0}^{\mathrm{CW}} = \frac{\int_{\mathbb{R}} f(z)\exp(\frac{1}{2}n^{2\lambda-1}z^2) Q_n(d(z/n^{1-\lambda}))}{\int_{\mathbb{R}} \exp(\frac{1}{2}n^{2\lambda-1}z^2) Q_n(d(z/n^{1-\lambda}))}.$$

It is tempting to replace $Q_n(d(z/n^{1-\lambda}))$ by $\exp[-nI_\rho^{(1)}(z/n^{1-\lambda})]dz$, where $I_\rho^{(1)}$ is the level-1 entropy function of ρ. Consider the Taylor expansion of $I_\rho^{(1)}(z)$ around $z = 0$:

$$I_\rho^{(1)}(z) = \frac{1-z}{2}\log(1-z) + \frac{1+z}{2}\log(1+z) = \frac{z^2}{2} + \frac{z^4}{12} + O(z^6).$$

Hence for large n one expects that

$$\begin{aligned} \exp\left[n^{2\lambda-1}\frac{z^2}{2}\right] Q_n\left(d\left(\frac{z}{n^{1-\lambda}}\right)\right) &\approx \exp\left[n^{2\lambda-1}\frac{z^2}{2}\right]\exp\left[-n\left(\frac{1}{2}\left(\frac{z}{n^{1-\lambda}}\right)^2\right.\right. \\ &\qquad \left.\left. + \frac{1}{12}\left(\frac{z}{n^{1-\lambda}}\right)^4 + \gamma_n(z)\right)\right] dz \\ &= \exp(-\tfrac{1}{12}n^{4\lambda-3}z^4 + \gamma_n(z))dz, \end{aligned}$$

where $\gamma_n(z) = nO((z/n^{1-\lambda})^6) = O(n^{6\lambda-5}z^6)$. If we pick $\lambda = \frac{3}{4}$, then $\gamma_n(z) \to 0$ for each z, and formally the right-hand side of (5.69) converges to $\int_{\mathbb{R}} f(z) \exp(-\frac{1}{12}z^4)dz / \int \exp(-\frac{1}{12}z^4)dz$. This leads to our next theorem, which was discovered by Simon and Griffiths (1973). The proof given here is due to Ellis and Newman (1978b).[13]

Theorem V.9.5. *Fix $\beta = \beta_c^{\text{CW}} = 1$. Then there exists a random variable X with density proportional to $\exp(-\frac{1}{12}x^4)$ such that as $n \to \infty$*

$$\frac{S_n}{n^{3/4}} \xrightarrow{\mathscr{D}} X \quad \text{w.r.t. the finite-volume Gibbs states } \{P_{n,1,0}^{\text{CW}}\}.$$

Proof. While large deviations and the entropy function motivated the theorem, the proof involves the free energy function $c_\rho(t) = \log \int_{\mathbb{R}} e^{tx} \rho(dx) = \log\cosh t$ of the measure ρ. We will prove that for any real number r

$$\lim_{n\to\infty} \left\langle \exp\left(r \frac{S_n}{n^{3/4}} \right) \right\rangle_{n,1,0}^{\text{CW}} = \frac{\int_{\mathbb{R}} \exp(rx - \frac{1}{12}x^4)dx}{\int_{\mathbb{R}} \exp(-\frac{1}{12}x^4)dx}. \tag{5.70}$$

Theorem A.8.7(a) then yields the theorem. The expression $\langle \exp(rS_n/n^{3/4}) \rangle_{n,1,0}^{\text{CW}}$ equals

$$\frac{\displaystyle\int_\Omega \exp\left[r \frac{\sum_{j=1}^n \omega_j}{n^{3/4}} + \frac{1}{2}\left(\frac{\sum_{j=1}^n \omega_j}{\sqrt{n}} \right)^2 \right] P_\rho(d\omega)}{\displaystyle\int_\Omega \exp\left[\frac{1}{2}\left(\frac{\sum_{j=1}^n \omega_j}{\sqrt{n}} \right)^2 \right] P_\rho(d\omega)}. \tag{5.71}$$

We replace the quantities $\exp[\frac{1}{2}(\sum_{j=1}^n \omega_j/\sqrt{n})^2]$ using the identity $\exp(\frac{1}{2}y^2) = \int_{\mathbb{R}} \exp(yx - \frac{1}{2}x^2)dx/\sqrt{2\pi}$, then do the P_ρ integration. After a short calculation, one finds that the ratio in (5.71) equals

$$\exp(-\tfrac{1}{2}r^2/\sqrt{n}) \int_{\mathbb{R}} \exp[rx - nG(x/n^{1/4})]dx \bigg/ \int_{\mathbb{R}} \exp[-nG(x/n^{1/4})]dx, \tag{5.72}$$

where $G(x) = \frac{1}{2}x^2 - c_\rho(x) = \frac{1}{2}x^2 - \log\cosh x$. There exist positive real numbers A and ε such that

(a) $G(x) = \frac{1}{12}x^4 + O(x^6)$ for $|x| \le A$;
(b) $G(x) \ge \varepsilon x^4$ for $|x| \le A$;
(c) $G(x) \ge \varepsilon x^2$ for $|x| \ge A$.

We split each integral in (5.72) into an integral over $|x/n^{1/4}| \le A$ and an integral over $|x/n^{1/4}| > A$. Property (c) implies that the integrals over $|x/n^{1/4}| > A$ converge to 0 as $n \to \infty$. Given this, properties (a) and (b) and the Lebesgue dominated convergence theorem imply that the ratio in (5.72) converges to $\int_{\mathbb{R}} \exp(rx - \frac{1}{12}x^4)dx / \int_{\mathbb{R}} \exp(-\frac{1}{12}x^4)dx$. This completes the proof. □

We compare this theorem with Theorem V.8.6. The latter stated that if for a general ferromagnet on $\mathbb{Z}^D$ the critical exponents η and δ satisfy $2 - \eta = D(\delta - 1)/(\delta + 1)$ (and other hypotheses hold), then a scaling limit of S_n/Σ_n is non-Gaussian and its tail is bounded by $\exp(-\text{const} \cdot x^{\delta+1})$. For the Curie–Weiss model, we have seen that δ equals 3. Although according to the Gaussian approximation η is defined to have the value $\eta = 0$, we consider another internally consistent definition of η for the Curie–Weiss model. For general ferromagnets, the critical exponent η was defined by the relation $\langle Y_0 ; Y_k \rangle_{\beta_c,0} \sim \|k\|^{-(D-2+\eta)}$, and this implies that $\Sigma_n^2 = \langle S_n^2 \rangle_{\beta_c,0} \sim n^{D+2-\eta}$ [Lemma V.8.3(c)]; n is the number of sites in each edge of the symmetric hypercube Λ and $|\Lambda|$ equals n^D. In comparing the Curie–Weiss model with a general ferromagnet on $\mathbb{Z}^D$, we should consider the model on a set Λ consisting of n^D sites (rather than n sites, as we have done in this section). Let S_n denote the corresponding spin on Λ and $\langle - \rangle_{n^D,1,0}^{\text{CW}}$ expectation with respect to the finite-volume Gibbs state on Λ, $P_{n^D,1,0}^{\text{CW}}$ [see (5.64)]. The asymptotic behavior of $\langle S_n^2 \rangle_{n^D,1,0}^{\text{CW}}$ follows from the proof of Theorem V.9.5. Indeed (5.70) and Theorem A.8.7(b) imply that

$$(5.73) \quad \lim_{n\to\infty} \langle [S_n/(n^D)^{3/4}]^2 \rangle_{n^D,1,0}^{\text{CW}} = \int_{\mathbb{R}} x^2 \exp(-\tfrac{1}{12}x^4)dx \bigg/ \int_{\mathbb{R}} \exp(-\tfrac{1}{12}x^4)dx.$$

Thus $\langle S_n^2 \rangle_{n^D,1,0}^{\text{CW}} \sim n^{3D/2}$. A comparison of the relations $\langle S_n^2 \rangle_{n^D,1,0}^{\text{CW}} \sim n^{3D/2}$ and $\Sigma_n^2 \sim n^{D+2-\eta}$ gives $\eta = 2 - \frac{1}{2}D$. With this value of η and with the value $\delta = 3$, the equation $2 - \eta = D(\delta - 1)/(\delta + 1)$ holds. It is also consistent with Theorem V.8.6 that the density of the Curie–Weiss scaling limit is proportional to $\exp(-\frac{1}{12}x^4) = \exp(-\frac{1}{12}x^{1+\delta})$. Notice that the value $\eta = 2 - \frac{1}{2}D$ agrees with the (Gaussian approximation) value $\eta = 0$ for the upper critical dimension $D = 4$.

This completes our discussion of the Curie–Weiss model. In the next section, we consider a generalization.

V.10. The Circle Model and Random Waves

Let Θ denote the unit interval $[0, 1]$ with its endpoints identified and let Λ_n denote the subset $\{\theta \in \Theta : \theta = j/n, j = 1, 2, \ldots, n\}$; n is a positive integer that eventually tends to ∞. The circle model is a magnetic system on the subsets $\{\Lambda_n\}$ which generalizes the Curie–Weiss model. The circle model is interesting because it exhibits a new kind of phase transition described in terms of random waves.[14] The present section is based upon the paper by Eisele and Ellis (1983).

We are given a continuous symmetric function J of period 1 on $\mathbb{R}$. J is not necessarily non-negative. We define a Hamiltonian

$$(5.74) \qquad H_n(\omega) = -\frac{1}{2n} \sum_{i,j=1}^{n} J\left(\frac{i}{n} - \frac{j}{n}\right) \omega_i \omega_j$$

for each configuration $\omega = (\omega_1, \omega_2, \ldots, \omega_n) \in \Omega_n = \{1, -1\}^{\Lambda_n}$. The quantity ω_i denotes the value of a spin at the site i/n in Λ_n. The circle model is defined by the probability measure $P_{n,\beta}$ on $\mathscr{B}(\Omega_n)$ which assigns to each $\{\omega\}$, $\omega \in \Omega_n$, the probability

$$(5.75) \qquad P_{n,\beta}\{\omega\} = \exp[-\beta H_n(\omega)] \pi_n P_\rho\{\omega\} \cdot \frac{1}{Z(n.\beta)}.$$

In this formula β is positive, $\pi_n P_\rho$ is the product measure on $\mathscr{B}(\Omega_n)$ with identical one-dimensional marginals $\rho = \frac{1}{2}\delta_1 + \frac{1}{2}\delta_{-1}$, and $Z(n, \beta)$ is the partition function $\int_{\Omega_n} \exp[-\beta H_n(\omega)] \pi_n P_\rho(d\omega)$. The choice $J \equiv 1$ defines the Curie–Weiss model with external field $h = 0$. The circle model is more complicated than the Curie–Weiss model since the interaction strength between sites depends in general on the distance between the sites. Our main results describe the asymptotic behavior of the circle model for different interactions J.

The circle model differs from the general ferromagnetic models previously considered. The latter are defined on symmetric hypercubes Λ which expand to fill $\mathbb{Z}^D$ (infinite-volume limit) and their interactions are independent of Λ. By contrast, the circle model is defined on the subsets $\{\Lambda_n\}$ of Θ which become dense in Θ as $n \to \infty$ (a continuum limit). Also the interaction in the model, $n^{-1}J(i/n - j/n)$, depends on n. This interaction is reminiscent of the scaled interaction $J^{(\gamma)}(i-j) = \gamma J(\gamma(i-j))$ which arises in the Kac limit for models on $\mathbb{Z}$. We recall that if $J \not\equiv 0$ is a non-negative, continuous, symmetric function on $\mathbb{R}$ with bounded support and $h = 0$, then the corresponding specific Gibbs free energies $\psi^{(\gamma)}(\beta, 0)$ tend to the Curie–Weiss specific Gibbs free energy $\psi^{\text{CW}}(\mathscr{J}_0; \beta, 0)$ as $\gamma \to 0^+$, where $\mathscr{J}_0 = \int_{\mathbb{R}} J(x)dx$. Because of this analogy with the Kac limit, one might expect that in the limit $n \to \infty$ the circle model behaves like the Curie–Weiss model for any J for which $\mathscr{J}_0 = \int_\Theta J(\theta)d\theta$ is positive. Our results are consistent with this expectation. For a positive interaction, the asymptotic behavior of the circle model coincides with that of the Curie–Weiss model. By contrast, we present an example of a J with variable sign and $\mathscr{J}_0 = \int_\Theta J(\theta)d\theta < 0$ for which the circle model exhibits features absent in the Curie–Weiss case.

Let $Y_j(\omega) = \omega_j$ be the coordinate functions on Ω_n. Given a nonempty interval Δ in Θ of length $|\Delta|$, define

$$S_n(\Delta)(\omega) = \frac{1}{|\Delta|} \sum_{\{j \in \Theta: j/n \in \Delta\}} Y_j(\omega).$$

This sum, which is the spin in Δ, involves approximately $n|\Delta|$ summands. We shall describe the asymptotic behavior of the circle model by giving the limiting joint distribution, with respect to the finite-volume Gibbs states $\{P_{n.\beta}\}$, of the vector $(S_n(\Delta_1)/n, \ldots, S_n(\Delta_r)/n)$; r is a positive integer and $\Delta_1, \ldots, \Delta_r$ are any r nonempty intervals in Θ. The limiting joint distribution depends strongly on the sign of J. For $J > 0$, it is described completely in terms of the Curie–Weiss model. For a specific choice of J with variable sign, it is described in terms of random waves on Θ.

In order to motivate the main results, we define a limiting random field $\{Y(\theta); \theta \in \Theta\}$ by the expectations

$$(5.76) \quad \langle f(Y(\theta_1), \ldots, Y(\theta_r))\rangle_\beta = \lim_{\Delta_i \downarrow \{\theta_i\}} \lim_{n\to\infty} \left\langle f\left(\frac{S_n(\Delta_1)}{n}, \ldots, \frac{S_n(\Delta_r)}{n}\right)\right\rangle_{n,\beta}$$

for $f \in \mathscr{C}(\mathbb{R}^r)$. In this formula $\theta_1, \ldots, \theta_r$ are r distinct points in Θ and $\langle - \rangle_{n,\beta}$ denotes expectation with respect to $P_{n,\beta}$. We also introduce the set of ground states of the Hamiltonian H_n, which is defined as the set of configurations $\bar{\omega} \in \Omega_n$ which minimize H_n. Since $H_n(\bar{\omega}) < H_n(\omega)$ is equivalent to $P_{n,\beta}\{\bar{\omega}\} > P_{n,\beta}\{\omega\}$, the ground states are the configurations to which $P_{n,\beta}$ assigns maximal probability. In other words, they are the most likely configurations for the circle model. According to formula (5.76), the distribution of $Y(\theta)$ for $\theta \in \Theta$ fixed is approximately the same as the distribution of $S_n(\Delta)/n$ for large n and Δ a small interval centered at θ. As $\beta \to \infty$, the latter distribution is determined by the values $\{S_n(\Delta)(\bar{\omega})/n;\ \bar{\omega}$ a ground state$\}$. Indeed, if ω is not a ground state, then $P_{n,\beta}\{\omega\} \to 0$ as $\beta \to \infty$.

We first consider an interaction J which is positive.* Such an interaction is called *ferromagnetic*. For any $\beta > 0$ and fixed n, the ground states are the totally aligned configurations $\bar{\omega}_+$ and $\bar{\omega}_-$ in Ω_n ($\bar{\omega}_{+,j} = 1$, $\bar{\omega}_{-,j} = -1$ for each $j \in \{1, \ldots, n\}$). By symmetry, as $\beta \to \infty$ the measures $\{P_{n,\beta}; \beta > 0\}$ converge weakly to the measure $\frac{1}{2}\delta_{\bar{\omega}_+} + \frac{1}{2}\delta_{\bar{\omega}_-}$ on Ω_n. Thus for any $f \in \mathscr{C}(\mathbb{R})$

$$(5.77) \quad \lim_{\beta\to\infty} \left\langle f\left(\frac{S_n(\Delta)}{n}\right)\right\rangle_{n,\beta} = \frac{1}{2} f\left(\frac{S_n(\Delta)(\bar{\omega}_+)}{n}\right) + \frac{1}{2} f\left(\frac{S_n(\Delta)(\bar{\omega}_-)}{n}\right).$$

If Δ is centered at a point θ, then the right-hand side is approximately $\frac{1}{2}f(1) + \frac{1}{2}f(-1)$, and this holds independently of the choice of θ. It is consistent with (5.77) to hope that for all sufficiently large $\beta > 0$ there exists a number $m(\beta) > 0$ such that $\langle f(S_n(\Delta)/n)\rangle_{n,\beta} \to \frac{1}{2}f(m(\beta)) + \frac{1}{2}f(-m(\beta))$ as $n \to \infty$. Then the limiting random field Y at each point θ will equals $m(\beta)$ or $-m(\beta)$ with probability $\frac{1}{2}$ each. The limit (5.78) below with $r = 1$ shows that this is the case. We write $\mathbf{0}$ and $\mathbf{1}$ for the constant vectors $(0, \ldots, 0)$ and $(1, \ldots, 1)$ in $\mathbb{R}^r$ and 0 and 1 for the constant functions on Θ.

Theorem V.10.1. *Suppose that $J > 0$ is a continuous symmetric function of period* 1 *on $\mathbb{R}$ and set $\mathscr{J}_0 = \int_\Theta J(\theta)d\theta$. For $\beta > \beta_c^{\mathrm{CW}} = \mathscr{J}_0^{-1}$, define $m(\beta)$ to be the Curie–Weiss spontaneous magnetization $m^{\mathrm{CW}}(\mathscr{J}_0; \beta, +)$; $m(\beta)$ is the unique positive solution of the equation $z = \tanh(\beta\mathscr{J}_0 z)$ [see* (5.52)]. *Then for any positive integer r, nonempty intervals $\Delta_1, \ldots, \Delta_r$ in Θ, and function $f \in \mathscr{C}(\mathbb{R}^r)$*

$$(5.78) \quad \lim_{n\to\infty} \left\langle f\left(\frac{S_n(\Delta_1)}{n}, \ldots, \frac{S_n(\Delta_r)}{n}\right)\right\rangle_{n,\beta} = \begin{cases} f(\mathbf{0}) & \text{for } 0 < \beta \le \beta_c^{\mathrm{CW}}, \\ \frac{1}{2}f(m(\beta)\mathbf{1}) + \frac{1}{2}f(-m(\beta)\mathbf{1}) & \text{for } \beta > \beta_c^{\mathrm{CW}}. \end{cases}$$

*The same results hold for non-negative interactions which are irreducible on Θ.

Thus the limiting random field $\{Y(\theta); \theta \in \theta\}$ *equals* 0 *for* $0 < \beta \le \beta_c^{\text{CW}}$ *and equals* $\kappa m(\beta) 1$ *for* $\beta > \beta_c^{\text{CW}}$, *where* $\text{Prob}\{\kappa = 1\} = \text{Prob}\{\kappa = -1\} = \frac{1}{2}$.

The form of $Y(\theta)$ for $\beta > \beta_c^{\text{CW}}$ represents a (discrete) symmetry-breaking of the sign-invariance of the model ($H_n(-\omega) = H_n(\omega)$ for each $\omega \in \Omega_n$). It corresponds to the symmetry-breaking of the infinite-volume Gibbs states $P_{\beta,0,+} \neq P_{\beta,0,-}$ for general ferromagnetic models on $\mathbb{Z}^D$ [page 116].

We now consider an interaction J which is negative. Such an interaction is called *antiferromagnetic*.[15] In order to find a ground state, one may try to minimize $H_n(\omega)$ by making each summand $J(i/n - j/n)\omega_i\omega_j$ positive, or since J is negative, by making each product $\omega_i\omega_j$ negative. But this is impossible since each ω_i can take only the values 1 or -1. The form of the ground state must represent a compromise with this incompatible ordering. Specifically, the spins in each ground state cluster into alternating islands of plus spins and minus spins, where the number of islands and their sizes are determined by J. Since the interaction is translation invariant, the phase shifts of the islands are not determined. If $\bar{\omega}$ is a ground state, then define the translated configurations $\bar{\omega}^{(\alpha)}$, $\alpha = 1, \ldots, n$, by

$$\bar{\omega}_j^{(\alpha)} = \begin{cases} \bar{\omega}_{j+\alpha} & \text{if } j = 1, \ldots, n - \alpha, \\ \bar{\omega}_{j+\alpha-n} & \text{if } j = n - \alpha + 1, \ldots, n. \end{cases}$$

The configurations $\{\bar{\omega}^{(\alpha)}\}$ are also ground states.

Let us consider in more detail the case where there is a unique ground state $\bar{\omega}$ up to translations. As $\beta \to \infty$ the measures $\{P_{n,\beta}; \beta > 0\}$ converge weakly to the measure $n^{-1} \sum_{\alpha=1}^{n} \delta_{\bar{\omega}^{(\alpha)}}$ on Ω_n, and for any $f \in \mathscr{C}(\mathbb{R})$

$$\lim_{\beta \to \infty} \left\langle f\left(\frac{S_n(\Delta)}{n}\right)\right\rangle_{n,\beta} = n^{-1} \sum_{\alpha=1}^{n} f\left(\frac{S_n(\Delta)(\bar{\omega}^{(\alpha)})}{n}\right). \tag{5.79}$$

Let us think of $\bar{\omega}_j$ as the value at j/n of some function $\bar{g}$ on Θ and take Δ to be a small interval centered at a point θ in Θ. The right-hand side of (5.79) is approximately $n^{-1} \sum_{\alpha=1}^{n} f(\bar{g}(\theta + \alpha/n))$. If $\bar{g}$ is smooth, then for large n the latter sum in turn is approximately $\int_\Theta f(\bar{g}(\theta + \lambda))d\lambda$. In this case it is consistent to hope that for all sufficiently large $\beta > 0$ the limiting random field $Y(\theta)$ is described by a wave on the circle Θ whose phase shift λ is random and whose shape is given by a function $\bar{g}_\beta$. As $\beta \to \infty$, $\bar{g}_\beta$ should converge to the above function $\bar{g}$.

We are unable to treat the antiferromagnetic case in any generality. The next theorem verifies the above intuitive picture for the particular interaction $J(\theta) = -1 + b\cos(2\pi p\theta)$, where b is a nonzero real number and p is a positive integer. This function is antiferromagnetic only for $|b| \le 1$, but interestingly the same wave structure is valid for any $b \neq 0$.

Theorem V.10.2. *Given* $b \neq 0$ *and a positive integer* p, *define the interaction* $J(\theta) = -1 + b\cos(2\pi p\theta)$, $\theta \in \Theta$. *For each* $\beta > 2/|b|$ *the equation*

$$\gamma = \int_\Theta \cos(2\pi p\theta) \tanh[\beta b\gamma \cos 2\pi p\theta] d\theta \tag{5.80}$$

has a unique positive solution $\gamma = \gamma(\beta, b, p)$. *Define the function*

(5.81)
$$\bar{g}_\beta(\theta) = \tanh[\beta b\gamma \cos(2\pi p\theta)],$$

and for each point λ *and nonempty interval* Δ *in* Θ, *let* $\bar{g}_\beta(\Delta;\lambda)$ *denote the average* $|\Delta|^{-1}\int_\Delta \bar{g}_\beta(\theta+\lambda)d\theta$. *Then for any positive integer r, nonempty intervals* $\Delta_1, \ldots, \Delta_r$ *in* Θ, *and function* $f \in \mathscr{C}(\mathbb{R}^r)$

(5.82)
$$\lim_{n\to\infty}\left\langle f\left(\frac{S_n(\Delta_1)}{n}, \ldots, \frac{S_n(\Delta_r)}{n}\right)\right\rangle_{n,\beta} = \begin{cases} f(\mathbf{0}) & \text{for } 0 \le \beta \le 2/|b|, \\ \displaystyle\int_\Theta f(\bar{g}_\beta(\Delta_1;\lambda), \ldots, \bar{g}_\beta(\Delta_r;\lambda))d\lambda & \text{for } \beta > 2/|b|. \end{cases}$$

Thus the limiting random field $\{Y(\theta); \theta \in \Theta\}$ *equals* 0 *for* $0 < \beta \le 2/|b|$ *and equals* $\bar{g}_\beta(\theta + \lambda)$ *for* $\beta > 2/|b|$, *where* λ *is uniformly distributed in* Θ.*

Since for $\beta > 2/|b|$, $\bar{g}_\beta$ is a periodic function with period $1/p$ and $2p$ nodes, the form of $Y(\theta)$ validates the island picture of antiferromagnetism presented before the theorem. The form of $Y(\theta)$ for $\beta > 2/|b|$ also represents a continuous symmetry-breaking of the translation invariance of the model. This contrasts with the discrete symmetry-breaking in the ferromagnetic case.

The proofs of Theorems V.10.1 and V.10.2 generalize the large deviation analysis of the Curie–Weiss model given in Section IV.4. There are three main steps: derive a Gibbs variational formula for the specific Gibbs free energy of the circle model; relate this formula to the limits in the two theorems; determine where the supremum in the Gibbs variational formula is attained. Since the proofs of these steps are technical, all details will be omitted. Instead, we will motivate the main ideas of the proofs by stressing analogies with the Curie–Weiss case. Full details are given in the paper by Eisele and Ellis (1983).

For the Curie–Weiss model we wrote the Hamiltonian $H_{n,h}(\omega)$ as $-n\frac{1}{2}\mathscr{J}_0(\sum_{j=1}^n \omega_j/n)^2 - nh(\sum_{j=1}^n \omega_j/n)$, then expressed the partition function as

$$Z^{\text{CW}}(n,\beta,h) = \int_{\Omega_n} \exp\left[n\left(\tfrac{1}{2}\beta\mathscr{J}_0\left(\frac{\sum_{j=1}^n \omega_j}{n}\right)^2 + h\frac{\sum_{j=1}^n \omega_j}{n}\right)\right]\pi_n P_\rho(d\omega).$$

Level-1 large deviations for the distributions of $\sum_{j=1}^n \omega_j/n$ gave the variational formula for the specific Gibbs free energy:

$$-\beta\psi^{\text{CW}}(\beta,h) = \lim_{n\to\infty}\frac{1}{n}\log Z^{\text{CW}}(n,\beta,h) = \sup_{z\in\mathbb{R}}\{\tfrac{1}{2}\beta\mathscr{J}_0 z^2 + \beta h z - I_\rho^{(1)}(z)\},$$

where

$$I_\rho^{(1)}(z) = \frac{1-z}{2}\log(1-z) + \frac{1+z}{2}\log(1+z) \qquad \text{for } |z| \le 1$$

and $I_\rho^{(1)}(z) = \infty$ for $|z| > 1$ [see Remark IV.4.2].

*As $\Delta \downarrow \{\theta\}$, $\bar{g}_\beta(\Delta;\lambda) \to \bar{g}_\beta(\theta+\lambda)$.

The specific Gibbs free energy $\psi(\beta)$ of the circle model is defined by the limit

$$-\beta\psi(\beta) = \lim_{n\to\infty} \frac{1}{n} \log Z(n, \beta),$$

where $Z(n, \beta)$ equals $\int_{\Omega_n} \exp[-\beta H_n(\omega)]\pi_n P_\rho(d\omega)$. The Hamiltonian H_n is defined in (5.74). We seek a stochastic process which does for the circle model what the sum $\sum_{j=1}^n \omega_j/n$ did for the Curie–Weiss model. Let $\mathscr{X}$ be the space of measurable functions g on $\mathbb{R}$ which are of period 1 and which satisfy $-1 \leq \operatorname{ess\,inf} g \leq \operatorname{ess\,sup} g \leq 1$. $\mathscr{X}$ is a subspace of the real Hilbert space $L^2(\Theta)$. Denote by $\langle -, - \rangle$ the inner product of $L^2(\Theta)$. We say that a sequence $\{g_n; n = 1, 2, \ldots\}$ in $\mathscr{X}$ converges to an element g in $\mathscr{X}$ if $\langle g_n, f\rangle \to \langle g, f\rangle$ for each $f \in L^2(\Theta)$. With this topology, $\mathscr{X}$ is metrizable as a compact metric space. For $g \in \mathscr{X}$, define

$$F(g) = -\frac{1}{2}\int_\Theta \int_\Theta J(\theta - v)g(\theta)g(v)d\theta dv. \tag{5.83}$$

F is a bounded continuous functional on $\mathscr{X}$. The following facts (proofs omitted) allow one to derive a variational formula for the circle model free energy.

(a) There exists a stochastic process $\xi_n(\omega, \theta)$, $n = 1, 2, \ldots$, $\omega \in \Omega_n$, $\theta \in \Theta$, taking values in $\mathscr{X}$ such that $\sup_{\omega\in\Omega_n} |H_n(\omega) - nF(\xi_n(\omega, \cdot))| = o(n)$ as $n \to \infty$.[16]

(b) For A a Borel subset of $\mathscr{X}$, define the distribution

$$Q_n\{A\} = \pi_n P_\rho\{\omega \in \Omega_n : \xi_n(\omega, \cdot) \in A\}.$$

Then the sequence $\{Q_n; n = 1, 2, \ldots\}$ has a large deviation property with $a_n = n$ and entropy function

$$I(g) = \int_\Theta I_\rho^{(1)}(g(\theta))d\theta. \tag{5.84}$$

Fact (a) and the comparison lemma, Lemma II.7.4, imply that the functions

$$Z(n, \beta) = \int_{\Omega_n} \exp[-\beta H_n(\omega)]\pi_n P_\rho(d\omega),$$

$$\bar{Z}(n, \beta) = \int_{\Omega_n} \exp[-n\beta F(\xi_n(\omega, \cdot)]\pi_n P_\rho(d\omega)$$

have the same leading order asymptotic behavior as n $\to \infty$. Because of the large deviation property of $\{Q_n\}$, we may apply Varadhan's theorem, Theorem II.7.1, to evaluate

$$\begin{aligned} -\beta\psi(\beta) &= \lim_{n\to\infty} \frac{1}{n}\log Z(n, \beta) = \lim_{n\to\infty} \frac{1}{n}\log \bar{Z}(n, \beta) \\ &= \lim_{n\to\infty} \frac{1}{n}\log \int_{\mathscr{X}} \exp[-n\beta F(g)]Q_n(dg). \end{aligned} \tag{5.85}$$

The following theorem states the Gibbs variational formula for the circle model.

Theorem V.10.3. *The specific Gibbs free energy $\psi(\beta)$ for the circle model is given by*

$$-\beta\psi(\beta) = \sup_{g \in \mathscr{X}} \{-\beta F(g) - I(g)\}. \tag{5.86}$$

If in (5.86) the functions g are restricted to constants $z \in [-1, 1]$, then we obtain the Curie–Weiss lower bound*

$$\begin{aligned} -\beta\psi(\beta) \geq \sup_{g \equiv z} \{-\beta F(g) - I(g)\} &= \sup_{z \in [-1,1]} \{\tfrac{1}{2}\beta \mathscr{J}_0 z^2 - I_\rho^{(1)}(z)\} \\ &= -\beta\psi^{\text{CW}}(\mathscr{J}_0; \beta, 0). \end{aligned} \tag{5.87}$$

Theorem V.10.4(a) below shows that for $J > 0$ the supremum in (5.86) is attained only at constant functions g. Hence in this case $\psi(\beta)$ equals $\psi^{\text{CW}}(\mathscr{J}_0; \beta, 0)$.

We now relate the Gibbs variational formula (5.86) to the limits in Theorems V.10.1 and V.10.2. Again the key idea can be seen in our analysis of the Curie–Weiss model. For $f \in \mathscr{C}(\mathbb{R})$, the heuristic relation (4.13) states that

$$\left\langle f\left(\frac{S_n}{n}\right)\right\rangle_{n,\beta,h}^{\text{CW}} \approx \int_{\mathbb{R}} f(z) \exp[-n i_{\beta,h}(z)]\,dz \cdot \frac{1}{\displaystyle\int_{\mathbb{R}} \exp[-n i_{\beta,h}(z)]\,dz}, \tag{5.88}$$

where $i_{\beta,h} = -(\frac{1}{2}\beta\mathscr{J}_0 z^2 + \beta h z) + I_\rho^{(1)}(z)$; $-\beta\psi^{\text{CW}}(\mathscr{J}_0; \beta, h)$ equals $\sup_{z \in \mathbb{R}}\{-i_{\beta,h}(z)\}$ or $-\inf_{z \in \mathbb{R}} i_{\beta,h}(z)$. Theorem II.7.2(b) showed that the limit of $\langle f(S_n/n)\rangle_{n,\beta,h}^{\text{CW}}$ is determined by the points z at which the infimum of $i_{\beta,h}(z)$ is attained. The limit is given in Theorem IV.4.1(a).

For the circle model, we now look for an expression analogous to (5.88). Let Δ be a nonempty interval in Θ and define the function $G_\Delta(\theta) = |\Delta|^{-1}\chi_\Delta(\theta)$. A key observation (proof omitted) is that for any integer r, nonempty intervals $\Delta_1, \ldots, \Delta_r$ and function $f \in \mathscr{C}(\mathbb{R}^r)$, the quantity $\langle f(S_n(\Delta_1)/n, \ldots, S_n(\Delta_r)/n)\rangle_{n,\beta}$ has the same limit as $n \to \infty$ as the quantity

$$\frac{\displaystyle\int_{\Omega_n} f(\langle G_{\Delta_1}, \xi_n(\omega,\cdot)\rangle, \ldots, \langle G_{\Delta_r}, \xi_n(\omega,\cdot)\rangle) \exp[-n\beta F(\xi_n(\omega,\cdot))]\pi_n P_\rho(d\omega)}{\displaystyle\int_{\Omega_n} \exp[-n\beta F(\xi_n(\omega,\cdot))]\pi_n P_\rho(d\omega)} \tag{5.89}$$

This ratio can be expressed in terms of the distribution $Q_n(dg)$ of $\xi_n(\omega, \cdot)$. Since the distributions $\{Q_n\}$ have a deviation property with entropy function

*Inequality (5.87) is analogous to the lower bound in Theorem V.9.2.

$I(g)$, heuristically we may write $Q_n(dg)$ as $\exp(-nI(g))dg$ and the ratio in (5.89) as

$$\int_{\mathscr{X}} f(\langle G_{\Delta_1}, g\rangle, \ldots, \langle G_{\Delta_r}, g\rangle)\exp[-ni_\beta(g)]dg \cdot \frac{1}{\int_{\mathscr{X}} \exp[-ni_\beta(g)]dg},$$

where $i_\beta(g) = \beta F(g) + I(g)$. This is the analog of the Curie–Weiss expression in (5.88).

We have remarked that $\langle f(S_n(\Delta_1)/n, \ldots, S_n(\Delta_r)/n\rangle_{n,\beta}$ and the ratio in (5.89) have the same limits as $n \to \infty$. According to Theorem II.7.2(b), the latter limit is determined by the functions $\bar{g}$ in $\mathscr{X}$ at which $-i_\beta(g)$ attains its supremum. The next theorem lists the maximizing functions $\bar{g}$ for the interactions in Theorems V.10.1 and V.10.2. The limits in these theorems are direct consequences (proof omitted).

Theorem V.10.4. *Let J be a continuous symmetric function of period* 1 *on* $\mathbb{R}$. *For each $\beta > 0$, let $\mathscr{G}_\beta$ denote the set of functions $\bar{g}$ in $\mathscr{X}$ for which $-\beta F(\bar{g}) - I(\bar{g})$ equals* $\sup_{g \in \mathscr{X}}\{-\beta F(g) - I(g)\}$. *Then the following conclusions hold.*

(a) *$\mathscr{G}_\beta$ is nonempty. Any element in $\mathscr{G}_\beta$ equals* a.s. *a solution of the nonlinear integral equation**

$$(5.90) \qquad (I_\rho^{(1)})'(\bar{g}(\theta)) = \beta \int_\Theta J(\theta - v)\bar{g}(v)dv.$$

(b) *Suppose that J is positive and set $\mathscr{J}_0 = \int_\Theta J(\theta)d\theta$. Then*

$$(5.91) \qquad \mathscr{G}_\beta = \begin{cases} \{0\} & \text{for } 0 < \beta \le \beta_c^{\text{CW}} = \mathscr{J}_0^{-1}, \\ \{m(\beta)1, -m(\beta)1\} & \text{for } \beta > \beta_c^{\text{CW}}, \end{cases}$$

where $m(\beta)$ is the Curie–Weiss spontaneous magnetization $m^{\text{CW}}(\mathscr{J}_0; \beta, +)$.

(c) *Suppose that $J(\theta) = -1 + b\cos(2\pi p\theta)$, $b \ne 0$, $p \ge 1$ an integer. Then*

$$(5.92) \qquad \mathscr{G}_\beta = \begin{cases} \{0\} & \text{for } 0 < \beta \le 2/|b|, \\ \{\bar{g}_\beta(\cdot + \lambda); \lambda \in \Theta\} & \text{for } \beta < 2/|b|, \end{cases}$$

where $\bar{g}_\beta$ is defined by (5.80)–(5.81).

It is not hard to check the conclusions of parts (b) and (c) of Theorem V.10.4 in the limits $\beta \to 0^+$ and $\beta \to \infty$. First take β small. Then in the Gibbs variational formula the entropy term dominates, and $\sup_{g \in \mathscr{X}}\{-I(g)\} = -\inf_{g \in \mathscr{X}} \int_\Theta I_\rho^{(1)}(g(\theta))d\theta$ is attained at the unique function $\bar{g} \equiv 0$. This is consistent with the first lines of (5.91) and (5.92). For large $\beta > 0$ the energy term dominates. The functions at which $-F$ attains its supremum, or F its infimum, depend on the form of J. For $J > 0$, F attains its infimum at the constant functions $\bar{g} \equiv 1$ and $\bar{g} \equiv -1$. This is consistent with the second

* A.s. denotes almost surely with respect to Lebesgue measure on Θ. Equation (5.90) is the analog of the Curie–Weiss equation (4.14).

line of (5.91) since $m^{\text{CW}}(\mathscr{J}_0;\beta,h)$ converges to 1 as $\beta\to\infty$. On the other hand, for $J(\theta)=-1+b\cos(2\pi p\theta)$, F attains its infimum at the function $\bar{g}(\theta)=\operatorname{sgn}(\cos(2\pi p\theta))$ ($\bar{g}(\theta)=0$ when $\cos(2\pi p\theta)=0$) and at the translates $\{\bar{g}(\theta+\lambda);\lambda\in\Theta\}$ of this function. This is consistent with the second line of (5.92) since $\bar{g}_\beta(\theta)\to\bar{g}(\theta)$ as $\beta\to\infty$.

We have now completed our discussion of the circle model. The study of critical phenomena for the model is an interesting open problem.

V.11. A Postscript on Magnetic Models

The probabilistic models of magnetism which were analyzed in Chapters IV and V span a range of difficulty and exhibit a number of different phenomena. Three models were considered. The simplest is the Curie–Weiss model, which studies spin random variables on the subsets $\Lambda=\{1,\ldots,n\}$ of $\mathbb{Z}$. The interaction strength between each pair of spins is equal (proportional to n^{-1}), and thus the model has no geometry. Of intermediate difficulty is the circle model, which studies spin random variables on the subsets $\Lambda_n=\{\theta:\theta=j/n, j=1,\ldots,n\}$ of $[0,1]$. It replaces the Curie–Weiss interaction by the interaction $n^{-1}J(i/n-j/n)$, where J is a continuous symmetric function of period 1 on $\mathbb{R}$. The most complicated of the models are the ferromagnetic models on $\mathbb{Z}^D$. They study spin random variables on the symmetric hypercubes Λ of $\mathbb{Z}^D$ and include, as a special case, the Ising model. The interaction strength between each pair of spins i,j depends only on the vector $i-j$, not on the subset Λ.

Large deviation techniques allowed us to analyze the Curie–Weiss model and the circle model. For the Curie–Weiss model, this approach showed that spontaneous magnetization occurs at all $\beta>\beta_c^{\text{CW}}=\mathscr{J}_0^{-1}$. It also showed that the law of large numbers for the spin is valid for $\beta>0$, $h\neq 0$ and $0<\beta\le\beta_c^{\text{CW}}$, $h=0$ and that the law of large numbers fails for $\beta>\beta_c^{\text{CW}}$, $h=0$. Finally, it suggested the correct form for the block spin scaling limit at the critical point. Large deviations showed that the asymptotic behavior of the circle model depends strongly on the choice of interaction J. For $J>0$ this behavior coincides with that of the Curie–Weiss model while for a specific J with variable sign the model is described asymptotically by random waves.

In analyzing ferromagnetic models on $\mathbb{Z}^D$, we used large deviations to prove exponential convergence properties of Gibbs states and to derive the Gibbs variational formula for $D=1$. Using convexity and moment inequalities, we found that the phase transitions associated with these models exist on two levels: spontaneous magnetization on level-1 and nonuniqueness of infinite-volume Gibbs states on level-3. The level-1 phase transition represents a sensitivity of the specific magnetization $m(\beta,h)$ to the limit $h\to 0^+$ versus $h\to 0^-$. For each model there exists a number $\beta_c\in(0,\infty]$ such that for $\beta>\beta_c$, $m(\beta,+)=\lim_{h\to 0^+}m(\beta,h)>0>m(\beta,-)=$

Table V.4. Gibbs Variational Formulas

Model	Formula	Physical significance of points at which supremum is attained
Curie–Weiss	$-\beta\psi^{\mathrm{CW}}(\beta,h)=\sup_{z\in\mathbb{R}}\left\{\frac{\beta}{2}J_0z^2+\beta hz-I_\rho^{(1)}(z)\right\}$	Value of specific magnetization $m^{\mathrm{CW}}(J_0;\beta,h)$
Circle	$-\beta\psi(\beta)=\sup_{g\in\mathscr{X}}\{-\beta F(g)-I(g)\}$	Realization of limiting random field $\{Y(\theta);\theta\in\Theta\}$
Ferromagnetic on $\mathbb{Z}^D$	$-\beta\psi(\beta,h)=\sup_{P\in\mathscr{M}_s(\Omega)}\{-\beta u(h;P)-I_\rho^{(3)}(P)\}$	Translation invariant infinite-volume Gibbs state

$\lim_{h\to 0^-} m(\beta,h)$. The level-3 phase transition represents a sensitivity of the infinite-volume Gibbs states to the choice of plus versus minus external condition. For $\beta>\beta_c$ and $h=0$, two distinct infinite-volume Gibbs states $P_{\beta,0,+}$ and $P_{\beta,0,-}$ exist. Since the mean of the one-dimensional marginal of $P_{\beta,0,+}$ is $m(\beta,+)$, spontaneous magnetization is a contraction of the level-3 phase transition down to level-1. Among the most interesting problems discussed were critical phenomena of ferromagnetic models. These include the form of block spin scaling limits and the relation among these limits, critical exponents, mean field theory, and dimensionality. The Curie–Weiss model plays a central role here. It gives universal bounds on the specific magnetization and the critical inverse temperature for ferromagnetic models on $\mathbb{Z}^D$, and the critical exponents of these models are believed to coincide with their mean field/Curie–Weiss values for dimension $D>4$.

A common theme shared by the three models analyzed in Chapters IV and V is the Gibbs variational formula. For each model, this formula expresses the specific Gibbs free energy as the supremum, over some complete separable metric space, of β times an energy functional minus an entropy functional. Table V.4 lists the Gibbs variational formulas and the physical significance of the points at which the respective suprema are attained. The choice of metric space ($\mathbb{R}$, function space $\mathscr{X}$, space of measures $\mathscr{M}_s(\Omega)$) reflects the complexity of the model. The Gibbs variational formulas make explicit the energy-entropy competition underlying the phase transitions which we studied. For each model, the Gibbs variational formula was explicitly solved in the limits $\beta\to 0^+$ (where entropy dominates) and $\beta\to\infty$ (where energy dominates).

Chapters IV and V have treated a number of topics in an area of mathematical physics which is currently under very active study. The notes and problems in the next two sections discuss some other developments.

V.12. Notes

1 (page 142). The moment inequalities presented in Section V.3 are valid for ferromagnetic models with other single-site distributions besides $\rho=\frac{1}{2}\delta_1+\frac{1}{2}\delta_{-1}$ [Problems V.13.4 and V.13.7]. The GKS-1 and GKS-2 inequalities extend to models with many-body interactions [Problem V.13.5]. For

generalizations of the FKG inequality, see Battle and Rosen (1980), Newman (1980, 1984), Eaton (1982), Graham (1983), and the references listed in these papers. Other moment inequalities have been proved by many people, including Griffiths (1967c, 1969), Ginibre (1970, 1972), Lebowitz (1974), Newman (1975b, 1975c), Simon (1980a), Lieb (1980), Aizenman (1982), Brydges, Fröhlich, and Spencer (1982), and Brydges, Fröhlich, and Sokal (1983). Monroe and Pearce (1979) is a review of moment inequalities for vector-spin systems.

2 (page 152) Yang and Lee (1952) proposed a theory of phase transitions based on the location of the zeroes of the partition function $Z(\Lambda, \beta, h)$ as a function of complex h. Lee and Yang (1952) showed that for the ferromagnetic models in Section V.2 all the zeroes lie on the imaginary h-axis. It follows from theorems of Vitali and Hurwitz that since the interaction is summable, for $\beta > 0$ $\psi(\beta, h)$ is an analytic function of h for $\operatorname{Re} h \neq 0$. Theorem V.4.4 follows. For discussions and generalizations, see Ruelle (1969, Section 5.1), Griffiths (1972, Section IV), Newman (1974, 1975a, 1976b), Simon (1974, Section IX.3), Dunlop (1977), Glimm and Jaffe (1981, Sections 4.5 and 4.6), and Lieb and Sokal (1981). Lebowitz and Penrose (1968) prove other analyticity properties of $\psi(\beta, h)$.

3 (page 153). The critical inverse temperature for the Ising model on $\mathbb{Z}^2$ was first located by Kramers and Wannier (1941) by a duality argument. Onsager's formula (5.18) for $m(\beta, +)$ was also derived by Yang (1952) and by Montroll, Potts, and Ward (1963) using heuristically equivalent definitions of spontaneous magnetization. The values of β_c and $m(\beta, +)$ given in (5.18) were established rigorously by Lebowitz (1972) and Benettin et al. (1973), respectively. Also see Abraham and Martin-Löf (1973). McCoy and Wu (1973) treat the Ising model on $\mathbb{Z}^2$ in great detail. Baxter (1982) discusses other exactly solved models in statistical mechanics.

4 (page 154). The argument of Peierls (1936) was not rigorous. Rigorous versions were given independently by Griffiths (1964) and Dobrushin (1965). For proving phase transitions in systems with continuous symmetry, infrared bounds (Fourier methods) are useful [Fröhlich, Simon, and Spencer (1976), Fröhlich, Israel, Lieb, and Simon (1978)]. For treatments of phase transitions in spin systems, see the references listed in Note 2a of Chapter IV together with Fröhlich (1978, 1980), Lieb (1978), Fröhlich and Spencer (1982b), and Sinai (1982).

5 (page 163). Newman (1980) proves a central limit theorem for the spins per site over finitely many disjoint hypercubes in $\mathbb{Z}^D$. He shows convergence to independent Gaussians as the hypercubes tend to infinity. Theorem V.7.2(a) is a special case. Newman (1983) is a further generalization. Iagolnitzer and Souillard (1979) prove a central limit theorem for the spin per site as an application of the Lee–Yang theorem. Also see Martin-Löf (1973). Martin-Löf (1979, Section 3.5) presents a heuristic approach to the central limit theorem based on large deviations. See Newman (1980) for a central limit theorem with respect to $P_{\beta,0,+}$ and $P_{\beta,0,-}$ for $\beta > \beta_c$.

6 (page 169). A proof of Theorem V.7.6(a) based on Griffiths (1967c) is sketched in Lebowitz (1975). There is a large literature on decay of correlations for spin systems. For example, see Penrose and Lebowitz (1974), Iagolnitzer and Souillard (1977), Gross (1979), and Künsch (1982).

7 (page 170). A somewhat different scaling limit is studied for the Ising model on $\mathbb{Z}^2$ by Palmer and Tracy (1981). Their work and the work of other authors listed in the references of their paper reveal deep connections among the Ising model on $\mathbb{Z}^2$, relativistic quantum field theory, and differential equations.

8 (page 171). Let J be a finite-range ferromagnetic interaction on $\mathbb{Z}^D$, $D \geq 2$, for which the critical inverse temperature β_c is finite. Define $\mathscr{J}_0 = \sum_{k \in \mathbb{Z}^D} J(k)$. According to Theorem V.7.6(a), there exists a number $\beta_0 \in [\mathscr{J}_0^{-1}, \beta_c]$ such that for $\beta > 0$, $h \neq 0$ and $0 < \beta < \beta_0$, $h = 0$, $\langle Y_0; Y_k \rangle_{\beta,h}$ decays exponentially as $\|k\| \to \infty$. We assume that β_0 is the largest number with this property. Let us relate β_0 to other quantities of interest. Define the *correlation length* $\xi(\beta, h)$ by the formula

$$\frac{1}{\xi(\beta,h)} = -\limsup_{\|k\|\to\infty} \frac{1}{\|k\|} \log \langle Y_0; Y_k \rangle_{\beta,h}.$$

Theorem V.7.6 implies that $\xi(\beta, h)$ is finite for $\beta > 0$, $h \neq 0$ and $0 < \beta < \beta_0$, $h = 0$ and that β_0 is characterized by the formula

$$\beta_0 = \sup\{\beta \in (0, \beta_c] : \xi(\beta, 0) < \infty\}.$$

According to Simon (1980a), β_0 equals the number

$$\beta_1 = \sup\{\beta \in (0, \beta_c] : \chi(\beta, 0) < \infty\},$$

where $\chi(\beta, h)$ is the specific magnetic susceptibility. Property (c) of Definition V.8.1 of a normal critical point requires that β_1, or equivalently β_0, equal β_c.

9 (page 172). Surveys of critical phenomena and the renormalization group are given in Fisher (1965, 1967b, 1974, 1983), Stanley (1971, 1985), Wilson (1974, 1979, 1983), Wilson and Kogut (1974), Wegner (1975), Ma (1976), Wallace and Zia (1978), and Lang (1981). Jona-Lasinio (1975) and Cassandro and Jona-Lasinio (1978) discuss connections between the renormalization group and limit theorems in probability. Large deviations and the renormalization group are discussed by Jona-Lasinio (1983). Also see Collet and Eckmann (1978), Dobrushin (1980), Newman (1981a), Aizenman (1983), De Coninck (1984), and the references listed in these works.

10 (page 178). If $k = (m, n) \in \mathbb{Z}^2$, then define $f(m, n) = \langle Y_0; Y_k \rangle_{\beta_c, 0}$. Suppose that $m \geq n > 0$. Inequalities (10) and (11) in Messager and Miracle-Sole (1977) yield

$$f\left(\frac{m+n-1}{2}, \frac{m+n-1}{2}\right) > f(m,n) \geq f(m,m) \qquad \text{if } m+n \text{ is odd,}$$

$$f\left(\frac{m+n-2}{2}, \frac{m+n-2}{2}\right) \geq f(m,n) \geq f(m,m) \qquad \text{if } m+n \text{ is even.}$$

By symmetry and by the asymptotic relation (5.50), it follows that for any $(m,n) \in \mathbb{Z}^2$ $f(m,n)$ is bounded above and below by const$(\sqrt{m^2+n^2})^{-1/4}$ as $\sqrt{m^2+n^2} \to \infty$. Hence η equals 1/4.

11 (page 178). One can show that δ equals 15 by combining several inequalities for critical exponents given in Griffiths (1972, page 102). This was pointed out to me by M. E. Fisher and R. B. Griffiths; also see Abraham (1978, page 352). Inequality 2 together with the known values of the critical exponents α' and β (0 and 1/8, respectively) yields $\delta \geq 15$. Inequality 9 together with the known values of the critical exponents α and γ (0 and 7/4, respectively) yields $\delta_s = \delta \leq 15$. However, the derivations of the critical exponent inequalities involve a number of "tacit" assumptions which though extremely reasonable have not all been verified (see page 103 of Griffiths (1972)).

12 (page 185). The Kac limit for continuous systems of particles was proved by Lebowitz and Penrose (1966). Hemmer and Lebowitz (1976) is a review of the Kac limit and related results.

13 (page 189). The Simon and Griffiths (1973) proof of Theorem V.9.5 is closely related to Khinchin's classical result on large deviations for sums of independent Bernoulli random variables (1929). Dunlop and Newman (1975) generalize Theorem V.9.5 to vector-spin models using a multidimensional local limit theorem for large deviations due to Richter (1958). Ellis and Newman (1978b) prove Theorems V.9.4 and V.9.5 for other single-site distributions ρ on $\mathbb{R}$. See Ellis and Newman (1978a, 1978d), Jeon (1983), Ellis, Newman, and Rosen (1980), and Chaganty and Sethuraman (1982) for related results. Dawson (1983) studies critical dynamics of Curie–Weiss models.

14 (page 190). Messer and Spohn (1982) and Kusuoka and Tamura (1984) discuss models related to the circle model.

15 (page 193). Antiferromagnets are discussed in Griffiths (1972, Section V.C.1).

16 (page 195). In order to simplify the discussion, the exposition on page 195 avoids an important point. The process $\xi_n(\omega,\theta)$ is actually an array $\xi_{p,n}(\omega,\theta)$, $p = 1, 2, \ldots,$ $n = 1, 2, \ldots$. We have

$$\sup_{\omega \in \Omega_n} |H_n(\omega) - nF(\xi_{p,n}(\omega,\cdot))| = o(n) \qquad \text{as } n \to \infty, p \to \infty,$$

$$\limsup_{p\to\infty} \limsup_{n\to\infty} \frac{1}{n} \log \pi_n P_\rho\{\xi_{p,n} \in K\} \leq -\inf_{g \in K} I(g) \quad \text{for each closed set } K \text{ in } \mathscr{X},$$

$$\liminf_{p\to\infty} \liminf_{n\to\infty} \frac{1}{n} \log \pi_n P_\rho\{\xi_{p,n} \in \mathrm{G}\} \geq -\inf_{g \in \mathrm{G}} I(g) \qquad \text{for each open set G in } \mathscr{X}.$$

A straightforward extension of Varadhan's theorem yields Theorem V.10.3:

$$
\begin{aligned}
-\beta\psi(\beta) &= \lim_{n\to\infty}\frac{1}{n}\log Z(n,\beta) \\
&= \lim_{p\to\infty}\lim_{n\to\infty}\frac{1}{n}\log\int_{\Omega_n}\exp[-n\beta F(\xi_{p,n}(\omega,\cdot))]\pi_n P_\rho(d\omega) \\
&= \sup_{g\in\mathscr{X}}\{-\beta F(g) - I(g)\}.
\end{aligned}
$$

Similar adjustments must be made on pages 196–197.

V.13. Problems

V.13.1. (a) Consider a model on $\mathbb{Z}^D$ whose interaction function J is identically 0. Show that the specific magnetization $m(\beta, h)$ equals $\tanh \beta h$ for all $\beta > 0$, h real.

(b) Let J be a summable ferromagnetic interaction on $\mathbb{Z}^D$ and $m(\beta, h)$ the corresponding specific magnetization. Prove that $m(\beta, h) \geq \tanh \beta h$ for all $\beta > 0$, $h \geq 0$. This implies that $m(\beta, h) > 0$ for $\beta > 0$, $h > 0$; for $\beta > 0$, $m(\beta, h) \to 1$ as $h \to \infty$; and for $h > 0$, $m(\beta, h) \to 1$ as $\beta \to \infty$.

(c) Assume that there exist two linearly independent unit vectors i and j in $\mathbb{Z}^D$ such that $J(i)$ and $J(j)$ are positive. Then β_c is finite by Theorem V.5.1(b). Prove that $m(\beta, +) \to 1$ and $m(\beta, -) \to -1$ as $\beta \to \infty$. [*Hint:* (5.23)].

(d) For $\beta > 0$, $h \neq 0$ and $0 < \beta < \beta_c$, $h = 0$, there exists a unique infinite-volume Gibbs state $P_{\beta,h}$ [Theorem V.6.1]. Prove the following.

(i) For any h real, $P_{\beta,h} \Rightarrow P_\rho$ as $\beta \to 0^+$. [*Hint:* Use the FKG inequality to compare $\int_\Omega f_B dP_{\beta,h}$ with $\int_\Omega f_B dP_{\Lambda,\beta,h,+}$ and $\int_\Omega f_B dP_{\Lambda,\beta,h,-}$.]

(ii) For $h > 0$ $P_{\beta,h} \Rightarrow P_{\delta_1}$ as $\beta \to \infty$ and for $h < 0$ $P_{\beta,h} \Rightarrow P_{\delta_{-1}}$ as $\beta \to \infty$. [*Hint:* For $h > 0$, show that as $\beta \to \infty$ $P_{\beta,h}\{\omega \in \Omega: \omega_i = 1\} \to 1$ for each site $i \in \mathbb{Z}^D$ and thus that $P_{\beta,h}\{\Sigma\} \to P_{\delta_1}\{\Sigma\}$ for each cylinder set Σ.]

(iii) If the assumption of part (c) holds (so that β_c is finite), then $P_{\beta,0,+} \Rightarrow P_{\delta_1}$ and $P_{\beta,0,-} \Rightarrow P_{\delta_{-1}}$ as $\beta \to \infty$.

V.13.2. Supply the symmetry argument which shows that the integral (5.6) equals 0 unless all the integers m_i are even. This will complete the proof of Lemma V.3.1(b).

V.13.3. Supply the symmetry argument which shows that the integral (5.9) equals 0 unless the integers k, l, m, n all have the same parity. This will complete the proof of Lemma V.3.2.

For the next four problems, we denote by $\mathscr{E}$ the set of nondegenerate symmetric probability measures ρ on $\mathbb{R}$ for which $\int_{\mathbb{R}} e^{\sigma x^2}\rho(dx)$ is finite for all $\sigma > 0$.

V.13.4. Let Λ be an arbitrary nonempty finite subset of $\mathbb{Z}^D$, $\{J_{ij}; i,j\in\Lambda\}$ a set of non-negative real numbers, and $\{h_i; i\in\Lambda\}$ a set of real numbers. For $\rho\in\mathscr{E}$, define the probability measure on $\mathscr{B}(\mathbb{R}^\Lambda)$

$$P_{\Lambda,\rho}(d\omega) = \exp[-H_\Lambda(\omega)]\pi_\Lambda P_\rho(d\omega)\cdot\frac{1}{Z}, \tag{5.93}$$

where $H_\Lambda(\omega) = -\frac{1}{2}\sum_{i,j\in\Lambda} J_{ij}\omega_i\omega_j - \sum_{i\in\Lambda} h_i\omega_i$ and Z is a normalization. The measure ρ is called the *single-site distribution*. Prove the GKS-1, GKS-2, and Percus inequalities for the measure $P_{\Lambda,\rho}$.

V.13.5 [Griffiths (1967a, 1967b), Kelly and Sherman (1968)]. This problem concerns many-body interactions [see Section C.3]. Let Λ be an arbitrary nonempty finite subset of $\mathbb{Z}^D$ and $\{J(A)\}$ a set of non-negative real numbers defined for each nonempty subset A of Λ. For $\rho\in\mathscr{E}$, define the probability measure on $\mathscr{B}(\mathbb{R}^\Lambda)$

$$P_{\Lambda,\rho}(d\omega) = \exp[-H_\Lambda(\omega)]\pi_\Lambda P_\rho(d\omega)\cdot\frac{1}{Z},$$

where $H_\Lambda(\omega) = -\sum_{A\subset\Lambda} J(A)\omega_A$ and Z is a normalization. Prove the GKS-1 and GKS-2 inequalities for the measure $P_{\Lambda,\rho}$.

V.13.6. (a) For $\rho\in\mathscr{E}$ define the free energy function $c_\rho(t) = \log\int_{\mathbb{R}}\exp(tx)\,\rho(dx)$. Prove that $c_\rho(t)$ is finite for all real t and that $c_\rho(t) = o(t^2)$ as $t\to\infty$.

(b) [Simon and Griffiths (1973)] The function $-\beta^{-1}c_\rho(\beta h)$ defines the specific Gibbs free energy in a single-site system with interaction $J(0) = 0$, inverse temperature $\beta > 0$, external field h, and single-site distribution ρ. The derivative $c_\rho'(\beta h)$ gives the specific magnetization.* For $0\le\lambda<1$, define $\rho_\lambda = \lambda\delta_0 + \frac{1}{2}(1-\lambda)(\delta_1+\delta_{-1})$. Prove that for $\frac{2}{3}<\lambda<1$, $c_{\rho_\lambda}'''(\beta h)$ is not non-positive for all $h\ge 0$. Conclude that the (one-site) GHS inequality is not valid for ρ_λ.

V.13.7. For $\rho\in\mathscr{E}$, let $P_{\Lambda,\rho}$ be the probability measure on $\mathscr{B}(\mathbb{R}^\Lambda)$ defined in (5.93). Let GHS denote the set of measures $\rho\in\mathscr{E}$ such that the GHS inequality is valid for the measure $P_{\Lambda,\rho}$ for all nonempty finite subsets Λ in $\mathbb{Z}^D$. In Section V.3, we showed that $\rho = \frac{1}{2}\delta_1 + \frac{1}{2}\delta_{-1}$ belongs to GHS. GHS also contains the following measures: $(k+1)^{-1}\sum_{j=0}^k \delta_{k-2j}$ for k a positive integer [Griffiths (1969)]; the measures $\rho_\lambda = \lambda\delta_0 + \frac{1}{2}(1-\lambda)(\delta_1+\delta_{-1})$ for $0\le\lambda\le\frac{2}{3}$; and all probability measures of the form const $\cdot\exp(-V(x))dx$, where $V(x)$ is even and C^1, $V(x)\to\infty$ as $|x|\to\infty$, and $V'(x)$ is convex on $[0,\infty)$ [Ellis, Monroe, and Newman (1976)]. By Problem V.13.6(b), ρ_λ does not belong to GHS for $\frac{2}{3}<\lambda<1$. The GHS and related inequalities are discussed further in Ellis and Newman (1976, 1978c).

(a) Prove that if $\rho\in$ GHS, then $c_\rho'''(t)\le 0$ for all $t\ge 0$ and $c_\rho(t)\le\frac{1}{2}\sigma_\rho^2 t^2$

*By Theorem VII.5.1, $c_\rho(t)$ is real analytic and $c_\rho'(t) = \int_{\mathbb{R}} xe^{tx}\rho(dx)/\int_{\mathbb{R}} e^{tx}\rho(dx)$.

for all t real, where $\sigma_\rho^2 = \int_{\mathbb{R}} x^2 \rho(dx)$. Measures $\rho \in \mathscr{E}$ satisfying $c_\rho(t) \leq \frac{1}{2}\sigma_\rho^2 t^2$ for all t real are called *sub-Gaussian* [cf. (5.68)].

(b) Let $\langle - \rangle$ denote expectation with respect to the measure $P_{\Lambda,\rho}$ in (5.93). For any $\rho \in \mathscr{E}$ and sites i, j, k, l in Λ, prove that

$$(5.94) \quad \frac{\partial^3}{\partial h_i \partial h_j \partial h_k} \langle \omega_l \rangle \Big|_{\text{all}\{h_i\}=0}$$
$$= \langle \omega_i \omega_j \omega_k \omega_l \rangle - \langle \omega_i \omega_j \rangle \langle \omega_k \omega_l \rangle - \langle \omega_i \omega_k \rangle \langle \omega_j \omega_l \rangle - \langle \omega_i \omega_l \rangle \langle \omega_j \omega_k \rangle.$$

Prove that if $\rho \in \text{GHS}$ (e.g., $\rho = \frac{1}{2}\delta_1 + \frac{1}{2}\delta_{-1}$), then (5.94) is non-positive [Lebowitz (1974)]. [*Hint:* $\partial^2 \langle \omega_l \rangle / \partial h_j \partial h_k = 0$ when all $\{h_i\}$ equal 0.]

V.13.8. Let J be a summable ferromagnetic interaction on $\mathbb{Z}^D$ and set $\mathscr{J}_0 = \sum_{k \in \mathbb{Z}^D} J(k)$. Let β_c be the corresponding critical inverse temperature.

(a) By modifying Problem IV.9.7, prove that $\beta_c \geq \mathscr{J}_0^{-1}$. [*Hint:* Redefine J_Λ.]

(b) Assume that there exist two linearly independent vectors i and j in $\mathbb{Z}^D$, not necessarily unit vectors, such that $J(i)$ and $J(j)$ are positive. By modifying the proof of Theorem V.5.1(b), prove that β_c is finite. [*Hint:* Consider a "nearest-neighbor" model on the sublattice $A = \{ai + bj: a, b \in \mathbb{Z}\}$.]

(c) Let $m_D(\beta, +)$ and $\beta_{c,D}$ denote the spontaneous magnetization and the critical inverse temperature, respectively, for the Ising model on $\mathbb{Z}^D$, $D \geq 2$. Prove that

$$m_2(\beta, +) \leq m_3(\beta, +) \leq \cdots \leq m_D(\beta, +),$$
$$\beta_{c,2} \geq \beta_{c,3} \geq \cdots \geq \beta_{c,D} \geq (2D\mathscr{J})^{-1},$$

where $\mathscr{J}$ is the nearest neighbor interaction strength.

V.13.9. Let J be a summable ferromagnetic interaction on $\mathbb{Z}^D$. Let $P_{L \times N, \beta, h, +}$ be the corresponding finite-volume Gibbs state on a rectangle Λ in $\mathbb{Z}^2$ with side lengths L and N. Let A be a nonempty finite subset of Λ and define $\omega_A = \prod_{\omega \in A} \omega_i$. Prove that

$$\langle \omega_A \rangle_{\beta,h,+} = \lim_{N\to\infty} \langle \omega_A \rangle_{N \times N, \beta, h, +} = \lim_{L\to\infty} \lim_{N\to\infty} \langle \omega_A \rangle_{L \times N, \beta, h, +}$$
$$= \lim_{N\to\infty} \lim_{L\to\infty} \langle \omega_A \rangle_{L \times N, \beta, h, +}.$$

V.13.10. Let J be a summable ferromagnetic interaction on $\mathbb{Z}^D$. In Theorems IV.6.5(a) and V.6.1(a) we stated the existence of infinite-volume Gibbs states $P_{\beta,h,+}$ and $P_{\beta,h,-}$. Prove that these measures are nondegenerate and are not product measures. [*Hint:* Consider $\langle Y_0; Y_k \rangle_{\beta,h,+}$ and $\langle Y_0; Y_k \rangle_{\beta,h,-}$.]

V.13.11. Let J be a summable ferromagnetic interaction on $\mathbb{Z}^D$ and $P_{\Lambda,\beta,h,f}$ the corresponding finite-volume Gibbs state on a symmetric hypercube Λ with the free external condition.

(a) Prove that for any $\beta > 0$ and h real, $P_{\beta,h,f} = w\text{-}\lim_{\Lambda \uparrow \mathbb{Z}^D} P_{\Lambda,\beta,h,f}$ exists. [*Hint:* GKS-2 inequality.]

(b) Prove that for $\beta > 0$, $h \neq 0$ and $0 < \beta < \beta_c$, $h = 0$, $P_{\beta,h,f}$ equals $P_{\beta,h,+} = P_{\beta,h,-} = P_{\beta,h}$. [*Hint:* FKG inequality.]

V.13.12 [Sokal (1982a)]. Let J be an exponentially decaying ferromagnetic interaction on $\mathbb{Z}^D$. Set $\mathscr{J}_0 = \sum_{k \in \mathbb{Z}^D} J(k)$. This problem proves that for $0 < \beta < \mathscr{J}_0^{-1}$ and $h = 0$, the correlations $\langle Y_i Y_j \rangle_{\beta,0}$ decay exponentially [Theorem V.7.6(a)]. We consider a finite-volume Gibbs state on a symmetric hypercube Λ with $\beta > 0$, $h = 0$, interaction $\lambda J(k)$ $(\lambda > 0)$, and the free external condition. Denote by $\langle - \rangle_{\Lambda,\lambda}$ expectation with respect to this measure.

(a) For any sites i, j in Λ, prove that

$$\begin{aligned} \frac{\partial}{\partial \lambda} \langle Y_i Y_j \rangle_{\Lambda,\lambda} &= \frac{\beta}{2} \sum_{k,l \in \Lambda} J(k-l)[\langle Y_i Y_j Y_k Y_l \rangle_{\Lambda,\lambda} - \langle Y_i Y_j \rangle_{\Lambda,\lambda} \langle Y_k Y_l \rangle_{\Lambda,\lambda}] \\ &\leq \beta \sum_{k,l \in \Lambda} \langle Y_i Y_k \rangle_{\Lambda,\lambda} J(k-l) \langle Y_l Y_j \rangle_{\Lambda,\lambda}. \end{aligned} \tag{5.95}$$

[*Hint:* Problem V.13.7(b).]

(b) If $F = \{f(i); i \in \Lambda\}$ is an element of $\mathbb{R}^\Lambda$, then define operators I_Λ and F_Λ on $\mathbb{R}^\Lambda$ by $(I_\Lambda f)(i) = f(i)$ and $(F_\Lambda f)(i) = \sum_{j \in \Lambda} J(i-j) f(j)$, $i \in \Lambda$. For all sufficiently small $\lambda > 0$ the inverse operator $(I_\Lambda - \lambda\beta F_\Lambda)^{-1}$ exists. By Szarski (1965, Theorem 9.3, page 25) the solution $\langle Y_i Y_j \rangle_{\Lambda,\lambda}$ of the differential inequality (5.95) is bounded above by the solution of the corresponding differential equality with the same initial condition at $\lambda = 0$. Using this, show that for all sufficiently small $\lambda > 0$ $\langle Y_i Y_j \rangle_{\Lambda,\lambda} \leq (I_\Lambda - \lambda\beta F_\Lambda)_{ij}^{-1}$. The quantity $(I_\Lambda - \lambda\beta F_\Lambda)_{ij}^{-1}$ is defined by the formula

$$((I_\Lambda - \lambda\beta F_\Lambda)^{-1} f)(i) = \sum_{j \in \Lambda} (I_\Lambda - \lambda\beta F_\Lambda)_{ij}^{-1} f(j), \qquad f \in \mathbb{R}^\Lambda.$$

(c) For $\varepsilon > 0$, let l_ε denote the Banach space of functions $f \in \mathbb{R}^{\mathbb{Z}^D}$ for which $\|f\|_\varepsilon = \sum_{j \in \mathbb{Z}^D} |f(j)| e^{\varepsilon|j|}$ is finite ($|j| = |j_1| + \ldots + |j_D|$). Since all norms on $\mathbb{Z}^D$ are equivalent, we may assume that $0 \leq J(k) \leq c_1 e^{-c_2|k|}$ for all $k \in \mathbb{Z}^D$. Define an operator F on l_ε by $(Ff)(i) = \sum_{j \in \mathbb{Z}^D} J(i-j) f(j)$. Prove that for any $\delta > 0$ there exists a sufficiently small $\varepsilon > 0$ such that

$$\|Ff\|_\varepsilon \leq (\mathscr{J}_0 + \delta)\|f\|_\varepsilon \qquad \text{for all } f \in l_\varepsilon.$$

Conclude that if $0 < \beta < (\mathscr{J}_0 + \delta)^{-1}$, then $(I - \beta F)^{-1}$ exists and is a bounded operator from l_ε to l_ε. [*Hint:* Consider the Neumann series of $(I - \beta F)^{-1}$.]

(d) Show that if $0 < \beta < \mathscr{J}_0^{-1}$, then there exist $\varepsilon > 0$ and $a > 0$ such that for any symmetric hypercube Λ

$$\langle Y_i Y_j \rangle_{\Lambda,\lambda=1} \leq (I_\Lambda - \beta F_\Lambda)_{ij}^{-1} \leq a e^{-\varepsilon|i-j|}, \qquad i, j \in \Lambda.$$

Complete the proof of Theorem V.7.6(a). [*Hint:* Problem V.13.11(b).]

V.13.13. (a) Let f be a non-negative, nonincreasing, continuous radial function on $\mathbb{R}^D$, $D \in \{1, 2, \ldots\}$. Prove that there exist constants $c_1 > c_2 > 0$

such that for all sufficiently large $R > 0$

$$c_2 \sum_{\{k \in \mathbb{Z}^D: \|k\| \leq R\}} f(\|k\|) \leq \int_{\|x\| \leq R} f(\|x\|) d^D x \leq c_1 \sum_{\{k \in \mathbb{Z}^D: \|k\| \leq R\}} f(\|k\|).$$

[*Hint:* Divide the sphere $\{\|x\| \leq R\}$ into shells $\{(k-1)a < \|x\| \leq ka\}$ for suitable a; similarly for $\{k \in \mathbb{Z}^D: \|k\| \leq R\}$.]

(b) In Lemma V.8.3(c), prove $G(R) \sim R^{2-\eta}$.

V.13.14 [Eisele and Ellis (1983, Appendix B)]. Let ρ be a probability measure in the set GHS [see Problem V.13.7]. Define the Curie–Weiss probability measure on $\mathscr{B}(\mathbb{R}^n)$

$$(5.96)\quad P^{\mathrm{CW}}_{n,\beta,h}(d\omega) = \exp\left(\beta\left[\frac{1}{2n}\left(\sum_{j=1}^{n} \omega_j\right)^2 + h \sum_{j=1}^{n} \omega_j\right]\right) \pi_n P_\rho(d\omega) \cdot \frac{1}{Z^{\mathrm{CW}}(n,\beta,h)},$$

where

$$Z^{\mathrm{CW}}(n,\beta,h) = \int_{\mathbb{R}^n} \exp\left(\beta\left[\frac{1}{2n}\left(\sum_{j=1}^{n} \omega_j\right)^2 + h \sum_{j=1}^{n} \omega_j\right]\right) \pi_n P_\rho(d\omega).$$

(a) Define $\psi^{\mathrm{CW}}(\beta,h) = -\beta^{-1} \lim_{n\to\infty} n^{-1} \log Z^{\mathrm{CW}}(n,\beta,h)$ and let $I^{(1)}_\rho$ be the level-1 entropy function of ρ. Prove that

$$-\beta\psi^{\mathrm{CW}}(\beta,h) = \sup_{z \in \mathbb{R}} \{\tfrac{1}{2}\beta z^2 + \beta h z - I^{(1)}_\rho(z)\} \leq \log \int_{\mathbb{R}} \exp(\tfrac{1}{2}\beta x^2 + \beta h x)\rho(dx).$$

(b) Let S_ρ be the support of ρ and $\operatorname{conv} S_\rho$ the smallest closed interval containing S_ρ. $I^{(1)}_\rho(z)$ is differentiable for $z \in \operatorname{int}(\operatorname{conv} S_\rho)$ [Theorem VIII.3.4(a)]. Show that the supremum in part (a) is attained at some point $z \in \operatorname{int}(\operatorname{conv} S_\rho)$ for which $\beta(z+h) = (I^{(1)}_\rho)'(z)$. [*Hint:* If $\operatorname{conv} S_\rho = [-L, L]$, $L < \infty$, then $(I^{(1)}_\rho)'(z) \to \infty$ as $|z| \to L$ [Theorem VIII.3.4(a)]. If $\operatorname{conv} S_\rho = \mathbb{R}$, prove using Problem V.13.6(a) that $\frac{1}{2}\beta z^2 + \beta h z - I^{(1)}_\rho(z) \to -\infty$ as $|z| \to \infty$.]

(c) For $\rho \in$ GHS, one may show that $(I^{(1)}_\rho)'(z)$ is strictly convex for $z \geq 0$. Using this fact, prove that spontaneous magnetization occurs at all $\beta > 1/\sigma^2_\rho$, where $\sigma^2_\rho = \int_{\mathbb{R}} x^2 \rho(dx)$.

V.13.15. This problem generalizes Theorem V.9.4. Let ρ be a probability measure in the set GHS and pick $0 < \beta < 1/\sigma^2_\rho$, where $\sigma^2_\rho = \int_{\mathbb{R}} x^2 \rho(dx)$. Prove that with respect to the measures $\{P^{\mathrm{CW}}_{n,\beta,0}\}$ in (5.96)

$$\frac{S_n}{\sqrt{n}} \xrightarrow{\mathscr{D}} N(0, \sigma^2_\rho(\beta)), \qquad \text{where } \sigma^2_\rho(\beta) = (\sigma^{-2}_\rho - \beta)^{-1}.$$

[*Hint:* Problem V.13.7(a).]

Part II

Convexity and Proofs of Large Deviation Theorems

Chapter VI

Convex Functions and the Legendre–Fenchel Transform

VI.1. Introduction

The first part of this book illustrated the use of entropy concepts in analyzing stochastic and statistical mechanical systems. Convexity was a recurring theme. Suppose that ρ is a probability measure on $\mathbb{R}^d$ such that

$$c_\rho(t) = \log \int_{\mathbb{R}^d} \exp\langle t, x\rangle \rho(dx)$$

is finite for all t in $\mathbb{R}^d$. The function $c_\rho(t)$, called the free energy function of ρ, is a convex function on $\mathbb{R}^d$ [Example VII.1.2]. The Legendre–Fenchel transform of $c_\rho(t)$ is given by

$$I_\rho^{(1)}(z) = \sup_{t \in \mathbb{R}^d} \{\langle t, z\rangle - c_\rho(t)\}, \qquad z \in \mathbb{R}^d.$$

According to Theorem II.4.1, $I_\rho^{(1)}(z)$ is convex and is the level-1 entropy function for i.i.d. random vectors distributed by ρ. In Chapters IV and V, the concavity of the specific Gibbs free energy $\psi(\beta, h)$ and of the specific magnetization $m(\beta, h)$ as functions of real h played an important role in analyzing spin systems.

The present chapter summarizes the theory of convex functions on $\mathbb{R}^d$.[1] For proofs, we refer to Rockafellar (1970) by page number. This chapter will provide background material for those parts of Chapters IV and V which depended on convexity arguments. The results will be applied later to derive the large deviation theorems and the properties of entropy functions which were stated without proof in the first part of the book.

VI.2. Basic Definitions

A subset C of $\mathbb{R}^d$ is said to be *convex* if $\lambda x + (1 - \lambda)y$ is in C for every $x \in C$, $y \in C$, and $0 < \lambda < 1$. Let f be an extended real-valued function defined on a subset C of $\mathbb{R}^d$. We say that f is *convex*, or *convex on* C, if C is convex, $f(x)$ is finite for at least one $x \in C$, f does not take the value $-\infty$, and

(6.1) $$f(\lambda x + (1-\lambda)y) \le \lambda f(x) + (1-\lambda)f(y)$$

for every x and y in C and $0 < \lambda < 1$.* The function f may take the value ∞. In order to make sense of (6.1) when f equals ∞, we define

(6.2) $$\alpha + \infty = \infty + \alpha = \infty \qquad \text{for any } -\infty < \alpha \le \infty.$$

A convex function f on C can always be extended to a convex function on all of $\mathbb{R}^d$ by defining $f(x) = \infty$ for all $x \notin C$. Because of this, we need consider only convex functions on $\mathbb{R}^d$.

Let f be an extended real-valued function on $\mathbb{R}^d$. We define the *effective domain* of f to be the set

$$\operatorname{dom} f = \{x \in \mathbb{R}^d : f(x) < \infty\}.$$

If f is convex, then $\operatorname{dom} f$ is clearly a convex subset of $\mathbb{R}^d$. We define the *epigraph* of f to be the set

$$\operatorname{epi} f = \{(x, \alpha) : x \in \mathbb{R}^d, \alpha \in \mathbb{R}, \alpha \ge f(x)\}.$$

It is not hard to prove that f is convex if and only if f does not take the value $-\infty$ and $\operatorname{epi} f$ is a nonempty convex subset of $\mathbb{R}^{d+1}$ [Problem VI.7.1].

Let f be an extended real-valued function defined on a subset C of $\mathbb{R}^d$. We say that f is *concave on* C if its negative $-f$ is convex on C. Unless otherwise noted, all definitions and theorems concerning convex functions carry over to concave functions with little or no change, and we shall not bother to restate them.

A real-valued function f defined on a convex set C is said to be *affine on* C if

$$f(\lambda x + (1-\lambda)y) = \lambda f(x) + (1-\lambda)f(y)$$

for every x and y in C and $0 < \lambda < 1$. Thus, an affine function on C is both convex and concave on C.

A real-valued function f on a convex set C is said to be *strictly convex on* C if

(6.3) $$f(\lambda x + (1-\lambda)y) < \lambda f(x) + (1-\lambda)f(y)$$

for any two distinct points x and y in C and every $0 < \lambda < 1$. If $C = \operatorname{dom} f$, then f is said to be *strictly convex*. For example, the function on $\mathbb{R}$ which equals 0 for $x \le 0$ and x^2 for $x \ge 0$ is affine on $(-\infty, 0]$ and is strictly convex on $[0, \infty)$.

Before studying properties of convex functions, we need several more definitions. The Euclidean inner product on $\mathbb{R}^d$ is denoted by $\langle -, - \rangle$. A subset H of $\mathbb{R}^d$ is called a *hyperplane* if it has the form

$$H = \{x \in \mathbb{R}^d : \langle x, y \rangle = b\},$$

*The definition of convex function coincides with the notion of proper convex function in Rockafellar (page 24).

where γ is a nonzero vector in $\mathbb{R}^d$ and b is a real number. The vector γ is called a *normal* to the hyperplane H. The sets $\{x \in \mathbb{R}^d : \langle x, \gamma \rangle \leq b\}$ and $\{x \in \mathbb{R}^d : \langle x, \gamma \rangle \geq b\}$ are called the *closed half-spaces determined by H*. H is said to be a *vertical hyperplane* in $\mathbb{R}^d$ if the component γ_d of γ equals 0. Let C be a convex set in $\mathbb{R}^d$ and z a point of C which lies in its boundary. H is said to be a *supporting hyperplane* to C at z if z lies in H and C lies in one of the closed half-spaces determined by H.

Our study of properties of convex functions begins in the next section.

VI.3. Properties of Convex Functions

We first state several results which show that convex functions are well behaved over large subsets of their effective domains. Let C be a convex set in $\mathbb{R}^d$. The *closure* of C is denoted by $\operatorname{cl} C$ and the *interior* of C by $\operatorname{int} C$. The *boundary* of C is defined as the set difference $\operatorname{cl} C \backslash \operatorname{int} C$ and is denoted by $\operatorname{bd} C$. The concept of interior can be generalized to the more convenient concept of relative interior. It is motivated by the fact that a line segment or a triangle embedded in $\mathbb{R}^3$ has a natural interior of sorts which is not a real interior in the sense of the whole space $\mathbb{R}^3$. A subset M of $\mathbb{R}^d$ is said to be *affine* if $\lambda x + (1 - \lambda)y$ is in M for every $x \in M$, $y \in M$, and real λ. If C is a convex set in $\mathbb{R}^d$, then the *affine hull* of C, $\operatorname{aff} C$, is defined to be the intersection of all the affine sets which contain C. We define the *relative interior* of C, $\operatorname{ri} C$, as the interior which results when C is regarded as a subset of $\operatorname{aff} C$. The *relative boundary* of C, $\operatorname{rbd} C$, is defined as the set difference $\operatorname{cl} C \backslash \operatorname{ri} C$. C is said to be *relatively open* if $\operatorname{ri} C = C$. For any nonempty convex set C in $\mathbb{R}^d$, $\operatorname{ri} C$ is nonempty [Rockafellar (page 45)].

The first theorem is a basic continuity result.

Theorem VI.3.1 [Rockafellar (page 82)]. *A convex function f on $\mathbb{R}^d$ is continuous relative to* $\operatorname{ri}(\operatorname{dom} f)$; *i.e., if $\{x_n; n = 1, 2, \ldots\}$ is a sequence in* $\operatorname{ri}(\operatorname{dom} f)$ *and $x_n \to x \in \operatorname{ri}(\operatorname{dom} f)$, then $f(x_n) \to f(x)$.*

A convex function f need not be continuous up to the relative boundary of $\operatorname{dom} f$. For example, the function which equals 0 for $-\infty < x < 1$, 1 for $x = 1$, and ∞ for $x > 1$ is convex but is discontinuous at $x = 1$.

In many situations, convex functions f arise which are lower semicontinuous on $\mathbb{R}^d$; i.e., if $x_n \to x$, then $\liminf_{n\to\infty} f(x_n) \geq f(x)$. Such a function is called a *closed* convex function on $\mathbb{R}^d$. A reason for the term "closed" is that an extended real-valued function f on $\mathbb{R}^d$ is lower semicontinuous on $\mathbb{R}^d$ if and only if the epigraph of f is a closed set in $\mathbb{R}^{d+1}$ [Problem VI.7.1]. According to Theorem VI.3.1, any convex function f is continuous, and thus lower semicontinuous, relative to $\operatorname{ri}(\operatorname{dom} f)$. Hence the property of being closed involves only the behavior of f at the relative boundary of $\operatorname{dom} f$. In fact, it translates into a continuity property at the relative boundary. Let y be a point in $\mathbb{R}^d$ and x a point in dom f. Since f is convex,

$$\limsup_{\lambda\to 0^+} f(y+\lambda(x-y)) \le \limsup_{\lambda\to 0^+}[\lambda f(x)+(1-\lambda)f(y)] = f(y).$$

On the other hand, if f is closed, then

$$\liminf_{\lambda\to 0^+} f(y+\lambda(x-y)) \ge f(y).$$

These inequalities yield the following result.

Theorem VI.3.2. *Let f be a closed convex function on $\mathbb{R}^d$, y a point in* rbd(dom f), *and x a point in* ri(dom f). *Then for each $0<\lambda\le 1$, the point $\lambda x+(1-\lambda)y$ lies in* ri(dom f) *and*

$$\lim_{\lambda\to 0^+} f(\lambda x+(1-\lambda)y) = f(y).$$

The first assertion is proved in Rockafellar (page 45). The second assertion is a consequence of the two displays which appear before the theorem.

For $d=1$ the closed convex functions can be easily characterized. A convex function f on $\mathbb{R}$ is closed if and only if it is continuous* at each endpoint of dom f that is in dom f and $f(x)\to\infty$ as x approaches any finite endpoint not in dom f. For example, the convex function mentioned after Theorem VI.3.1 is not closed.

We next state a basic convergence theorem that has been applied several times in Chapters IV and V.

Theorem VI.3.3 [Rockafellar (page 90)]. *Let C be a relatively open convex set in $\mathbb{R}^d$ and $\{f_n; n=1,2,\ldots\}$ a sequence of finite convex functions on C. Suppose that the sequence converges pointwise on a dense subset of C. Then the following conclusions hold.*

(a) *The limit $f(x)=\lim_{n\to\infty} f_n(x)$ exists for every $x\in C$ and defines a finite convex function on C.*

(b) *The sequence $\{f_n\}$ converges to f uniformly on each compact subset of C.*

This completes the discussion of continuity properties. Convex functions also have many useful differentiability properties. Let us first consider the one-dimensional case. If f is an extended real-valued function on $\mathbb{R}$ and x is a point where f is finite, then we define the *right-derivative* $f'_+(x)$ and the *left-derivative* $f'_-(x)$ by the formulas

$$f'_+(x) = \lim_{y\to x^+}\frac{f(y)-f(x)}{y-x}, \qquad f'_-(x) = \lim_{y\to x^-}\frac{f(y)-f(x)}{y-x}.$$

f is differentiable at x if and only if $f'_+(x)=f'_-(x)$. Now assume that f is convex. Then the difference quotient $(f(y)-f(x))/(y-x)$ is a monotone function of y as $y\to x^+$ or as $y\to x^-$, and so $f'_+(x)$ and $f'_-(x)$ both exist as

* Relative to dom f.

extended real numbers. The following properties are easy to see: if x is in $\mathrm{int}(\mathrm{dom} f)$, then $f'_+(x)$ and $f'_-(x)$ are both finite; if $\mathrm{dom} f$ is a closed bounded interval $[\alpha, \beta]$, then $f'_+(\alpha)$ exists but may equal $-\infty$ and $f'_-(\beta)$ exists but may equal ∞ [Rockafellar (pages 214, 228)].

We now turn to functions on $\mathbb{R}^d$. If f is an extended real-valued function on $\mathbb{R}^d$ and x is a point where f is finite, then f is said to be *differentiable at* x if there exists a vector z with the property that

$$f(w) = f(x) + \langle z, w - x\rangle + o(\|w - x\|) \qquad \text{as } w \to x.^*$$

Such a z, if it exists, is called the *gradient* of f at x and is denoted by $\nabla f(x)$. Clearly, if f is differentiable at x, then the partial derivatives $\partial f(x)/\partial x_1, \ldots, \partial f(x)/\partial x_d$ exist and

$$\nabla f(x) = \left(\frac{\partial f(x)}{\partial x_1}, \ldots, \frac{\partial f(x)}{\partial x_d}\right).$$

If the function f is convex on $\mathbb{R}^d$, then this statement has a useful converse.

Theorem VI.3.4 [Rockafellar (page 244)]. *Let f be a convex function on $\mathbb{R}^d$ such that* $\mathrm{int}(\mathrm{dom} f)$ *is nonempty. Then f is differentiable at* $x \in \mathrm{int}(\mathrm{dom} f)$ *if and only if the d partial derivatives $\partial f(x)/\partial x_1, \ldots, \partial f(x)/\partial x_d$ exist at x and are finite.*

Given f a convex function on $\mathbb{R}^d$, define $A(f)$ to be the subset of $\mathrm{int}(\mathrm{dom} f)$ where f is differentiable. Then the complement of $A(f)$ in $\mathrm{int}(\mathrm{dom} f)$ is a set of Lebesgue measure zero and ∇f is continuous on $A(f)$. If $d = 1$, then $A(f)$ contains all but at most countably many points of $\mathrm{int}(\mathrm{dom} f)$ [Rockafellar (pages 244–246)].

Let f be a convex function on $\mathbb{R}^d$. Then for every x and y in $\mathbb{R}^d$ such that $f(y)$ is finite and every $0 < \lambda < 1$

$$\lambda(f(x) - f(y)) \ge f(y + \lambda(x - y)) - f(y).$$

If f is differentiable at y, then as $\lambda \to 0^+$

$$\begin{aligned} \lambda(f(x) - f(y)) &\ge f(y + \lambda(x - y)) - f(y) \\ &= \lambda\langle \nabla f(y), x - y\rangle + o(\lambda\|x - y\|). \end{aligned} \tag{6.4}$$

Dividing both sides of (6.4) by $\lambda > 0$ and taking λ to 0, we obtain the inequality

$$f(x) \ge f(y) + \langle \nabla f(y), x - y\rangle \qquad \text{for all } x \in \mathbb{R}^d. \tag{6.5}$$

The notion of subdifferential extends the derivative concept to points at which f is not differentiable. If y is a point in $\mathbb{R}^d$, then a vector $z \in \mathbb{R}^d$ is called a *subgradient* of f at y if

$$f(x) \ge f(y) + \langle z, x - y\rangle \qquad \text{for all } x \in \mathbb{R}^d. \tag{6.6}$$

* $\|-\|$ denotes the Euclidean norm on $\mathbb{R}^d$.

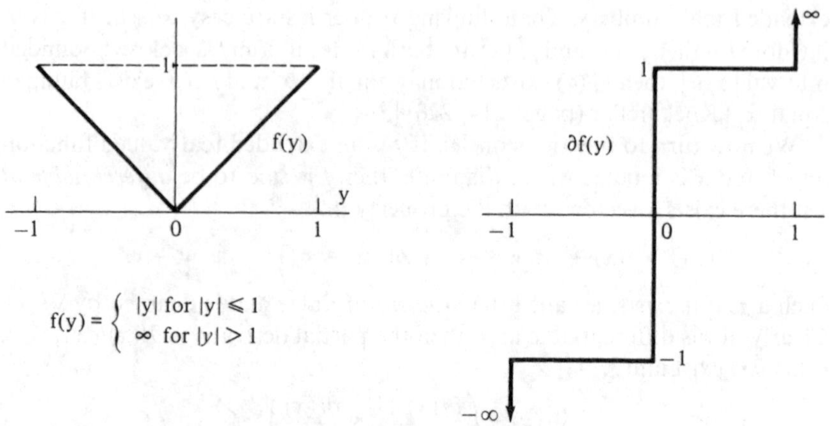

Figure VI.1. A function f and its subdifferential for $d = 1$.

For example, if f is differentiable at y, then by (6.5) $z = \nabla f(y)$ is a subgradient of f at y. We define the *subdifferential* of f at y to be the set

$$\partial f(y) = \{z \in \mathbb{R}^d : z \text{ is a subgradient of } f \text{ at } y\}.$$

It is easily checked that $\partial f(y)$ is a closed convex set. If $\partial f(y)$ is not empty, then f is said to be *subdifferentiable* at y. Note that f is not subdifferentiable at any point y not in $\operatorname{dom} f$. Indeed, since $f(y) = \infty$, the subgradient inequality (6.6) cannot be satisfied for $x \in \operatorname{dom} f$ by any vector z. The set of all y at which f is subdifferentiable is called the *effective domain* of ∂f and is denoted by $\operatorname{dom} \partial f$.

Condition (6.6) has a simple geometric interpretation when f is finite at y. It says that the graph of the affine function $g(x) = f(y) + \langle z, x - y \rangle$ is a non-vertical supporting hyperplane in $\mathbb{R}^{d+1}$ to the convex set $\operatorname{epi} f$ at the point $(y, f(y))$.

If $d = 1$ and $\operatorname{dom} f$ is a closed bounded interval $[\alpha, \beta]$, then the subdifferential is easily determined [see Figure VI.1]:

$$\partial f(y) = \begin{cases} (-\infty, f'_+(\alpha)] & \text{for } y = \alpha, \\ [f'_-(y), f'_+(y)] & \text{for } \alpha < y < \beta, \\ [f'_-(\beta), \infty) & \text{for } y = \beta. \end{cases} \tag{6.7}$$

Thus, in general, the mapping $y \to \partial f(y)$ is multivalued. The set $\partial f(y)$ is empty if $y < \alpha$ or $y > \beta$.

The next theorem relates properties of f with topological properties of the subdifferential. The proof for $d = 1$ is immediate from (6.7).

Theorem VI.3.5 [Rockafellar (pages 217, 242)]. *Let f be a convex function on $\mathbb{R}^d$. Then the following conclusions hold.*

(a) *For y not in* $\text{dom} f$, $\partial f(y)$ *is empty.*
(b) *For y in* $\text{ri}(\text{dom} f)$, $\partial f(y)$ *is a nonempty, closed, convex set.*
(c) $\partial f(y)$ *is a nonempty bounded set if and only if y is in* $\text{int}(\text{dom} f)$.
(d) *f is differentiable at y if and only if f has a unique subgradient at y. If unique, this subgradient is* $\nabla f(y)$.

Parts (a)–(c) of the theorem show that the effective domain of ∂f, $\text{dom}\,\partial f$, lies between $\text{ri}(\text{dom} f)$ and $\text{dom} f$. For $d = 1$, this is easy to check [see (6.7)].

As an application of the concept of subdifferential, we prove *Jensen's inequality*.

Theorem VI.3.6. *Let f be a finite convex function on an open interval A of $\mathbb{R}$. Let X be a random variable on a probability space $(\Omega, \mathcal{F}, P)$ whose essential range* is a subset of A.*

(a) *If $E\{X\}$ is finite, then*

$$f(E\{X\}) \leq E\{f(X)\} \leq \infty. \tag{6.8}$$

(b) *Suppose that in addition f is strictly convex on A and $E\{f(X)\}$ is finite. Then equality holds in* (6.8) *if and only if X is a constant P-a.s.*

Proof. (a) Define ρ to be the P-distribution of X and let C be the smallest closed interval of $\mathbb{R}$ containing the support of ρ. By hypothesis on X, C is a subset of A, and so the point $\langle\rho\rangle = \int_{\mathbb{R}} x\rho(dx)$ lies in A. For all real x and any $z \in \partial f(\langle\rho\rangle)$,

$$f(x) \geq f(\langle\rho\rangle) + z(x - \langle\rho\rangle). \tag{6.9}$$

Integrating this inequality with respect to ρ yields (6.8).

(b) X is a constant P-a.s. if and only if ρ is a unit point measure. If ρ is a unit point measure, then clearly equality holds in (6.8). Conversely, suppose that equality holds in (6.8). According to the proof of part (a), if $\int_{\mathbb{R}} f(x)\rho(dx)$ is finite, then for all points y in the support of ρ and $z \in \partial f(\langle\rho\rangle)$

$$f(y) = f(\langle\rho\rangle) + z(y - \langle\rho\rangle). \tag{6.10}$$

Any point y in C can be expressed as $\lambda y_1 + (1 - \lambda)y_2$, where y_1 and y_2 are in the support of ρ and $0 \leq \lambda \leq 1$. The convexity of f, (6.9), and (6.10) imply that

$$\begin{aligned} f(\langle\rho\rangle) + z(y - \langle\rho\rangle) \leq f(y) &\leq \lambda f(y_1) + (1 - \lambda) f(y_2) \\ &= f(\langle\rho\rangle) + z(y - \langle\rho\rangle). \end{aligned}$$

Thus (6.10) holds for all y in C, and f coincides with an affine function on this set. Unless ρ is a unit point measure (so that C reduces to a point), we have a contradiction to the strict convexity of f on A. □

* This is the smallest closed set B which satisfies $P\{X \in B\} = 1$.

The main concepts introduced in this section were those of a closed convex function and of the subdifferential of a convex function at a point. These concepts will be explored further when we consider the Legendre–Fenchel transform of a convex function on $\mathbb{R}^d$. But first, in order to motivate the Legendre–Fenchel transform, we look at a simple one-dimensional example.

VI.4. A One-Dimensional Example of the Legendre–Fenchel Transform.

Let g be a real-valued, increasing, continuous function on $\mathbb{R}$ which satisfies $g(0) = 0$, $g(x) \to -\infty$ as $x \to -\infty$, and $g(x) \to \infty$ as $x \to \infty$. Its inverse function g^{-1} is well defined and has the same properties as g. Hence, if we define functions

$$f(x) = \int_0^x g(s)ds \quad \text{for } x \in \mathbb{R}, \qquad f^*(y) = \int_0^y g^{-1}(t)dt \quad \text{for } y \in \mathbb{R}, \tag{6.11}$$

then f and f^* are both strictly convex on $\mathbb{R}$ [Problem VI.7.3(b)]. Figure VI.2 shows the graph of $t = g(s)$. The shaded portions represent $f(x)$ and $f^*(y)$.

A number of results follow from the definitions of f and f^* or can be seen directly from the figure. Part (a) is known as *Young's inequality*.

Theorem VI.4.1.

(a) $xy \le f(x) + f^*(y)$ *for all* x *and* y *in* $\mathbb{R}$.
(b) $xy = f(x) + f^*(y)$ *if and only if* $y = f'(x)$.
(c) $(f^*)' = (f')^{-1}$.
(d) $f^*(y) = \sup_{x \in \mathbb{R}} \{xy - f(x)\}$ *for* $y \in \mathbb{R}$.
(e) $f(x) = \sup_{y \in \mathbb{R}} \{xy - f^*(y)\}$ *for* $x \in \mathbb{R}$.

Part (d) gives an alternate definition of $f^*(y)$ as $\sup_{x \in \mathbb{R}} \{xy - f(x)\}$. We call f^* the *Legendre–Fenchel transform* of f.[2] If we write $f^{**}(x)$ for $(f^*)^*(x)$, then part (e) states that $f^{**}(x) = f(x)$; i.e., performing the Legendre–Fenchel transform twice gives back the original function. Theorem VI.4.1 is generalized in the next section to functions that do not necessarily have the form (6.11) (in fact, to all closed convex functions on $\mathbb{R}^d$).

As a corollary of Theorem VI.4.1, we derive *Hölder's inequality*.

Corollary VI.4.2. *Let* $(\Omega, \mathscr{F}, P)$ *be a measure space and let* p *and* q *be real numbers satisfying* $p > 1$, $q > 1$, $1/p + 1/q = 1$. *Suppose that* h *and* k *are real-valued, measurable functions on* Ω *and that* $\|h\|_p = \{\int_\Omega |h|^p dP\}^{1/p}$ *and*

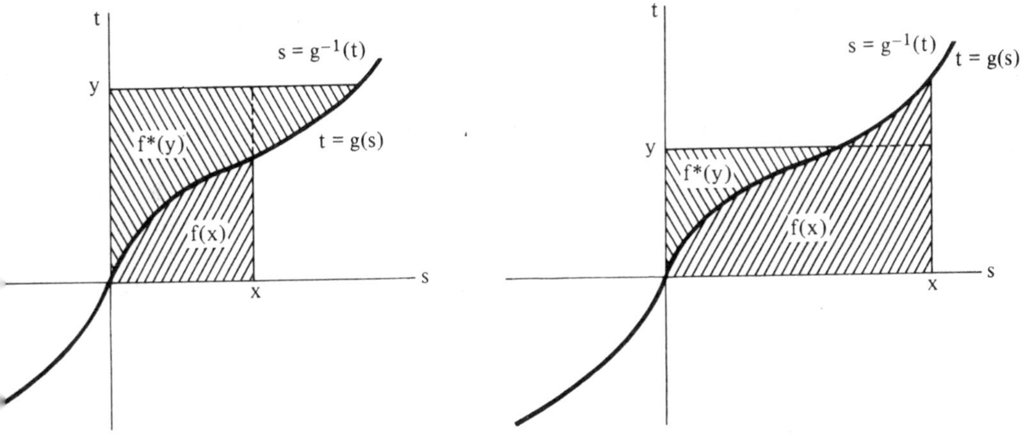

Figure VI.2. The functions f and f^* defined in (6.11).†

$\|k\|_q = \{\int_\Omega |k|^q dP\}^{1/q}$ *are finite. Then*

$$\int_\Omega |hk|\,dP \le \|h\|_p \, \|k\|_q \tag{6.12}$$

with equality if and only if $\left(\frac{|h|}{\|h\|_p}\right)^p = \left(\frac{|k|}{\|k\|_q}\right)^q$ P-a.s.

Proof. The inequality is trivial if either $\|h\|_p$ or $\|k\|_q$ equals 0, so we assume that $\|h\|_p$ and $\|k\|_q$ are positive. We need consider only $x \ge 0$ and $y \ge 0$ in (6.11). If $g(x) = x^{p-1}$, then $f(x) = p^{-1}x^p$ and $f^*(y) = q^{-1}y^q$ and therefore

$$xy \le \frac{1}{p}x^p + \frac{1}{q}y^q \qquad \text{with equality iff } y = x^{p-1}.$$

In particular,

$$\frac{|h(\omega)|}{\|h\|_p}\frac{|k(\omega)|}{\|k\|_q} \le \frac{1}{p}\frac{|h(\omega)|^p}{\|h\|_p^p} + \frac{1}{q}\frac{|k(\omega)|^q}{\|k\|_q^q}$$

for all $\omega \in \Omega$ such that $h(\omega)$ and $k(\omega)$ are defined; i.e., for P-almost all ω. Integrating both sides of this inequality with respect to P, we find that

$$\frac{1}{\|h\|_p\|k\|_q}\int |hk|\,dP \le \frac{1}{p} + \frac{1}{q} = 1$$

$$\text{with equality iff } \frac{|k|}{\|k\|_q} = \left(\frac{|h|}{\|h\|_p}\right)^{p-1} \quad P\text{-a.s.}$$

Since $p - 1 = p/q$, (6.12) follows. □

† Figure VI.2 is adapted from Figure 15.1 in Roberts and Varberg (1973, page 29).

VI.5. The Legendre–Fenchel Transform for Convex Functions on $\mathbb{R}^d$

In the previous section, we considered the Legendre–Fenchel transform f^* for convex functions f on $\mathbb{R}$ which can be written as the integral of another function g having special properties. In this case, the formula $f^{**} = f$ is valid. Our aim now is to extend the definition of the Legendre–Fenchel transform to convex functions f on $\mathbb{R}^d$ while at the same time preserving the property $f^{**} = f$. To do this, we take as the definition of the Legendre–Fenchel transform

$$(6.13) \qquad f^*(y) = \sup_{x \in \mathbb{R}^d} \{\langle x, y\rangle - f(x)\}, \qquad y \in \mathbb{R}^d.$$

The supremum over $\mathbb{R}^d$ may be replaced by the supremum over $\operatorname{dom} f$ since $f(x)$ equals ∞ for $x \notin \operatorname{dom} f$. Formula (6.13) reduces to the formula in part (d) of Theorem VI.4.1 when $d = 1$. The function f^* in (6.13) is also known as the *conjugate* of f or as the *conjugate function*. For f a concave function on $\mathbb{R}^d$, the definition of the Legendre–Fenchel transform changes to $f^*(y) = \inf_{x \in \mathbb{R}^d} \{\langle x, y\rangle - f(x)\}$.

Example VI.5.1. Let V be a symmetric, positive definite $d \times d$ matrix ($\langle Vx, x\rangle > 0$ for all nonzero $x \in \mathbb{R}^d$) and define the function $f(x) = \frac{1}{2}\langle Vx, x\rangle$. f is strictly convex by Problem VI.7.4(b). To calculate f^*, let $V^{1/2}$ be the unique symmetric, positive definite square root of V. Substituting $w = V^{1/2}x$, we have for $y \in \mathbb{R}^d$

$$\begin{aligned} f^*(y) &= \sup_{x \in \mathbb{R}^d} \{\langle x, y\rangle - \tfrac{1}{2}\langle Vx, x\rangle\} = \sup_{w \in \mathbb{R}^d} \{\langle w, V^{-1/2}y\rangle - \tfrac{1}{2}\langle w, w\rangle\} \\ &= \tfrac{1}{2}\langle V^{-1}y, y\rangle - \tfrac{1}{2} \inf_{w \in \mathbb{R}^d} \langle w - V^{-1/2}y, w - V^{-1/2}y\rangle = \tfrac{1}{2}\langle V^{-1}y, y\rangle. \end{aligned}$$

In this example, both f and f^* are finite on all of $\mathbb{R}^d$. However, it is not hard to find examples of convex functions which are everywhere finite but whose Legendre–Fenchel transforms are not [see Theorem VIII.3.3]. A relationship between closed convex functions f and the effective domains of f^* is given in formula (6.19) below.

Before generalizing Theorems VI.4.1, we note some simple properties of f^* for convex functions f on $\mathbb{R}^d$. Recall that f is said to be closed if it is lower semicontinuous on $\mathbb{R}^d$.

Lemma VI.5.2. *If f is a convex function on $\mathbb{R}^d$, then* $\operatorname{dom} f^*$ *is nonempty and f^* is a closed convex function on $\mathbb{R}^d$.*

Proof. If $\operatorname{dom} f$ consists of one point x_0, then for any $y \in \mathbb{R}^d$ $f(x) \geq f(x_0) + \langle y, x - x_0\rangle$ for all x. Hence for all x, $\langle x, y\rangle - f(x) \leq \langle x_0, y\rangle - f(x_0) < \infty$, and so $f^*(y)$ is finite. If $\operatorname{dom} f$ consists of more than one point, then $\operatorname{ri}(\operatorname{dom} f)$

is nonempty and $\partial f(x_0)$ is nonempty for any $x_0 \in \mathrm{ri}(\mathrm{dom} f)$ [Theorem VI.3.5(b)]. For any $y \in \partial f(x_0)$, $f(x) \geq f(x_0) + \langle y, x - x_0 \rangle$ for all x. Hence as above $f^*(y)$ is finite. We conclude that in all cases $\mathrm{dom} f^*$ is nonempty.

To prove that f^* is convex, note that since $f \not\equiv \infty$, f^* cannot take the value $-\infty$. Pick $0 < \lambda < 1$ and y_1 and y_2 in $\mathbb{R}^d$. Writing $\alpha_x(y)$ for the affine function $\langle x, y \rangle - f(x)$, we have

$$\begin{aligned} f^*(\lambda y_1 + (1-\lambda) y_2) &= \sup_{x \in \mathbb{R}^d} \{\lambda \alpha_x(y_1) + (1-\lambda)\alpha_x(y_2)\} \\ &\leq \lambda \sup_{x \in \mathbb{R}^d} \alpha_x(y_1) + (1-\lambda) \sup_{x \in \mathbb{R}^d} \alpha_x(y_2) \\ &= \lambda f^*(y_1) + (1-\lambda) f^*(y_2). \end{aligned}$$

To prove that f^* is closed, let $\{y_n; n = 1, 2, \ldots\}$ be a sequence of points in $\mathbb{R}^d$ converging to some point y. Then for any point $x \in \mathbb{R}^d$

$$f^*(y_n) \geq \langle x, y_n \rangle - f(x), \qquad \liminf_{n \to \infty} f^*(y_n) \geq \langle x, y \rangle - f(x).$$

Since x is arbitrary, we conclude that $\liminf_{n \to \infty} f^*(y_n) \geq f^*(y)$. □

Lemma VI.5.2 shows that for any convex function f on $\mathbb{R}^d$, $f^{**} = (f^*)^*$ is automatically closed. Hence we cannot preserve the property $f^{**} = f$ unless we start with a closed convex function. The important fact is that for such functions Theorem VI.4.1 can be generalized. Part (b) of the next theorem is known as *Fenchel's inequality*. In part (d) the subdifferential $\partial f^*(y)$ is well defined since f^* is a convex function.[3]

Theorem VI.5.3. *Let f be a closed convex function on $\mathbb{R}^d$. Then the following conclusions hold.*

(a) *f^* is a closed convex function on $\mathbb{R}^d$.*
(b) *$\langle x, y \rangle \leq f(x) + f^*(y)$ for all x and y in $\mathbb{R}^d$.*
(c) *$\langle x, y \rangle = f(x) + f^*(y)$ if and only if $y \in \partial f(x)$.*
(d) *$y \in \partial f(x)$ if and only if $x \in \partial f^*(y)$.*
(e) *$f^{**} = f$; i.e., $f(x) = \sup_{y \in \mathbb{R}^d} \{\langle x, y \rangle - f^*(y)\}$ for all $x \in \mathbb{R}^d$.*

In order to prove part (e), we need additional information. Consider a hyperplane in $\mathbb{R}^d$ of the form $H = \{x \in \mathbb{R}^d : \langle x, \gamma \rangle = b\}$, where γ is a nonzero vector in $\mathbb{R}^d$ and b is a real number. The sets

$$\{x \in \mathbb{R}^d : \langle x, \gamma \rangle \leq b\} \qquad \text{and} \qquad \{x \in \mathbb{R}^d : \langle x, \gamma \rangle \geq b\}$$

are called the *closed half-spaces determined by H*. The sets

$$\{x \in \mathbb{R}^d : \langle x, \gamma \rangle < b\} \qquad \text{and} \qquad \{x \in \mathbb{R}^d : \langle x, \gamma \rangle > b\}$$

are called the *open half-spaces determined by H*. Let C_1 and C_2 be nonempty sets in $\mathbb{R}^d$. A hyperplane H is said to *separate C_1 and C_2* if C_1 is contained in one of the closed half-spaces determined by H and C_2 lies in the opposite closed half-space. Let B denote the closed unit ball $\{x \in \mathbb{R}^d : \|x\| \le 1\}$. A hyperplane H is said to *separate C_1 and C_2 strongly* if there exists some $\varepsilon > 0$ such that $C_1 + \varepsilon B$ is contained in one of the open half-spaces determined by H and $C_2 + \varepsilon B$ is contained in the opposite open half-space.

Lemma VI.5.4 [Rockafellar (1970, pages 95–98)]. *Let C_1 and C_2 be non-empty convex sets in $\mathbb{R}^d$.*

(a) *If* ri C_1 *and* ri C_2 *have no point in common, then there exists a hyperplane $H = \{x \in \mathbb{R}^d : \langle x, y \rangle = b\}$ which separates C_1 and C_2 and*

$$\sup\{\langle x, y \rangle : x \in C_1\} \le b \le \inf\{\langle x, y \rangle : x \in C_2\}.$$

(b) *If* $\inf\{\|x_1 - x_2\| : x_1 \in C_1, x_2 \in C_2\}$ *is positive, then there exists a hyperplane $H = \{x \in \mathbb{R}^d : \langle x, y \rangle = b\}$ which separates C_1 and C_2 strongly and*

$$\sup\{\langle x, y \rangle : x \in C_1\} < b < \inf\{\langle x, y \rangle : x \in C_2\}.$$

Let f be a closed convex function on $\mathbb{R}^d$. We have defined the epigraph of f to be the set

$$\operatorname{epi} f = \{(x, \alpha) : x \in \mathbb{R}^d, \alpha \in \mathbb{R}, \alpha \ge f(x)\}.$$

We will need the fact that if a point (x_0, α_0) in $\mathbb{R}^{d+1}$ does not belong to epi f, then (x_0, α_0) may be separated from epi f by a non-vertical hyperplane in $\mathbb{R}^{d+1}$. This fact, due to Fenchel (1949), is proved in the next lemma. We follow the presentation of Gross (1982).

Lemma VI.5.5. *Let f be a closed convex function on $\mathbb{R}^d$. Let x_0 be a point in $\mathbb{R}^d$ and α_0 a real number satisfying $\alpha_0 < f(x_0)$. Then there exists a point $y_0 \in \mathbb{R}^d$ such that*

$$\alpha_0 + \langle y_0, x - x_0 \rangle \le f(x) \quad \textit{for all } x \in \mathbb{R}^d. \tag{6.14}$$

Proof. It suffices to prove (6.14) for $x \in \operatorname{dom} f$. Since f is a closed convex function, epi f is a closed convex set in $\mathbb{R}^{d+1}$ [Problem VI.7.1]. Since the point (x_0, α_0) does not belong to epi f, there exists a hyperplane in $\mathbb{R}^{d+1}$ which separates (x_0, α_0) and epi f strongly. Thus we can find a vector y in $\mathbb{R}^d$ and a real number c such that

$$\langle y, x_0 \rangle + c\alpha_0 < \langle y, x \rangle + c\alpha \tag{6.15}$$

for all $x \in \operatorname{dom} f$ and $\alpha \in \mathbb{R}$ satisfying $\alpha \ge f(x)$. Suppose first that x_0 is in dom f. With $x = x_0$, (6.15) states that $c(\alpha - \alpha_0) > 0$ for all $\alpha \ge f(x_0)$. Since $f(x_0) > \alpha_0$, c must be positive. In this case, we may put $y_0 = -c^{-1}y$ in (6.15) to obtain (6.14). Suppose now that x_0 is not in dom f. If c were negative, then taking α in (6.15) to be sufficiently large and positive would give a contradiction. Thus, c must be non-negative. If c is positive, then we proceed as before to obtain (6.14). If c is zero, then

$$0 < \langle y, x - x_0 \rangle \quad \text{for all } x \in \operatorname{dom} f. \tag{6.16}$$

To deduce (6.14), we must work harder. Let x_1 be any point in dom$f$. If $\alpha_1 < f(x_1)$, then there exists a vector γ_1 and a real number c_1 such that (6.15) holds with $\gamma = \gamma_1$, $c = c_1$, $x_0 = x_1$, and $\alpha_0 = \alpha_1$. As above c_1 must be positive, and it follows that

$$\text{(6.17)} \qquad \langle y_1, x \rangle + a \le f(x) \qquad \text{for all } x \in \operatorname{dom} f,$$

where $y_1 = -c_1^{-1}\gamma_1$ and $a = \alpha_1 + c_1^{-1}\langle \gamma_1, x_1 \rangle$. We put $y_0 = y_1 - s\gamma$ for some positive real number s. If s is sufficiently large, then (6.16) and (6.17) yield (6.14) for all x in dom$f$:

$$\begin{aligned} \alpha_0 + \langle y_0, x - x_0 \rangle &= \alpha_0 + \langle y_1, x \rangle - \langle y_1, x_0 \rangle - s\langle \gamma, x - x_0 \rangle \\ &\le f(x) + \alpha_0 - a - \langle y_1, x_0 \rangle - s\langle \gamma, x - x_0 \rangle \\ &\le f(x). \qquad \square \end{aligned}$$

Proof of Theorem VI.5.3. (a) Lemma VI.5.2.

(b) Definition of f^*.

(c) The subgradient inequality defining the condition $y \in \partial f(x)$ can be written as $\langle y, x \rangle - f(x) \ge \langle y, z \rangle - f(z)$ for all z. But this is the same as saying that the function $\langle y, z \rangle - f(z)$ attains its supremum at $z = x$. Since by definition this supremum is $f^*(y)$, we have proved part (c).

(d) By part (c) applied to f^* and f^{**}, the equation $\langle y, x \rangle = f^*(y) + f^{**}(x)$ holds iff $x \in \partial f^*(y)$ for some $y \in \operatorname{dom} f^*$. By part (e), this equation is the same as $\langle y, x \rangle = f^*(y) + f(x)$, and again by part (c) this holds iff $y \in \partial f(x)$ for some $x \in \operatorname{dom} f$. Hence $y \in \partial f(x)$ iff $x \in \partial f^*(y)$.

(e) For any x and y, $f(x) \ge \langle x, y \rangle - f^*(y)$. This implies that $f(x) \ge f^{**}(x)$. Let x_0 be any point in $\mathbb{R}^d$. We prove that $f^{**}(x_0) \ge f(x_0)$ using Lemma VI.5.5. If α_0 is any real number satisfying $\alpha_0 < f(x_0)$, then there exists a point $y_0 \in \mathbb{R}^d$ such that

$$\langle y_0, x \rangle - f(x) \le \langle y_0, x_0 \rangle - \alpha_0 \qquad \text{for all } x \in \mathbb{R}^d.$$

This implies that $f^*(y_0) \le \langle y_0, x_0 \rangle - \alpha_0$ or that

$$\alpha_0 \le \langle y_0, x_0 \rangle - f^*(y_0) \le \sup_{y \in \mathbb{R}^d} \{\langle y, x_0 \rangle - f^*(y)\} = f^{**}(x_0).$$

Letting α_0 converge to $f(x_0)$, we conclude that $f(x_0) \le f^{**}(x_0)$. $\square$

Theorem VI.5.3 shows that the Legendre–Fenchel transform $f \to f^*$ defines a one-to-one correspondence in the class of all closed convex functions on $\mathbb{R}^d$. We investigate what additional property f^* satisfies if f is differentiable on int(dom$f$). Suppose that $d = 1$ and f is a differentiable convex function on all of $\mathbb{R}$. Then for each x in $\mathbb{R}$, the subdifferential $\partial f(x)$ consists of the single point $f'(x)$. Since $y = f'(x)$ if and only if $x \in \partial f^*(y)$ [Theorem VI.5.3(d)], we see that whenever y_1 and y_2 are different points, the intersection $\partial f^*(y_1) \cap \partial f^*(y_2)$ is empty. Clearly, the only way for this to hold is for f^* to be strictly convex. A similar argument shows that if f is finite and strictly convex on all of $\mathbb{R}$, then the set $\text{dom} f^*$ has nonempty interior and f^* is differentiable on this interior.

In order to generalize these statements, we need some definitions. A convex function f on $\mathbb{R}^d$ is said to be *essentially smooth* if it satisfies the following three conditions.

(a) $C = \text{int}(\text{dom} f)$ is nonempty.
(b) f is differentiable on C.
(c) $\lim_{n\to\infty} \|\nabla f(x_n)\| = \infty$ whenever $\{x_n; n = 1, 2, \dots\}$ is a sequence in C converging to a boundary point of C.

Note that a convex function which is finite and differentiable on all of $\mathbb{R}^d$ is essentially smooth since condition (c) holds vacuously.

Let f be a convex function on $\mathbb{R}^d$. The subdifferential mapping ∂f defined in Section VI.3 assigns to each $y \in \mathbb{R}^d$ a certain closed set $\partial f(y)$ in $\mathbb{R}^d$. The effective domain of ∂f, which is the set

$$\text{dom}\,\partial f = \{y \in \mathbb{R}^d : \partial f(y) \text{ is non-empty}\},$$

is not necessarily convex.[†] However, it differs very little from being convex, in the sense that

$$\text{ri}(\text{dom} f) \subseteq \text{dom}\,\partial f \subseteq \text{dom} f \tag{6.18}$$

[Theorem VI.3.5]. The function f is said to be *essentially strictly convex* if f is strictly convex on every convex subset of $\text{dom}\,\partial f$. Since $\text{ri}(\text{dom} f)$ is a convex subset of $\text{dom}\,\partial f$, this condition implies that f is strictly convex on $\text{ri}(\text{dom} f)$. Hence for $d = 1$, the notions of essentially strictly convex and strictly convex coincide. Rockafellar (page 253) gives an example of a closed convex function f on $\mathbb{R}^d$ which is essentially strictly convex but is not strictly convex on the entire set $\text{dom} f$. On the other hand, since $\partial f(x)$ is empty for x not in $\text{dom} f$, it follows that if f is strictly convex on all of $\text{dom} f$, then f is essentially strictly convex.

Theorem VI.5.6 [Rockafellar (page 253)]. *A closed convex function is essentially strictly convex if and only if its Legendre–Fenchel transform is essentially smooth.*

Let f be a closed convex function on $\mathbb{R}^d$ which is differentiable on the set $\text{int}(\text{dom} f)$ (assumed nonempty). If $\text{ran}\,\nabla f$ denotes the range of the gradient mapping $\nabla f(x)$, $x \in \text{int}(\text{dom} f)$, then according to Theorem VI.5.3(c), $\text{ran}\,\nabla f$ is a subset of the essential domain of f^*. If in addition f is essentially smooth, then this relation can be strengthened.

Theorem VI.5.7. *Let f be an essentially smooth, closed convex function on $\mathbb{R}^d$; in particular, a convex function which is finite and differentiable on all of $\mathbb{R}^d$.*

[†] See the example in Rockafellar (page 218) for $d = 2$.

Then

$$\text{ri}(\text{dom} f^*) \subseteq \text{ran}\, \nabla f \subseteq \text{dom} f^*.$$

Sketch of Proof. According to (6.18), $\text{ri}(\text{dom} f^*) \subseteq \text{dom}\, \partial f^* \subseteq \text{dom} f^*$. The range of ∂f is defined by $\bigcup_{y \in \mathbb{R}^d} \partial f(y)$ and is denoted by $\text{ran}\, \partial f$. By Theorem VI.5.3(d), the range of ∂f equals the effective domain of ∂f^*, so that

$$\text{ri}(\text{dom} f^*) \subseteq \text{ran}\, \partial f \subseteq \text{dom} f^*. \tag{6.19}$$

The proof is done once we show that $\text{ran}\, \partial f$ equals $\text{ran}\, \nabla f$. Since f is essentially smooth, f is differentiable on $\text{int}(\text{dom} f)$. Hence $\partial f(x)$ consists of the vector $\nabla f(x)$ alone when x is in $\text{int}(\text{dom} f)$. The set $\partial f(x)$ is empty when x is not in $\text{dom} f$. Rockafellar (page 252) shows that condition (c) in the definition of essential smoothness guarantees that $\partial f(x)$ is also empty when x is in $\text{bd}(\text{dom} f)$. It follows that $\text{ran}\, \partial f$ equals $\text{ran}\, \nabla f$. This completes the proof. □

In closing this section, we note that the definition (6.13) of the Legendre–Fenchel transform f^* makes sense even if f is not convex. In this case, we wish to describe f^{**}. Let f be an extended real-valued function on $\mathbb{R}^d$ which is finite on at least one point in $\mathbb{R}^d$ and majorizes at least one closed convex function. The *closed convex hull* of f is defined as the largest closed convex function majorized by f and is denoted by $\text{cl}(\text{conv} f)$. This function is well defined [Rockafellar (pages 36, 52)].

Theorem VI.5.8 [Rockafellar (page 104)]. *Let f be an extended real-valued function on $\mathbb{R}^d$ which is finite on at least one point in $\mathbb{R}^d$ and majorizes at least one closed convex function. Then f^* is a closed convex function on $\mathbb{R}^d$, and*

$$f^* = (\text{cl}(\text{conv} f))^*, \qquad f^{**} = \text{cl}(\text{conv} f).$$

This completes our discussion of convex functions on $\mathbb{R}^d$. The theory will now be applied to derive the large deviation theorems stated in Chapter II.

VI.6. Notes

1 (page 211). For the theory of convex functions on infinite dimensional spaces, see Moreau (1962, 1966), Ioffe and Tikhomirov (1968), Roberts and Varberg (1973), and Ekeland and Temam (1976, Chapter I). Chapter I of Roberts and Varberg (1973) is an elementary treatment of convex functions on $\mathbb{R}$.

2 (page 218). The Legendre–Fenchel transforms in Section VI.4 are important in the study of Birnbaum–Orlicz spaces. See Krasnosel'skiĭ and Rutickiĭ (1961) and Hewitt and Stromberg (1965, page 203). The function

$f(x)$ defined in (6.11) has a derivative for all x which is invertible. Hence its Legendre–Fenchel transform f^* is an example of the classical Legendre transformation [Rockafellar (page 256)]. The latter is an important tool in the theory of differential equations [Kamke (1930, 1974)], the calculus of variations [Courant and Hilbert (1962, pages 32–39)], and Hamiltonian mechanics [Goldstein (1959, Section 7.1)].

3 (page 221). The formula $f^{**} = f$ for closed convex functions on $\mathbb{R}^d$ [Theorem VI.5.3(e)] was discovered by Fenchel (1949) and was extended to infinite dimensional spaces by Moreau (1962) and Brøndsted (1964). Relationships between the convergence of a sequence of closed convex functions and the convergence of the corresponding Legendre–Fenchel transforms are discussed by Wijsman (1966) and Wets (1980).

VI.7. Problems

VI.7.1. Let f be an extended real-valued function on $\mathbb{R}^d$.

(a) [Rockafellar (page 25)]. Prove that f is convex if and only if f does not take the value $-\infty$ and epi f is a nonempty convex subset of $\mathbb{R}^{d+1}$.

(b) Prove that f is lower semicontinuous if and only if epi f is a closed set in $\mathbb{R}^{d+1}$.

VI.7.2 [Rockafellar (page 26)]. Let f be a real-valued, twice continuously differentiable function on an open interval (α, β) of $\mathbb{R}$.

(a) Prove that f is convex on (α, β) if and only if $f''(x) \geq 0$ for all $x \in (\alpha, \beta)$.

(b) Prove that f is strictly convex on (α, β) if $f''(x) > 0$ for all $x \in (\alpha, \beta)$. Show that the converse is false.

(c) Prove the strict convexity of the following functions.

(i) $f(x) = e^{\alpha x}$, where α is real;
(ii) $f(x) = x^p$ for $x \geq 0, f(x) = \infty$ for $x < 0$, where $1 < p < \infty$;
(iii) $f(x) = -x^p$ for $x \geq 0, f(x) = \infty$ for $x < 0$, where $0 < p < 1$;
(iv) $f(x) = -\log x$ for $x > 0, f(x) = \infty$ for $x \leq 0$;
(v) $f(x) = x \log x$ for $x > 0, f(x) = 0$ for $x = 0, f(x) = \infty$ for $x < 0$.

VI.7.3. Let f be a real-valued, continuously differentiable function on an open interval (α, β) of $\mathbb{R}$.

(a) Prove that f is convex on (α, β) if and only if f' is non-decreasing on (α, β).

(b) Prove that f is strictly convex on (α, β) if f' is increasing on (α, β).

VI.7.4. Let f be a real-valued, twice continuously differentiable function on an open convex set C in $\mathbb{R}^d$.

(a) [Rockafellar (page 27)]. Prove that f is convex on C if and only if its Hessian matrix $V_f(x) = \{\partial^2 f(x)/\partial x_i \partial x_j; i, j = 1, \ldots, d\}$ is non-negative definite for all $x \in C$ ($\langle V_f(x)z, z\rangle \geq 0$ for all $z \in \mathbb{R}^d$). [*Hint:* The convexity of f on C is equivalent to the convexity of the function $g(\lambda) = f(y + \lambda z)$ on the open real interval $\{\lambda \in \mathbb{R} : y + \lambda z \in C\}$ for each $y \in C$ and $z \in \mathbb{R}^d$.]

(b) Prove that f is strictly convex on C if $V_f(x)$ is positive definite for all $x \in C$ ($\langle V_f(x)z, z\rangle > 0$ for all nonzero $z \in \mathbb{R}^d$).

VI.7.5. Let $f(x)$ be defined by formula (6.11). Given $y > 0$, let $x_0 \geq 0$ be the unique point such that $f'(x_0) = y$. Let b denote the ordinate of the point at which the tangent line to the graph of $f(x)$ at x_0 intersects the line $x = 0$. Prove that $f^*(y) = -b$.

VI.7.6 Prove Theorem VI.4.1 as a special case of Theorem VI.5.3. [*Hint:* Reduce the proof to showing that for all $y \in \mathbb{R}$

$$\int_0^y g^{-1}(t)dt = \sup_{x \in \mathbb{R}} \{xy - f(x)\} = g^{-1}(y) \cdot y - \int_0^{g^{-1}(y)} g(s)ds.]$$

VI.7.7. Let T_1 and T_2 be positive numbers and pick $0 < \lambda < 1$. Using the strict concavity of the logarithm, prove that

$$T_1^{\lambda} T_2^{(1-\lambda)} \leq \lambda T_1 + (1 - \lambda) T_2$$

with equality if and only if $T_1 = T_2$. Deduce Hölder's inequality [Corollary VI.4.2].

VI.7.8 [Rockafellar (page 108)]. Calculate the Legendre–Fenchel transform of the convex function $\frac{1}{2}\langle Vx, x\rangle$ on $\mathbb{R}^d$, where V is a $d \times d$ symmetric non-negative definite matrix with nonempty nullspace [cf., Example VI.5.1].

VI.7.9 [Rockafellar (page 106)]. Let f be a convex function on $\mathbb{R}^d$. Prove that $f = f^*$ if and only if f is the function $\frac{1}{2}\langle x, x\rangle$, $x \in \mathbb{R}^d$.

VI.7.10 [Rockafellar (page 110) and Ekeland and Temam (1976, page 19)]. Let f be an even, closed, convex function on $\mathbb{R}$ which is non-decreasing on $[0, \infty)$. Define $F(x) = f(\|x\|)$ for $x \in \mathbb{R}^d$.

(a) Prove that F is closed and convex.

(b) Prove that $F^*(y) = f^*(\|y\|)$ for all $y \in \mathbb{R}^d$. [*Hint:* For $r \geq 0$ $\sup\{\langle y, x\rangle : \|x\| = r\} = r$.]

VI.7.11. Let C be a nonempty subset of $\mathbb{R}^d$. The indicator function of C, $\delta(x|C)$, and the support function of C, $\sigma(y|C)$, are defined by

$$\delta(x|C) = \begin{cases} 0 & \text{for } x \in C, \\ \infty & \text{for } x \notin C, \end{cases} \qquad \sigma(y|C) = \sup_{x \in C} \langle x, y\rangle.$$

(a) Prove that C is a convex set if and only if $\delta(\cdot|C)$ is a convex function on $\mathbb{R}^d$.

(b) Prove that if C is a convex set, then $\sigma(\cdot|C)$ is the Legendre–Fenchel transform of $\delta(\cdot|C)$.

(c) Prove that for any subset C of $\mathbb{R}^d$,

$$\sigma(\cdot|C) = \sigma(\cdot|\mathrm{cl}(\mathrm{conv}\, C)),$$

where $\mathrm{cl}(\mathrm{conv}\, C)$ denotes the closed convex hull of C (the intersection of all the closed convex sets containing C).

(d) Let C be a subspace of $\mathbb{R}^d$ and $C^\perp$ its orthogonal complement. Prove that $\sigma(\cdot|C) = \delta(\cdot|C^\perp)$.

VI.7.12. (a) Prove that the intersection of an arbitrary collection of convex sets is convex.

(b) Let f be a closed convex function on $\mathbb{R}^d$. Prove that the Legendre–Fenchel transform f^* is a closed convex function by showing that $\mathrm{epi}\, f^*$ is a nonempty closed convex subset of $\mathbb{R}^{d+1}$ [Problem VI.7.1].

VI.7.13. The functions in Problem VI.7.2(c) are each closed and convex. In each case, calculate f^* and f^{**} and verify that $f^{**} = f$.

VI.7.14 [Eisele and Ellis (1983, Appendix C)]. Let f and g be closed convex functions on $\mathbb{R}^d$. Prove that

$$\sup_{x \in \mathrm{dom}\, g} \{f(x) - g(x)\} = \sup_{y \in \mathrm{dom}\, f^*} \{g^*(y) - f^*(y)\}.$$

VI.7.15. Prove Theorem VI.5.8. [*Hint*: $f \geq \mathrm{cl}(\mathrm{conv} f)$, so $f^* \leq (\mathrm{cl}(\mathrm{conv} f))^*$.]

Chapter VII

Large Deviations for Random Vectors

VII.1. Statement of Results

In Section II.4, we stated three levels of large deviation properties for i.i.d. random vectors taking values in $\mathbb{R}^d$. In Section II.6 we also stated a large deviation theorem for random vectors [Theorem II.6.1] which generalized the level-1 property. In this chapter, Theorem II.6.1 will be proved [Sections VII.2–VII.4] and the level-1 large deviation property will be derived as a corollary [Section VII.5]. The results on exponential convergence of random vectors stated in Theorem II.6.3 will be proved in Section VII.6.

Let us recall the definitions in Section II.6. $\mathscr{W} = \{W_n; n = 1, 2, \ldots\}$ is a sequence of random vectors which are defined on probability spaces $\{(\Omega_n, \mathscr{F}_n, P_n); n = 1, 2, \ldots\}$ and which take values in $\mathbb{R}^d$. We define functions

$$(7.1) \qquad c_n(t) = \frac{1}{a_n} \log E_n\{\exp\langle t, W_n\rangle\}, \qquad n = 1, 2, \ldots, \quad t \in \mathbb{R}^d,$$

where $\{a_n; n = 1, 2, \ldots\}$ is a sequence of positive numbers tending to infinity, E_n denotes expectation with respect to P_n, and $\langle -,- \rangle$ denotes the Euclidean inner product on $\mathbb{R}^d$. The following hypotheses are assumed to hold.

(a) Each function $c_n(t)$ is finite for all $t \in \mathbb{R}^d$.

(b) $c_{\mathscr{W}}(t) = \lim_{n\to\infty} c_n(t)$ exists for all $t \in \mathbb{R}^d$ and is finite.

The function $c_{\mathscr{W}}(t)$ is called the *free energy function of* $\mathscr{W}$.

Proposition VII.1.1. *$c_n(t)$ and $c_{\mathscr{W}}(t)$ are convex functions on $\mathbb{R}^d$.*

Proof. For every t_1 and t_2 in $\mathbb{R}^d$ and $0 < \lambda < 1$, Hölder's inequality with $p = 1/\lambda$ and $q = 1/(1-\lambda)$ implies that

$$(7.2) \qquad \begin{aligned} &E_n\{\exp\langle \lambda t_1 + (1-\lambda)t_2, W_n\rangle\} \\ &\qquad \le [E_n\{\exp\langle t_1, W_n\rangle\}]^{\lambda} \cdot [E_n\{\exp\langle t_2, W_n\rangle\}]^{1-\lambda}. \end{aligned}$$

It follows that $c_n(t)$ is convex. Since convexity is preserved under pointwise limits, $c_{\mathscr{W}}(t)$ is convex. $\square$

The next example discusses an important special case of free energy functions.

Example VII.1.2. Let W_n be the nth partial sum of i.i.d. random vectors $X_1, X_2, \ldots$ on a probability space $(\Omega, \mathscr{F}, P)$. Assume that X_1 has distribution ρ and that $\int_{\mathbb{R}^d} \exp\langle t, x\rangle \rho(dx)$ is finite for all $t \in \mathbb{R}^d$. Then

$$c_{\mathscr{W}}(t) = \lim_{n\to\infty} \frac{1}{n} \log E\{\exp\langle t, W_n\rangle\} = \log E\{\exp\langle t, X_1\rangle\}$$

$$= \log \int_{\mathbb{R}^d} \exp\langle t, x\rangle \rho(dx).$$

This function is the free energy function of ρ, $c_\rho(t)$. By Proposition VII.1.1, c_ρ is convex on $\mathbb{R}^d$. Properties of c_ρ and of its Legendre–Fenchel transform (the level-1 entropy function of the distributions of $\{n^{-1}\sum_{i=1}^n X_i\}$) will be determined in Section VII.5.

We return to the general case. Since the convex function $c_{\mathscr{W}}(t)$ is finite for all $t \in \mathbb{R}^d$, $c_{\mathscr{W}}$ is a continuous function on $\mathbb{R}^d$ [Theorem VI.3.1]. Hence $c_{\mathscr{W}}$ is closed. We define the function $I_{\mathscr{W}}(z)$ by the Legendre–Fenchel transform

(7.3) $$I_{\mathscr{W}}(z) = \sup_{t \in \mathbb{R}^d} \{\langle t, z\rangle - c_{\mathscr{W}}(t)\}, \qquad z \in \mathbb{R}^d.$$

The large deviation theorem is stated next.[1]

Theorem II.6.1. *Assume hypotheses* (a) *and* (b) *on page* 229. *Let* Q_n *be the distribution of* W_n/a_n *on* $\mathbb{R}^d$. *Then the following conclusions holds.*

(a) $I_{\mathscr{W}}(z)$ *is convex, closed* (*lower semicontinuous*), *and non-negative.* $I_{\mathscr{W}}(z)$ *has compact level sets and* $\inf_{z \in \mathbb{R}^d} I_{\mathscr{W}}(z) = 0$.

(b) *The upper large deviation bound is valid:*

(7.4) $$\limsup_{n\to\infty} \frac{1}{a_n} \log Q_n\{K\} \le -\inf_{z \in K} I_{\mathscr{W}}(z) \qquad \textit{for each closed set } K \textit{ in } \mathbb{R}^d.*$$

(c) *Assume in addition that* $c_{\mathscr{W}}(t)$ *is differentiable for all* t. *Then the lower large deviation bound is valid:*

(7.5) $$\liminf_{n\to\infty} \frac{1}{a_n} \log Q_n\{G\} \ge -\inf_{z \in G} I_{\mathscr{W}}(z) \qquad \textit{for each open set } G \textit{ in } \mathbb{R}^d.$$

Hence, if $c_{\mathscr{W}}(t)$ *is differentiable for all* t, *then* (7.5) *and parts* (a) *and* (b) *imply that* $\{Q_n; n = 1, 2, \ldots\}$ *has a large deviation property with entropy function* $I_{\mathscr{W}}$.

The properties of $I_{\mathscr{W}}$ stated in part (a) will be proved in the next section as direct consequences of the theory of Legendre–Fenchel transforms. The large deviation bounds will be proved first in Section VII.3 for $d = 1$. We save for Section VII.4 the much harder proofs of these bounds for $d \ge 1$.

The large deviation theorem is closely related to convergence properties

* See footnote on page 36.

of $\{W_n/a_n\}$. Let z_0 be a point in $\mathbb{R}^d$. We say that W_n/a_n *converges exponentially to* z_0, and write $W_n/a_n \overset{\exp}{\to} z_0$, if for any $\varepsilon > 0$ there exists a number $N = N(\varepsilon) > 0$ such that

$$P_n\{\|W_n/a_n - z_0\| \geq \varepsilon\} \leq e^{-a_n N} \qquad \text{for all sufficiently large } n.$$

Theorem II.6.3. *Assume hypotheses* (a) *and* (b) *on page* 229. *Then the following statements are equivalent.*

(a) $W_n/a_n \overset{\exp}{\to} z_0$.
(b) $c_{\mathscr{W}}(t)$ *is differentiable at* $t = 0$ *and* $\nabla c_{\mathscr{W}}(0) = z_0$.
(c) $I_{\mathscr{W}}(z)$ *attains its infimum over* $\mathbb{R}^d$ *at the unique point* $z = z_0$.

If part (a) holds, then $W_n(\omega)/a_n$ is close to z_0 for all but at most an exponentially small set of microstates ω. We may think of z_0 as the (necessarily unique) equilibrium state corresponding to the distributions $\{Q_n\}$ of $\{W_n/a_n\}$. Theorem II.6.3 will be proved in Section VII.6.

VII.2. Properties of $I_{\mathscr{W}}$

As a preliminary to proving the upper and lower large deviation bounds in Theorem II.6.1, we establish properties of the function $I_{\mathscr{W}}$. We write $c(t)$ for $c_{\mathscr{W}}(t)$ and $I(z)$ for $I_{\mathscr{W}}(z)$.

Theorem VII.2.1. *Assume hypotheses* (a) *and* (b) *on page* 229. *Then the following conclusions hold.*

(a) $I(z)$ *is a closed convex function on* $\mathbb{R}^d$.
(b) $\langle t, z\rangle \leq c(t) + I(z)$ *for all* t *and* z *in* $\mathbb{R}^d$.
(c) $\langle t, z\rangle = c(t) + I(z)$ *if and only if* $z \in \partial c(t)$.
(d) $z \in \partial c(t)$ *if and only if* $t \in \partial I(z)$.
(e) $c(t) = \sup_{z \in \mathbb{R}^d}\{\langle t, z\rangle - I(z)\}$ *for all* $t \in \mathbb{R}^d$.
(f) $I(z)$ *has compact level sets.*
(g) *The infimum of* $I(z)$ *over* $\mathbb{R}^d$ *is* 0, *and* $I(z_0) = 0$ *if and only if* z_0 *is in* $\partial c(0)$. *The latter is a nonempty, closed, bounded, convex subset of* $\mathbb{R}^d$.
(h) *The function* $c(t)$ *is differentiable for all* t *if and only if* $I(z)$ *is essentially strictly convex. In particular, if* $c(t)$ *is differentiable for all* t, *then for* $d = 1$, $I(z)$ *is strictly convex and for* $d \geq 2$, $I(z)$ *is strictly convex on* ri(dom I).

Proof. (a)–(e) Since $c(t)$ is a closed convex function, Theorem VI.5.3 may be applied.

(f) Consider a level set $K_b = \{z \in \mathbb{R}^d : I(z) \leq b\}$, where b is a real number. K_b is closed since $I(z)$ is lower semicontinuous. If z is in K_b, then for any $t \in \mathbb{R}^d$

$$\langle t, z\rangle \leq c(t) + I(z) \leq c(t) + b.$$

Hence for $R > 0$, there exists a constant A independent of $z \in K_b$ such that

$$\sup_{\|t\| \le R} \langle t, z \rangle = R\|z\| \le \sup_{\|t\| \le R} c(t) + b \le A < \infty.$$

This implies that K_b is bounded.

(g) By the definition of subdifferential, $I(z)$ attains its infimum at z_0 if and only if 0 is in $\partial I(z_0)$. The point 0 is in $\partial I(z_0)$ if and only if z_0 is in $\partial c(0)$ [part (d)] and $I(z_0) = 0$ for any $z_0 \in \partial c(0)$ [part (c)]. The subdifferential $\partial c(0)$ is a nonempty, closed, bounded, convex set [Theorem VI.3.5(b)–(c)].

(h) This follows from Theorem VI.5.6 and the fact that for $d = 1$, the notions of essentially strictly convex and strictly convex coincide. □

Part (a) of Theorem II.6.1 states that $I_{\mathscr{W}}(z)$ is convex, closed, and nonnegative, $I_{\mathscr{W}}(z)$ has compact level sets, and $\inf_{z \in \mathbb{R}^d} I_{\mathscr{W}}(z) = 0$. These assertions follow from parts (a), (f), and (g) of Theorem VII.2.1.

VII.3. Proof of the Large Deviation Bounds for $d = 1$

We write $c(t)$ for $c_{\mathscr{W}}(t)$ and $I(z)$ for $I_{\mathscr{W}}(z)$. If A is a nonempty subset of $\mathbb{R}$, then $I(A)$ denotes the infimum of I over A. $I(\phi)$ equals ∞.

Proof of the upper large deviation bound. We show that if K is any nonempty closed subset of $\mathbb{R}$, then

$$\limsup_{n \to \infty} \frac{1}{a_n} \log Q_n\{K\} \le -I(K). \tag{7.6}$$

For the empty set, (7.6) holds trivially. Let $c'_+(0)$ and $c'_-(0)$ denote the right-hand and left-hand derivatives, respectively, of c at 0. We first prove (7.6) for the closed half intervals $[\alpha, \infty)$, $\alpha > c'_+(0)$. For any $t > 0$ and $\alpha > c'_+(0)$, Chebyshev's inequality implies

$$Q_n\{[\alpha, \infty)\} \le \exp(-a_n t\alpha) E_n\{\exp(tW_n)\} = \exp[-a_n(t\alpha - c_n(t))],$$

where $c_n(t) = a_n^{-1} \log E_n\{\exp(tW_n)\}$. Since $t > 0$ is arbitrary,

$$\limsup_{n \to \infty} \frac{1}{a_n} \log Q_n\{[\alpha, \infty)\} \le \inf_{t > 0}\{-(t\alpha - c(t))\} = -\sup_{t>0}\{t\alpha - c(t)\}. \tag{7.7}$$

Lemma VII.3.1. *If* $\alpha > c'_+(0)$, *then* $\sup_{t>0}\{t\alpha - c(t)\} = I(\alpha) = \inf_{z \ge \alpha} I(z)$.

Proof. Since $c(t)$ is continuous at $t = 0$,

$$I(\alpha) = \sup_{t \in \mathbb{R}}\{t\alpha - c(t)\} = \sup_{t \ne 0}\{t\alpha - c(t)\}.$$

Since $c(t)$ is convex, $c'_-(0) \ge c(t)/t$ for any $t < 0$. Therefore, for any $t < 0$,

$$t\alpha - c(t) = t(\alpha - c(t)/t) \le t(\alpha - c'_-(0)) < 0 = 0 \cdot \alpha - c(0).$$

The second inequality holds since $\alpha > c'_+(0) \ge c'_-(0)$. From this display, we see that the supremum in the formula for $I(\alpha)$ cannot occur for $t < 0$. It follows that

$$I(\alpha) = \sup_{t>0}\{t\alpha - c(t)\}.$$

$I(z)$ is a non-negative, closed, convex function which is 0 on the interval $\partial c(0) = [c'_-(0), c'_+(0)]$. Thus $I(z)$ is nonincreasing for $z \leq c'_-(0)$ and is nondecreasing for $z \geq c'_+(0)$. This means that $I(\alpha) = \inf\{I(z): z \geq \alpha\}$. □

Inequality (7.7) and the lemma imply that if K is the closed interval $[\alpha, \infty)$, $\alpha > c'_+(0)$, then (7.6) holds:

$$\limsup_{n\to\infty} \frac{1}{a_n} \log Q_n\{K\} \leq -I(\alpha) = -I(K).$$

A similiar proof yields (7.6) if $K = (-\infty, \alpha]$, $\alpha < c'_-(0)$.

Let K be an arbitrary nonempty closed set. If $K \cap [c'_-(0), c'_+(0)]$ is nonempty, then $I(K)$ equals 0 and (7.6) holds trivially since $\log Q_n\{K\}$ is always non-positive. If $K \cap [c'_-(0), c'_+(0)]$ is empty, then let (α_1, α_2) be the largest open interval which contains $[c'_-(0), c'_+(0)]$ and which has empty intersection with K. K is a subset of $(-\infty, \alpha_1] \cup [\alpha_2, \infty)$ and by the first part of the proof

$$\begin{aligned}\limsup_{n\to\infty} \frac{1}{a_n} \log Q_n\{K\} &\leq \limsup_{n\to\infty} \frac{1}{a_n} \log [Q_n\{(-\infty, \alpha_1]\} + Q_n\{[\alpha_2, \infty)\}] \\ &= -\min\{I(\alpha_1), I(\alpha_2)\}.\end{aligned}$$

If $\alpha_1 = -\infty$ or $\alpha_2 = \infty$, then the corresponding term is missing. From the monotonicity properties of $I(z)$ on $(-\infty, c'_-(0)]$ and on $[c'_+(0), \infty)$,

$$I(K) = \min(I(\alpha_1), I(\alpha_2)).$$

We conclude that $\limsup_{n\to\infty} a_n^{-1} \log Q_n\{K\} \leq -I(K)$. This completes the proof of the upper large deviation bound for $d = 1$.

Proof of the lower large deviation bound. Under the hypothesis that $c(t)$ is differentiable for all t real, we want to prove that

$$\text{(7.8)} \quad \liminf_{n\to\infty} \frac{1}{a_n} \log Q_n\{G\} \geq -I(G) \quad \text{for each nonempty open set } G \text{ in } \mathbb{R}.$$

For the empty set, (7.8) holds trivially. In order to simplify the proof, we shall also assume that the range of $c'(t)$ is all of $\mathbb{R}$.* Let z be any point in G. Pick $\varepsilon > 0$ such that the interval $A_{z,\varepsilon} = (z - \varepsilon, z + \varepsilon)$ is a subset of G. For any t real, we define probability measures

$$\text{(7.9)} \qquad Q_{n,t}(dx) = \exp(a_n t x) Q_n(dx) \cdot \frac{1}{Z_n(t)}, \qquad n = 1, 2, \ldots,$$

where $Z_n(t)$ is the normalization $\int_{\mathbb{R}} \exp(a_n t x) Q_n(dx) = \exp[a_n c_n(t)]$. If x is a point in $A_{z,\varepsilon}$, then $-tx \geq -tz - |t|\varepsilon$. Hence we have

$$\begin{aligned}Q_n\{G\} \geq Q_n\{A_{z,\varepsilon}\} &= Z_n(t) \int_{A_{z,\varepsilon}} \exp(-a_n t x) Q_{n,t}(dx) \\ &\geq \exp[a_n(c_n(t) - tz) - a_n|t|\varepsilon] \cdot Q_{n,t}\{A_{z,\varepsilon}\},\end{aligned}$$

*This hypothesis will be removed in the next section.

$$\text{(7.10)} \quad \liminf_{n\to\infty} \frac{1}{a_n} \log Q_n\{G\} \geq c(t) - tz - |t|\varepsilon + \liminf_{n\to\infty} \frac{1}{a_n} \log Q_{n,t}\{A_{z,\varepsilon}\}.$$

By hypothesis on the range of $c'(t)$, there exists a point t such that $c'(t) = z$. Substituting this t in (7.10), we have $c(t) - tz = -I(z)$ [Theorem VII.2.1(c)]. Below we prove that

$$\lim_{n\to\infty} Q_{n,t}\{A_{z,\varepsilon}\} = 1,$$

from which it follows that the last term in (7.10) equals 0. Taking $\varepsilon \to 0$ yields

$$\liminf_{n\to\infty} \frac{1}{a_n} \log Q_n\{G\} \geq -I(z).$$

Since $z \in G$ is arbitrary, we may replace $-I(z)$ by $\sup\{-I(z): z \in G\} = -I(G)$. This yields (7.8).

Lemma VII.3.2. *If $c'(t) = z$ and ε is positive, then $Q_{n,t}\{A_{z,\varepsilon}\} \to 1$ as $n \to \infty$.*

Proof. This is a consequence of Theorem II.6.3. Let $\mathscr{W}_t = \{W_{n,t}; n = 1, 2, \ldots\}$ be a sequence of random variables such that $W_{n,t}/a_n$ is distributed by $Q_{n,t}$. We calculate the free energy function of $\mathscr{W}_t$. For any real number s

$$\begin{aligned} c_{\mathscr{W}_t}(s) &= \lim_{n\to\infty} \frac{1}{a_n} \log \int_{\mathbb{R}} \exp(a_n s x) Q_{n,t}(dx) \\ &= \lim_{n\to\infty} \frac{1}{a_n} \log \left\{ \int_{\mathbb{R}} \exp[a_n(s+t)x] Q_n(dx) \cdot \frac{1}{\exp[a_n c_n(t)]} \right\} \\ &= c(s+t) - c(t). \end{aligned}$$

Since c is differentiable at t, $c_{\mathscr{W}_t}(s)$ is differentiable at $s = 0$ and $c'_{\mathscr{W}_t}(0) = c'(t) = z$. By the implication (b) $\Rightarrow$ (a) in Theorem II.6.3,

$$W_{n,t}/a_n \overset{\exp}{\to} c'(0) = z.$$

This implies that $Q_{n,t}\{A_{z,\varepsilon}\} \to 1$ as $n \to \infty$. □

The proof of the large deviation bounds for $d = 1$ indicates a general pattern of proof for many other large deviation estimates. The proof of the upper bound relies on a global estimate which is Chebyshev's inequality. By contrast, the proof of the lower bound is a local estimate, and it may be interpreted physically as follows. According to Theorem II.6.3, if c is differentiable at $t = 0$, then the macrostate $z_0 = c'(0)$ is the unique equilibrium state corresponding to the distributions $\{Q_n\}$. We have $W_n/a_n \overset{\exp}{\to} z_0$, and so in particular $Q_n\{A_{z_0,\varepsilon}\} \to 1$ for any $\varepsilon > 0$. If z differs from the equilibrium state $z_0 = c'(0)$, then $Q_n\{A_{z,\varepsilon}\}$ decays exponentially for $0 < \varepsilon < |z - z_0|$. According to Lemma VII.3.2, if t satisfies $c'(t) = z$, then z is the equilibrium state corresponding to the distributions $\{Q_{n,t}\}$. Since

$$\lim_{n\to\infty} \frac{1}{a_n} \log \frac{dQ_{n,t}}{dQ_n}(z) = tz - c(t) = I(z),$$

$I(z)$ measures the error made in replacing $\{Q_n\}$ by $\{Q_{n,t}\}$.

VII.4. Proof of the Large Deviation Bounds for $d \geq 1$

Proof of the upper large deviation bound. The same notation as in the previous section is used. We show that if K is any nonempty closed set in $\mathbb{R}^d$, then

$$\limsup_{n\to\infty} \frac{1}{a_n} \log Q_n\{K\} \leq -I(K). \tag{7.11}$$

If $I(K) = 0$, then the bound is obvious, so we assume that $I(K)$ is positive. For $d = 1$, we proved the upper bound first for closed half intervals, then for arbitrary closed sets. For $d \geq 1$, the half intervals are replaced by certain half-spaces. The following lemma is the key.

Lemma VII.4.1. *Let K be a nonempty closed set in $\mathbb{R}^d$. Given a nonzero point t in $\mathbb{R}^d$ and a real number α, define the closed half-space*

$$H_+(t, \alpha) = \{z \in \mathbb{R}^d : \langle t, z\rangle - c(t) \geq \alpha\}.$$

(a) *If $0 < I(K) < \infty$, then for any $0 < \varepsilon < I(K)$, there exist finitely many nonzero points $t_1, \ldots, t_r$ in $\mathbb{R}^d$ such that*

$$K \subseteq \bigcup_{i=1}^{r} H_+(t_i, I(K) - \varepsilon). \tag{7.12}$$

(b) *If $I(K) = \infty$, then for any $R > 0$ there exist finitely many nonzero points $t_1, \ldots, t_r$ in $\mathbb{R}^d$ such that*

$$K \subseteq \bigcup_{i=1}^{r} H_+(t_i, R). \tag{7.13}$$

The lemma will be proved in a moment. If $0 < I(K) < \infty$, then the upper large deviation bound (7.11) follows from (7.12). Indeed Chebyshev's inequality implies that

$$\begin{aligned} Q_n\{K\} &\leq \sum_{i=1}^{r} P_n\{W_n/a_n \in H_+(t_i, I(K) - \varepsilon)\} \\ &= \sum_{i=1}^{r} P_n\{\langle t_i, W_n\rangle \geq a_n[c(t_i) + I(K) - \varepsilon]\} \\ &\leq \sum_{i=1}^{r} \exp\{a_n[c_n(t_i) - c(t_i) - I(K) + \varepsilon]\}. \end{aligned}$$

This yields (7.11) since $c_n(t_i) \to c(t_i)$ as $n \to \infty$ and $\varepsilon \in (0, I(K))$ is arbitrary.

Similarly, if $I(K) = \infty$, then (7.11) follows from (7.13) since $R > 0$ is arbitrary.

Proof of Lemma VII.4.1. (a) Define the set $A = \{z \in \mathbb{R}^d : I(z) \le I(K) - \varepsilon\}$, where $0 < \varepsilon < I(K)$. A is nonempty, closed, and bounded [Theorem VII.2.1(f)–(g)], and it is disjoint from K. Let B be a closed ball containing A such that the boundary of B (bd B) and A are disjoint. Define $C = (K \cap B) \cup \text{bd}\, B$. We shall find finitely many nonzero points $t_1, \ldots, t_r$ in $\mathbb{R}^d$ such that

$$C \subseteq \bigcup_{i=1}^{r} H_+(t_i, I(K) - \varepsilon). \tag{7.14}$$

Afterwards we prove that (7.14) implies (7.12).

Write $H_+(t)$ for the closed half space $H_+(t, I(K) - \varepsilon)$ and $H_-(t)$ for the opposite closed half-space

$$H_-(t) = \{z \in \mathbb{R}^d : \langle t, z \rangle - c(t) \le I(K) - \varepsilon\}.$$

Since $c(t)$ is continuous at $t = 0$, we have for any $z \in \mathbb{R}^d$

$$I(z) = \sup_{t \in \mathbb{R}^d} \{\langle t, z \rangle - c(t)\} = \sup_{t \neq 0} \{\langle t, z \rangle - c(t)\}.$$

It follows that $A = \bigcap_{t \neq 0} H_-(t)$ and thus that $A^c = \bigcup_{t \neq 0} \text{int}\, H_+(t)$. Since C is a subset of A^c and C is compact, there exist finitely many nonzero points $t_1, \ldots, t_r$ such that

$$C \subseteq \bigcup_{i=1}^{r} \text{int}\, H_+(t_i) \subseteq \bigcup_{i=1}^{r} H_+(t_i).$$

We now prove (7.12). The set K equals $(K \cap B) \cup (K \cap B^c)$, and this is a subset of $C \cup B^c$. Because of (7.14), it suffices to prove that any point $x \in B^c$ belongs to $\bigcup_{i=1}^r H_+(t_i)$. We argue by contradiction. If there exists a point $x \in B^c$ which does not belong to $\bigcup_{i=1}^r H_+(t_i)$, then x lies in $\bigcap_{i=1}^r \text{int}\, H_-(t_i)$. Pick any point $\theta \in A$. Since A is a subset of $\bigcap_{i=1}^r H_-(t_i)$ and since each int $H_-(t_i)$ is convex, the set

$$(\theta, x] = \{z \in \mathbb{R}^d : z = \lambda\theta + (1 - \lambda)x, 0 \le \lambda < 1\}$$

is a subset of $\bigcap_{i=1}^r \text{int}\, H_-(t_i)$ [Rockafellar (1970, page 45)]. B was defined as a closed ball containing A whose boundary is disjoint from A. Since x lies in B^c and θ lies in A, the set $(\theta, x]$ must intersect bd B in some point b, and thus b belongs to $\bigcap_{i=1}^r \text{int}\, H_-(t_i)$. As this contradicts (7.14), the proof of part (a) is done.

(b) Repeat the proof of part (a) with $A = \{z \in \mathbb{R}^d : I(z) \le R\}$ and closed half-spaces $H_+(t, R) = [z \in \mathbb{R}^d : \langle t, z \rangle - c(t) \ge R\}$, $H_-(t, R) = \{z \in \mathbb{R}^d : \langle t, z \rangle - c(t) \le R\}$. □

Proof of the lower large deviation bound. We need the analog of Lemma VII.3.2, which follows from the implication (b) ⇒ (a) in Theorem II.6.3. This implication is proved first. The proof depends on the upper large deviation bound which has just been proved.

Lemma VII.4.2. (a) *If $c(t)$ is differentiable at $t = 0$, then $W_n/a_n \overset{\exp}{\to} z_0 = \nabla c(0)$.*

(b) *For $t \in \mathbb{R}^d$, define the probability measures*

$$Q_{n,t}(dx) = \exp(a_n\langle t, x\rangle)Q_n(dx) \cdot \frac{1}{\exp[a_n c_n(t)]}, \qquad n = 1, 2, \ldots.$$

Assume that c is differentiable at t and let $A_{z,\varepsilon}$ be the open ball of radius $\varepsilon > 0$ centered at $z = \nabla c(t)$. Then

$$Q_{n,t}\{A_{z,\varepsilon}\} \to 1 \qquad \text{as } n \to \infty.$$

Proof. (a) Since $c(t)$ is differentiable at $t = 0$, the set $\partial c(0)$ consists of the unique point $z_0 = \nabla c(0)$ [Theorem VI.3.5(d)]. By Theorem VII.2.1(g), z_0 is the unique minimum point of I ($I(z) > I(z_0) = 0$ for $z \neq z_0$). We have shown that I is a lower semicontinuous function on $\mathbb{R}^d$, that it has compact level sets, and that the upper large deviation bound is valid. These are hypotheses (a)–(c) of large deviation property. Hence Theorem II.3.3 implies that $W_n/a_n \overset{\exp}{\to} z_0$.

(b) Let $\mathscr{W}_t = \{W_{n,t}; n = 1, 2, \ldots\}$ be a sequence of random vectors such that $W_{n,t}/a_n$ is distributed by $Q_{n,t}$. As in the proof of Lemma VII.3.2, the free energy function $c_{\mathscr{W}_t}(s)$ of $\mathscr{W}_t$ equals $c(s + t) - c(t)$. Since c is differentiable at t, $c_{\mathscr{W}_t}(s)$ is differentiable at $s = 0$ and

$$\nabla c_{\mathscr{W}_t}(0) = \nabla c(t) = z.$$

By part (a), $W_{n,t}/a_n \overset{\exp}{\to} z$. This implies that $Q_{n,t}\{A_{z,\varepsilon}\} \to 1$ as $n \to \infty$. □

We show that if $c(t)$ is differentiable for all t, then for any nonempty open set G in $\mathbb{R}^d$

$$\liminf_{n\to\infty} \frac{1}{a_n} \log Q_n\{G\} \geq -I(G). \tag{7.15}$$

The effective domain of I, $\operatorname{dom} I$, has nonempty relative interior $\operatorname{ri}(\operatorname{dom} I)$. Since $I(z)$ equals ∞ for $z \notin \operatorname{dom} I$, $I(G)$ equals ∞ if $G \cap \operatorname{dom} I$ is empty. In this case the lower bound (7.15) is valid. If $G \cap \operatorname{dom} I$ is nonempty, then $I(G)$ equals $I(G \cap \operatorname{dom} I)$. The set $G \cap \operatorname{ri}(\operatorname{dom} I)$ is also nonempty* and by the continuity property of I expressed in Theorem VI.3.2,

$$I(G \cap \operatorname{dom} I) = I(G \cap \operatorname{ri}(\operatorname{dom} I)).$$

Hence it suffices to prove that

$$\liminf_{n\to\infty} \frac{1}{a_n} \log Q_n\{G\} \geq -I(G \cap \operatorname{ri}(\operatorname{dom} I)), \tag{7.16}$$

* See Rockafellar (1970, page 46).

where $G \cap \text{ri}(\text{dom } I)$ is nonempty. Let z be any point in $G \cap \text{ri}(\text{dom } I)$. Pick $\varepsilon > 0$ such that $A_{z,\varepsilon}$, the open ball of radius ε centered at z, is contained in G, and pick any $t \in \mathbb{R}^d$. For any point $x \in A_{z,\varepsilon}$, $-\langle t, x\rangle \geq -\langle t, z\rangle - \|t\|\varepsilon$. Hence, if $Q_{n,t}(dx)$ denotes the measure in Lemma VII.4.2(b), then

$$Q_n\{G\} \geq Q_n\{A_{z,\varepsilon}\} = \exp(a_n c_n(t)) \int_{A_{z,\varepsilon}} \exp(-a_n\langle t, x\rangle) Q_{n,t}(dx),$$

$$\liminf_{n\to\infty} \frac{1}{a_n} \log Q_n\{G\} \geq c(t) - \langle t, z\rangle - \|t\|\varepsilon + \liminf_{n\to\infty} \frac{1}{a_n} \log Q_{n,t}\{A_{z,\varepsilon}\}.$$

(7.17)

Since $c(t)$ is finite and differentiable on all of $\mathbb{R}^d$, ri(dom I) is a subset of the range of $\nabla c(t)$, $t \in \mathbb{R}^d$ [Theorem VI.5.7]. Thus we can find a point t such that $\nabla c(t) = z$. Substituting this t in (7.17), we have $c(t) - \langle t, z\rangle = -I(z)$ and $Q_{n,t}\{A_{z,\varepsilon}\} \to 1$ as $n \to \infty$. Hence the last term in (7.17) equals 0. Taking $\varepsilon \to 0$ yields

$$\liminf_{n\to\infty} \frac{1}{a_n} \log Q_n\{G\} \geq -I(z). \tag{7.18}$$

Since z is an arbitrary point in $G \cap \text{ri}(\text{dom } I)$, we may replace $-I(z)$ by $-I(G \cap \text{ri}(\text{dom } I))$, and we obtain (7.16). This completes the proof of the lower large deviation bound (7.15).

Here is another proof when dom I consists of a single point z_0. $I(z_0)$ equals 0, $c(t)$ equals $\langle t, z_0\rangle$ for all t, and $\nabla c(0)$ equals z_0. By Lemma VII.4.2(a), $W_n/a_n \overset{\exp}{\to} z_0$. Thus for any open set G containing z_0, $Q_n\{G\} \to 1$ as $n \to \infty$ and

$$\liminf_{n\to\infty} \frac{1}{a_n} \log Q_n\{G\} = 0 = -I(G).$$

If G is a nonempty open set which does not contain z_0, then

$$\liminf_{n\to\infty} \frac{1}{a_n} \log Q_n\{G\} \geq -\infty = -I(G).$$

VII.5. Level-1 Large Deviations for I.I.D. Random Vectors[2]

As a consequence of Theorem II.6.1, we will derive the level-1 large deviation property stated in Theorem II.4.1(a).

Theorem II.4.1. (a) *Let $X_1, X_2, \ldots,$ be a sequence of* i.i.d. *random vectors taking values in $\mathbb{R}^d$ and let $\rho \in \mathcal{M}(\mathbb{R}^d)$ be the distribution of X_1. For $t \in \mathbb{R}^d$, define the free energy function*

$$c_\rho(t) = \log E\{\exp\langle t, X_1\rangle\} = \log \int_{\mathbb{R}^d} \exp\langle t, x\rangle \rho(dx).$$

Assume that $c_\rho(t)$ is finite for all t and define

$$I_\rho^{(1)}(z) = \sup_{t\in\mathbb{R}^d} \{\langle t, z\rangle - c_\rho(t)\} \qquad \text{for } z \in \mathbb{R}^d.$$

Let S_n denote the nth *partial sum* $\sum_{j=1}^{n} X_j$, $n = 1, 2, \ldots$. *Then* $\{Q_n^{(1)}\}$, *the distributions of* $\{S_n/n\}$ *on* $\mathbb{R}^d$, *have a large deviation property with* $a_n = n$ *and entropy function* $I_\rho^{(1)}(z)$.

Below we will prove that $c_\rho(t)$ is a real analytic function of $t \in \mathbb{R}^d$. This implies that $c_\rho(t)$ is differentiable for all t.* Hence Theorem II.4.1(a) is a consequence of Theorem II.6.1. After proving the real analyticity of $c_\rho(t)$, we will derive properties of the level-1 entropy function.

Proof of the real analyticity of $c_\rho(t)$. Let $f(x)$ be a real-valued function on an open set A in $\mathbb{R}^d$. We say that f is *real analytic on* A if in some open ball about each point $\tilde{x}$ in A, $f(x)$ is the sum of an absolutely convergent power series

$$f(x) = \sum_{\mathbf{n}} b_{\mathbf{n}}(x_1 - \tilde{x}_1)^{n_1} \cdots (x_d - \tilde{x}_d)^{n_d},$$

where the sum runs over all d-tuples $\mathbf{n} = (n_1, \ldots, n_d)$ of non-negative integers.

Theorem VII.5.1. *Assume that* $c_\rho(t)$ *is finite for all* $t \in \mathbb{R}^d$. *Then the following conclusions hold.*

(a) c_ρ *is a real analytic, closed, convex function on* $\mathbb{R}^d$.

(b) c_ρ *has mixed partial derivatives of all orders which are likewise real analytic functions on* $\mathbb{R}^d$. *These derivatives can be calculated by differentiation under the integral sign. In particular,*

$$\frac{\partial c_\rho(t)}{\partial t_i} = \int_{\mathbb{R}^d} x_i \exp\langle t, x\rangle \rho(dx) \cdot \frac{1}{\int_{\mathbb{R}^d} \exp\langle t, x\rangle \rho(dx)} \qquad \text{for } t \in \mathbb{R}^d.$$

In order to prove the theorem, we need some facts about analytic functions. $\mathbb{C}$ denotes the complex numbers and $\mathbb{C}^d$ the set of all vectors $\xi = (\xi_1, \ldots, \xi_d)$ with each $\xi_i \in \mathbb{C}$. For $\xi \in \mathbb{C}^d$, define $\|\xi\|^2 = \sum_{j=1}^{d} |\xi_j|^2$. We consider $\mathbb{R}^d$ as a subset of $\mathbb{C}^d$ in the obvious way. An open ball in $\mathbb{C}^d$ is a set of the form $B(\tilde{\xi}, r) = \{\xi \in \mathbb{C}^d : \|\xi - \tilde{\xi}\| < r\}$ for some $\tilde{\xi} \in \mathbb{C}^d$ and $r > 0$. $B(\tilde{\xi}, r)$ is called an open ball about $\tilde{\xi}$. A subset U of $\mathbb{C}^d$ is open if U equals a union of open balls. Let $f(\xi)$ be a complex-valued function on an open set U in $\mathbb{C}^d$. We say that f is *analytic on* U if in some open ball about each point $\tilde{\xi}$ in U, $f(\xi)$ is the sum of an absolutely convergent power series

$$f(\xi) = \sum_{\mathbf{n}} b_{\mathbf{n}}(\xi_1 - \tilde{\xi}_1)^{n_1} \cdots (\xi_d - \tilde{\xi}_d)^{n_d},$$

where the sum runs over all d-tuples $\mathbf{n} = (n_1, \ldots, n_d)$ of non-negative integers. The next lemma is proved in Bochner and Martin (1948). Part (a) is Hartogs' theorem on analyticity in each variable.

*That $c_\rho(t)$ is differentiable for all t can be proved directly via the Lebesgue dominated convergence theorem.

Lemma VII.5.2. *Let $f(\xi)$ be a complex-valued function on an open set U in $\mathbb{C}^d$.*

(a) *f is an analytic function on U if and only if for every point ξ in U, each function*

$$f_j(\xi, \zeta_j) = f(\xi_1, \ldots, \xi_{j-1}, \zeta_j, \xi_{j+1}, \ldots, \xi_d), \qquad j = 1, \ldots, d,$$

is an analytic function of the single complex variable ζ_j on some open ball about ξ_j.

(b) *For $\xi \in U$ fixed, the analyticity of $f_j(\xi, \zeta_j)$ as a function of the complex variable ζ_j is equivalent to the one-time differentiability of $f_j(\xi, \zeta_j)$ as a function of ζ_j.*

(c) *If f is analytic on U, then f has mixed partial derivatives of all orders which are likewise analytic on U.*

(d) *Assume that U has nonempty intersection with $\mathbb{R}^d$ and that f maps $U \cap \mathbb{R}^d$ into an open set V in $\mathbb{R}$. If h is a real analytic function on V, then $h(f(x))$ is a real analytic function of $x \in U \cap \mathbb{R}^d$.*

Theorem VII.5.1 will be proved by first considering the *moment generating function* $m_\rho(t)$ of the measure ρ. This function is defined by the formula

$$m_\rho(t) = \int_{\mathbb{R}^d} \exp\langle t, x\rangle \rho(dx) \qquad \text{for } t \in \mathbb{R}^d.$$

We shall study m_ρ as a function of the complex vector $\xi = t + i\eta$, t and η in $\mathbb{R}^d$. The function is well defined since

$$|m_\rho(\xi)| \leq \int_{\mathbb{R}^d} |\exp\langle \xi, x\rangle| \rho(dx) = \int_{\mathbb{R}^d} \exp\langle t, x\rangle \rho(dx) = m_\rho(t).$$

The proof of the next lemma is taken from Barndorff–Nielsen (1978, page 105).

Lemma VII.5.3. *Assume that $m_\rho(t)$ is finite for all t in $\mathbb{R}^d$. Then the following conclusions hold.*

(a) *$m_\rho(\xi)$ is an analytic function on $\mathbb{C}^d$.*

(b) *$m_\rho(\xi)$ has mixed partial derivatives of all orders which are likewise analytic functions on $\mathbb{C}^d$. These derivatives can be calculated by differentiation under the integral sign; that is, if $j_1, \ldots, j_d$ are non-negative integers, then*

$$\frac{\partial^{j_1 + \cdots + j_d} m_\rho(\xi)}{\partial \xi_1^{j_1} \cdots \partial \xi_d^{j_d}} = \int_{\mathbb{R}^d} x_1^{j_1} \cdots x_d^{j_d} \exp\langle \xi, x\rangle \rho(dx) \quad \text{for } \xi \in \mathbb{C}^d. \tag{7.19}$$

Proof. (a) Let $\xi = t + i\eta$ be any point in $\mathbb{C}^d$. If u_j denotes the jth unit coordinate vector and h is a nonzero complex number, then

$$\frac{\partial m_\rho(\xi)}{\partial \xi_j} = \lim_{h \to 0} \frac{1}{h}(m(\xi + hu_j) - m(\xi)) = \lim_{h \to 0} \int_{\mathbb{R}^d} \frac{\exp(hx_j) - 1}{h} \cdot \exp\langle \xi, x\rangle \rho(dx).$$

The interchange of the limit $h \to 0$ and the integral is justified by the Lebesgue dominated convergence theorem and the inequality

$$\left|\frac{\exp(hx_j) - 1}{h}\right| \leq \sum_{m=1}^{\infty} \delta^{m-1}|x_j|^m/m! = \delta^{-1}\exp(\delta|x_j|)$$
$$\leq \delta^{-1}[\exp(\delta x_j) + \exp(-\delta x_j)], \qquad |h| \leq \delta,$$

which is valid for any $x_j \in \mathbb{R}$ and $\delta > 0$. It follows that

$$\frac{\partial m_\rho(\xi)}{\partial \xi_j} = \int_{\mathbb{R}^d} x_j \exp\langle \xi, x\rangle \rho(dx) \tag{7.20}$$

and that $m_\rho(\xi)$ is an analytic function of $\xi_i \in \mathbb{C}$ [Lemma VII.5.2(b)]. By Lemma VII.5.2(a), $m_\rho(\xi)$ is an analytic function of $\xi \in \mathbb{C}^d$.

(b) The first assertion follows from Lemma VII.5.2(c). Formula (7.20) shows that first-order derivatives of m_ρ are calculated by differentiation under the integral sign. The result for general order is proved by induction [Problem VII.8.14]. □

Proof of Theorem VII.5.1. (a) The real analyticity of c_ρ follows from Lemmas VII.5.3(a) and VII.5.2(d) and from the real analyticity of $\log x$ on $(0, \infty)$. The convexity follows from Proposition VII.1.1. Being finite on $\mathbb{R}^d$, c_ρ is continuous on $\mathbb{R}^d$ [Theorem VI.3.1] and is therefore closed.

(b) Lemma VII.5.3(b). □

The following remark was used on page 49.

Remark VII.5.4. The proof of Theorem VII.5.1 also applies to the functions $c_n(t)$ defined in (7.1). In particular, since $c_n(t)$ is finite for all $t \in \mathbb{R}^d$, $c_n(t)$ is a real analytic function on $\mathbb{R}^d$ and

$$\nabla c_n(t) = \frac{1}{a_n} E_n\{W_n \exp\langle t, W_n\rangle\} \cdot \frac{1}{E_n\{\exp\langle t, W_n\rangle\}}.$$

The differentiability of $c_\rho(t)$ combined with Theorem VII.2.1 yields properties of the level-1 entropy function $I_\rho^{(1)}(z) = \sup_{t \in \mathbb{R}^d}\{\langle t, z\rangle - c_\rho(t)\}$.

Theorem VII.5.5. *Assume that $c_\rho(t)$ is finite for all $t \in \mathbb{R}^d$. Then the following conclusions hold.*

(a) *$I_\rho^{(1)}(z)$ is a closed convex function on $\mathbb{R}^d$.*
(b) *$\langle t, z\rangle \leq c_\rho(t) + I_\rho^{(1)}(z)$ for all t and z in $\mathbb{R}^d$.*
(c) *$\langle t, z\rangle = c_\rho(t) + I_\rho^{(1)}(z)$ if and only if $z = \nabla c_\rho(t)$.*
(d) *$z = \nabla c_\rho(t)$ if and only if $t \in \partial I_\rho^{(1)}(z)$.*
(e) *$c_\rho(t) = \sup_{z \in \mathbb{R}^d}\{\langle t, z\rangle - I_\rho^{(1)}(z)\}$ for all $t \in \mathbb{R}^d$.*
(f) *$I_\rho^{(1)}(z)$ has compact level sets.*
(g) *Let $m_\rho = \int_{\mathbb{R}^d} x\rho(dx)$. Then for any $z \in \mathbb{R}^d$, $I_\rho^{(1)}(z) \geq 0$ and $I_\rho^{(1)}(z) = 0$ if and only if $z = m_\rho$.*

(h) $I_\rho^{(1)}(z)$ *is essentially strictly convex. In particular, for* $d = 1$, $I_\rho^{(1)}(z)$ *is strictly convex and for* $d \geq 2$, $I_\rho^{(1)}(z)$ *is strictly convex on* $\operatorname{ri}(\operatorname{dom} I_\rho^{(1)})$.

Proof. (a)–(f), (h) These parts follow from Theorem VII.2.1(a)–(f), (h), respectively.

(g) Since $c_\rho(t)$ is differentiable at $t = 0$, $I_\rho^{(1)}(z)$ attains its infimum over $\mathbb{R}^d$ at the unique point $z = \nabla c_\rho(0)$ [Theorem II.6.3]. According to Theorem VII.5.1(b),

$$\frac{\partial c_\rho(t)}{\partial t_i} = \frac{\int_{\mathbb{R}^d} x_i \exp\langle t, x\rangle \rho(dx)}{\int_{\mathbb{R}^d} \exp\langle t, x\rangle \rho(dx)}, \qquad \frac{\partial c_\rho(0)}{\partial t_i} = \int_{\mathbb{R}^d} x_i \rho(dx).$$

Hence $\nabla c_\rho(0) = m_\rho$. □

This completes our discussion of level-1 large deviations.

VII.6. Exponential Convergence and Proof of Theorem II.6.3

Theorem II.6.3 states that three properties are equivalent.

(a) $W_n/a_n \overset{\exp}{\to} z_0$.
(b) $c_{\mathscr{W}}(t)$ is differentiable at $t = 0$ and $\nabla c_{\mathscr{W}}(0) = z_0$.
(c) $I_{\mathscr{W}}(z)$ attains its infimum over $\mathbb{R}^d$ at the unique point $z = z_0$.

We now prove the theorem. We write $c(t)$ for $c_{\mathscr{W}}(t)$.

Proof. (b) ⇒ (a) Lemma VII.4.2(a).

(b) ⇔ (c) $c(t)$ is differentiable at $t = 0$ if and only if the subdifferential $\partial c(0)$ consists of the unique point $z_0 = \nabla c(0)$ [Theorem VI.3.5(d)]. By Theorem VII.2.1(g), $\partial c(0)$ consists of the unique point z_0 if and only if $I_{\mathscr{W}}(z)$ attains its infimum over $\mathbb{R}^d$ at the unique point $z = z_0$.

(a) ⇒ (b) Let s be an arbitrary unit vector in $\mathbb{R}^d$. Define $D_s^+ c(0)$ and $D_s^- c(0)$ to be the right-hand and left-hand derivatives at 0 of the convex function $\mu \to c(\mu s)$, $\mu \in \mathbb{R}$. By Theorem VI.3.4, it suffices to prove

$$D_s^+ c(0) = D_s^- c(0) = \langle s, z_0\rangle. \tag{7.21}$$

For any nonzero real number μ

$$\begin{aligned} \frac{c_n(\mu s)}{\mu} - \langle s, z_0\rangle &= \frac{1}{a_n \mu} \log E_n\{\exp[\mu\langle s, W_n\rangle]\} - \langle s, z_0\rangle \\ &= \frac{1}{a_n \mu} \log E_n\{\exp[a_n \mu\langle s, V_n\rangle]\}, \end{aligned}$$

where $V_n = a_n^{-1} W_n - z_0$. Given $\varepsilon > 0$, we divide the latter expectation into two parts: the first over the set where $\|V_n\| < \varepsilon$ and the second over the set where $\|V_n\| \geq \varepsilon$. The first is bounded above by $\exp(a_n|\mu|\varepsilon)$. For the second, we have

$$E_n\{\exp[a_n\mu\langle s, V_n\rangle]; \|V_n\| \geq \varepsilon\}$$
$$\leq \exp(-a_n\mu\langle s, z_0\rangle)[E_n\{\exp[2\mu\langle s, W_n\rangle]\}]^{1/2}[P_n\{\|V_n\| \geq \varepsilon\}]^{1/2}.$$

By hypotheses $W_n/a_n \overset{\exp}{\to} z_0$ or $V_n \overset{\exp}{\to} 0$. Putting these facts together, we conclude that for any $\varepsilon > 0$ there exists $N(\varepsilon) > 0$ such that for all $\mu > 0$ and all sufficiently large n

$$(7.22)\qquad \frac{c_n(\mu s)}{\mu} - \langle s, z_0\rangle \leq \frac{1}{a_n\mu}\log\{\exp(a_n|\mu|\varepsilon) + \exp[-a_n\mu\langle s, z_0\rangle + a_n\cdot\tfrac{1}{2}(c_n(2\mu s) - N(\varepsilon))]\}.$$

For $\mu < 0$, the sense of the inequality is reversed.

Below we prove that for any $\varepsilon > 0$ there exist positive numbers $\bar{\mu} = \bar{\mu}(\varepsilon)$ and $\bar{n} = \bar{n}(\varepsilon, \bar{\mu})$ such that for all $|\mu| \leq \bar{\mu}$ and all $n \geq \bar{n}$

$$(7.23)\qquad |\mu|\varepsilon \geq -\mu\langle s, z_0\rangle + \tfrac{1}{2}(c_n(2\mu s) - N(\varepsilon)).$$

Let us accept this. Then for $\varepsilon > 0$, $0 < \mu \leq \bar{\mu}$, and all sufficiently large n, we have from (7.22)

$$\frac{c_n(\mu s)}{\mu} - \langle s, z_0\rangle \leq \varepsilon + \frac{\log 2}{a_n\mu}$$

while for $\varepsilon > 0$, $-\bar{\mu} \leq \mu < 0$, and all sufficiently large n, we have

$$\frac{c_n(\mu s)}{\mu} - \langle s, z_0\rangle \geq -\varepsilon + \frac{\log 2}{a_n\mu}.$$

Taking $n \to \infty$, then $\mu \to 0$, then $\varepsilon \to 0$, we obtain (7.21):

$$\langle s, z_0\rangle \leq D_s^- c(0) \leq D_s^+ c(0) \leq \langle s, z_0\rangle.$$

The function $\mu \to c(2\mu s)$ is continuous and $N(\varepsilon)$ is positive. Hence for any $\varepsilon > 0$ there exists $\bar{\mu} = \bar{\mu}(\varepsilon) > 0$ such that for all $|\mu| \leq \bar{\mu}$

$$|\mu|\varepsilon \geq -\mu\langle s, z_0\rangle + \tfrac{1}{2}c(2\mu s) - \tfrac{1}{4}N(\varepsilon).$$

This implies the inequality (7.23) since the convex functions $\{c_n(2\mu s); n = 1, 2, \ldots\}$ converge uniformly to $c(2\mu s)$ for $|\mu| \leq \bar{\mu}$ [Theorem VI.3.3(b)]. The proof of Theorem II.6.3 is complete. □

This completes our discussion of large deviations for random vectors taking values in $\mathbb{R}^d$. In the next two chapters, we will apply the large deviation result in Theorem II.6.1 to derive the level-2 and level-3 large deviation properties for i.i.d. random variables with a finite state space.

VII.7. Notes

1 (page 230). In Ellis (1984), Theorem II.6.1, II.6.3, and VII.2.1 are proved under less restrictive hypotheses than those given on page 229. Other authors

have studied large deviations for random vectors, but their results differ from Theorem II.6.1. Dacunha–Castelle (1979) treats the case $d = 1$. Steinebach (1978) treats $d \geq 1$, but he is interested in different large deviation estimates and his hypotheses are not the same as ours. See this paper for earlier related references. Gärtner (1977) contains results related to Theorem II.6.1, but he does not prove the large deviation bounds for arbitrary closed and open subsets of $\mathbb{R}^d$. Our proofs of these bounds depend upon Lemmas 1.1 and 1.2 in Gärtner (1977). Section 1 of Orey (1985) contains results related to Theorem II.6.1 as does Section 5.1 of Freidlin and Wentzell (1984). De Acosta (1984) and Ellis (1984) extend the upper large deviation bound (7.4) to certain infinite dimensional problems.

2 (page 238). Let $X_1, X_2, \ldots$ be a sequence of i.i.d. random variables with distribution ρ, mean 0, and variance 1. Define $S_n = X_1 + \cdots + X_n$ and $F_n(x) = P\{S_n/\sqrt{n} \leq x\}$, $n = 1, 2, \ldots$, and let $\Phi(x)$ be the $N(0, 1)$-Gaussian distribution. The central limit theorem implies that for fixed $x > 0$

$$\text{(7.24)} \qquad \frac{1 - F_n(x)}{1 - \Phi(x)} \to 1, \qquad \frac{F_n(-x)}{\Phi(-x)} \to 1 \qquad \text{as } n \to \infty.$$

Khinchin (1929), Cramér (1938, Theorem 1), and Petrov in 1954 [see Petrov (1975, Chapter VIII)] studied the behavior of these ratios when $x = x_n$ tends to infinity with n and $x_n = o(\sqrt{n})$. Linnik (1961) and Feller (1966, Section XVI.6) also made contributions to this problem. Richter (1957, 1958) derived local limit theorems related to the Cramér–Petrov results. Richter's work has been generalized by Chaganty and Sethuraman (1984, 1985) and by other authors listed in the references of these papers.

Large deviations, which correspond to the choice $x = \sqrt{n}y$ in (7.24), are not covered by the results in the previous paragraph. For suitable ρ, Cramér (1938, Theorem 6) proved the limit

$$\lim_{n \to \infty} \frac{1}{n} \log P\{S_n/n \geq y\} = -I_\rho^{(1)}(y) \qquad \text{for} \qquad y > \int_{\mathbb{R}} x\rho(dx).$$

Extensions and related results have been found by many people, including Chernoff (1952), Blackwell and Hodges (1959), Bahadur and Rao (1960), Efron and Truax (1968), Petrov (1975), Höglund (1979), Martin–Löf (1982), Ney (1983) and Bolthausen (1984a). Lanford (1973, Section A.4) proves large deviation limits by using superadditivity properties of the probabilities. His results have been greatly generalized by Bahadur and Zabell (1979). The level-1 large deviation property for i.i.d. random vectors taking values in a Banach space has been proved by Donsker and Varadhan (1976a). Azencott (1980) and Stroock (1984) give another proof which is based upon the ideas of Bahadur and Zabell (1979). Also see de Acosta (1985).

Theorem II.4.1(a) is true under less restrictive hypotheses on c_ρ (see, e.g., Ellis (1984)).

VII.8. Problems

VII.8.1. In Section VII.3, we proved the upper large deviation bound (7.6) for the closed half interval $[\alpha, \infty)$, $\alpha > c'_+(0)$. Prove the upper bound (7.6) for the closed half interval $(-\infty, \alpha]$, $\alpha < c'_-(0)$.

VII.8.2. Let $\{Q_n; n = 1, 2, \ldots\}$ be a sequence of Borel probability measures on $\mathbb{R}^d$ which have a large deviation property with constants $\{a_n\}$ and entropy function $I(z)$. Assume that $c(t) = \lim_{n\to\infty} a_n^{-1} \log \int_{\mathbb{R}^d} \exp(a_n\langle t, z\rangle) Q_n(dz)$ exists and is finite for all $t \in \mathbb{R}^d$. Prove that the Legendre–Fenchel transform of $c(t)$ equals the closed convex hull of $I(z)$. [*Hint:* Theorems II.7.1 and VI.5.8.]

In the next three problems, assume that $\mathscr{W} = \{W_n; n = 1, 2, \ldots\}$ is a sequence of random vectors which satisfy hypotheses (a) and (b) on page 229:

$$c_n(t) = \frac{1}{a_n} \log E_n\{\exp\langle t, W_n\rangle\}$$

is finite for all $t \in \mathbb{R}^d$ and $c_{\mathscr{W}}(t) = \lim_{n\to\infty} c_n(t)$ exists for all $t \in \mathbb{R}^d$ and is finite.

VII.8.3 [Ellis (1981)]. (a) Prove that there exists a random vector X such that $W_{n'}/a_{n'} \xrightarrow{\mathscr{D}} X$ for some subsequence $\{n'\}$ and that the support of X is contained in the set $\partial c_{\mathscr{W}}(0)$. [*Hint:* Prove that $\partial c_{\mathscr{W}}(0)$ is compact and that the distributions $\{Q_n\}$ of $\{W_n/a_n\}$ are tight [Billingsley (1968, page 37)].]

(b) Let s be any nonzero vector in $\mathbb{R}^d$. Prove that for any $\mu > 0$

$$\sup_{n,\pm} E_n\{\exp[\pm\mu\langle s, W_n/a_n\rangle]\} = \sup_{n,\pm} \exp[a_n c_n(\pm\mu s/a_n)] < \infty.$$

[*Hint:* Modify the proof of Lemma V.7.5 to show that

$$\limsup_{n\to\infty} \frac{c_n(\mu s/a_n)}{\mu/a_n} \le D_s^+ c_{\mathscr{W}}(0) \quad \text{and} \quad \liminf_{n\to\infty} \frac{c_n(-\mu s/a_n)}{-\mu/a_n} \ge D_s^- c_{\mathscr{W}}(0),$$

where $D_s^+ c_{\mathscr{W}}(0)$ and $D_s^- c_{\mathscr{W}}(0)$ denote the right-hand and left-hand derivatives at 0 of the function $\lambda \to c_{\mathscr{W}}(\lambda s)$, $\lambda \in \mathbb{R}$.]

(c) Prove that for any nonzero vector s in $\mathbb{R}^d$ and positive integer j

$$E_{n'}\{\langle s, W_{n'}/a_{n'}\rangle^j\} \to E\{\langle s, X\rangle^j\} \qquad \text{as } n' \to \infty,$$

where $\{n'\}$ is the subsequence in part (a).

VII.8.4 [Ellis (1981)]. (a) Suppose that $c_{\mathscr{W}}(t)$ is not differentiable at $t = 0$. Prove that if z is a boundary point of $\partial c_{\mathscr{W}}(0)$ and if $E_n\{W_n/a_n\} \to z$, then $W_n/a_n \xrightarrow{\mathscr{D}} z$ and that the convergence in distribution cannot be strengthened to exponential convergence.

(b) Parts (b) and (c) of Theorem IV.6.6 show that for $\beta > \beta_c$ the spin per site $S_\Lambda/|\Lambda|$ in symmetric intervals Λ of $\mathbb{Z}$ has an almost sure limit with respect to the infinite-volume Gibbs state $P_{\beta,0,+}$, $P_{\beta,0,-}$, or $P_{\beta,0}^{(\lambda)} = \lambda P_{\beta,0,+} +$

$(1-\lambda)P_{\beta,0,-}$ $(0<\lambda<1)$. Using part (a) of the present problem, prove the limits in Theorem IV.6.6(b)–(c) with $\xrightarrow{\text{a.s.}}$ replaced by $\xrightarrow{\mathcal{D}}$; i.e., for each choice of sign

$$S_\Lambda/|\Lambda| \xrightarrow{\mathcal{D}} m(\beta, \pm) \qquad \text{w.r.t.} \quad P_{\beta,0,\pm},$$

$$S_\Lambda/|\Lambda| \xrightarrow{\mathcal{D}} \lambda\delta_{m(\beta,+)} + (1-\lambda)\delta_{m(\beta,-)} \qquad \text{w.r.t.} \quad P_{\beta,0}^{(\lambda)}.$$

VII.8.5 [Ellis (1981)]. Let $d=1$. Assume the following.

(i) The random variables $\{W_n; n=1,2,\ldots\}$ are all defined on the same probability space $(\Omega, \mathscr{F}, P)$.

(ii) The free energy function $c_{\mathscr{W}}(t)$ is not differentiable at $t=0$.

(iii) There exists a random variable X on $(\Omega, \mathscr{F}, P)$ with the property that for any $\varepsilon>0$ there exists a number $N=N(\varepsilon)>0$ such that

$$P\{|W_n/a_n - X| \geq \varepsilon\} \leq e^{-a_n N} \qquad \text{for all sufficiently large } n.$$

Prove that for all sufficiently small $\delta>0$, X has support in the interval $[c'_-(0), c'_-(0)+\delta]$ and in the interval $[c'_+(0)-\delta, c'_+(0)]$.

VII.8.6. Let ρ be a Borel probability measure on $\mathbb{R}^d$. Assume that the set of $t \in \mathbb{R}^d$ where $c_\rho(t)$ is finite is a proper subspace of $\mathbb{R}^d$. Prove that $c_\rho(t)$ is a closed convex function on $\mathbb{R}^d$.

VII.8.7. Let $\{\rho_n; n=1,2,\ldots\}$ be a sequence of Borel probability measures on $\mathbb{R}^d$ which converges weakly to a measure ρ. Assume that $\sup_{n=1,2,\ldots} c_{\rho_n}(t) < \infty$ for each $t \in \mathbb{R}^d$. Let W_n be the sum of n i.i.d. random vectors with distribution ρ_n and define Q_n to be the distribution of W_n/n. Prove that $\{Q_n; n=1,2,\ldots\}$ has a large deviation property with entropy function $I_\rho^{(1)}$. This result has been generalized by Bolthausen (1984b) to measures on a Banach space.

VII.8.8 [Freidlin and Wentzell (1984, page 142)]. Let ρ be a Borel probability measure on $\mathbb{R}^d$ such that $c_\rho(t)$ is finite for all t in some neighborhood of $t=0$ and the Hessian matrix

$$\left\{\frac{\partial^2 c_\rho(0)}{\partial t_i \partial t_j}; \qquad i,j=1,\ldots,d\right\}$$

is nonsingular. Let $X_1, X_2, \ldots$ be a sequence of i.i.d. random vectors with distribution ρ and $\{b_n; n=1,2,\ldots\}$ a sequence of positive real numbers such that $b_n/\sqrt{n} \to \infty$ and $b_n/n \to 0$ as $n\to\infty$. Set $S_n = X_1 + \cdots + X_n$, $n=1,2,\ldots$. Prove that the distributions of $\{(S_n - nE\{X_1\})/b_n\}$ have a large deviation property with $a_n = b_n^2/n$ and entropy function $I(z) = \frac{1}{2}\langle Bz, z\rangle$, $z \in \mathbb{R}^d$, where B is the inverse matrix of

$$\left\{\frac{\partial^2 c_\rho(0)}{\partial t_i \partial t_j}; \qquad i,j=1,\ldots,d\right\}.$$

In the next four problems, S_n is the nth partial sum of i.i.d. random variables with given distribution ρ.

VII.8.9. Let ρ be a nondegenerate Borel probability measure on $\mathbb{R}$ (not a unit point measure) with mean 0. Assume that $c_\rho(t)$ is finite for all $t \in \mathbb{R}$. Let y be a positive real number which is in the range of $c'_\rho(t)$, $t \in \mathbb{R}$. Chernoff (1952) proved that

$$\lim_{n\to\infty} \frac{1}{n} \log P_\rho\{S_n/n \geq y\} = -I_\rho^{(1)}(y) = -\inf_{z \geq y} I_\rho^{(1)}(z). \tag{7.25}$$

The purpose of this problem is to prove (7.25) by a method due to Bahadur and Rao (1960).

(a) Denote by $t = t(y)$ the unique solution of $c'_\rho(t) = y$ [Theorem VIII.3.3(a)]. Prove that

$$P_\rho\left\{\frac{S_n}{n} \geq y\right\}$$
$$= \exp[-nI_\rho^{(1)}(y)] \int_{\{x_1,\ldots,x_n \text{ real}: \sum_{i=1}^n (x_i - y) \geq 0\}} \exp\left[-t \sum_{i=1}^n (x_i - y)\right] \prod_{i=1}^n \rho_t(dx_i)$$

where ρ_t is the Borel probability measure on $\mathbb{R}$ which is absolutely continuous with respect to ρ and whose Radon–Nikodym derivative is given by

$$\frac{d\rho_t}{d\rho}(x) = \exp(tx) \cdot \frac{1}{\int_{\mathbb{R}} \exp(tx)\rho(dx)}.$$

(b) Let $Y_1, Y_2, \ldots$ be a sequence of i.i.d. random variables with distribution $\rho_{t(y)}$ and set $W_i(y) = Y_i - y$. Prove that for any $\varepsilon > 0$

$$\begin{aligned} \gamma(n, y, \varepsilon) \cdot \exp[-nI_\rho^{(1)}(y) - \sqrt{n}\varepsilon t(y)] &\leq P_\rho\left\{\frac{S_n}{n} > y\right\} \leq P_\rho\left\{\frac{S_n}{n} \geq y\right\} \\ &\leq \exp[-nI_\rho^{(1)}(y)], \end{aligned} \tag{7.26}$$

where $\gamma(n, y, \varepsilon) = P\{0 < (W_1(y) + \cdots + W_n(y))/\sqrt{n} \leq \varepsilon\}$.

(c) Deduce the limit (7.25).

VII.8.10. Let ρ be a nondegenerate Borel probability measure on $\mathbb{R}$ with mean 0 and variance σ^2. Assume that $c_\rho(t)$ is finite for all $t \in \mathbb{R}$. By an *order of growth estimate* for S_n, we mean a sequence of positive real numbers $\{\alpha_n; n = 1, 2, \ldots\}$ tending to ∞ such that $P\{S_n > \alpha_n \text{ i.o.}\} = 0$.*

(a) By Theorem VIII.3.4(c), $I_\rho^{(1)}(z)$ is real analytic in a neighborhood of $z = 0$. Prove that $(I_\rho^{(1)})''(0) = 1/\sigma^2$. It follows that for $|z|$ sufficiently small,

$$I_\rho^{(1)}(z) = z^2/(2\sigma^2) + O(z^3).$$

(b) Let $\{\alpha_n; n = 1, 2, \ldots\}$ be a sequence of positive real numbers tending to ∞ such that $\alpha_n/n \to 0$. Using (7.26), prove that for any $\varepsilon > 0$

*i.o. means infinitely often. $P\{S_n > \alpha_n \text{ i.o.}\} = 0$ if and only if $P\left\{\limsup_{n\to\infty} \frac{S_n}{\alpha_n} \leq 1\right\} = 1$.

$$(7.27)\quad P_\rho\{S_n > \alpha_n\} \le p_n(\varepsilon) = \exp\left\{-\frac{\alpha_n^2}{2n\sigma^2}(1-\varepsilon)\right\} \qquad \text{as } n \to \infty.$$

(c) Deduce that $P\{S_n > \alpha_n \text{ i.o.}\} = 0$ if $\alpha_n/n \to 0$ and if for some $\varepsilon > 0 \sum_{n\ge 1} p_n(\varepsilon)$ converges.

The best order of growth estimates are due to Feller [see Chung (1968, page 213)].

VII.8.11. Let ρ be a nondegenerate Borel probability measure on $\mathbb{R}$ with mean 0 and variance σ^2. The law of the iterated logarithm states that

$$\limsup_{n\to\infty} S_n/\alpha_n = 1 \qquad \text{a.s.} \qquad \text{where } \alpha_n = (2\sigma^2 n \log\log n)^{1/2}.$$

According to Lamperti (1966, Section 11), one may prove the law of the iterated logarithm by showing that for arbitrary $\varepsilon > 0$

$$(7.28)\quad \begin{aligned} \exp\left\{-\frac{\alpha_n^2}{2n\sigma^2}(1+\varepsilon)\right\} &\le P\{S_n > \alpha_n\} \\ &\le \exp\left\{-\frac{\alpha_n^2}{2n\sigma^2}(1-\varepsilon)\right\} \qquad \text{as } n \to \infty. \end{aligned}$$

Under the extra assumption that $c_\rho(t)$ is finite for all $t \in \mathbb{R}$, one may deduce these bounds from the previous two problems. Since $\alpha_n \to \infty$ and $\alpha_n/n \to 0$, the upper bound in (7.28) follows from Problem VII.8.10(b). Parts (a)–(c) below yield the lower bound in (7.28).

(a) Let $t(y)$ be as in Problem VII.8.9(a). Prove that $\sqrt{n}t(\alpha_n/n) = O(\alpha_n^2/n)$ as $n \to \infty$.

(b) Let $\gamma(n, y, \varepsilon)$ be the probability $P\{0 < (W_1(y) + \cdots + W_n(y))/\sqrt{n} \le \varepsilon\}$ in (7.26). Prove that for any $\varepsilon > 0$

$$\lim_{n\to\infty} \gamma(n, \alpha_n/n, \varepsilon) = (2\pi\sigma^2)^{-1/2} \int_0^\varepsilon \exp(-x^2/2\sigma^2))dx.$$

[*Hint:* Consider the characteristic function of $(W_1(y) + \cdots + W_n(y))/\sqrt{n}$, $y = \alpha_n/n$.]

(d) Deduce the lower bound in (7.28) from (7.26).

VII.8.12. (a) [Feller (1957, page 166)]. For any $w > 0$ prove the bounds

$$\left(\frac{1}{w} - \frac{1}{w^2}\right)\exp\left(-\frac{1}{2}w^2\right) < \int_w^\infty \exp\left(-\frac{1}{2}x^2\right)dx < \frac{1}{w}\exp\left(-\frac{1}{2}w^2\right).$$

$$\left[\textit{Hint: } \int_w^\infty \left(1 - \frac{3}{x^4}\right)\exp\left(-\frac{1}{2}x^2\right)dx = \left(\frac{1}{w} - \frac{1}{w^3}\right)\exp\left(-\frac{1}{2}w^2\right).\right]$$

(b) Let ρ be a Gaussian probability measure on $\mathbb{R}$ with mean 0 and variance σ^2. For any $y > 0$, prove from part (a)

$$\lim_{n\to\infty}\frac{1}{n}\log P_\rho\left\{\frac{S_n}{n}\geq y\right\} = -I_\rho^{(1)}(y) = -\inf_{z\geq y} I_\rho^{(1)}(z).$$

VII.8.13. Let V be a symmetric, positive definite $d\times d$ matrix. Define on $\mathbb{R}^d$ the Gaussian probability measure

$$\rho(dx) = [(2\pi)^d \det V]^{-1/2}\exp(-\tfrac{1}{2}\langle V^{-1}x, x\rangle)dx.$$

(a) Prove that ρ has mean zero and that its covariance matrix is V ($\int_{\mathbb{R}^d} x_i x_j \rho(dx) = V_{ij}$, $i, j = 1, \ldots, d$). Verify that

$$c_\rho(t) = \tfrac{1}{2}\langle Vt, t\rangle \qquad \text{and} \qquad I_\rho^{(1)}(z) = \tfrac{1}{2}\langle V^{-1}z, z\rangle.$$

(b) For Borel subsets B of $\mathbb{R}^d$, define the measures

$$Q_n\{B\} = \rho\{x\in\mathbb{R}^d : x/\sqrt{n}\in B\}, \qquad n = 1, 2, \ldots.$$

Prove that $\{Q_n\}$ has a large deviation property with $a_n = n$ and entropy function $I_\rho^{(1)}(z) = \frac{1}{2}\langle V^{-1}z, z\rangle$.

(c) For F a bounded continuous function from $\mathbb{R}^d$ to $\mathbb{R}$, prove that

$$\lim_{n\to\infty}\frac{1}{n}\log\int_{\mathbb{R}^d}\exp[nF(x/\sqrt{n})]\rho(dx) = \sup_{z\in\mathbb{R}^d}\{F(z) - \tfrac{1}{2}\langle V^{-1}z, z\rangle\}.$$

For generalizations, see Schilder (1966), Pincus (1968), Donsker and Varadhan (1976a), Simon (1979a, Chapter 18), (1983), Azencott (1980), Ellis and Rosen (1981, 1982a, 1982b) Davies and Truman (1982), Chevet (1983), and Stroock (1984).

VII.8.14. In Lemma VII.5.3(b) prove that derivatives of the moment generating function $m_\rho(t)$ can be calculated by differentiation under the integral sign. Verify formula (7.19).

VII.8.15. Let ρ be a Borel probability measure on $\mathbb{R}^d$. Assume that there exist positive real numbers a and b such that $\int_{\mathbb{R}^d}\exp(a\|x\|^{1+b})\rho(dx) < \infty$. Prove that $c_\rho(t)$ is finite for all $t\in\mathbb{R}^d$ and that there exist positive real numbers α and β such that

$$I_\rho^{(1)}(z) \geq \alpha\|z\|^{1+b} - \beta \qquad \text{for all } z\in\mathbb{R}^d.$$

[*Hint:* Problem VI.7.10].

VII.8.16. Let ρ and ν be Borel probability measures on $\mathbb{R}^d$ such that $c_\rho(t)$ and $c_\nu(t)$ are finite for all $t\in\mathbb{R}^d$. Prove that if $I_\rho^{(1)}(z)$ equals $I_\nu^{(1)}(z)$ for all $z\in\mathbb{R}^d$, then ρ equals ν.

Chapter VIII

Level-2 Large Deviations for I.I.D. Random Vectors

VIII.1. Introduction

Theorem II.4.3 stated the level-2 large deviation property for i.i.d. random vectors taking values in $\mathbb{R}^d$. This theorem follows from the results contained in Donsker and Varadhan (1975a, 1976a), which prove level-2 large deviation properties for Markov processes taking values in a complete separable metric space.[1] In Chapter VIII, we will give an elementary, self-contained proof of Theorem II.4.3 in the special case of i.i.d. random variables with a finite state space. This version of the theorem was applied in Chapter III to study the exponential convergence of velocity observables for the discrete ideal gas with respect to the microcanonical ensemble [Theorem III.4.4].

Theorems II.5.1 and II.5.2 stated contraction principles relating levels-1 and 2 for i.i.d. random vectors taking values in $\mathbb{R}^d$. In the chapters on statistical mechanics, the contraction principles were applied only in the case $d = 1$. Proofs for this case are given in Section VIII.3. We save for Section VIII.4 the more difficult proofs of the contraction principles for $d \geq 2$. Section VIII.4 can be skipped with no loss in continuity.

VIII.2. The Level-2 Large Deviation Theorem

We consider a sequence of i.i.d. random variables $X_1, X_2, \ldots$ with a finite state space Γ. Γ is topologized by the discrete topology. The Borel σ-field $\mathscr{B}(\Gamma)$ of Γ coincides with the set of all subsets of Γ. The empirical measure $L_n(\omega, \cdot) = n^{-1}\sum_{j=1}^n \delta_{X_j(\omega)}(\cdot)$, $n = 1, 2, \ldots$, takes values in the space $\mathscr{M}(\Gamma)$, which is the set of probability measures on $\mathscr{B}(\Gamma)$. The topology on $\mathscr{M}(\Gamma)$ is the topology of weak convergence. The following theorem is Theorem II.4.3 for the case of a finite state space.

Theorem VIII.2.1. *Let $X_1, X_2, \ldots$ be a sequence of* i.i.d. *random variables which take values in a set $\Gamma = \{x_1, x_2, \ldots, x_r\}$ with $x_1 < x_2 < \cdots < x_r$. Let $\rho \in \mathscr{M}(\Gamma)$ be the distribution of X_1; we assume that each $\rho_i = \rho\{x_i\} > 0$. If $\nu = \sum_{i=1}^r \nu_i \delta_{x_i}$ is a probability measure on $\mathscr{B}(\Gamma)$, then define the relative* *entropy of* ν *with respect to* ρ *by the formula*

$$I_\rho^{(2)}(\nu) = \sum_{i=1}^{r} \nu_i \log \frac{\nu_i}{\rho_i}.$$

The following conclusions hold.

(a) $\{Q_n^{(2)}\}$, *the distributions on* $\mathcal{M}(\Gamma)$ *of the empirical measures* $\{L_n\}$, *have a large deviation property with* $a_n = n$ *and entropy function* $I_\rho^{(2)}$.

(b) $I_\rho^{(2)}(\nu)$ *is a convex function of* ν. $I_\rho^{(2)}(\nu)$ *measures the discrepancy between* ν *and* ρ *in the sense that* $I_\rho^{(2)}(\nu) \geq 0$ *with equality if and only if* $\nu = \rho$.

Part (b) of the theorem was proved in Proposition I.4.1. If A is a nonempty subset of $\mathcal{M}(\Gamma)$, then $I_\rho^{(2)}(A)$ denotes the infimum of $I_\rho^{(2)}$ over A. $I(\phi)$ equals ∞. In order to prove part (a) of the theorem, we must verify the following hypotheses.

(i) $I_\rho^{(2)}(\nu)$ is lower semicontinuous on $\mathcal{M}(\Gamma)$.
(ii) $I_\rho^{(2)}(\nu)$ has compact level sets in $\mathcal{M}(\Gamma)$.
(iii) $\limsup_{n\to\infty} n^{-1} \log Q_n^{(2)}\{K\} \leq -I_\rho^{(2)}(K)$ for each closed set K in $\mathcal{M}(\Gamma)$.
(iv) $\liminf_{n\to\infty} n^{-1} \log Q_n^{(2)}\{G\} \geq -I_\rho^{(2)}(G)$ for each open set G in $\mathcal{M}(\Gamma)$.

The set $\mathcal{M}(\Gamma)$ with the topology of weak convergence is homeomorphic to the compact convex subset of $\mathbb{R}^r$ consisting of all vectors $\nu = (\nu_1, \ldots, \nu_r)$ with $\nu_i \geq 0$ and $\sum_{i=1}^r \nu_i = 1$; that is, $\nu_n \Rightarrow \nu$ in $\mathcal{M}(\Gamma)$ if and only if the corresponding vectors converge in $\mathbb{R}^r$. Let $\mathcal{M}$ denote this compact convex subset of $\mathbb{R}^r$. Since $I_\rho^{(2)}(\nu)$ is continuous relative to $\mathcal{M}$, hypotheses (i) and (ii) follow. The vector in $\mathcal{M}$ corresponding to the empirical measure $L_n(\omega, \cdot)$ is the vector $L_n(\omega)$ whose ith component is $L_n(\omega, \{x_i\})$. The next theorem establishes a large deviation property for the distributions $\{\bar{Q}_n^{(2)}\}$ of $\{L_n(\omega)\}$ on $\mathbb{R}^r$. The entropy function $I(\nu)$ equals $I_\rho^{(2)}(\nu)$ for $\nu \in \mathcal{M}$ and equals ∞ for $\nu \in \mathbb{R}^r \backslash \mathcal{M}$. Hence

$$\limsup_{n\to\infty} \frac{1}{n} \log \bar{Q}_n^{(2)}\{K\} \leq -I(K) = -I_\rho^{(2)}(K \cap \mathcal{M}) \quad \text{for each closed set } K \text{ in } \mathbb{R}^r,$$

$$\liminf_{n\to\infty} \frac{1}{n} \log \bar{Q}_n^{(2)}\{G\} \geq -I(G) = -I_\rho^{(2)}(G \cap \mathcal{M}) \quad \text{for each open set } G \text{ in } \mathbb{R}^r.$$

Since the support of $\bar{Q}_n^{(2)}$ is contained in $\mathcal{M}$, $\bar{Q}_n^{(2)}\{A\}$ equals $\bar{Q}_n^{(2)}\{A \cap \mathcal{M}\}$ for any Borel subset A of $\mathbb{R}^r$. As the topology on $\mathcal{M}$ is its relative topology as a subset of $\mathbb{R}^r$, the last display implies that

$$\limsup_{n\to\infty} \frac{1}{n} \log \bar{Q}_n^{(2)}\{K\} \leq -I_\rho^{(2)}(K) \qquad \text{for each closed set } K \text{ in } \mathcal{M},$$

$$\liminf_{n\to\infty} \frac{1}{n} \log \bar{Q}_n^{(2)}\{G\} \geq -I_\rho^{(2)}(G) \qquad \text{for each open set } G \text{ in } \mathcal{M}.$$

These inequalities yield hypotheses (iii) and (iv) above.

Theorem VIII.2.2. *Let $\bar{Q}_n^{(2)}$ be the distribution of the random vector $L_n(\omega)$ on $\mathbb{R}^r$. Then the sequence $\{\bar{Q}_n^{(2)}; n = 1, 2, \ldots\}$ has a large deviation property with $a_n = n$. The free energy function $c(t)$ and the entropy function $I(v)$ are given by*

$$c(t) = \log \sum_{i=1}^{r} e^{t_i}\rho_i \quad \text{and} \quad I(v) = \sup_{t\in\mathbb{R}^r}\{\langle t, v\rangle - c(t)\} = \begin{cases} I_\rho^{(2)}(v) & \text{for } v \in \mathcal{M}, \\ \infty & \text{for } v \notin \mathcal{M}. \end{cases}$$

Proof. We apply the large deviation theorem for random vectors, Theorem II.6.1. If t is a vector in $\mathbb{R}^r$, then define the function $f_t(x_i) = t_i$ for $x_i \in \Gamma$. We have

$$\langle t, L_n(\omega)\rangle = \frac{1}{n}\sum_{i=1}^{r} t_i \sum_{j=1}^{n} \delta_{X_j(\omega)}\{x_i\} = \frac{1}{n}\sum_{j=1}^{n} f_t(X_j(\omega)),$$

which is a sum of i.i.d. random variables. In the notation of Theorem II.6.1, W_n equals nL_n and a_n equals n. The free energy function of the sequence $\{nL_n; n = 1, 2, \ldots\}$ is given by

$$c(t) = \lim_{n\to\infty}\frac{1}{n}\log E\{\exp(n\langle t, L_n\rangle)\} = \log E\{\exp f_t(X_1)\} = \log \sum_{i=1}^{r} e^{t_i}\rho_i.$$

The function $c(t)$ is differentiable for all $t \in \mathbb{R}^r$. Hence the theorem is proved once we identify the Legendre–Fenchel transform $I(v)$ of $c(t)$. The lower large deviation bound states that

$$\liminf_{n\to\infty}\frac{1}{n}\log \bar{Q}_n^{(2)}\{G\} \geq -I(G) \qquad \text{for each open set } G \text{ in } \mathbb{R}^r.$$

Since the support of $\bar{Q}_n^{(2)}$ is contained in $\mathcal{M}$, $I(G)$ equals ∞ whenever G is open and $G \cap \mathcal{M}$ is empty. Thus $I(v)$ equals ∞ for $v \notin \mathcal{M}$. The form of $I(v)$ for $v \in \mathcal{M}$ is given by part (c) of the next lemma.

Lemma VIII.2.3. *Let $\mathcal{N}$ denote the set of vectors $t \in \mathbb{R}^r$ of the form $t_i = \log(z_i/\rho_i)$ for some vector $z \in \operatorname{ri}\mathcal{M}$.* Then the following conclusions hold.*

(a) *$c(t) = \log\sum_{i=1}^r e^{t_i}\rho_i$ equals 0 if and only if t is in $\mathcal{N}$.*

(b) *For any vector t in $\mathbb{R}^r$, $h(t) = t - c(t)\mathbf{1}$ is in $\mathcal{N}$, where $\mathbf{1}$ is the constant vector $(1, \ldots, 1)$.*

(c) *$I(v)$ equals $I_\rho^{(2)}(v) = \sum_{i=1}^r v_i \log(v_i/\rho_i)$ for any $v \in \mathcal{M}$.*

Proof (a) If t is in $\mathcal{N}$, then $\log\sum_{i=1}^r e^{t_i}\rho_i = \log\sum_{i=1}^r z_i = 0$. If $\log\sum_{i=1}^r e^{t_i}\rho_i = 0$, then $e^{t_i} = z_i/\rho_i$, where $z_i = \rho_i e^{t_i} > 0$. Hence t is in $\mathcal{N}$.

(b) $\log\sum_{i=1}^r e^{h(t)_i}\rho_i = \log\sum_{i=1}^r e^{t_i}\rho_i - c(t) = 0$. By part (a), $h(t)$ is in $\mathcal{N}$.

(c) For any $v \in \mathcal{M}$,

$$I(v) = \sup_{t\in\mathbb{R}^r}\{\langle t, v\rangle - c(t)\} \geq \sup_{t\in\mathcal{N}}\{\langle t, v\rangle - c(t)\} = \sup_{t\in\mathcal{N}}\langle t, v\rangle.$$

* $\operatorname{ri}\mathcal{M}$ denotes the relative interior of $\mathcal{M}$, which is the set of $z = (z_1, \ldots, z_r)$ satisfying $z_i > 0$ and $\sum_{i=1}^r z_i = 1$ [Rockafellar (1970, page 48)].

For any $\nu \in \mathcal{M}$, $\langle t, \nu \rangle - c(t) = \langle t - c(t)\mathbf{1}, \nu \rangle = \langle h(t), \nu \rangle$, and since $h(t)$ belongs to $\mathcal{N}$,

$$I(\nu) = \sup_{t \in \mathbb{R}^r} \langle h(t), \nu \rangle \le \sup_{t^0 \in \mathcal{N}} \langle t^0, \nu \rangle.$$

It follows that for any $\nu \in \mathcal{M}$

$$I(\nu) = \sup_{t \in \mathcal{N}} \langle t, \nu \rangle = \sup_{z \in \operatorname{ri} \mathcal{M}} \sum_{i=1}^{r} \nu_i \log(z_i/\rho_i).$$

Since $I(\nu)$ is defined as a Legendre–Fenchel transform, $I(\nu)$ is a closed convex function on $\mathbb{R}^r$. Suppose we show that $I(\nu)$ equals $I_\rho^{(2)}(\nu)$ for all ν in $\operatorname{ri} \mathcal{M}$. Since $I_\rho^{(2)}(\nu)$ is continuous relative to $\mathcal{M}$, the continuity property in Theorem VI.3.2 will imply that $I(\nu)$ equals $I_\rho^{(2)}(\nu)$ for all ν in $\mathcal{M}$. For any ν and z in $\operatorname{ri} \mathcal{M}$

$$\sum_{i=1}^{r} \nu_i \log \frac{z_i}{\rho_i} = \sum_{i=1}^{r} \nu_i \log \frac{\nu_i}{\rho_i} + \sum_{i=1}^{r} \nu_i \log \frac{z_i}{\nu_i} = I_\rho^{(2)}(\nu) - I_\nu^{(2)}(z).$$

Since $I_\nu^{(2)}(z) \ge 0$ with equality if and only if $z = \nu$, it follows that $I(\nu)$ equals $I_\rho^{(2)}(\nu)$ for all $\nu \in \operatorname{ri} \mathcal{M}$. □

Lemma VIII.2.3 completes the proof of Theorem VIII.2.2. By our remarks earlier in the section, Theorem VIII.2.1 follows.

VIII.3. The Contraction Principle Relating Levels-1 and 2 ($d = 1$)

In Theorem II.5.2, we stated a contraction principle relating levels-1 and 2 for Borel probability measures ρ on $\mathbb{R}$ whose support is a finite set. In the present section, a generalization is proved for any Borel probability measure ρ on $\mathbb{R}$ which is nondegenerate (not a unit point measure) and for which $c_\rho(t) = \log \int_{\mathbb{R}} \exp(tx)\rho(dx)$ is finite for all $t \in \mathbb{R}$. We also obtain additional properties of the level-1 entropy function $I_\rho^{(1)}$, including a relationship between the effective domain of this function and the support of ρ. These results will be proved in the next section for Borel probability measures ρ on $\mathbb{R}^d$, $d \ge 2$.

Let ρ be a Borel probability measure on $\mathbb{R}$. A point $x \in \mathbb{R}$ is said to be a *support point* for ρ if every neighborhood of x has positive ρ-measure. The *support* of ρ is defined as the set of all support points for ρ and is denoted by S_ρ. The support may be characterized as the smallest closed set having ρ-measure 1. The *convex hull* of S_ρ is defined as the intersection of all the convex sets containing S_ρ and is denoted by $\operatorname{conv} S_\rho$. Clearly, $\operatorname{conv} S_\rho$ is the smallest closed interval containing S_ρ; $\operatorname{conv} S_\rho$ has nonempty interior whenever ρ is nondegenerate. If $c_\rho(t)$ is finite for all $t \in \mathbb{R}$, then we define ρ_t to be the Borel probability measure on $\mathbb{R}$ which is absolutely continuous with respect to ρ and whose Radon–Nikodym derivative is given by

$$\frac{d\rho_t}{d\rho}(x) = \exp(tx) \cdot \frac{1}{\int_{\mathbb{R}} \exp(tx)\rho(dx)}. \tag{8.1}$$

According to Theorem VII.5.1(b), $c_\rho(t)$ is differentiable for all t and

$$c'_\rho(t) = \int_{\mathbb{R}} x \exp(tx)\rho(dx) \cdot \frac{1}{\int_{\mathbb{R}} \exp(tx)\rho(dx)} = \int_{\mathbb{R}} x\rho_t(dx).$$

If ν is a Borel probability measure on $\mathbb{R}$, then we define the relative entropy of ν with respect to ρ by the formula

$$I_\rho^{(2)}(\nu) = \begin{cases} \int_{\mathbb{R}} \log \frac{d\nu}{d\rho}(x)\nu(dx) & \text{if } \nu \ll \rho \text{ and } \int_{\mathbb{R}} \left|\log \frac{d\nu}{d\rho}\right| d\nu < \infty, \\ \infty & \text{otherwise.} \end{cases} \tag{8.2}$$

The next theorem states the contraction principle relating levels-1 and 2. The theorem shows that for each $z \in \mathbb{R}$

$$I_\rho^{(1)}(z) = \inf\left\{I_\rho^{(2)}(\nu) : \nu \in \mathscr{M}(\mathbb{R}), \int_{\mathbb{R}} x\nu(dx) = z\right\},$$

and it determines for $z \in \operatorname{conv} S_\rho$ where the infimum is attained. A related contraction principle was used in Chapter III to show the existence of the Maxwell–Boltzmann distribution for the discrete ideal gas [Lemma III.4.5].[2]

Theorem VIII.3.1. *Let ρ be a nondegenerate Borel probability measure on $\mathbb{R}$ such that $c_\rho(t)$ is finite for all $t \in \mathbb{R}$. Then the following conclusions hold.*

(a) *For each point $z \in \operatorname{int}(\operatorname{conv} S_\rho)$, there exists a unique point $t \in \mathbb{R}$ such that $\int_{\mathbb{R}} x\rho_t(dx) = z$. $I_\rho^{(2)}(\nu)$ attains its infimum over the set $\{\nu \in \mathscr{M}(\mathbb{R}) : \int_{\mathbb{R}} x\rho(dx) = z\}$ at the unique measure ρ_t and*

$$I_\rho^{(1)}(z) = I_\rho^{(2)}(\rho_t) = \inf\left\{I_\rho^{(2)}(\nu) : \nu \in \mathscr{M}(\mathbb{R}), \int_{\mathbb{R}} x\nu(dx) = z\right\} < \infty. \tag{8.3}$$

(b) *For $z \notin \operatorname{conv} S_\rho$,*

$$I_\rho^{(1)}(z) = \inf\left\{I_\rho^{(2)}(\nu) : \nu \in \mathscr{M}(\mathbb{R}), \int_{\mathbb{R}} x\nu(dx) = z\right\} = \infty. \tag{8.4}$$

(c) *Suppose that $\operatorname{conv} S_\rho$ has a finite endpoint α. If ρ has an atom at α ($\rho\{\alpha\} > 0$), then for $z = \alpha$ (8.3) is valid with ρ_t replaced by δ_α. If ρ does not have an atom at α, then (8.4) is valid for $z = \alpha$.*

Part (a) of the theorem states that for each point $z \in \operatorname{int}(\operatorname{conv} S_\rho)$, there exists a unique real number t such that $\int_{\mathbb{R}} x\rho_t(dx) = z$. Since $c'_\rho(t) = \int_{\mathbb{R}} x\rho_t(dx)$, we can prove the statement in part (a) by showing that the function c'_ρ defines a one-to-one mapping of $\mathbb{R}$ onto $\operatorname{int}(\operatorname{conv} S_\rho)$. Lemma VIII.3.2 and Theorem VIII.3.3 establish this fact together with other useful information. Theorem VIII.3.1 will be proved afterwards.

The following lemma is due to Lanford (1973, page 42).

Lemma VIII.3.2. *If the point* 0 *belongs to* $\text{int}(\text{conv}\, S_\rho)$, *then* $c'_\rho(t) = 0$ *for some* $t \in \mathbb{R}$.

Proof. A continuous function on $\mathbb{R}$ with compact level sets attains its infimum over $\mathbb{R}$ at some point t. If the function is differentiable, then the derivative vanishes at t. Hence it suffices to prove that the level sets $K_b = \{t \in \mathbb{R}: c_\rho(t) \leq b\}$, b real, are compact. K_b is closed since c_ρ is a closed convex function. Since 0 belongs to $\text{int}(\text{conv}\, S_\rho)$, there exist numbers $\varepsilon > 0$ and $0 < \delta < 1$ such that

$$\rho\{x \in \mathbb{R}: x \geq \varepsilon\} \geq \delta > 0 \qquad \text{and} \qquad \rho\{x \in \mathbb{R}: x \leq -\varepsilon\} \geq \delta > 0.$$

If t is non-negative, then

$$c_\rho(t) \geq \log \int_{\{x \geq \varepsilon\}} \exp(tx)\rho(dx) \geq t\varepsilon + \log \delta.$$

Similarly, if t is negative, then $c_\rho(t) \geq |t|\varepsilon + \log \delta$. Thus K_b is a subset of the interval $\{t \in \mathbb{R}: |t| \leq (b - \log \delta)/\varepsilon\}$, and so K_b is bounded. □

Lemma VIII.3.2 will be used in the proof of part (a) of the next theorem.

Theorem VIII.3.3. *Let* ρ *be a nondegenerate Borel probability measure on* $\mathbb{R}$ *such that* $c_\rho(t)$ *is finite for all* $t \in \mathbb{R}$. *Denote the range of the function* $c'_\rho(t)$, $t \in \mathbb{R}$, *by* $\text{ran}\, c'_\rho$. *Then the following conclusions hold.*

(a) c'_ρ *defines a one-to-one mapping of* $\mathbb{R}$ *onto* $\text{int}(\text{conv}\, S_\rho)$ *and*

$$\text{int}(\text{dom}\, I_\rho^{(1)}) = \text{ran}\, c'_\rho = \text{int}(\text{conv}\, S_\rho).$$

(b) $\text{dom}\, I_\rho^{(1)} \subseteq \text{conv}\, S_\rho$; *i.e.*, $I_\rho^{(1)}(z) = \infty$ *if* $z \notin \text{conv}\, S_\rho$.

(c) *Suppose that* $\text{conv}\, S_\rho$ *has a finite endpoint* α. *Then* α *belongs to* $\text{dom}\, I_\rho^{(1)}$ *if and only if* ρ *has an atom at* α; *in this case* $I_\rho^{(1)}(\alpha) = -\log \rho\{\alpha\}$. *In particular, if the support of* ρ *is a finite set, then* $\text{dom}\, I_\rho^{(1)}$ *equals* $\text{conv}\, S_\rho$ *and* $I_\rho^{(1)}$ *is continuous relative to* $\text{conv}\, S_\rho$.

Proof. (a), (b). A short calculation shows that

$$c''_\rho(t) = \int_{\mathbb{R}} (x - \langle x \rangle_t)^2 \rho_t(dx), \tag{8.5}$$

where $\langle x \rangle_t = \int_{\mathbb{R}} x\rho_t(dx)$ and the measure ρ_t is defined in (8.1). Since ρ is nondegenerate, (8.5) implies that $c''_\rho(t)$ is positive. Thus $c_\rho(t)$ is strictly convex [Problem VI.7.2(b)] and the derivative $c'_\rho(t)$ is an increasing function on $\mathbb{R}$. In particular, if z is given, then the equation $c'_\rho(t) = z$ has a unique solution t whenever a solution exists. We now show that $\text{ran}\, c'_\rho = \text{int}(\text{conv}\, S_\rho)$.

Step 1: $\text{ran}\, c'_\rho \subseteq \text{int}(\text{conv}\, S_\rho)$. Since the support S_{ρ_t} of ρ_t equals the support of ρ, $c'_\rho(t) = \int_{S_{\rho_t}} x\rho_t(dx)$ must lie in $\text{conv}\, S_\rho$ for each real t. If $c'_\rho(t)$ were to lie in $\text{bd}(\text{conv}\, S_\rho)$, then ρ_t and ρ would have to be point measures. This would contradict the nondegeneracy of ρ.

Step 2: $\text{int}(\text{conv}\, S_\rho) \subseteq \text{ran}\, c'_\rho$. Let z be a point in $\text{int}(\text{conv}\, S_\rho)$ and consider the translated measure $\rho(\cdot + z)$. The point 0 belongs to the interior of the set $\text{conv}\, S_{\rho(\cdot+z)}$ and

$$c'_{\rho(\cdot+z)}(t) = \int_{\mathbb{R}} x\rho_t(dx) - z = c'_\rho(t) - z.$$

Thus finding a point t which satisfies $c'_\rho(t) = z$ is equivalent to finding a point t which satisfies $c'_{\rho(\cdot+z)}(t) = 0$. In other words, without loss of generality, it suffices to prove that if 0 belongs to $\text{int}(\text{conv}\, S_\rho)$, then there exists a point t which satisfies $c'_\rho(t) = 0$. This implication was proved in Lemma VIII.3.2.

We have shown that $\text{ran}\, c'_\rho = \text{int}(\text{conv}\, S_\rho)$ and thus that c'_ρ defines a one-to-one mapping of $\mathbb{R}$ onto $\text{int}(\text{conv}\, S_\rho)$. We now describe the effective domain of the function

$$I_\rho^{(1)}(z) = \sup_{t\in\mathbb{R}} \{tz - c_\rho(t)\}, \qquad z \in \mathbb{R}.$$

Since $c_\rho(t)$ is convex, the supremum is attained at some point t if and only if $c'_\rho(t) = z$. In this case,

$$I_\rho^{(1)}(z) = t(z)\cdot z - c_\rho(t)) \qquad \text{where } c'_\rho(t(z)) = z.$$

Hence all points z in the range of c'_ρ are in the effective domain of $I_\rho^{(1)}$. Combining this with the previous result, we have

$$\text{ran}\, c'_\rho = \text{int}(\text{conv}\, S_\rho) \subseteq \text{dom}\, I_\rho^{(1)}.$$

We now show that $\text{dom}\, I_\rho^{(1)} \subseteq \text{conv}\, S_\rho$. This will yield part (b) of the theorem, and together with the last display it will show that

$$\text{int}(\text{dom}\, I_\rho^{(1)}) = \text{ran}\, c'_\rho = \text{int}(\text{conv}\, S_\rho).$$

Let z be a point outside $\text{conv}\, S_\rho$. If z lies to the right of $\text{conv}\, S_\rho$, then there exist real numbers $\delta > 0$ and b such that

$$\sup\{x\in\mathbb{R} : x \in \text{conv}\, S_\rho\} \le b - \delta < b + \delta \le z.$$

Since S_ρ is a subset of $\text{conv}\, S_\rho$,

$$\begin{aligned} I_\rho^{(1)}(z) &\ge \sup_{t>0}\left\{tz - \log\int_{S_\rho} \exp(tx)\rho(dx)\right\} \\ &\ge \sup_{t>0}\{t(b+\delta) - t(b-\delta)\} = \infty. \end{aligned}$$

A similar analysis shows that $I_\rho^{(1)}(z) = \infty$ if z lies to the left of $\text{conv}\, S_\rho$. This proves that $\text{dom}\, I_\rho^{(1)} \subseteq \text{conv}\, S_\rho$.

(c) Suppose that α is a right-hand endpoint of $\text{conv}\, S_\rho$. For fixed $x \in S_\rho$, $\exp[t(x-\alpha)]$ is a nonincreasing function of t which converges to $\chi_{\{\alpha\}}(x)$

as $t \to \infty$. Hence

$$I_\rho^{(1)}(\alpha) = -\inf_{t \in \mathbb{R}} \log \int_{S_\rho} \exp[t(x-\alpha)]\rho(dx) = -\log \int_{\mathbb{R}} \chi_{\{\alpha\}}(x)\rho(dx).$$

$I_\rho^{(1)}(\alpha)$ equals $-\log \rho\{\alpha\}$ or ∞ according to whether or not ρ has an atom at α. A similar proof works for a left-hand endpoint of $\operatorname{conv} S_\rho$. If the support of ρ is a finite set $\Gamma = \{x_1, x_2, \ldots, x_r\}$ with $x_1 < x_2 < \cdots < x_r$, then $\operatorname{conv} S_\rho = [x_1, x_r]$. Since ρ has atoms at x_1 and at x_r, $\operatorname{dom} I_\rho^{(1)}$ equals $\operatorname{conv} S_\rho$. $I_\rho^{(1)}$ is continuous relative to $\operatorname{conv} S_\rho$ by Theorems VI.3.1 and VI.3.2. □

Proof of the contraction principle, Theorem VIII.3.1. (a) Let z be a point in $\operatorname{int}(\operatorname{conv} S_\rho)$. Part (a) of Theorem VIII.3.3 implies that there exists a unique point $t \in \mathbb{R}$ such that $\int_{\mathbb{R}} x\rho_t(dx) = z$. We have

$$\begin{aligned} I_\rho^{(2)}(\rho_t) &= \int_{\mathbb{R}} \left[tx - \log \int_{\mathbb{R}} \exp(tx)\rho(dx) \right] \rho_t(dx) \\ &= tz - c_\rho(t) = I_\rho^{(1)}(z). \end{aligned}$$

Now let $\nu \neq \rho_t$ be any other Borel probability measure on $\mathbb{R}$ which has mean z and for which $I_\rho^{(2)}(\nu)$ is finite. Then ν is absolutely continuous with respect to ρ and to ρ_t and

$$\begin{aligned} I_\rho^{(2)}(\nu) &= \int_{\mathbb{R}} \log \frac{d\nu}{d\rho}\, d\nu = \int_{\mathbb{R}} \log \frac{d\nu}{d\rho_t}\, d\nu + \int_{\mathbb{R}} \log \frac{d\rho_t}{d\rho}\, d\nu \\ &= I_{\rho_t}^{(2)}(\nu) + \int_{\mathbb{R}} \left[tx - \log \int_{\mathbb{R}} \exp(tx)\rho(dx) \right] \nu(dx) \\ &= I_{\rho_t}^{(2)}(\nu) + tz - c_\rho(t) = I_{\rho_t}^{(2)}(\nu) + I_\rho^{(1)}(z). \end{aligned}$$

Since $\nu \neq \rho_t$, $I_{\rho_t}^{(2)}(\nu)$ is positive and so $I_\rho^{(2)}(\nu) > I_\rho^{(1)}(z)$. We conclude that for $\nu \in \mathscr{M}(\mathbb{R})$, $I_\rho^{(2)}(\nu) \geq I_\rho^{(1)}(z) = I_\rho^{(2)}(\rho_t)$ and that equality holds if and only if $\nu = \rho_t$.

(b) If z does not belong to $\operatorname{conv} S_\rho$, then any Borel probability measure ν on $\mathbb{R}$ which has mean z is not absolutely continuous with respect to ρ. Therefore

$$\inf\left\{ I_\rho^{(2)}(\nu) : \nu \in \mathscr{M}(\mathbb{R}), \int_{\mathbb{R}} x\nu(dx) = z \right\} = \infty.$$

$I_\rho^{(1)}(z)$ equals ∞ by Theorem VIII.3.3(b).

(c) If α is a finite endpoint of $\operatorname{conv} S_\rho$ and if ρ has an atom at α, then the measure δ_α is absolutely continuous with respect to ρ, δ_α has mean α, and δ_α is the only Borel probability measure on $\mathbb{R}$ with these properties. By Theorem VIII.3.3(c)

$$I_\rho^{(1)}(\alpha) = -\log \rho\{\alpha\} = I_\rho^{(2)}(\delta_\alpha) = \inf\left\{ I_\rho^{(2)}(\nu) : \nu \in \mathscr{M}(\mathbb{R}), \int_{\mathbb{R}} x\nu(dx) = \alpha \right\} < \infty.$$

If ρ does not have an atom at α, then

$$I_\rho^{(1)}(\alpha) = \inf\left\{I_\rho^{(2)}(\nu): \nu \in \mathscr{M}(\mathbb{R}), \int_{\mathbb{R}} x\nu(dx) = \alpha\right\} = \infty. \qquad \square$$

Our final result describes smoothness and mapping properties of the entropy function $I_\rho^{(1)}(z)$. We recall from Theorem VIII.3.3(a) that

$$\operatorname{int}(\operatorname{dom} I_\rho^{(1)}) = \operatorname{int}(\operatorname{conv} S_\rho).$$

Theorem VIII.3.4. *Let ρ be a nondegenerate Borel probability measure on $\mathbb{R}$ such that $c_\rho(t)$ is finite for all $t \in \mathbb{R}$. Then the following conclusions hold.*

(a) *$I_\rho^{(1)}(z)$ is essentially smooth; that is, $I_\rho^{(1)}(z)$ is differentiable for $z \in \operatorname{int}(\operatorname{dom} I_\rho^{(1)}) = \operatorname{int}(\operatorname{conv} S_\rho)$ and $|(I_\rho^{(1)})'(z)| \to \infty$ as $z \in \operatorname{int}(\operatorname{conv} S_\rho)$ converges to a boundary point of* $\operatorname{int}(\operatorname{conv} S_\rho)$.

(b) *The function $(I_\rho^{(1)})'$ defines a one-to-one mapping of* $\operatorname{int}(\operatorname{conv} S_\rho)$ *onto $\mathbb{R}$ with inverse c'_ρ.*

(c) *$I_\rho^{(1)}(z)$ is a real analytic function of $z \in \operatorname{int}(\operatorname{conv} S_\rho)$.*

Proof. (a) Since ρ is nondegenerate, $c_\rho(t)$ is strictly convex [see (8.5)] and so $I_\rho^{(1)}(z)$ is essentially smooth [Theorem VI.5.6].

(b) This follows from Theorem VIII.3.3(a) and the fact that for $z \in \operatorname{int}(\operatorname{conv} S_\rho)$, $t = (I_\rho^{(1)})'(z)$ if and only if $z = c'_\rho(t)$ [Theorem VII.5.5(d)].

(c) If z is a point in $\operatorname{int}(\operatorname{conv} S_\rho)$, then there exists a unique solution $t = t(z)$ of $c'_\rho(t) = z$ and

$$I_\rho^{(1)}(z) = t(z) \cdot z - c_\rho(t(z)).$$

It suffices to prove that the function $z \to t(z)$ is real analytic in some neighborhood of each fixed z. The function $c'_\rho(t)$ can be continued into the complex plane $\mathbb{C}$ to be an analytic function. Since $c''_\rho(t(z))$ is positive, this continuation defines a one-to-one analytic mapping of a complex open neighborhood of $t(z)$ onto a complex open neighborhood U of z. The inverse function is also analytic [Rudin (1974, page 231)]. Hence the restriction of the inverse function to points $z \in U \cap \mathbb{R}$ is real analytic. The restriction is the function $z \to t(z)$. $\square$

VIII.4. The Contraction Principle Relating Levels-1 and 2 ($d \geq 2$)

Let ρ be a nondegenerate Borel probability measure on $\mathbb{R}$ with support S_ρ and assume that $c_\rho(t)$ is finite for all $t \in \mathbb{R}$. In the previous section, we proved that the infimum in the contraction principle relating levels-1 and 2 is attained for all z in the interior of $\operatorname{conv} S_\rho$ (convex hull of S_ρ) and that the effective domain of $I_\rho^{(1)}$ is a subset of $\operatorname{conv} S_\rho$. These results will now be extended to nondegenerate Borel probability measures ρ on $\mathbb{R}^d$, $d \geq 2$. The sets S_ρ and $\operatorname{conv} S_\rho$ are defined as on page 253.

The role played by the set $\operatorname{conv} S_\rho$ for $d = 1$ is played, for $d \geq 2$, by the closed convex hull of S_ρ. The latter is defined as the intersection of all the closed convex sets containing S_ρ and is denoted by $\operatorname{cc} S_\rho$.* We suppose that $c_\rho(t) = \log \int_{\mathbb{R}^d} \exp\langle t, x\rangle \rho(dx)$ is finite for all $t \in \mathbb{R}^d$, and we define ρ_t to be the Borel probability measure on $\mathbb{R}^d$ which is absolutely continuous with respect to ρ and whose Radon–Nikodym derivative is given by

$$\text{(8.6)} \qquad \frac{d\rho_t}{d\rho}(x) = \exp\langle t, x\rangle \cdot \frac{1}{\int_{\mathbb{R}^d} \exp\langle t, x\rangle \rho(dx)}.$$

According to Theorem VII.5.1(b), $c_\rho(t)$ is differentiable for all t and

$$\text{(8.7)} \qquad \frac{\partial c_\rho(t)}{\partial t_i} = \int_{\mathbb{R}^d} x_i \exp\langle t, x\rangle \rho(dx) \cdot \frac{1}{\int_{\mathbb{R}^d} \exp\langle t, x\rangle \rho(dx)}.$$

Thus $\nabla c_\rho(t) = \int_{\mathbb{R}^d} x \rho_t(dx)$.

Let ρ be a Borel probability measure on $\mathbb{R}^d$ such that $c_\rho(t)$ is finite for all $t \in \mathbb{R}^d$. The relative entropy $I_\rho^{(2)}(v)$ of $v \in \mathcal{M}(\mathbb{R}^d)$ with respect to ρ is defined as in (8.2). Donsker and Varadhan (1976a, page 425) prove that

$$\text{(8.8)} \qquad I_\rho^{(1)}(z) = \inf\left\{I_\rho^{(2)}(v) : v \in \mathcal{M}(\mathbb{R}^d), \int_{\mathbb{R}^d} x v(dx) = z\right\}$$

for each $z \in \mathbb{R}^d$.[3] We will prove (8.8) only for z in the relative interior of $\operatorname{cc} S_\rho$ and for $z \notin \operatorname{cc} S_\rho$. For z in the relative interior of $\operatorname{cc} S_\rho$, we will also determine where the infimum in (8.8) is attained.

Theorem VIII.4.1. *Let ρ be a nondegenerate Borel probability measure on $\mathbb{R}^d$, $d \geq 2$, such that $c_\rho(t)$ is finite for all $t \in \mathbb{R}^d$. Then the following conclusions hold.*

(a) *For $z \in \operatorname{ri}(\operatorname{cc} S_\rho)$, let A_z denote the set of points t for which $\int_{\mathbb{R}^d} x \rho_t(dx) = z$. Then A_z is nonempty,*† *the measures $\{\rho_t; t \in A_z\}$ are all equal, and $I_\rho^{(2)}(v)$ attains its infimum over the set $\{v \in \mathcal{M}(\mathbb{R}^d) : \int_{\mathbb{R}^d} x v(dx) = z\}$ at the unique measure ρ_t, $t \in A_z$. Furthermore*

$$\text{(8.9)} \qquad I_\rho^{(1)}(z) = I_\rho^{(2)}(\rho_t) = \inf\left\{I_\rho^{(2)}(v) : v \in \mathcal{M}(\mathbb{R}^d), \int_{\mathbb{R}^d} x v(dx) = z\right\} < \infty.$$

(b) *For $z \notin \operatorname{cc} S_\rho$,*

$$\text{(8.10)} \qquad I_\rho^{(1)}(z) = \inf\left\{I_\rho^{(2)}(v) : v \in \mathcal{M}(\mathbb{R}^d), \int_{\mathbb{R}^d} x v(dx) = z\right\} = \infty.$$

* Although S_ρ is closed, $\operatorname{conv} S_\rho$ need not be closed for $d \geq 2$; e.g., if $S_\rho = \{x \in \mathbb{R}^2 : x_1 \in \mathbb{R}, x_2 = 0 \text{ or } x_1 = 0, x_2 = 1\}$, then $\operatorname{conv} S_\rho = \{x \in \mathbb{R}^2 : x_1 \in \mathbb{R}, 0 \leq x_2 < 1 \text{ or } x_1 = 0, x_2 = 1\}$. In general, $\operatorname{ri}(\operatorname{cc} S_\rho) = \operatorname{ri}(\operatorname{conv} S_\rho)$.

† A_z consists of a unique point if and only if ρ is maximal [see Theorems VIII.4.3(a) and VIII.4.4(a)].

This theorem generalizes parts (a) and (b) of Theorem VIII.3.1 to $\mathbb{R}^d$, $d \geq 2$. The proof of the latter depended on the relation between the range of the function $c'_\rho(t)$ and the support of ρ expressed in Theorem VIII.3.3(a). A similar proof will yield Theorem VIII.4.1 once we establish an analogous relation between the range of the mapping $\nabla c_\rho(t)$, $t \in \mathbb{R}^d$, and the support of ρ.

It is convenient first to consider those measures ρ for which the convex function $c_\rho(t)$ is a strictly convex function on $\mathbb{R}^d$. The next proposition gives a simple criterion for this strict convexity. Let $\operatorname{aff} S_\rho$ be the affine hull of the support of ρ. A Borel probability measure ρ on $\mathbb{R}^d$ is said to be *maximal* if $\operatorname{aff} S_\rho$ equals all of $\mathbb{R}^d$. This condition is equivalent to S_ρ not being a subset of any hyperplane in $\mathbb{R}^d$. For example, if $d = 1$, then any nondegenerate measure is maximal. For $d \geq 1$, $\operatorname{cc} S_\rho$ has nonempty interior whenever ρ is maximal, and $\operatorname{int}(\operatorname{cc} S_\rho)$ and $\operatorname{int}(\operatorname{conv} S_\rho)$ coincide.

Proposition VIII.4.2. *Let ρ be a Borel probability measure on $\mathbb{R}^d$ such that $c_\rho(t)$ is finite for all $t \in \mathbb{R}^d$. Then $c_\rho(t)$ is a strictly convex function on $\mathbb{R}^d$ if and only if ρ is maximal.*

Proof. Hölder's inequality implies that for every t_1 and t_2 in $\mathbb{R}^d$ and $0 < \lambda < 1$

$$\int_{\mathbb{R}^d} \exp\langle \lambda t_1 + (1-\lambda) t_2, x\rangle \rho(dx) \leq \left\{\int_{\mathbb{R}^d} \exp\langle t_1, x\rangle \rho(dx)\right\}^{\lambda}$$

$$\cdot \left\{\int_{\mathbb{R}^d} \exp\langle t_2, x\rangle \rho(dx)\right\}^{(1-\lambda)} \tag{8.11}$$

and that equality holds if and only if

$$\frac{\exp\langle t_1, x\rangle}{\int_{\mathbb{R}^d} \exp\langle t_1, x\rangle \rho(dx)} = \frac{\exp\langle t_2, x\rangle}{\int_{\mathbb{R}^d} \exp\langle t_2, x\rangle \rho(dx)} \qquad \rho\text{-a.s.} \tag{8.12}$$

If $c_\rho(t)$ is not strictly convex on $\mathbb{R}^d$, then there exist distinct points t_1 and t_2 and some $0 < \lambda < 1$ for which equality holds in (8.11). Hence by (8.12)

$$\langle t_1, x\rangle = \langle t_2, x\rangle + c_\rho(t_1) - c_\rho(t_2) \qquad \rho\text{-a.s.} \tag{8.13}$$

This equality implies that the support of ρ is a subset of the hyperplane

$$\{x \in \mathbb{R}^d : \langle x, t_1 - t_2\rangle = c_\rho(t_1) - c_\rho(t_2)\}.$$

Hence ρ is not maximal. Conversely, suppose that ρ is not maximal, but that its support is a subset of the hyperplane $H = \{x \in \mathbb{R}^d : \langle x, \gamma\rangle = b\}$, where γ is a unit vector. Let t_0 be a fixed element of H and let A be the set of vectors of the form

$$t = t_0 + \alpha\gamma, \qquad \alpha \text{ real} \quad (\alpha = \langle t - t_0, \gamma\rangle).$$

Since $\langle x, \gamma\rangle = b$ for all x in the support of ρ, we have for $t \in A$

$$c_\rho(t) = b\alpha + \log \int_{\mathbb{R}^d} \exp\langle t_0, x\rangle \rho(dx) = b\langle t - t_0, \gamma\rangle + c_\rho(t_0).$$

This shows that c_ρ is affine on A and thus is not strictly convex on $\mathbb{R}^d$. □

The next theorem generalizes Theorem VIII.3.3 to maximal probability measures on $\mathbb{R}^d$, $d \geq 2$. It is due to Barndorff-Nielsen (1970, 1978).[4] Measures which are not maximal will be considered later.

Theorem VIII.4.3. *Let ρ be a maximal Borel probability measure on $\mathbb{R}^d$, $d \geq 2$, such that $c_\rho(t)$ is finite for all $t \in \mathbb{R}^d$. Denote the range of the mapping $\nabla c_\rho(t)$, $t \in \mathbb{R}^d$, by* ran ∇c_ρ. *Then the following conclusions hold.*

(a) *∇c_ρ defines a one-to-one mapping of $\mathbb{R}^d$ onto* $\operatorname{int}(\operatorname{cc} S_\rho)$ *and*

$$\operatorname{int}(\operatorname{dom} I_\rho^{(1)}) = \operatorname{ran} \nabla c_\rho = \operatorname{int}(\operatorname{cc} S_\rho).$$

(b) $\operatorname{dom} I_\rho^{(1)} \subseteq \operatorname{cc} S_\rho$; *i.e., $I_\rho^{(1)}(z) = \infty$ if $z \notin \operatorname{cc} S_\rho$.*

(c) *Let z be a boundary point of* cc S_ρ *and let $\mathscr{H}_z$ be the set of all supporting hyperplanes to* cc S_ρ *at z. If $\rho(H) = 0$ for some $H \in \mathscr{H}_z$, then z does not belong to* $\operatorname{dom} I_\rho^{(1)}$.

Proof of parts (*a*) *and* (*b*). It is convenient to divide the proof of parts (a) and (b) into four steps. The proof of part (c) is omitted [see Barndorff-Nielsen (1978, pages 140–143) and Problem VIII.6.8.]

Step 1: *∇c_ρ defines a one-to-one mapping.* Let t_1 and t_2 be any two distinct points in $\mathbb{R}^d$ and define the function

$$f(\lambda) = c_\rho(t_1 + \lambda(t_2 - t_1)), \qquad \lambda \in \mathbb{R}.$$

Since ρ is maximal, $c_\rho(t)$ is strictly convex on $\mathbb{R}^d$, and so $f(\lambda)$ is strictly convex on $\mathbb{R}$. Thus

$$f'(0) = \langle \nabla c_\rho(t_1), t_2 - t_1\rangle < f'(1) = \langle \nabla c_\rho(t_2), t_2 - t_1\rangle.$$

It follows that $\nabla c_\rho(t_1) \neq \nabla c_\rho(t_2)$.

Step 2: $\operatorname{ran} \nabla c_\rho \subseteq \operatorname{int}(\operatorname{cc} S_\rho)$. We first prove that $\operatorname{ran} \nabla c_\rho \subseteq \operatorname{cc} S_\rho$. Suppose that $z = \nabla c_\rho(t)$ lies in the complement of cc S_ρ. By Lemma VI.5.4(b) there exists a hyperplane $H = \{x \in \mathbb{R}^d : \langle x, \gamma\rangle = b\}$ which separates cc S_ρ and $\{z\}$ strongly and

$$\sup\{\langle x, \gamma\rangle : x \in \operatorname{cc} S_\rho\} < b < \langle z, \gamma\rangle.$$

Since the support S_{ρ_t} of the measure ρ_t equals S_ρ and S_ρ is a subset of cc S_ρ,

$$b < \langle z, \gamma\rangle = \langle \operatorname{grad} c_\rho(t), \gamma\rangle = \int_{S_{\rho_t}} \langle x, \gamma\rangle \rho_t(dx) < b.$$

This contradiction proves that $\operatorname{ran}\nabla c_\rho \subseteq \operatorname{cc} S_\rho$. Now suppose that $z = \nabla c_\rho(t)$ lies in $\operatorname{bd}(\operatorname{cc} S_\rho)$. By Lemma VI.5.4(a) there exists a hyperplane $H = \{x \in \mathbb{R}^d : \langle x, \gamma \rangle = b\}$ which separates $\operatorname{cc} S_\rho$ and $\{z\}$, and

$$\sup\{\langle x, \gamma \rangle : x \in \operatorname{cc} S_\rho\} \leq b \leq \langle z, \gamma \rangle.$$

As above, $b \leq \langle z, \gamma \rangle = \int_{S_{\rho_t}} \langle x, \gamma \rangle \rho_t(dx) \leq b$. This implies that $\langle x, \gamma \rangle$ equals b for all $x \in S_{\rho_t} = S_\rho$, or that S_ρ is a subset of the hyperplane H. As this contradicts the maximality of ρ, the proof of Step 2 is done.

Step 3: $\operatorname{int}(\operatorname{cc} S_\rho) \subseteq \operatorname{ran}\nabla c_\rho$. As in the proof of the analogous fact for $d = 1$ [page 256], it suffices to prove that if the point 0 belongs to $\operatorname{int}(\operatorname{cc} S_\rho)$, then there exists a point t which satisfies $\nabla c_\rho(t) = 0$. For $d = 1$, this was proved in Lemma VIII.3.2. The same proof applies to $d \geq 2$ once we show that the level sets $K_b = \{t \in \mathbb{R}^d : c_\rho(t) \leq b\}$, b real, are compact.* K_b is closed since c_ρ is a closed convex function on $\mathbb{R}^d$. If $\operatorname{cc} S_\rho = \mathbb{R}^d$, then let $\varepsilon = R$, where R is any fixed positive number. Otherwise, let $\varepsilon > 0$ be half the distance from 0 to the complement of $\operatorname{cc} S_\rho$. For any unit vector γ, define the half-space $A(\gamma) = \{x \in \mathbb{R}^d : \langle x, \gamma \rangle > \varepsilon\}$. We prove below that there exists a number $0 < \delta \leq 1$ such that

$$\inf_{\|\gamma\|=1} \rho\{A(\gamma)\} \geq \delta. \tag{8.14}$$

From this it will follow that if t is nonzero, then

$$c_\rho(t) \geq \log \int_{A(t/\|t\|)} \exp\langle \|t\| \cdot t/\|t\|, x \rangle \rho(dx) \geq \varepsilon\|t\| + \log \delta.$$

The same inequality holds for $t = 0$ ($c_\rho(0) = 0 \geq \log \delta$). We conclude that K_b is a subset of the ball $\{t \in \mathbb{R}^d : \|t\| \leq \varepsilon^{-1}(b - \log \delta)\}$ and thus that K_b is bounded.

If (8.14) were not true for some $\delta > 0$, then there would exist a sequence of unit vectors $\gamma_1, \gamma_2, \ldots$ such that $\rho\{A(\gamma_n)\} \to 0$ as $n \to \infty$. By the compactness of the unit sphere in $\mathbb{R}^d$, there would exist a subsequence $\{\gamma_{n'}\}$ and a unit vector γ such that $\gamma_{n'} \to \gamma$. Fatou's lemma would imply $\rho\{A(\gamma)\} = 0$. On the other hand, by the definiton of ε, the open half-space $A(\gamma)$ contains a point in $\operatorname{conv} S_\rho$, and therefore $A(\gamma)$ contains a point in S_ρ. It follows that $\rho\{A(\gamma)\}$ must be positive. This contradiction proves that (8.14) must hold for some $\delta > 0$ ($\delta \leq 1$ since ρ is a probability measure).

Step 4: $\operatorname{int}(\operatorname{dom} I_\rho^{(1)}) = \operatorname{int}(\operatorname{cc} S_\rho) \subseteq \operatorname{dom} I_\rho^{(1)} \subseteq \operatorname{cc} S_\rho$. According to Theorem VI.5.3(c) $\operatorname{ran}\nabla c_\rho \subseteq \operatorname{dom} I_\rho^{(1)}$, and Steps 2 and 3 imply that $\operatorname{ran}\nabla c_\rho = \operatorname{int}(\operatorname{cc} S_\rho)$. Hence Step 4 is proved once we show that $\operatorname{dom} I_\rho^{(1)} \subseteq \operatorname{cc} S_\rho$; i.e., $I_\rho^{(1)}(z) = \infty$ for $z \notin \operatorname{cc} S_\rho$. If $z \notin \operatorname{cc} S_\rho$, then there exists a hyperplane $H = \{x \in \mathbb{R}^d : \langle x, \gamma \rangle = b\}$ which separates $\operatorname{cc} S_\rho$ and $\{z\}$ strongly, and

$$\sup\{\langle x, \gamma \rangle : x \in \operatorname{cc} S_\rho\} \leq b - \delta < b + \delta \leq \langle z, \gamma \rangle$$

for some $\delta > 0$. Since S_ρ is a subset of $\operatorname{cc} S_\rho$,

*The following proof is due to Lanford (1973, page 42).

$$I_\rho^{(1)}(z) = \sup_{t \in \mathbb{R}^d} \{\langle t, z\rangle - c_\rho(t)\} \geq \sup_{t=r\gamma, r>0} \left\{\langle t, z\rangle - \log \int_{S_\rho} \exp\langle t, x\rangle \rho(dx)\right\}$$
$$\geq \sup_{r>0} \{r(b+\delta) - r(b-\delta)\} = \infty.$$

This completes the proof of Step 4 and the proof of parts (a) and (b) of Theorem VIII.4.3. □

The previous theorem described the range of the mapping $\nabla c_\rho(t)$, $t \in \mathbb{R}^d$, and the effective domain of $I_\rho^{(1)}$ for maximal measures ρ. We now consider nondegenerate measures ρ which are not maximal. As a special case, assume that the affine hull of S_ρ equals the subspace $A = \{x \in \mathbb{R}^d : x_{m+1} = \cdots = x_d = 0\}$ of dimension $m < d$. Then $c_\rho(t) = \log \int_{\mathbb{R}^d} \exp(\sum_{i=1}^m t_i x_i)\rho(dx)$ is a function of $t \in A$. Since as a measure on A ρ is maximal, the previous theorem implies that $\operatorname{grad} c_\rho$ defines a one-to-one mapping of A onto $\operatorname{ri}(\operatorname{cc} S_\rho)$ (the interior of $\operatorname{cc} S_\rho$ relative to A) and that $\operatorname{ri}(\operatorname{dom} I_\rho^{(1)})$ equals $\operatorname{ri}(\operatorname{cc} S_\rho)$. In order to analyze an arbitrary nonmaximal measure, one reduces to this special case.

Theorem VIII.4.4. *Let ρ be a nondegenerate Borel probability measure on $\mathbb{R}^d$, $d \geq 2$, which is not maximal. Assume that $c_\rho(t)$ is finite for all $t \in \mathbb{R}^d$. Then the following conclusions hold.*

(a) *∇c_ρ defines a many-to-one mapping of $\mathbb{R}^d$ onto* $\operatorname{ri}(\operatorname{cc} S_\rho)$ *and*

$$\operatorname{ri}(\operatorname{dom} I_\rho^{(1)}) = \operatorname{ran} \nabla c_\rho = \operatorname{ri}(\operatorname{cc} S_\rho).$$

If $\nabla c_\rho(t_1) = \nabla c_\rho(t_2)$, then the measures ρ_{t_1} and ρ_{t_2} are equal.

(b) *∇c_ρ defines a one-to-one mapping of* $\operatorname{aff} S_\rho$ *onto* $\operatorname{ri}(\operatorname{cc} S_\rho)$.

(c) $\operatorname{dom} I_\rho^{(1)} \subseteq \operatorname{cc} S_\rho$.

We prove the assertion in part (a) that if $\nabla c_\rho(t_1) = \nabla c_\rho(t_2)$, then $\rho_{t_1} = \rho_{t_2}$. Suppose that $t_1 \neq t_2$ and define the convex function

$$f(\lambda) = c_\rho(\lambda t_1 + (1-\lambda)t_2), \qquad \lambda \in \mathbb{R}.$$

Since $\operatorname{grad} c_\rho(t_1) = \operatorname{grad} c_\rho(t_2)$, it follows that $f'(0) = f'(1)$ and thus that $f(\lambda)$ is affine on the interval $0 \leq \lambda \leq 1$. This implies that

$$c_\rho(\lambda t_1 + (1-\lambda)t_2) = \lambda c_\rho(t_1) + (1-\lambda)c_\rho(t_2) \qquad \text{for all } 0 \leq \lambda \leq 1.$$

The latter is equivalent to condition (8.12), which implies that $\rho_{t_1} = \rho_{t_2}$. The proof of the rest of the theorem is Problem VIII.6.9.

Proof of the contraction principle, Theorem VIII.4.1. (a) Since $\int_{\mathbb{R}^d} x\rho_t(dx) = \nabla c_\rho(t)$, the set A_z is the set of t for which $\nabla c_\rho(t) = z$. For $z \in \operatorname{ri}(\operatorname{cc} S_\rho)$, A_z is nonempty by Theorems VIII.4.3(a) and VIII.4.4(a). A_z consists of a unique point t if and only if ρ is maximal. We have just proved that if ρ is not maximal, then the measures $\{\rho_t; t \in A_z\}$ are all equal. The rest of part (a) is proved exactly like Theorem VIII.3.1(a).

(b) If z does not belong to cc S_ρ, then any Borel probability measure ν on $\mathbb{R}^d$ which has mean z is not absolutely continuous with respect to ρ. Therefore

$$\inf\left\{I_\rho^{(2)}(\nu): \nu \in \mathscr{M}(\mathbb{R}^d), \int_{\mathbb{R}^d} x\nu(dx) = z\right\} = \infty.$$

$I_\rho^{(1)}(z)$ equals ∞ by Theorem VIII.4.3(b). □

For nondegenerate probability measures ρ on $\mathbb{R}$, Theorem VIII.3.4 described smoothness and mapping properties of the entropy function $I_\rho^{(1)}(z)$. This has a direct generalization to maximal probability measures ρ on $\mathbb{R}^d$, $d \geq 2$ [Problem VIII.6.10].

We have completed our discussion of level-2 large deviations. The level-3 problem is the topic of the next chapter.

VIII.5. Notes

1 (page 250). We describe the results in Donsker and Varadhan (1975a). Let $X_0, X_1, X_2, \ldots$ be a stationary Markov process with transition probabilities $\gamma(x, dy)$ taking values in a compact metric space $\mathscr{X}$. Assume that $X_0 = x$ and that

(a) $\gamma(x, dy)$ is a Feller transition probability;

(b) $\gamma(x, dy)$ has a density $\gamma(x, y)$ relative to some reference probability measure $\beta(dy)$;

(c) $\gamma(x, y)$ is uniformly bounded away from 0 and ∞.

Let $\mathscr{U}$ be the set of positive functions $u \in \mathscr{C}(\mathscr{X})$. Define for any Borel probability measure ν on $\mathscr{X}$

$$I_\gamma(\nu) = -\inf_{u \in \mathscr{U}} \int_{\mathscr{X}} \log\left(\frac{\gamma u}{u}\right)(x)\,\nu(dx),$$

where $\gamma u(x) = \int_{\mathscr{X}} u(y)\gamma(x, dy)$. It is then proved that the distributions on $\mathscr{M}(\mathscr{X})$ of the empirical measures $L_{n,x}(\omega, \cdot) = n^{-1}\sum_{j=0}^{n-1}\delta_{X_j(\omega)}(\cdot), n = 1, 2, \ldots,$ have a large deviation property with $a_n = n$ and entropy function $I_\gamma(\nu)$.

The paper also treats continuous parameter, stationary Markov processes $\{X_t; t \geq 0\}$ taking values in a compact metric space $\mathscr{X}$. Let $p(t, x, dy)$ be the transition probabilities. Assume that $X_0 = x$ and that for each $t > 0$ $\gamma(x, dy) = p(t, x, dy)$ satisfies hypotheses (a)–(c) above [$\beta(dy)$ fixed for all $t > 0$]. Let L be the infinitesimal generator of the semigroup associated with $p(t, x, dy)$ and $\mathscr{D}$ the domain of L. Define for any Borel probability measure ν on $\mathscr{X}$

$$I(\nu) = -\inf_{u>0, u \in \mathscr{D}} \int_{\mathscr{X}} \left(\frac{Lu}{u}\right)(x)\,\nu(dx).$$

It is then proved that the distributions on $\mathscr{M}(\mathscr{X})$ of the empirical measures $L_t(\omega, \cdot) = t^{-1} \int_0^t \delta_{X_s(\omega)}(\cdot) ds$, $t > 0$, have a large deviation property with entropy function $I(v)$; that is, I is lower semicontinuous, I has compact level sets,

$$\limsup_{t\to\infty} \frac{1}{t} \log P\{L_t(\omega, \cdot) \in K\} \le -\inf_{v \in K} I(v) \qquad \text{for each closed set } K \text{ in } \mathscr{X},$$

$$\liminf_{t\to\infty} \frac{1}{t} \log P\{L_t(\omega, \cdot) \in G\} \ge -\inf_{v \in G} I(v) \qquad \text{for each open set } G \text{ in } \mathscr{X}.$$

In Donsker and Varadhan (1976a), a large deviation property is proved for Markov processes taking values in a complete separable metric space $\mathscr{X}$. The transition probabilities must satisfy strong transitivity and recurrence properties. As a corollary, one obtains the level-2 large deviation property for i.i.d. random vectors taking values in $\mathscr{X}$. Theorem II.4.3 is a special case.

Donsker and Varadhan have applied these large deviation theorems to a number of interesting problems. See references at the end of the book. Large deviation results for empirical measures have been obtained by many people, including Sanov (1957), Sethuraman (1964), Gärtner (1977), Bahadur and Zabell (1979), Bretagnolle (1979), Groeneboom, Oosterhoff, and Ruymgaart (1979), Kac (1980), Chiang (1982), Jain (1982), Csiszár (1984), Stroock (1984), and Pinsky (1985). Luttinger (1982) presents a formal approach.

2 (page 254) Csiszár (1975) studies a large class of minimization problems involving relative entropy. Formula (8.3) is an example as is Problem VIII.6.2 below.

3 (page 259) Donsker and Varadhan (1976a, page 425) prove the contraction principle (8.8) for ρ a Borel probability measure on a Banach space $\mathscr{X}$ which satisfies $\int_{\mathscr{X}} e^{\sigma\|x\|}\rho(dx) < \infty$ for all $\sigma > 0$. If $\mathscr{X} = \mathbb{R}^d$, then this condition is equivalent to $c_\rho(t) < \infty$ for all $t \in \mathbb{R}^d$.

4 (page 261) Theorem VIII.4.3 is proved in Barndorff-Nielsen (1978, Section 9.1) for maximal measures ρ under less restrictive hypotheses on c_ρ.

VIII.6. Problems

VIII.6.1. Let ρ be a maximal Borel probability measure on $\mathbb{R}^d$ such that $c_\rho(t)$ is finite for all $t \in \mathbb{R}^d$.

(a) Show that the Hessian matrix of $c_\rho(t)$ is positive definite for all $t \in \mathbb{R}^d$ and thus that c_ρ is strictly convex on $\mathbb{R}^d$.

(b) Prove that $I_\rho^{(1)}$ is essentially strictly convex.

VIII.6.2 [Kagan, Linnik and Rao (1973, Section 13.2.2)]. We are given a Borel measure ρ on $\mathbb{R}$ (not necessarily a probability measure); bounded, measurable, real-valued functions $h_1, \ldots, h_r$ on $\mathbb{R}$; and real numbers $\alpha_1, \ldots, \alpha_r$. For v a Borel probability measure on $\mathbb{R}$, define

$$I_\rho^{(2)}(\nu) = \begin{cases} \displaystyle\int_{\mathbb{R}} \log \frac{d\nu}{d\rho}(x)\nu(dx) & \text{if } \nu \ll \rho \text{ and } \displaystyle\int_{\mathbb{R}} \left|\log \frac{d\nu}{d\rho}\right| d\nu < \infty, \\ \infty & \text{otherwise.} \end{cases}$$

Let $\mathscr{P}$ denote the set of all Borel probability measures ν on $\mathbb{R}$ which are absolutely continuous with respect to ρ, for which $d\nu/d\rho$ has the form

$$\frac{d\nu}{d\rho}(x) = \exp(\beta_0 + \beta_1 h_1(x) + \cdots + \beta_r h_r(x)), \qquad \beta_0, \ldots, \beta_r \text{ real,}$$

and which satisfy $\int_{\mathbb{R}} h_i(x)\nu(dx) = \alpha_i$, $i = 1, \ldots, r$. Prove that if $\mathscr{P}$ is not empty, then $\mathscr{P}$ equals the set of measures at which

$$\text{(8.15)} \qquad \inf\left\{I_\rho^{(2)}(\nu): \nu \in \mathscr{M}(\mathbb{R}), \int_{\mathbb{R}} h_i(x)\nu_i(dx) = \alpha_i, \qquad i = 1, \ldots, r\right\}$$

is attained.

VIII.6.3. Let ρ be Lebesgue measure on $\mathscr{B}(\mathbb{R})$. For each of the following probability densities, find functions $h_1, \ldots, h_r$ such that the probability measure on $\mathbb{R}$ with the given density solves the minimization problem (8.15) for some constants $\alpha_1, \ldots, \alpha_r$.

(i) $f(x) = x^{a-1}(1-x)^{b-1}/\beta(a, b)$ on $(0, 1)$, $a > 0$, $b > 0$ (beta density).
(ii) $f(x) = ae^{-ax}$ on $(0, \infty)$, $a > 0$ (exponential density).
(iii) $f(x) = a^b x^{b-1} e^{-ax}/\Gamma(b)$ on $(0, \infty)$, $a > 0$, $b > 0$ (gamma density).
(iv) $f(x) = (2\pi\sigma^2)^{-1/2} \exp[-(x-m)^2/2\sigma^2]$ on $(-\infty, \infty)$, m real, $\sigma^2 > 0$ (Gaussian density).
(v) $f(x) = \frac{1}{2}ae^{-a|x|}$ on $(-\infty, \infty)$, $a > 0$ (Laplace density).

VIII.6.4. Let ρ be the probability measure $\frac{1}{3}(\delta_{(0,0)} + \delta_{(1,0)} + \delta_{(0,1)})$ on $\mathbb{R}^2$. Prove that for each $z \in \text{cc}\, S_\rho$

$$\text{(8.16)} \qquad I_\rho^{(1)}(z) = \inf\left\{I_\rho^{(2)}(\nu): \nu \in \mathscr{M}(\mathbb{R}^2), \int_{\mathbb{R}^2} x\nu(dx) = z\right\} < \infty$$

and determine the measure ν at which the infimum is attained. Theorem VIII.4.1(a) states where the infimum in (8.16) is attained only for points $z \in \text{int}(\text{cc}\, S_\rho)$.

VIII.6.5. Let ρ be a Borel probability measure on $\mathbb{R}^d$. Prove that $I_\rho^{(2)}$ is a strictly convex function on $\mathscr{M}(\mathbb{R}^d)$; i.e., if μ and ν are distinct Borel probability measures on $\mathbb{R}^d$, then for all $0 < \lambda < 1$

$$I_\rho^{(2)}(\lambda\mu + (1-\lambda)\nu) < \lambda I_\rho^{(2)}(\mu) + (1-\lambda)I_\rho^{(2)}(\nu).$$

VIII.6.6. (a) Let ρ be a Borel probability measure on $\mathbb{R}^d$ and A a nonempty compact convex subset of $\mathscr{M}(\mathbb{R}^d)$. Prove that if $\inf_{\nu \in A} I_\rho^{(2)}(\nu)$ is finite, then $I_\rho^{(2)}(\nu)$ attains its infimum over A at a unique measure.

(b) Let ρ be a nondegenerate Borel probability measure on $\mathbb{R}^d$ such that $c_\rho(t)$ is finite for all $t \in \mathbb{R}^d$. Let z be a point in rbd(cc S_ρ) such that

$$\inf\left\{ I_\rho^{(2)}(v): v \in \mathscr{M}(\mathbb{R}^d), \int_{\mathbb{R}^d} xv(dx) = z \right\} < \infty. \tag{8.17}$$

Prove that the infimum is attained at a unique measure $\bar{\rho}$ and that $\bar{\rho}$ is the weak limit of measures $\{\rho_{t_n}; n = 1, 2, \ldots\}$ for some sequence $\{t_n; n = 1, 2, \ldots\}$ in $\mathbb{R}^d$ such that $\|t_n\| \to \infty$ [ρ_{t_n} is defined in (8.6)]. Theorem VIII.4.1(a) states where the infimum in (8.17) is attained only for points $z \in$ ri(cc S_ρ). [*Hint:* Use (8.8) and the following properties. $I_\rho^{(2)}(v)$ is strictly convex; $I_\rho^{(2)}(v)$ is lower semicontinuous; $I_\rho^{(2)}(v)$ has compact level sets; if $\sup_n I_\rho^{(2)}(v_n)$ is finite for some sequence $\{v_n\}$, then $\lim_{A \to \infty} \sup_n \int_{\|x\| \geq A} \|x\| v_n(dx) = 0$. The last three properties are proved in Donsker and Varadhan (1975a, 1976a).]

VIII.6.7 [Barndorff-Nielsen (1978, page 105)]. Let ρ be a Borel probability measure on $\mathbb{R}^d$ such that $c_\rho(t)$ is finite for all $t \in \mathbb{R}^d$. S_ρ denotes the support of ρ. Prove that for any nonzero $t \in \mathbb{R}^d$

$$\lim_{\lambda \to \infty} [-\lambda \sigma(t|S_\rho) + c_\rho(\lambda t)] = \log \rho\{x: \langle t, x\rangle = \sigma(t|S_\rho)\},$$

where $\sigma(t|S_\rho) = \sup_{x \in S_\rho} \langle t, x\rangle$ is the support function of S_ρ [see Problem VI.7.11].

VIII.6.8 [Barndorff-Nielsen (1978, pages 140–143)]. Let ρ be a maximal Borel probability measure on $\mathbb{R}^d$ such that $c_\rho(t)$ is finite for all $t \in \mathbb{R}^d$. Let z be a boundary point of cc S_ρ and $\mathscr{H}_z$ the set of all supporting hyperplanes to cc S_ρ at z. This problem shows that z does not belong to dom $I_\rho^{(1)}$ if $\rho(H) = 0$ for some $H \in \mathscr{H}_z$. [Theorem VIII.4.3(c)].

(a) Any $H \in \mathscr{H}_z$ can be written as $\{x \in \mathbb{R}^d: \langle x - z, \gamma\rangle = 0\}$, where γ is a unit vector and $\langle x - z, \gamma\rangle \leq 0$ for all $x \in$ cc S_ρ. Prove this statement.

(b) Suppose that $H \in \mathscr{H}_z$ has unit normal vector γ. Prove that

$$I_\rho^{(1)}(z) \geq \sup_{\lambda \in \mathbb{R}} \{\lambda\langle\gamma, z\rangle - c_\rho(\lambda\gamma)\} = \lim_{\lambda \to \infty} \{\lambda\langle\gamma, z\rangle - c_\rho(\lambda\gamma)\}.$$

Conclude that if $\rho(H) = 0$, then z does not belong to dom $I_\rho^{(1)}$. [*Hint:* Problem VIII.6.7.]

(c) Prove that if ρ is absolutely continuous with respect to Lebesgue measure, then dom $I_\rho^{(1)} =$ int(cc S_ρ).

VIII.6.9. The purpose of this problem is to prove Theorem VIII.4.4. Assume the hypotheses of that theorem.

(a) Suppose that the dimension m of aff S_ρ is less than d. Prove that there exist an orthogonal matrix U and a vector γ such that aff $S_\rho = UA + \gamma$, where $A = \{x \in \mathbb{R}^d: x_{m+1} = \cdots = x_d = 0\}$. For Borel subsets B of A, define $\bar{\rho}\{B\} = \rho\{UB + \gamma\}$. Prove that for suitable choice of γ

$$c_\rho(t) = \langle t, \gamma \rangle + c_{\bar{\rho}}(U^T(t - \gamma)), \qquad t \in \operatorname{aff} S_\rho,$$

where U^T is the transpose of U. Relate the range of $\nabla c_\rho(t)$, $t \in \operatorname{aff} S_\rho$, to the range of $\nabla c_\rho(s)$, $s \in A$, and prove parts (a) and (b) of Theorem VIII.4.4.

(b) Prove part (c) of Theorem VIII.4.4.

VIII.6.10. The purpose of this problem is to generalize Theorem VIII.3.4 to higher dimensions. Let ρ be a maximal Borel probability measure on $\mathbb{R}^d$, $d \geq 2$, such that $c_\rho(t)$ is finite for all $t \in \mathbb{R}^d$.

(a) Prove that $I_\rho^{(1)}(z)$ is essentially smooth.

(b) Prove that $\nabla I_\rho^{(1)}$ defines a one-to-one mapping of $\operatorname{int}(\operatorname{cc} S_\rho)$ onto $\mathbb{R}^d$ with inverse ∇c_ρ.

(c) Using the implicit function theorem for analytic functions [Bochner and Martin (1948, page 39)], prove that $I_\rho^{(1)}(z)$ is a real analytic function of $z \in \operatorname{int}(\operatorname{cc} S_\rho)$. [*Hint:* For any point $z \in \operatorname{int}(\operatorname{cc} S_\rho)$, there exists a unique point $t(z) \in \mathbb{R}^d$ such that $\nabla c_\rho(t(z)) = z$ [Theorem VIII.4.3(a)]. The Hessian matrix of c_ρ at $t(z)$ is positive-definite [Problem VIII.6.1(a)]. The mapping $\nabla c_\rho(t)$ can be continued into the complex space $\mathbb{C}^d$ to be an analytic mapping [Bochner and Martin (1948, page 34)].]

Chapter IX

Level-3 Large Deviations for I.I.D. Random Vectors

IX.1. Statement of Results

Theorem II.4.4 stated the level-3 large deviation property for i.i.d. random vectors taking values in $\mathbb{R}^d$. In this chapter, we prove Theorem II.4.4 in the special case of i.i.d. random variables with a finite state space. This version of the theorem covers the applications of level-3 large deviations which were made in Chapters III, IV, and V to the Gibbs variational principle. Theorem II.4.4 can also be proved via the methods of Donsker and Varadhan (1983a). The main result in that paper is a level-3 theorem for continuous parameter Markov processes taking values in a complete separable metric space.[1]

Let ρ be a Borel probability measure on $\mathbb{R}$ whose support is a finite set Γ. We topologize Γ by the discrete topology and $\Gamma^{\mathbb{Z}}$ by the product topology. With respect to the probability space $(\Gamma^{\mathbb{Z}}, \mathscr{B}(\Gamma^{\mathbb{Z}}), P_\rho)$, the coordinate representation process $X_j(\omega) = \omega_j$ is a sequence of i.i.d. random variables distributed by ρ. $\mathscr{M}_s(\Gamma^{\mathbb{Z}})$ denotes the set of strictly stationary probability measures on $\mathscr{B}(\Gamma^{\mathbb{Z}})$ with the topology of weak convergence. The empirical process is defined as

$$R_n(\omega, \cdot) = \frac{1}{n}\sum_{k=0}^{n-1} \delta_{T^k X(n,\omega)}(\cdot), \qquad n = 1, 2, \ldots, \quad \omega \in \Gamma^{\mathbb{Z}},$$

where T is the shift mapping on $\Gamma^{\mathbb{Z}}$ and $X(n, \omega)$ is the periodic point in $\Gamma^{\mathbb{Z}}$ obtained by repeating $(X_1(\omega), X_2(\omega), \ldots, X_n(\omega))$ periodically. For each Borel subset B of $\Gamma^{\mathbb{Z}}$, $R_n(\omega, B)$ is the relative frequency with which $X(n, \omega)$, $TX(n, \omega), \ldots, T^{n-1}X(n, \omega)$ is in B. Thus $R_n(\omega, \cdot)$ is for each ω an element of $\mathscr{M}_s(\Gamma^{\mathbb{Z}})$. For $P \in \mathscr{M}_s(\Gamma^{\mathbb{Z}})$, $\tilde{P}_\omega$ denotes a regular conditional distribution, with respect to P, of X_1 given the σ-field $\mathscr{F}\{X_j; j \le 0\}$. The level-3 entropy function is defined as

$$I_\rho^{(3)}(P) = \int_{\Gamma^{\mathbb{Z}}} I_\rho^{(2)}(\tilde{P}_\omega) P(d\omega), \tag{9.1}$$

where $I_\rho^{(2)}(\tilde{P}_\omega)$ is the relative entropy of $\tilde{P}_\omega$ with respect to ρ.

The following theorem is Theorem II.4.4 for the case of a finite state space.

Theorem IX.1.1. *Let ρ be a Borel probability measure on $\mathbb{R}$ whose support is a finite set Γ. Then the following conclusions hold.*

(a) $\{Q_n^{(3)}\}$, *the* P_ρ-*distributions on* $\mathcal{M}_s(\Gamma^{\mathbb{Z}})$ *of the empirical processes* $\{R_n\}$, *have a large deviation property with* $a_n = n$ *and entropy function* $I_\rho^{(3)}$.

(b) $I_\rho^{(3)}(P)$ *is an affine function of* P. $I_\rho^{(3)}(P)$ *measures the discrepancy between* P *and the infinite product measure* P_ρ *in the sense that* $I_\rho^{(3)}(P) \geq 0$ *with equality if and only if* $P = P_\rho$.

If A is a nonempty subset of $\mathcal{M}_s(\Gamma^{\mathbb{Z}})$, then $I_\rho^{(3)}(A)$ denotes the infimum of $I_\rho^{(3)}$ over A. $I_\rho^{(3)}(\phi)$ equals ∞. In order to prove part (a) of the theorem, we must verify the following hypotheses.

(i) $I_\rho^{(3)}(P)$ is lower semicontinuous.
(ii) $I_\rho^{(3)}(P)$ has compact level sets.
(iii) $\limsup_{n\to\infty} n^{-1} \log Q_n^{(3)}\{K\} \leq -I_\rho^{(3)}(K)$ for each closed set K in $\mathcal{M}_s(\Gamma^{\mathbb{Z}})$.
(iv) $\liminf_{n\to\infty} n^{-1} \log Q_n^{(3)}\{G\} \geq -I_\rho^{(3)}(G)$ for each open set G in $\mathcal{M}_s(\Gamma^{\mathbb{Z}})$.

Hypotheses (i) and (ii) and part (b) of the theorem will be proved in Section IX.2. We prove hypotheses (iii) and (iv) by first showing the large deviation bounds for finite-dimensional sets in $\mathcal{M}_s(\Gamma^{\mathbb{Z}})$. The proof of the bounds for such sets depends on the following facts.

(v) For each $\alpha \geq 1$ the distributions of the α-dimensional marginals of $\{R_n\}$ have a large deviation property with entropy function denoted by $I_{\rho,\alpha}^{(3)}$.
(vi) $I_{\rho,\alpha}^{(3)}$ is related to $I_\rho^{(3)}$ by the contraction principle

$$\inf\{I_\rho^{(3)}(P) \colon P \in \mathcal{M}_s(\Gamma^{\mathbb{Z}}), \pi_\alpha P = \tau\} = I_{\rho,\alpha}^{(3)}(\tau),$$

where $\tau = \pi_\alpha P$ is the fixed α-dimensional marginal of P.

Item (vi) is proved in Section IX.3; (v), (iii), and (iv) are proved in Section IX.4.

IX.2. Properties of the Level-3 Entropy Function

Let ρ be a Borel probability measure on $\mathbb{R}$ whose support is a finite set $\Gamma = \{x_1, x_2, \ldots, x_r\}$ with $x_1 < x_2 < \cdots < x_r$. Set $\rho_i = \rho\{x_i\} > 0$. Let α be a positive integer and π_α the projection of $\Gamma^{\mathbb{Z}}$ onto Γ^α defined by $\pi_\alpha \omega = (\omega_1, \ldots, \omega_\alpha)$. If P is a strictly stationary probability measure on $\mathscr{B}(\Gamma^{\mathbb{Z}})$, then define a probability measure $\pi_\alpha P$ on $\mathscr{B}(\Gamma^\alpha)$ by requiring

$$\pi_\alpha P\{F\} = P\{\pi_\alpha^{-1} F\} = P\{\omega \in \Gamma^{\mathbb{Z}} \colon (\omega_1, \ldots, \omega_\alpha) \in F\}$$

for subsets F of Γ^α. The measure $\pi_\alpha P$ is called the α-*dimensional marginal* of P. We consider the quantity

$$I_{\pi_\alpha P_\rho}^{(2)}(\pi_\alpha P) = \sum_{\omega \in \Gamma^\alpha} \pi_\alpha P\{\omega\} \log \frac{\pi_\alpha P\{\omega\}}{\pi_\alpha P_\rho\{\omega\}},$$

which is the relative entropy of $\pi_\alpha P$ with respect to $\pi_\alpha P_\rho$. In Chapter I, the level-3 entropy function $I_\rho^{(3)}(P)$ was defined as $\lim_{\alpha\to\infty} \alpha^{-1} I^{(2)}_{\pi_\alpha P_\rho}(\pi_\alpha P)$ [see (1.37)]. We will show in Theorem IX.2.3 that this limit exists and coincides with the quantity $I_\rho^{(3)}(P)$ defined in (9.1).

Given elements $x_{i_1}, \ldots, x_{i_\alpha}$ in Γ, set

$$p(x_{i_1}, \ldots, x_{i_\alpha}) = P\{X_1 = x_{i_1}, \ldots, X_\alpha = x_{i_\alpha}\}$$

and provided $P\{X_1 = x_{i_1}, \ldots, X_{\alpha-1} = x_{i_{\alpha-1}}\} > 0$, let $p(x_{i_\alpha}|x_{i_1}, \ldots, x_{i_{\alpha-1}})$ denote the conditional probability $P\{X_\alpha = x_{i_\alpha}|X_1 = x_{i_1}, \ldots, X_{\alpha-1} = x_{i_{\alpha-1}}\}$. We define

$$H_{\rho,\alpha}(P) = \begin{cases} \sum_{i_1=1}^{r} p(x_{i_1}) \log[p(x_{i_1})/\rho_{i_1}] & \text{for } \alpha = 1, \\ \sum p(x_{i_1}, \ldots, x_{i_\alpha}) \log[p(x_{i_\alpha}|x_{i_1}, \ldots, x_{i_{\alpha-1}})/\rho_{i_\alpha}] & \text{for } \alpha \geq 2. \end{cases}$$

For $\alpha \geq 2$ the sum defining $H_{\rho,\alpha}(P)$ runs over all $i_1, \ldots, i_{\alpha-1}, i_\alpha$ for which $p(x_{i_1}, \ldots, x_{i_{\alpha-1}}) > 0$. Since $0 \log 0 = 0$, the sum may be restricted to all $i_1, \ldots, i_\alpha$ for which $p(x_{i_1}, \ldots, x_{i_\alpha}) > 0$. The quantity $-\sum p(x_{i_1}, \ldots, x_{i_\alpha}) \cdot \log p(x_{i_\alpha}|x_{i_1}, \ldots, x_{i_{\alpha-1}})$ is known as the *conditional entropy of X_α given $X_1, \ldots, X_{\alpha-1}$*.

Lemma IX.2.1 (a) *Let $p_1, \ldots, p_N$ and $q_1, \ldots, q_N$ be non-negative real numbers such that $\sum_{i=1}^N p_i = 1$, $\sum_{i=1}^N q_i = 1$, and $p_i = 0$ whenever $q_i = 0$. Then*

$$\sum_{i=1}^{N} p_i \log p_i \geq \sum_{i=1}^{N} p_i \log q_i$$

with equality if and only if $p_i = q_i$ for all i.

(b) $I^{(2)}_{\pi_\alpha P_\rho}(\pi_\alpha P) = H_{\rho,1}(P) + H_{\rho,2}(P) + \cdots + H_{\rho,\alpha-1}(P) + H_{\rho,\alpha}(P)$.

(c) $0 \leq H_{\rho,1}(P) \leq H_{\rho,2}(P) \leq \cdots \leq H_{\rho,\alpha-1}(P) \leq H_{\rho,\alpha}(P)$.

(d) $H_{\rho,2}(P) = H_{\rho,1}(P)$ *if and only if X_1 and X_2 are independent.*

(e) *For $\alpha \geq 3$, $H_{\rho,\alpha}(P) = H_{\rho,\alpha-1}(P)$ if and only if X_1 and X_α are conditionally independent given $X_2, \ldots, X_{\alpha-1}$; that is, if and only if*

$$p(x_{i_\alpha}|x_{i_1}, x_{i_2}, \ldots, x_{i_{\alpha-1}}) = p(x_{i_\alpha}|x_{i_2}, \ldots, x_{i_{\alpha-1}})$$
$$\text{whenever } p(x_{i_1}, x_{i_2}, \ldots, x_{i_{\alpha-1}}) > 0.*$$

Proof. (a) We have

$$\sum_{i=1}^{N} p_i \log p_i - \sum_{i=1}^{N} p_i \log q_i = \sum_{q_i>0} p_i \log \frac{p_i}{q_i}.$$

The latter is non-negative and equals 0 if and only if $p_i = q_i$ whenever $q_i > 0$ [Proposition I.4.1(b)]. Since by hypothesis $p_i = q_i$ whenever $q_i = 0$, the proof of part (a) is done.

* See page 301.

(b) If $\alpha \geq 2$ and $p(x_{i_1}, \ldots, x_{i_\alpha}) > 0$, then

$$p(x_{i_1}, \ldots, x_{i_\alpha}) = p(x_{i_1}) \cdot \prod_{\beta=2}^{\alpha} p(x_{i_\beta} | x_{i_1}, \ldots, x_{i_{\beta-1}}).$$

Hence for $\alpha \geq 2$

$$I^{(2)}_{\pi_\alpha P_\rho}(\pi_\alpha P) = \sum p(x_{i_1}, \ldots, x_{i_\alpha}) \left[\log \frac{p(x_{i_1})}{\rho_{i_1}} + \sum_{\beta=2}^{\alpha} \log \frac{p(x_{i_\beta} | x_{i_1}, \ldots, x_{i_{\beta-1}})}{\rho_{i_\beta}} \right]$$

$$= \sum_{\beta=1}^{\alpha} H_{\rho,\beta}(P).$$

(c)–(e) $H_{\rho,1}(P)$ equals the relative entropy of $\pi_1 P$ with respect to ρ and so is non-negative. Since P is strictly stationary,

$$H_{\rho,1}(P) = \sum_{i_1, i_2 = 1}^{r} p(x_{i_1}, x_{i_2}) \log \frac{p(x_{i_2})}{\rho_{i_2}}.$$

The sum may be restricted to i_1, i_2 for which $p(x_{i_1}) > 0$. For if $p(x_{i_1}) = 0$, then $p(x_{i_1}, x_{i_2}) = 0$ for all i_2, and there is no contribution to the sum. Thus

$$H_{\rho,2}(P) - H_{\rho,1}(P) = \sum_{p(x_{i_1})>0} p(x_{i_1}) \left\{ \sum_{i_2=1}^{r} p(x_{i_2} | x_{i_1}) \log p(x_{i_2} | x_{i_1}) - \sum_{i_2=1}^{r} p(x_{i_2} | x_{i_1}) \log p(x_{i_2}) \right\}.$$

By part (a), $H_{\rho,2}(P) \geq H_{\rho,1}(P)$ with equality if and only if $p(x_{i_2} | x_{i_1}) = p(x_{i_2})$ whenever $p(x_{i_1}) > 0$. The latter condition is equivalent to the independence of X_1 and X_2. Similarly, for $\alpha \geq 3$,

$$H_{\rho,\alpha}(P) - H_{\rho,\alpha-1}(P)$$
$$= \sum p(x_{i_1}, \ldots, x_{i_{\alpha-1}}) \left\{ \sum_{i_\alpha=1}^{r} p(x_{i_\alpha} | x_{i_1}, \ldots, x_{i_{\alpha-1}}) \log p(x_{i_\alpha} | x_{i_1}, \ldots, x_{i_{\alpha-1}}) - \sum_{i_\alpha=1}^{r} p(x_{i_\alpha} | x_{i_1}, \ldots, x_{i_{\alpha-1}}) \log p(x_{i_\alpha} | x_{i_2}, \ldots, x_{i_{\alpha-1}}) \right\},$$

where the outer sum runs over all $i_1, \ldots, i_{\alpha-1}$ for which $p(x_{i_1}, \ldots, x_{i_{\alpha-1}}) > 0$. By part (a), $H_{\rho,\alpha}(P) \geq H_{\rho,\alpha-1}(P)$ with equality if and only if

$$p(x_{i_\alpha} | x_{i_1}, \ldots, x_{i_{\alpha-1}}) = p(x_{i_\alpha} | x_{i_2}, \ldots, x_{i_{\alpha-1}}) \quad \text{whenever } p(x_{i_1}, \ldots, x_{i_{\alpha-1}}) > 0. \qquad \square$$

We also need the following lemma, which arises in the proof of the Shannon–McMillan–Breiman theorem in information theory [Billingsley (1965, page 131)].

Lemma IX.2.2. *For each* $i_1 \in \{1, \ldots, r\}$,

$$0 \leq \int_{\{X_1 = x_{i_1}\}} \sup_{n \geq 0} [-\log P\{X_1 = x_{i_1} | X_{-n}, \ldots, X_0\}(\omega)] P(d\omega) < \infty.$$

Proof. For n a non-negative integer, $i \in \{1, \ldots, r\}$, and $\omega \in \Gamma^{\mathbb{Z}}$, define

$$f_n^{(i)}(\omega) = -\log P\{X_1 = x_i | X_{-n}, \ldots, X_0\}(\omega),$$

$$g_n(\omega) = \sum_{i=1}^{r} f_n^{(i)}(\omega) \cdot \chi_{\{X_1 = x_i\}}(\omega).$$

Since $0 \leq P\{X_1 = x_i | X_{-n}, \ldots, X_0\}(\omega) \leq 1$ P-a.s., it suffices to prove that

$$\int_{\Gamma^{\mathbb{Z}}} \sup_{n \geq 0} g_n(\omega) P(d\omega) < \infty. \tag{9.2}$$

If $A_n = \{\omega \in \Gamma^{\mathbb{Z}} : \max_{0 \leq j < n} g_j(\omega) \leq \lambda < g_n(\omega)\}$, $\lambda \geq 0$, then

$$P\{A_n\} = \sum_{i=1}^{r} P\{\{X_1 = x_i\} \cap A_n\} = \sum_{i=1}^{r} P\{\{X_1 = x_i\} \cap B_n^{(i)}\},$$

where $B_n^{(i)} = \{\omega \in \Gamma^{\mathbb{Z}} : \max_{0 \leq j < n} f_j^{(i)}(\omega) \leq \lambda < f_n^{(i)}(\omega)\}$. $B_n^{(i)}$ is in the σ-field generated by $X_{-n}, \ldots, X_0$, and so

$$\begin{aligned} P\{\{X_1 = x_i\} \cap B_n^{(i)}\} &= \int_{B_n^{(i)}} P\{X_1 = x_i | X_{-n}, \ldots, X_0\}(\omega) P(d\omega) \\ &= \int_{B_n^{(i)}} \exp(-f_n^{(i)}(\omega)) P(d\omega) \leq e^{-\lambda} P\{B_n^{(i)}\}. \end{aligned}$$

Since the sets $\{B_n^{(i)}; n = 0, 1, \ldots\}$ are disjoint,

$$\sum_{n=0}^{\infty} P\{A_n\} \leq e^{-\lambda} \sum_{i=1}^{r} \sum_{n=0}^{\infty} P\{B_n^{(i)}\} \leq r e^{-\lambda}.$$

Hence $P\{\omega \in \Gamma^{\mathbb{Z}} : \sup_{n \geq 0} g_n(\omega) > \lambda\} \leq r e^{-\lambda}$, and (9.2) follows. □

Theorem IX.2.3. *For* $P \in \mathcal{M}_s(\Gamma^{\mathbb{Z}})$, *define* $I_\rho^{(3)}(P)$ *by* (9.1). *Then* $\lim_{\alpha \to \infty} \alpha^{-1} I_{\pi_\alpha P_\rho}^{(2)}(\pi_\alpha P)$ *exists,* $\lim_{\alpha \to \infty} H_{\rho,\alpha}(P)$ *exists, and*

$$\lim_{\alpha \to \infty} \frac{1}{\alpha} I_{\pi_\alpha P_\rho}^{(2)}(\pi_\alpha P) = \lim_{\alpha \to \infty} H_{\rho,\alpha}(P) = I_\rho^{(3)}(P) < \infty.*$$

Proof. By Lemma IX.2.1(b), it suffices to prove that $\lim_{\alpha \to \infty} H_{\rho,\alpha}(P)$ exists and equals $I_\rho^{(3)}(P)$. Since P is strictly stationary, we have for $n \geq 0$

$$H_{\rho,n+2}(P) = \sum_{i_1=1}^{r} \int_{\{X_1 = x_{i_1}\}} \log[P\{X_1 = x_{i_1} | X_{-n}, \ldots, X_0\}/\rho_{i_1}] dP.$$

*There is an elementary proof of the existence of $\lim_{\alpha \to \infty} \alpha^{-1} I_{\pi_\alpha P_\rho}^{(2)}(\pi_\alpha P)$ which does not identify the form of the limit. See Note 2.

As $n \to \infty$, $P\{X_1 = x_{i_1} | X_{-n}, \ldots, X_0\}$ converges to $P\{X_1 = x_{i_1} | \{X_j; j \leq 0\}\}$ almost surely [Theorem A.6.2(a)]. By the Lebesgue dominated convergence theorem and Lemma IX.2.2, we conclude that $\lim_{\alpha\to\infty} H_{\rho,\alpha}(P)$ exists and*

$$\begin{aligned}
\lim_{\alpha\to\infty} H_{\rho,\alpha}(P) &= \sum_{i_1=1}^{r} \int_{\{X_1 = x_{i_1}\}} \log[P\{X_1 = x_{i_1} | \{X_j; j \leq 0\}\}/\rho_{i_1}] dP \\
&= \sum_{i_1=1}^{r} \int_{\Gamma^{\mathbb{Z}}} \log[P\{X_1 = x_{i_1} | \{X_j; j \leq 0\}\}/\rho_{i_1}] \\
&\quad \cdot P\{X_1 = x_{i_1} | \{X_j; j \leq 0\}\} dP \\
&= \int_{\Gamma^{\mathbb{Z}}} \int_{\Gamma} \log \frac{\tilde{P}_\omega\{x\}}{\rho\{x\}} \tilde{P}_\omega(dx) P(d\omega) \\
&= \int_{\Gamma^{\mathbb{Z}}} I_\rho^{(2)}(\tilde{P}_\omega) P(d\omega) = I_\rho^{(3)}(P) < \infty. \qquad \square
\end{aligned}$$

In order to prove properties of $I_\rho^{(3)}(P)$, we need a standard lemma.

Lemma IX.2.4. *Let $b_1, b_2, \ldots$ be a superadditive sequence of real numbers; i.e., $b_{m+n} \geq b_m + b_n$ for all positive integers m and n. Then $\lim_{n\to\infty} b_n/n = \sup_{n\geq 1} b_n/n$.*

Proof. Define $s = \sup_{n\geq 1} b_n/n$ and suppose that s is not ∞. Given $\varepsilon > 0$ choose a positive integer k such that $b_k/k > s - \varepsilon$. Any positive integer n has the form $n = mk + l$, where $m \geq 0$ and $0 \leq l < k$. By hypothesis

$$b_n = b_{mk+l} \geq mb_k + lb_1.$$

Hence $s \geq \liminf_{n\to\infty} b_n/n \geq \liminf_{n\to\infty} mb_k/n = b_k/k > s - \varepsilon$. Since $\varepsilon > 0$ is arbitrary, it follows that $\lim_{n\to\infty} b_n/n = s$ if s is finite. The proof is similar if s is ∞. $\square$

$I_\rho^{(3)}(P)$ is lower semicontinuous. We first prove that for $P \in \mathcal{M}_s(\Gamma^{\mathbb{Z}})$ the sequence of relative entropies $I_\alpha = I_{\pi_\alpha P_\rho}^{(2)}(\pi_\alpha P)$ is superadditive. Let α and β be arbitrary positive integers and introduce the notation

$$\omega \circ \bar{\omega} = (\omega_1, \omega_2, \ldots, \omega_\alpha, \bar{\omega}_1, \bar{\omega}_2, \ldots, \bar{\omega}_\beta) \in \Gamma^{\alpha+\beta} \qquad \text{for } \omega \in \Gamma^\alpha, \bar{\omega} \in \Gamma^\beta.$$

Since $\pi_{\alpha+\beta} P_\rho\{\omega \circ \bar{\omega}\} = \pi_\alpha P_\rho\{\omega\} \cdot \pi_\beta P_\rho\{\bar{\omega}\}$,

$$I_{\alpha+\beta} = \sum_{\omega\in\Gamma^\alpha} \sum_{\bar{\omega}\in\Gamma^\beta} \pi_{\alpha+\beta} P\{\omega \circ \bar{\omega}\} \log \frac{\pi_{\alpha+\beta} P\{\omega \circ \bar{\omega}\}}{\pi_\alpha P_\rho\{\omega\} \cdot \pi_\beta P_\rho\{\bar{\omega}\}}.$$

Since P is strictly stationary, $\sum_{\omega\in\Gamma^\alpha} \pi_{\alpha+\beta} P\{\omega \circ \bar{\omega}\} = \pi_\beta P\{\bar{\omega}\}$, and so

* The second equality in the display uses (A.3) [page 300].

$$I_{\alpha+\beta} - I_\alpha - I_\beta = \sum_{\omega\in\Gamma^\alpha}\sum_{\bar\omega\in\Gamma^\beta} \pi_{\alpha+\beta}P\{\omega\circ\bar\omega\}\log\pi_{\alpha+\beta}P\{\omega\circ\bar\omega\}$$
$$- \sum_{\omega\in\Gamma^\alpha}\sum_{\bar\omega\in\Gamma^\beta} \pi_{\alpha+\beta}P\{\omega\circ\bar\omega\}\log[\pi_\alpha P\{\omega\}\cdot\pi_\beta P\{\bar\omega\}].$$

By Lemma IX.2.1(a) $I_{\alpha+\beta} - I_\alpha - I_\beta \geq 0$. It follows from Lemma IX.2.4 that

$$I_\rho^{(3)}(P) = \lim_{\alpha\to\infty}\frac{1}{\alpha}I^{(2)}_{\pi_\alpha P_\rho}(\pi_\alpha P) = \sup_{\alpha\geq 1}\frac{1}{\alpha}I^{(2)}_{\pi_\alpha P_\rho}(\pi_\alpha P).$$

Since the mapping $P \to \pi_\alpha P$ is continuous, each function $I^{(2)}_{\pi_\alpha P_\rho}(\pi_\alpha P)$ is a continuous function of $P\in\mathscr{M}_s(\Gamma^{\mathbb{Z}})$. As the supremum of a sequence of continuous functions, $I_\rho^{(3)}(P)$ is a lower semicontinuous function of $P\in\mathscr{M}_s(\Gamma^{\mathbb{Z}})$.

$I_\rho^{(3)}(P)$ has compact level sets. $\mathscr{M}_s(\Gamma^{\mathbb{Z}})$ is a compact metric space [Theorem A.9.2(c)]. Since $I_\rho^{(3)}(P)$ is lower semicontinuous, its level sets are closed subsets of $\mathscr{M}_s(\Gamma^{\mathbb{Z}})$ and thus are compact subsets of $\mathscr{M}_s(\Gamma^{\mathbb{Z}})$.

$I_\rho^{(3)}(P)$ is affine. We show that if P and Q are measures in $\mathscr{M}_s(\Gamma^{\mathbb{Z}})$, then for every $0 < \lambda < 1$

$$I_\rho^{(3)}(\lambda P + (1-\lambda)Q) = \lambda I_\rho^{(3)}(P) + (1-\lambda)I_\rho^{(3)}(Q). \tag{9.3}$$

Write $p(\omega)$ for $\pi_\alpha P\{\omega\}/\pi_\alpha P_\rho\{\omega\}$ and $q\{\omega\}$ for $\pi_\alpha Q\{\omega\}/\pi_\alpha P_\rho\{\omega\}$, $\omega\in\Gamma^\alpha$. Since $x\log x$, $x\geq 0$, is convex and $\log x$, $x>0$, is nondecreasing,

$$\begin{aligned}
\lambda I^{(2)}_{\pi_\alpha P_\rho}(P) + (1-\lambda)I^{(2)}_{\pi_\alpha P_\rho}(Q) &= \sum_{\omega\in\Gamma^\alpha}\pi_\alpha P_\rho\{\omega\}[\lambda p(\omega)\log p(\omega) \\
&\qquad + (1-\lambda)q(\omega)\log q(\omega)] \\
&\geq \sum_{\omega\in\Gamma^\alpha}\pi_\alpha P_\rho\{\omega\}[\lambda p(\omega) + (1-\lambda)q(\omega)] \\
&\qquad \cdot\log[\lambda p(\omega) + (1-\lambda)q(\omega)] \\
&\geq \sum_{\omega\in\Gamma^\alpha}\pi_\alpha P_\rho\{\omega\}[\lambda p(\omega)\log(\lambda p(\omega)) \\
&\qquad + (1-\lambda)q(\omega)\log((1-\lambda)q(\omega))] \\
&= \lambda\sum_{\omega\in\Gamma^\alpha}\pi_\alpha P_\rho\{\omega\}\cdot p(\omega)\log p(\omega) \\
&\qquad + (1-\lambda)\sum_{\omega\in\Gamma^\alpha}\pi_\alpha P_\rho\{\omega\}\cdot q(\omega)\log q(\omega) \\
&\qquad + \lambda\log\lambda + (1-\lambda)\log(1-\lambda) \\
&\geq \lambda I^{(2)}_{\pi_\alpha P_\rho}(P) + (1-\lambda)I^{(2)}_{\pi_\alpha P_\rho}(Q) - \log 2.
\end{aligned}$$

The sum on the third and fourth lines is $I^{(2)}_{\pi_\alpha P_\rho}(\lambda P + (1-\lambda)Q)$. Dividing each term in the display by α and letting α tend to ∞, we obtain (9.3).

$I_\rho^{(3)}(P)\geq 0$ with equality if and only if $P = P_\rho$. Given $P\in\mathscr{M}_s(\Gamma^{\mathbb{Z}})$, let ν equal $\pi_1 P$. If $\nu = \rho$, then by Theorem IX.3.1 below

$$I_\rho^{(3)}(P) \geq I_\rho^{(2)}(\rho) = 0 \qquad \text{with equality if and only if } P = P_\rho.$$

If $\nu \neq \rho$, then by Theorems IX.3.1 and VIII.2.1(b)

$$I_\rho^{(3)}(P) \geq I_\rho^{(2)}(\nu) > I_\rho^{(2)}(\rho) = 0.$$

This completes the proofs of properties of $I_\rho^{(3)}$.

IX.3. Contraction Principles

The first theorem is a contraction principle relating levels-2 and 3.

Theorem IX.3.1. *Let ν be a probability measure on $\mathscr{B}(\Gamma)$ and P_ν the infinite product measure on $\mathscr{B}(\Gamma^{\mathbb{Z}})$ with $\pi_1 P = \nu$. Then $I_\rho^{(3)}(P)$ attains its infimum over the set $\{P \in \mathscr{M}_s(\Gamma^{\mathbb{Z}}) : \pi_1 P = \nu\}$ at the unique measure P_ν and*

$$\inf\{I_\rho^{(3)}(P) : P \in \mathscr{M}_s(\Gamma^{\mathbb{Z}}),\ \pi_1 P = \nu\} = I_\rho^{(3)}(P_\nu) = I_\rho^{(2)}(\nu).$$

Proof. Let $\mathscr{A}$ be the set $\{P \in \mathscr{M}_s(\Gamma^{\mathbb{Z}}) : \pi_1 P = \nu\}$ and P any measure in $\mathscr{A}$. $H_{\rho,1}(P)$ equals $\sum_{i=1}^r \nu_i \log(\nu_i/\rho_i) = I_\rho^{(2)}(\nu)$ $(\nu_i = \nu\{x_i\})$. By Lemma IX.2.1(c) and Theorem IX.2.3, for any $k \geq 2$

$$0 \leq H_{\rho,1}(P) = I_\rho^{(2)}(\nu) \leq \cdots \leq H_{\rho,k}(P) \leq H_{\rho,k+1}(P) \uparrow I_\rho^{(3)}(P). \tag{9.4}$$

If $P = P_\nu$, then for any $k \geq 1$ $H_{\rho,k}(P_\nu) = I_\rho^{(2)}(\nu) = I_\rho^{(3)}(P_\nu)$. We see that $I_\rho^{(3)}(P)$ attains its infimum over $\mathscr{A}$ at the measure $P = P_\nu$ and that the infimum equals $I_\rho^{(2)}(\nu)$. The proof is done once we show that $P = P_\nu$ is the unique measure in $\mathscr{A}$ for which $I_\rho^{(3)}(P)$ equals $I_\rho^{(2)}(\nu)$.

Suppose that $I_\rho^{(3)}(P)$ equals $I_\rho^{(2)}(\nu)$ for some P in $\mathscr{A}$. We prove by induction that for all $k \geq 1$

$$p(x_{i_1}, \ldots, x_{i_k}) = \nu_{i_1} \cdots \nu_{i_k}. \tag{9.5}$$

This will imply that P equals P_ν. Formula (9.5) holds for $k = 1$ since $\pi_1 P$ equals ν. Since $I_\rho^{(3)}(P) = I_\rho^{(2)}(\nu)$, (9.4) shows that $H_{\rho,k}(P)$ equals $H_{\rho,1}(P)$ for all $k \geq 2$. With respect to P, X_1 and X_2 are independent [Lemma IX.2.1(d)], and so (9.5) holds for $k = 2$. Assume that (9.5) has been shown for $k = 1, 2, \ldots, c-1$, some $c \geq 3$. With respect to P, X_1 and X_c are conditionally independent given $X_2, \ldots, X_{c-1}$ [Lemma IX.2.1(e)]. If $p(x_{i_1}, \ldots, x_{i_{c-1}}) > 0$, then by the induction hypothesis and strict stationarity

$$p(x_{i_1}, \ldots, x_{i_c}) = p(x_{i_1}, \ldots, x_{i_{c-1}}) p(x_{i_c} | x_{i_2}, \ldots, x_{i_{c-1}}) = \nu_{i_1} \cdots \nu_{i_{c-1}} \nu_{i_c}.$$

If $p(x_{i_1}, \ldots, x_{i_{c-1}}) = 0$, then $p(x_{i_1}, \ldots, x_{i_c}) = 0 = \nu_{i_1} \cdots \nu_{i_c}$. Thus (9.5) holds for $k = c$. The proof of the theorem is complete. □

We now generalize the contraction principle just proved by calculating the infimum of $I_\rho^{(3)}(P)$ over all measures $P \in \mathscr{M}_s(\Gamma^{\mathbb{Z}})$ with fixed marginal $\tau = \pi_\alpha P$, $\alpha \in \{2, 3, \ldots\}$. Denote by $\mathscr{M}_s(\Gamma^\alpha)$ the set of probability measures τ

on $\mathscr{B}(\Gamma^\alpha)$ which have the form $\tau = \pi_\alpha P$ for some $P \in \mathscr{M}_s(\Gamma^{\mathbb{Z}})$. For $\tau \in \mathscr{M}_s(\Gamma^\alpha)$, set $\tau_{i_1 \cdots i_\alpha} = \tau\{x_{i_1}, \ldots, x_{i_\alpha}\}$ and $(v_\tau)_{i_1 \cdots i_{\alpha-1}} = \sum_{i_\alpha=1}^r \tau_{i_1 \cdots i_\alpha}$ and define

$$(9.6) \qquad I_{\rho,\alpha}^{(3)}(\tau) = \sum \tau_{i_1 \cdots i_\alpha} \log \frac{\tau_{i_1 \cdots i_\alpha}}{(v_\tau)_{i_1 \cdots i_{\alpha-1}} \rho_{i_\alpha}},$$

where the sum runs over all $i_1, \ldots, i_{\alpha-1}, i_\alpha$ for which $(v_\tau)_{i_1 \cdots i_{\alpha-1}} > 0$. $I_{\rho,\alpha}^{(3)}(\tau)$ is well defined ($0 \log 0 = 0$) and equals the relative entropy of τ with respect to $\{(v_\tau)_{i_1 \cdots i_{\alpha-1}} \rho_{i_\alpha}\}$. We have encountered the function $I_{\rho,2}^{(3)}$ as the entropy function in Theorem I.5.1, which was a large deviation result for the quantities $\{M_n(\omega, \cdot)\}$ (empirical pair measures). The latter measures are related to the empirical processes by the formula $M_n(\omega, \cdot) = \pi_2 R_n(\omega, \cdot)$ [see (1.35)].

In the next section, the distributions of $\{\pi_\alpha R_n(\omega, \cdot)\}$ on $\mathscr{M}_s(\Gamma^\alpha)$ will be shown to have a large deviation property with entropy function $I_{\rho,\alpha}^{(3)}$. Hence $I_{\rho,\alpha}^{(3)}$ is called the *level*-3,α *entropy function.* We now prove that for fixed $\tau \in \mathscr{M}_s(\Gamma^\alpha)$ $I_{\rho,\alpha}^{(3)}(\tau)$ equals the infimum of $I_\rho^{(3)}(P)$ over the set $\mathscr{A}_\alpha = \{P \in \mathscr{M}_s(\Gamma^{\mathbb{Z}}): \pi_\alpha P = \tau\}$. We also show that $I_\rho^{(3)}(P)$ attains its infimum over $\mathscr{A}_\alpha$ at an $(\alpha - 1)$-dependent Markov chain. The latter measures are defined for $\alpha = 2$ in Example A.7.3(b) and for $\alpha \geq 3$ in Appendix A.10.

Lemma IX.3.2. *Let $\alpha \geq 2$ be an integer.*

(a) *A probability measure τ on $\mathscr{B}(\Gamma^\alpha)$ belongs to $\mathscr{M}_s(\Gamma^\alpha)$ if and only if*

$$(9.7) \qquad \sum_{j=1}^r \tau_{i_1 \cdots i_{\alpha-1} j} = \sum_{k=1}^r \tau_{k i_1 \cdots i_{\alpha-1}} \qquad \text{for each } i_1, \ldots, i_{\alpha-1}.$$

(b) *Let τ be a measure in $\mathscr{M}_s(\Gamma^\alpha)$. Define $\mathscr{M}_\tau$ to be the subset of $\mathscr{M}_s(\Gamma^{\mathbb{Z}})$ consisting of $(\alpha - 1)$-dependent Markov chains which satisfy $\pi_\alpha P = \tau$. Then $\mathscr{M}_\tau$ is nonempty. If $(v_\tau)_{i_1 \cdots i_{\alpha-1}} > 0$ for each $i_1, \ldots, i_{\alpha-1}$, then $\mathscr{M}_\tau$ consists of a unique measure.*

Proof. (a) If τ belongs to $\mathscr{M}_s(\Gamma^\alpha)$, then τ equals $\pi_\alpha P$ for some $P \in \mathscr{M}_s(\Gamma^{\mathbb{Z}})$, and for each $i_1, \ldots, i_{\alpha-1}$,

$$\sum_{j=1}^r \tau_{i_1 \cdots i_{\alpha-1} j} = P\{\omega \in \Gamma^{\mathbb{Z}} : \omega_1 = x_{i_1}, \ldots, \omega_{\alpha-1} = x_{i_{\alpha-1}}\}$$

$$= \sum_{k=1}^r P\{\omega \in \Gamma^{\mathbb{Z}} : \omega_0 = x_k, \omega_1 = x_{i_1}, \ldots, \omega_{\alpha-1} = x_{i_{\alpha-1}}\}.$$

The last sum equals $\sum_{k=1}^r \tau_{k i_1 \cdots i_{\alpha-1}}$, and so (9.7) holds. Now suppose that (9.7) holds. We write v for v_τ. For each $i_1, \ldots, i_{\alpha-1} \in \{1, \ldots, r\}$, define

$$\gamma_{i_1 \cdots i_\alpha} = \frac{\tau_{i_1 \cdots i_\alpha}}{v_{i_1 \cdots i_{\alpha-1}}}, \qquad i_\alpha = 1, \ldots, r, \qquad \text{if } v_{i_1 \cdots i_{\alpha-1}} > 0.$$

If $v_{i_1 \cdots i_{\alpha-1}} = 0$, then define $\{\gamma_{i_1 \cdots i_\alpha}; i_\alpha = 1, \ldots, r\}$ to be any non-negative real numbers which sum to 1. We have $v_{i_1 \cdots i_{\alpha-1}} \geq 0$, $\sum_{i_1, \ldots, i_{\alpha-1}=1}^r v_{i_1 \cdots i_{\alpha-1}} = 1$, and by (9.7)

$$\sum_{j=1}^{r} v_{i_1\cdots i_{\alpha-2}j} = \sum_{j,l=1}^{r} \tau_{i_1\cdots i_{\alpha-2}jl} = \sum_{j,k=1}^{r} \tau_{ki_1\cdots i_{\alpha-2}j} = \sum_{k=1}^{r} v_{ki_1\cdots i_{\alpha-2}}.$$

Also $\gamma_{i_1\cdots i_\alpha} \geq 0$, $\sum_{i_\alpha=1}^{r} \gamma_{i_1\cdots i_\alpha} = 1$, $v_{i_1\cdots i_{\alpha-1}}\gamma_{i_1\cdots i_\alpha} = \tau_{i_1\cdots i_\alpha}$ and by (9.7),

$$\sum_{i_1=1}^{r} v_{i_1\cdots i_{\alpha-1}}\gamma_{i_1\cdots i_\alpha} = \sum_{i_1=1}^{r} \tau_{i_1\cdots i_\alpha} = v_{i_2\cdots i_\alpha}.$$

Hence assumptions (A.5) and (A.6) in Appendix A.10 are satisfied. By Theorem A.10.1, there exists a unique $(\alpha - 1)$-dependent Markov chain P in $\mathcal{M}_s(\Gamma^{\mathbb{Z}})$ with transition array $\gamma = \{\gamma_{i_1\cdots i_\alpha}\}$ and invariant measure $v = \sum_{i_1,\ldots,i_{\alpha-1}=1}^{r} v_{i_1\cdots i_{\alpha-1}} \delta_{\{x_{i_1},\ldots,x_{i_{\alpha-1}}\}}$. The marginal $\pi_\alpha P$ equals τ since

$$\begin{aligned}\pi_\alpha P\{x_{i_1},\ldots,x_{i_\alpha}\} &= P\{X_1 = x_{i_1},\ldots,X_\alpha = x_{i_\alpha}\}\\ &= v_{i_1\cdots i_{\alpha-1}}\gamma_{i_1\cdots i_\alpha} = \tau_{i_1\cdots i_\alpha}.\end{aligned}$$

(b) Part (a) shows that $\mathcal{M}_\tau$ is nonempty. Suppose that $v_{i_1\cdots i_{\alpha-1}}$ is positive for each $i_1,\ldots,i_{\alpha-1}$. If P is an $(\alpha - 1)$-dependent Markov chain in $\mathcal{M}_s(\Gamma^{\mathbb{Z}})$ satisfying $\pi_\alpha P = \tau$, then P must have invariant measure v and transition array $\{\tau_{i_1\cdots i_\alpha}/v_{i_1\cdots i_{\alpha-1}}\}$. We conclude that P is unique. □

Theorem IX.3.3. *Let τ be a measure in $\mathcal{M}_s(\Gamma^\alpha)$, $\alpha \in \{2, 3, \ldots\}$. Define $\mathcal{M}_\tau$ to be the subset of $\mathcal{M}_s(\Gamma^{\mathbb{Z}})$ consisting of $(\alpha - 1)$-dependent Markov chains which satisfy $\pi_\alpha P = \tau$. Then $I_\rho^{(3)}(P)$ attains its infimum over the set $\{P \in \mathcal{M}_s(\Gamma^{\mathbb{Z}})$: $\pi_\alpha P = \tau\}$ at all $P \in \mathcal{M}_\tau$ and only at such P. For any $P \in \mathcal{M}_\tau$*

$$\inf\{I_\rho^{(3)}(P) : P \in \mathcal{M}_s(\Gamma^{\mathbb{Z}}), \pi_\alpha P = \tau\} = I_\rho^{(3)}(P) = I_{\rho,\alpha}^{(3)}(\tau).$$

Proof. Let $\mathcal{A}_\alpha$ be the set $\{P \in \mathcal{M}_s(\Gamma^{\mathbb{Z}}) : \pi_\alpha P = \tau\}$ and P any measure in $\mathcal{A}_\alpha$. If $(v_\tau)_{i_1\cdots i_{\alpha-1}} = p(x_{i_1},\ldots,x_{i_{\alpha-1}}) > 0$, then $p(x_{i_\alpha}|x_{i_1},\ldots,x_{i_{\alpha-1}})$ equals $\tau_{i_1\cdots i_\alpha}/(v_\tau)_{i_1\cdots i_{\alpha-1}}$. Hence

$$H_{\rho,\alpha}(P) = \sum \tau_{i_1\cdots i_\alpha} \log \frac{\tau_{i_1\cdots i_\alpha}}{(v_\tau)_{i_1\cdots i_{\alpha-1}}\rho_{i_\alpha}} = I_{\rho,\alpha}^{(3)}(\tau).$$

By Lemma IX.2.1(c) and Theorem IX.2.3, for any $k \geq \alpha + 1$

$$0 \leq H_{\rho,\alpha}(P) = I_{\rho,\alpha}^{(3)}(\tau) \leq \cdots \leq H_{\rho,k}(P) \leq H_{\rho,k+1}(P) \uparrow I_\rho^{(3)}(P). \tag{9.8}$$

If P is an $(\alpha - 1)$-dependent Markov chain in $\mathcal{M}_\tau$, then for any $k \geq \alpha$ $H_{\rho,k}(P) = I_{\rho,\alpha}^{(3)}(\tau) = I_\rho^{(3)}(P)$. We see that $I_\rho^{(3)}(P)$ attains its infimum over $\mathcal{A}_\alpha$ at $P \in \mathcal{M}_\tau$ and that the infimum equals $I_{\rho,\alpha}^{(3)}(\tau)$. The proof is done once we show that the measures P in $\mathcal{M}_\tau$ are the unique measures in $\mathcal{A}_\alpha$ for which $I_\rho^{(3)}(P)$ equals $I_{\rho,\alpha}^{(3)}(\tau)$.

Suppose that $I_\rho^{(3)}(P)$ equals $I_{\rho,\alpha}^{(3)}(\tau)$ for some P in $\mathcal{A}_\alpha$. We prove by induction that for all $k \geq \alpha$

$$p(x_{i_1},\ldots,x_{i_k}) = (v_\tau)_{i_1\cdots i_{\alpha-1}}\gamma_{i_1\cdots i_\alpha}\cdots\gamma_{i_{k-\alpha+1}\cdots i_k}, \tag{9.9}$$

where the transition array γ is constructed as in Lemma IX.3.2(a). This will imply that P is in $\mathscr{M}_\tau$. Formula (9.9) holds for $k = \alpha$ since $\pi_\alpha P$ equals τ. Since $I_\rho^{(3)}(P) = I_{\rho,\alpha}^{(3)}(\tau)$, (9.8) shows that $H_{\rho,k}(P)$ equals $H_{\rho,\alpha}(P)$ for all $k \geq \alpha + 1$. Assume that (9.9) has been shown for $k = \alpha, \alpha + 1, \ldots, c - 1$, some $c \geq \alpha + 1$. With respect to P, X_1 and X_c are conditionally independent given $X_2, \ldots, X_{c-1}$ [Lemma IX.2.1(e)]. If $p(x_{i_1}, \ldots, x_{i_{c-1}}) > 0$, then by the induction hypothesis and strict stationarity,

$$p(x_{i_1}, \ldots, x_{i_c}) = p(x_{i_1}, \ldots, x_{i_{c-1}})p(x_{i_c}|x_{i_2}, \ldots, x_{i_{c-1}})$$
$$= (\nu_\tau)_{i_1 \cdots i_{\alpha-1}} \gamma_{i_1 \cdots i_\alpha} \cdots \gamma_{i_{c-\alpha} \cdots i_{c-1}} \gamma_{i_{c-\alpha+1} \cdots i_c}.$$

If $p(x_{i_1}, \ldots, x_{i_{c-1}}) = 0$, then

$$p(x_{i_1}, \ldots, x_{i_c}) = 0 = (\nu_\tau)_{i_1 \cdots i_{\alpha-1}} \gamma_{i_1 \cdots i_\alpha} \cdots \gamma_{i_{c-\alpha+1} \cdots i_c}.$$

Thus (9.9) holds for $k = c$. The proof of the theorem is complete. □

The contraction principles just proved will now be used in order to deduce the level-3 large deviation bounds.

IX.4. Proof of the Level-3 Large Deviation Bounds

In this section we prove that

$$\limsup_{n\to\infty} \frac{1}{n} \log Q_n^{(3)}\{K\} \leq -I_\rho^{(3)}(K) \qquad \text{for each closed set } K \text{ in } \mathscr{M}_s(\Gamma^{\mathbb{Z}}), \tag{9.10}$$

$$\liminf_{n\to\infty} \frac{1}{n} \log Q_n^{(3)}\{G\} \geq -I_\rho^{(3)}(G) \qquad \text{for each open set } G \text{ in } \mathscr{M}_s(\Gamma^{\mathbb{Z}}), \tag{9.11}$$

where $Q_n^{(3)}$ is the P_ρ-distribution of $R_n(\omega, \cdot)$ on $\mathscr{M}_s(\Gamma^{\mathbb{Z}})$. With these bounds, we will complete the proof of the level-3 large deviation property since $I_\rho^{(3)}$ has already been shown to be lower semicontinuous and to have compact level sets. Our strategy is first to show that for each $\alpha \geq 2$ the P_ρ-distributions of the α-dimensional marginals $\{\pi_\alpha R_n(\omega, \cdot)\}$ have a large deviation property with entropy function $I_{\rho,\alpha}^{(3)}$ defined in (9.6). The bounds (9.10) and (9.11) will follow by an approximation argument. We prove the large deviation property for $\{\pi_\alpha R_n(\omega, \cdot)\}$ by applying the large deviation theorem for random vectors, Theorem II.6.1. In order to calculate the corresponding free energy functions, we need some facts about non-negative matrices.

Let $B = \{B_{ij}\}$ be a real, square matrix. We say that B is *non-negative*, and write $B \geq 0$, if each B_{ij} is non-negative. We say that B is *positive*, and write $B > 0$, if each B_{ij} is positive. If B is non-negative, then B is said to be *primitive* if there exists a positive integer k such that B^k is positive. Clearly, a positive matrix is primitive. If B is non-negative, then B is said to be *stochastic* if $\sum_j B_{ij} = 1$ for each i. The next lemma is due to Perron and Frobenius.

Lemma IX.4.1. *Let $B = \{B_{ij}\}$ be a non-negative primitive matrix. Then there exists an eigenvalue $\lambda(B)$ of B with the following properties.*

(a) *$\lambda(B)$ is real and positive and $\lambda(B)$ exceeds in absolute value any other eigenvalue of B.*

(b) *$\lambda(B)$ is a simple root of the characteristic equation of B.*

(c) *With $\lambda(B)$ may be associated a positive left eigenvector u and a positive right eigenvector w. These eigenvectors are unique up to constant multiples.*

(d) $\lim_{n\to\infty} n^{-1} \log B_{ij}^n = \log \lambda(B)$ *for each i and j.**

(e) *If B is stochastic, then $\lambda(B)$ equals* 1.

(f) *If the entries of B are $\mathscr{C}^1$ functions of a parameter $t \in \mathbb{R}^d$, then $\lambda(B)$ is a $\mathscr{C}^1$ function of $t \in \mathbb{R}^d$; in particular, $\lambda(B)$ is a differentiable function of $t \in \mathbb{R}^d$.*

Proof. (a)–(c) See, e.g., Seneta (1981, Theorem 1.1).

(d) Let u and w be positive left and right eigenvectors associated with $\lambda(B)$ and normalized so that $\langle u, w\rangle = 1$ [part (c)]. Define $\gamma_{ij} = B_{ij}w_j/(\lambda(B)w_i)$. The matrix $\gamma = \{\gamma_{ij}\}$ is a stochastic matrix which is primitive since B is primitive. Since the vector $v_i = u_i w_i$ satisfies $\sum_i v_i = 1$ and $\sum_i v_i \gamma_{ij} = v_j$ for each j, it follows that $\gamma_{ij}^n \to v_j = u_j w_j$ for each i and j [Lemma A.9.5]. This limit and the fact that

$$\gamma_{ij}^n = B_{ij}^n w_j/(\lambda(B)^n w_i), \qquad n = 1, 2, \ldots,$$

imply that $n^{-1} \log B_{ij}^n \to \log \lambda(B)$ as $n \to \infty$.

(e) Since $\sum_j B_{ij} 1 = 1$, $\lambda(B)$ cannot be less than 1. If w is a positive right eigenvector corresponding to $\lambda(B)$, then pick an index i such that $w_i = \max_j w_j$. We have

$$\lambda(B) = \sum_j B_{ij} w_j/w_i \le \sum_j B_{ij} \max_j w_j/\max_j w_j = 1.$$

(f) The eigenvalue $\lambda = \lambda(B)$ is a simple root of the characteristic equation $\det(\lambda\delta_{ij} - B_{ij}(t)) = 0$, in which the entries $B_{ij}(t)$ are $\mathscr{C}^1$ functions of $t \in \mathbb{R}^d$. The implicit function theorem completes the proof. □

If $\alpha \ge 2$ is an integer, then $\mathscr{M}_s(\Gamma^\alpha)$ denotes the set of probability measures τ on $\mathscr{B}(\Gamma^\alpha)$ which are of the form $\tau = \pi_\alpha P$ for some P in $\mathscr{M}_s(\Gamma^{\mathbb{Z}})$. $\mathscr{M}_s(\Gamma^\alpha)$ with the topology of weak convergence is homeomorphic to a compact convex subset $\mathscr{M}_{s,\alpha}$ of $\mathbb{R}^{r^\alpha}$. $\mathscr{M}_{s,\alpha}$ consists of all vectors $\tau = \{\tau_{i_1\cdots i_\alpha}; i_1, \ldots, i_\alpha = 1, \ldots, r\}$ which satisfy $\tau_{i_1\cdots i_\alpha} \ge 0$ for each $i_1, \ldots, i_\alpha$, $\sum_{i_1,\ldots,i_\alpha=1}^r \tau_{i_1\cdots i_\alpha} = 1$, and

$$\sum_{j=1}^r \tau_{i_1\cdots i_{\alpha-1}j} = \sum_{k=1}^r \tau_{ki_1\cdots i_{\alpha-1}} \qquad \text{for each } i_1, \ldots, i_{\alpha-1}. \tag{9.12}$$

For $\tau = \{\tau_{i_1,\ldots,i_\alpha}\}$ a point in $\mathbb{R}^{r^\alpha}$, we define the function

$$\bar{I}_{\rho,\alpha}^{(3)}(\tau) = \begin{cases} I_{\rho,\alpha}^{(3)}(\tau) & \text{for } \tau \in \mathscr{M}_{s,\alpha}, \\ \infty & \text{for } \tau \notin \mathscr{M}_{s,\alpha}, \end{cases} \tag{9.13}$$

where $I_{\rho,\alpha}^{(3)}$ is defined in (9.6). $\bar{I}_{\rho,\alpha}^{(3)}$ is continuous relative to $\mathscr{M}_{s,\alpha}$.

* B_{ij}^n denotes the ij-entry of the product B^n.

Large deviation property of $\{\pi_1 R_n(\omega, \cdot)\}$. Denote by $I_{\rho,1}^{(3)}$ the relative entropy function $I_\rho^{(2)}$. The one-dimensional marginal $\pi_1 R_n(\omega, \cdot)$ equals the empirical measure $L_n(\omega, \cdot)$. The following theorem was proved in Section VIII.2.

Theorem IX.4.2. *The P_ρ-distributions of $\{\pi_1 R_n(\omega, \cdot); n = 1, 2, \ldots,\}$ on $\mathscr{M}(\Gamma)$ have a large deviation property with $a_n = n$ and entropy function $I_{\rho,1}^{(3)} = I_\rho^{(2)}$.*

This was proved by showing a large deviation property for the random vectors $L_n(\omega) = (L_n(\omega, \{x_1\}), \ldots, L_n(\omega, \{x_r\}))$. We now prove a generalization for the α-dimensional marginals $\{\pi_\alpha R_n(\omega, \cdot)\}$, $\alpha \geq 2$; $\pi_\alpha R_n(\omega, \cdot)$ takes values in the space $\mathscr{M}_s(\Gamma^\alpha)$.

Theorem IX.4.3. *For $\alpha \geq 2$, the P_ρ-distributions of $\{\pi_\alpha R_n(\omega, \cdot); n = 1, 2, \ldots\}$ on $\mathscr{M}_s(\Gamma^\alpha)$ have a large deviation property with $a_n = n$ and entropy function $I_{\rho,\alpha}^{(3)}$.*

Let $M_{n,\alpha}(\omega)$ denote the vector in $\mathbb{R}^{r^\alpha}$ with

$$(M_{n,\alpha}(\omega))_{i_1 \cdots i_\alpha} = \pi_\alpha R_n(\omega, \{x_{i_1}, \ldots, x_{i_\alpha}\}), \qquad i_1, \ldots, i_\alpha = 1, \ldots, r.$$

$M_{n,\alpha}(\omega)$ takes values in the compact convex subset $\mathscr{M}_{s,\alpha}$ of $\mathbb{R}^{r^\alpha}$. Exactly as in Section VIII.2, it suffices to prove that the distributions of $\{M_{n,\alpha}\}$ on $\mathbb{R}^{r^\alpha}$ have a large deviation property with entropy function $\bar{I}_{\rho,\alpha}^{(3)}(\tau)$. We apply Theorem II.6.1. It is convenient to divide the proof into three steps.

Step 1: *Evaluation of the free energy function.* Let us first consider $\alpha = 2$. We write

$$(M_{n,2}(\omega))_{ij} = \frac{1}{n} \sum_{\beta=1}^{n} \delta_{Y_\beta^{(n)}(\omega)}\{x_i, x_j\},$$

where $Y_\beta^{(n)}(\omega)$ equals the ordered pair $(X_\beta(\omega), X_{\beta+1}(\omega))$ for $\beta \in \{1, \ldots, n-1\}$ and equals the cyclic pair $(X_n(\omega), X_1(\omega))$ for $\beta = n$. If $t = \{t_{ij}; i, j = 1, \ldots, r\}$ is a point in $\mathbb{R}^{r^2}$, then define the function $f_t(x_i, x_j) = t_{ij}$ for $x_i, x_j \in \Gamma$. Thus $f_t(Y_\beta^{(n)}(\omega)) = t_{ij}$ if $Y_\beta^{(n)}(\omega) = (x_i, x_j)$ and

$$\langle t, M_{n,2}(\omega)\rangle = \frac{1}{n} \sum_{i,j=1}^{r} t_{ij} \sum_{\beta=1}^{n} \delta_{Y_\beta^{(n)}(\omega)}\{x_i, x_j\} = \frac{1}{n} \sum_{\beta=1}^{n} f_t(Y_\beta^{(n)}(\omega)).$$

In the notation of Theorem II.6.1, W_n equals $nM_{n,2}$ and a_n equals n. The free energy function $c_2(t)$ of the sequence $\{nM_{n,2}(\omega); n = 1, 2, \ldots\}$ equals $\lim_{n\to\infty} c_{n,2}(t)$, where $c_{n,2}(t)$ is given by

$$\begin{aligned}\frac{1}{n}\log E_\rho\{\exp(n\langle t, M_{n,2}\rangle)\} &= \frac{1}{n}\log E_\rho\left\{\exp \sum_{\beta=1}^{n} f_t(Y_\beta^{(n)})\right\} \\ &= \frac{1}{n}\log \sum_{i_1, \ldots, i_n=1}^{r} \exp(t_{i_1 i_2} + \cdots + t_{i_{n-1} i_n} \\ &\qquad + t_{i_n i_1}) \cdot \rho_{i_1} \cdots \rho_{i_n}.\end{aligned}$$

Define $B_2(t)$ to be the positive matrix $\{e^{t_{ij}}\rho_j; i,j = 1, \ldots, r\}$. Then

$$c_{n,2}(t) = \frac{1}{n}\log \sum_{i_1=1}^{r} B_2(t)^n_{i_1 i_1}.$$

By Lemma IX.4.1(d), $c_2(t) = \lim_{n\to\infty} c_{n,2}(t) = \log \lambda(B_2(t))$.

In order to evaluate the free energy function $c_\alpha(t)$ of the sequence $\{nM_{n,\alpha}\}$ for $\alpha \geq 3$, we need some notation. If $n \geq \alpha$ and if $i_1, i_2, \ldots, i_n \in \{1, \ldots, r\}$, then define the multi-index

$$i(n,\alpha,j) = \begin{cases} (i_{j+1}, i_{j+2}, \ldots, i_{j+\alpha}) & \text{for } 0 \leq j \leq n-\alpha, \\ (i_{j+1}, \ldots, i_n, i_1, \ldots, i_{j+\alpha-n}) & \text{for } n-\alpha+1 \leq j \leq n-1. \end{cases}$$

Given $t = \{t_{i_1\ldots i_\alpha}; i_1, \ldots, i_\alpha = 1, \ldots, r\}$ a point in $\mathbb{R}^{r^\alpha}$, let $B_\alpha(t)$ be the $r^{\alpha-1} \times r^{\alpha-1}$ matrix

$$(9.14)\quad \begin{aligned} &B_\alpha(t)_{i_1\ldots i_{\alpha-1}, j_1\ldots j_{\alpha-1}} \\ &= \begin{cases} \exp(t_{i_1 i_2\ldots i_{\alpha-1} j_{\alpha-1}})\rho_{j_{\alpha-1}} & \text{for } 1 \leq i_1, i_2 = j_1, i_3 = j_2, \ldots, \\ & \quad i_{\alpha-1} = j_{\alpha-2}, j_{\alpha-1} \leq r, \\ 0 & \text{otherwise.} \end{cases} \end{aligned}$$

Then $c_\alpha(t)$ equals $\lim_{n\to\infty} c_{n,\alpha}(t)$, where for $n \geq \alpha$ $c_{n,\alpha}(t)$ is given by

$$(9.15)\quad \begin{aligned} \frac{1}{n}\log E_\rho\{\exp(n\langle t, M_{n,\alpha}\rangle)\} &= \frac{1}{n}\log \sum_{i_1,\ldots,i_n=1}^{r} \exp\left(\sum_{j=0}^{n-1} t_{i(n,\alpha,j)}\right)\rho_{i_1}\cdots\rho_{i_n} \\ &= \frac{1}{n}\log \sum_{i_1,\ldots,i_{\alpha-1}=1}^{r} B_\alpha(t)^n_{i_1\ldots i_{\alpha-1}, i_1\ldots i_{\alpha-1}}. \end{aligned}$$

$B = B_\alpha(t)$ is primitive since

$$\begin{aligned} &B^{\alpha-1}_{i_1\ldots i_{\alpha-1}, j_1\ldots j_{\alpha-1}} \\ &= B_{i_1\ldots i_{\alpha-1}, i_2\ldots i_{\alpha-1}j_1} B_{i_2\ldots i_{\alpha-1}j_1, i_3\ldots i_{\alpha-1}j_1 j_2} \cdots B_{i_{\alpha-1}j_1\ldots j_{\alpha-2}, j_1\ldots j_{\alpha-2}j_{\alpha-1}} > 0. \end{aligned}$$

By Lemma IX.4.1(d), $c_\alpha(t) = \lim_{n\to\infty} c_{n,\alpha}(t) = \log \lambda(B_\alpha(t))$.

Step 2: Large deviation property. The free energy function of the sequence $\{nM_{n,\alpha}; n = 1, 2, \ldots\}$ equals $\log \lambda(B_\alpha(t))$. Since the entries of $B_\alpha(t)$ are $\mathscr{C}^1$ functions of $t \in \mathbb{R}^{r^\alpha}$, $\lambda(B_\alpha(t))$ is a differentiable function of $t \in \mathbb{R}^{r^\alpha}$ [Lemma IX.4.1(f)]; $\log \lambda(B_\alpha(t))$ has the same property. By Theorem II.6.1, the P_ρ-distributions of $\{M_{n,\alpha}\}$ on $\mathbb{R}^{r^\alpha}$ have a large deviation property with entropy function

$$(9.16)\quad I_{\rho,\alpha}(\tau) = \sup_{t\in\mathbb{R}^{r^\alpha}} \{\langle t, \tau\rangle - \log \lambda(B_\alpha(t))\}, \qquad \tau \in \mathbb{R}^{r^\alpha}.$$

If $Q_{n,\alpha}$ denotes the P_ρ-distribution of $M_{n,\alpha}$, then

$$(9.17)\quad \limsup_{n\to\infty} \frac{1}{n}\log Q_{n,\alpha}\{K\} \leq -I_{\rho,\alpha}(K) \qquad \text{for each closed set } K \text{ in } \mathbb{R}^{r^\alpha},$$

(9.18) $$\liminf_{n\to\infty}\frac{1}{n}\log Q_{n,\alpha}\{G\} \geq -I_{\rho,\alpha}(G) \qquad \text{for each open set } G \text{ in } \mathbb{R}^{r^\alpha}.$$

Step 3: Evaluation of the entropy function. We show that for $\alpha \geq 2$ and $\tau \in \mathbb{R}^{r^\alpha}$, $I_{\rho,\alpha}(\tau)$ defined in (9.16) equals $\bar{I}^{(3)}_{\rho,\alpha}(\tau)$ defined in (9.13). In other words, we prove that

(9.19) $$\bar{I}^{(3)}_{\rho,\alpha}(\tau) = \sup_{t\in\mathbb{R}^{r^\alpha}}\{\langle t,\tau\rangle - \log\lambda(B_\alpha(t))\} \qquad \text{for } \tau\in\mathbb{R}^{r^\alpha}.$$

If G is any open subset of $\mathbb{R}^{r^\alpha}$ which is disjoint from $\mathcal{M}_{s,\alpha}$, then $Q_{n,\alpha}\{G\}$ equals 0, and by the lower bound (9.18) $I_{\rho,\alpha}(G)$ equals ∞. Thus for $\tau \notin \mathcal{M}_{s,\alpha}$, $I_{\rho,\alpha}(\tau) = \infty = \bar{I}^{(3)}_{\rho,\alpha}(\tau)$.

Since $I_{\rho,\alpha}(\tau)$ is defined as a Legendre–Fenchel transform, $I_{\rho,\alpha}(\tau)$ is a closed convex function on $\mathbb{R}^{r^\alpha}$. Suppose we show that $I_{\rho,\alpha}(\tau)$ equals $\bar{I}^{(3)}_{\rho,\alpha}(\tau)$ for all τ in the relative interior of $\mathcal{M}_{s,\alpha}$.* Since $\bar{I}^{(3)}_{\rho,\alpha}$ is continuous relative to $\mathcal{M}_{s,\alpha}$, the continuity property in Theorem VI.3.2 will imply that $I_{\rho,\alpha}(\tau)$ equals $\bar{I}^{(3)}_{\rho,\alpha}(\tau)$ for all τ in $\mathcal{M}_{s,\alpha}$. This will complete the proof of the large deviation property of $\{\pi_\alpha R_n(\omega,\cdot)\}$ with entropy function $I^{(3)}_{\rho,\alpha}$.

In order to treat all values of α by a single proof, it is useful to introduce multi-indices $i = (i_1, \ldots, i_{\alpha-1})$ and $j = (j_1, \ldots, j_{\alpha-1})$, where $i_1, \ldots, i_{\alpha-1}, j_1, \ldots, j_{\alpha-1} \in \{1, \ldots, r\}$. For $\alpha > 2$, we write $i \sim j$ if all the equalities $i_2 = j_1$, $i_3 = j_2, \ldots, i_{\alpha-1} = j_{\alpha-2}$ hold. Otherwise we write $i \not\sim j$. For $\alpha = 2$, we write $i \sim j$ for any i and j. For $\tau \in \mathcal{M}_{s,\alpha}$, define τ_{ij} to be $\tau_{i_1\cdots i_{\alpha-1}j_{\alpha-1}}$ if $i \sim j$ and to be 0 if $i \not\sim j$. Set $(v_\tau)_i = \sum_j \tau_{ij}$. For $t \in \mathbb{R}^{r^\alpha}$, define t_{ij} like τ_{ij}. In this notation

$$\bar{I}^{(3)}_{\rho,\alpha}(\tau) = \sum \tau_{ij}\log\frac{\tau_{ij}}{(v_\tau)_i\rho_{j_{\alpha-1}}} \qquad \text{and} \qquad B_\alpha(t)_{ij} = \begin{cases} e^{t_{ij}}\rho_{j_{\alpha-1}} & \text{if } i\sim j,\\ 0 & \text{if } i\not\sim j.\end{cases}$$

The sum defining $\bar{I}^{(3)}_{\rho,\alpha}(\tau)$ runs over all i and j for which $(v_\tau)_i > 0$ and $i \sim j$. Now let τ^0 be any point in the relative interior of $\mathcal{M}_{s,\alpha}$. Then τ^0_{ij} is positive whenever $i \sim j$, and so each $(v_{\tau^0})_i$ is positive. If we define

$$t^0_{ij} = \log\frac{\tau^0_{ij}}{(v_{\tau^0})_i\rho_{j_{\alpha-1}}} \qquad \text{for all } i\sim j,$$

then $B_\alpha(t^0)_{ij} = \tau^0_{ij}/(v^0_\tau)_i$ if $i \sim j$ and $B_\alpha(t^0)_{ij} = 0$ if $i \not\sim j$. $B_\alpha(t^0)$ is a stochastic matrix. Hence $\log\lambda(B_\alpha(t^0))$ equals 0 [Lemma IX.4.1(e)] and

$$I_{\rho,\alpha}(\tau^0) \geq \langle t^0,\tau^0\rangle - \log\lambda(B_\alpha(t^0)) = \sum_{i\sim j}\tau^0_{ij}\log\frac{\tau^0_{ij}}{(v^0_\tau)_i\rho_{j_{\alpha-1}}} = \bar{I}^{(3)}_{\rho,\alpha}(\tau^0).$$

In order to prove that $\bar{I}^{(3)}_{\rho,\alpha}(\tau^0) \geq I_{\rho,\alpha}(\tau^0) = \sup_{t\in\mathbb{R}^{r^\alpha}}\{\langle t,\tau^0\rangle - \log\lambda(B_\alpha(t))\}$, it suffices to prove that $\log\lambda(B_\alpha(t)) \geq \langle t,\tau^0\rangle - \bar{I}^{(3)}_{\rho,\alpha}(\tau^0)$ for all $t\in\mathbb{R}^{r^\alpha}$. We will in fact show that for all $t\in\mathbb{R}^{r^\alpha}$

(9.20) $$\log\lambda(B_\alpha(t)) = \sup_{\tau\in \operatorname{ri}\mathcal{M}_{s,\alpha}}\{\langle t,\tau\rangle - \bar{I}^{(3)}_{\rho,\alpha}(\tau)\}.$$

*The relative interior of $\mathcal{M}_{s,\alpha}$ consists of all positive vectors τ in $\mathcal{M}_{s,\alpha}$ (each $\tau_{i_1\cdots i_\alpha} > 0$) [Rockafellar (1970, page 48)].

If we insert the definition of $\bar{I}^{(3)}_{\rho,\alpha}(\tau)$, then the right-hand side of (9.20) becomes

$$\sup_{\tau \in \mathrm{ri}\,\mathscr{M}_{s,\alpha}} \sum_{i\sim j} \tau_{ij} \log \frac{B_\alpha(t)_{ij}(v_\tau)_i}{\tau_{ij}}.$$

The matrix $B_\alpha(t)$ is primitive and $B_\alpha(t)_{ij} > 0$ if and only if $i \sim j$. For any τ belonging to the relative interior of $\mathscr{M}_{s,\alpha}$, we have $\tau_{ij} \geq 0$, $\tau_{ij} > 0$ if and only if $B_\alpha(t)_{ij} > 0$, $\sum_{i,j} \tau_{ij} = 1$, and $\sum_j \tau_{ij} = \sum_k \tau_{ki}$ for each i. Hence (9.20) is a consequence of the next theorem.

Theorem IX.4.4. *Let $B = \{B_{ij}\}$ be a non-negative, primitive, $m \times m$ matrix (some $m \geq 2$). Let $\mathscr{M}_B$ be the set of all $\tau = \{\tau_{ij}; i,j = 1, \ldots, m\}$ which satisfy $\tau_{ij} \geq 0$, $\tau_{ij} > 0$ if and only if $B_{ij} > 0$, $\sum_{i,j=1}^m \tau_{ij} = 1$, and $\sum_{j=1}^m \tau_{ij} = \sum_{k=1}^m \tau_{ki}$ for each i. Define $(v_\tau)_i = \sum_{j=1}^m \tau_{ij}$. Then*

$$\log \lambda(B) = \sup_{\tau \in \mathscr{M}_B} \sum\nolimits_B \tau_{ij} \log \frac{B_{ij}(v_\tau)_i}{\tau_{ij}}, \tag{9.21}$$

where $\sum_B$ denotes the sum over all i and j for which $B_{ij} > 0$.

Proof. Let u and w be positive left and right eigenvectors associated with $\lambda(B)$ and normalized so that $\langle u, w\rangle = 1$. Define $\tau^0_{ij} = u_i B_{ij} w_j / \lambda(B)$ and $v^0_i = \sum_{j=1}^m \tau^0_{ij}$. $\tau^0 = \{\tau^0_{ij}\}$ belongs to $\mathscr{M}_B$. We prove that the supremum in (9.21) is attained at the unique point τ^0. Since v^0_i equals $u_i w_i$,

$$\sum\nolimits_B \tau^0_{ij} \log \frac{B_{ij} v^0_i}{\tau^0_{ij}} = \sum\nolimits_B \tau^0_{ij} \log \frac{w_i \lambda(B)}{w_j} = \log \lambda(B). \tag{9.22}$$

Let $\tau \neq \tau^0$ be any other point in $\mathscr{M}_B$. Each $(v_\tau)_i$ is positive. Indeed, if $(v_\tau)_i = 0$, then $\tau_{ij} = 0 = \tau_{ki}$ for each j and k, and so $B_{ij} = 0 = B_{ki}$ for each j and k. The latter cannot hold since B is primitive. Set $v = v_\tau$, $\gamma_{ij} = \tau_{ij}/v_i$, and $\gamma^0_{ij} = \tau^0_{ij}/v^0_i$. We have

$$\sum\nolimits_B \tau_{ij} \log \frac{B_{ij} v_i}{\tau_{ij}} = \sum\nolimits_B \tau_{ij} \log \frac{B_{ij} v^0_i}{\tau^0_{ij}} - \sum\nolimits_B \tau_{ij} \log\left(\frac{\tau_{ij}}{v_i} \frac{v^0_i}{\tau^0_{ij}}\right).$$

By the same calculation as in (9.22), the first sum equals $\log \lambda(B)$. Since $x \log x \geq x - 1$ with equality iff $x = 1$,

$$-\sum\nolimits_B \tau_{ij} \log\left(\frac{\tau_{ij}}{v_i} \frac{v^0_i}{\tau^0_{ij}}\right) = -\sum\nolimits_B \frac{v_i \tau^0_{ij}}{v^0_i} \frac{\tau_{ij} v^0_i}{v_i \tau^0_{ij}} \log\left(\frac{\tau_{ij} v^0_i}{v_i \tau^0_{ij}}\right) \leq -\sum\nolimits_B \left(\tau_{ij} - \frac{v_i \tau^0_{ij}}{v^0_i}\right) = 0 \tag{9.23}$$

and equality holds iff $\gamma_{ij} = \gamma^0_{ij}$ for each i and j. Since $\{\gamma_{ij}\}$ and $\{\gamma^0_{ij}\}$ are primitive stochastic matrices, $\{\gamma_{ij}\} = \{\gamma^0_{ij}\}$ implies that v and v^0 are equal [Lemma A.9.5]. Hence equality holds in (9.23) iff $\tau = \tau^0$. We conclude that for $\tau \in \mathscr{M}_B$, $\sum_B \tau_{ij} \log(B_{ij} v_i / \tau_{ij}) \leq \log \lambda(B)$ with equality iff $\tau = \tau^0$. □

With the last theorem, we have completed the proof of Theorem IX.4.3 (large deviation property of $\{\pi_\alpha R_n(\omega, \cdot)\}$, $\alpha \geq 2$). We are ready to prove the large deviation bounds (9.10) and (9.11) for the P_ρ-distributions of $\{R_n(\omega,\cdot)\}$ on $\mathcal{M}_s(\Gamma^{\mathbb{Z}})$. These bounds will be derived by means of an approximation argument based on our previous work in this chapter.

If A is any subset of $\mathcal{M}_s(\Gamma^{\mathbb{Z}})$, then obviously $P\in A$ implies $\pi_\alpha P\in\pi_\alpha A$ for any $\alpha\geq 1$. We say that A is *finite dimensional* if for all sufficiently large α, $\pi_\alpha P\in\pi_\alpha A$ implies $P\in A$.

Example IX.4.5. Let $\Sigma_1, \ldots, \Sigma_l$ be cylinder sets, P an element of $\mathcal{M}_s(\Gamma^{\mathbb{Z}})$, and $\varepsilon>0$. Consider the open set

$$G_0=\{\bar{P}\in\mathcal{M}_s(\Gamma^{\mathbb{Z}}):|\bar{P}\{\Sigma_i\}-P\{\Sigma_i\}|<\varepsilon, i=1,\ldots,l\}.$$

Since we are dealing with strictly stationary measures, we may assume without loss of generality that each Σ_i has the form

$$\Sigma_i=\{\omega\in\Gamma^{\mathbb{Z}}:(\omega_1,\ldots,\omega_{\alpha_i})\in F_i\},$$

where α_i is a positive integer and F_i is a subset of Γ^{α_i}. If $\alpha\geq\max_{i=1,\ldots,l}\alpha_i$, then $\pi_\alpha^{-1}(\pi_\alpha\Sigma_i)=\Sigma_i$ and so

$$G_0=\{\bar{P}\in\mathcal{M}_s(\Gamma^{\mathbb{Z}}):|\pi_\alpha\bar{P}\{\pi_\alpha\Sigma_i\}-P\{\Sigma_i\}|<\varepsilon, i=1,\ldots,l\}.$$

This shows that G_0 is finite dimensional.

For such open sets G_0, we can derive the lower large deviation bound (9.11) using the large deviation property of $\{\pi_\alpha R_n(\omega,\cdot)\}$ with entropy function $I_{\rho,\alpha}^{(3)}$ [Theorems IX.4.2 and IX.4.3] and the contraction principle relating $I_{\rho,\alpha}^{(3)}$ and $I_\rho^{(3)}$ [Theorems IX.3.1 and IX.3.3]. Given such a set G_0, pick α such that $\pi_\alpha P\in\pi_\alpha G_0$ implies $P\in G_0$. Then

$$\liminf_{n\to\infty}\frac{1}{n}\log Q_n^{(3)}\{G_0\}$$

$$=\liminf_{n\to\infty}\frac{1}{n}\log P_\rho\{\pi_\alpha R_n\in\pi_\alpha G_0\} \qquad (G_0 \text{ finite dimensional})$$

$$\geq -I_{\rho,\alpha}^{(3)}(\pi_\alpha G_0) \qquad (\pi_\alpha G_0 \text{ open; large deviation property of } \pi_\alpha R_n)$$

$$=-\inf\{I_\rho^{(3)}(P): P\in\mathcal{M}_s(\Gamma^{\mathbb{Z}}), \pi_\alpha P\in\pi_\alpha G_0\} \qquad (\text{contraction principle})$$

$$=-I_\rho^{(3)}(G_0) \qquad (G_0 \text{ finite dimensional}).$$

This is (9.11). A similar proof yields the upper bound (9.10) for closed sets of the form $K_0=\{\bar{P}\in\mathcal{M}_s(\Gamma^{\mathbb{Z}}):|\bar{P}\{\Sigma_i\}-P\{\Sigma_i\}|\leq\varepsilon, i=1,\ldots,l\}$.

We now prove the lower large deviation bound (9.11) for any nonempty open set G in $\mathcal{M}_s(\Gamma^{\mathbb{Z}})$. For the empty set, (9.11) holds trivially. An open base for the topology on $\mathcal{M}_s(\Gamma^{\mathbb{Z}})$ is the class of sets of the form

$$\left\{\bar{P}\in\mathcal{M}_s(\Gamma^{\mathbb{Z}}):\left|\int_{\Gamma^{\mathbb{Z}}}f_i d\bar{P}-\int_{\Gamma^{\mathbb{Z}}}f_i dP\right|<\varepsilon, i=1,\ldots,k\right\}, \tag{9.24}$$

where $P \in \mathscr{M}_s(\Gamma^{\mathbb{Z}})$, $f_1, \ldots, f_k \in \mathscr{C}(\Gamma^{\mathbb{Z}})$, and $\varepsilon > 0$. Let $\mathscr{A}$ denote the subset of $\mathscr{C}(\Gamma^{\mathbb{Z}})$ consisting of all finite, real, linear combinations $\sum a_i \chi_{\Sigma_i}$, where $\{\Sigma_i\}$ are cylinder sets. By the Stone–Weierstrass theorem [Theorem A.11.1], $\mathscr{A}$ is dense in $\mathscr{C}(\Gamma^{\mathbb{Z}})$. Hence each set (9.24) contains a set of the form

$$(9.25) \qquad G_0 = \{\bar{P} \in \mathscr{M}_s(\Gamma^{\mathbb{Z}}): |\bar{P}\{\Sigma_i\} - P\{\Sigma_i\}| < \varepsilon, i = 1, \ldots, l\},$$

where $\Sigma_1, \ldots, \Sigma_l$ are cylinder sets. Since each set (9.25) is also a set of the form (9.24) (with $k = l$ and $f_i = \chi_{\Sigma_i}$), it follows that the class of sets (9.25) is an open base for the topology on $\mathscr{M}_s(\Gamma^{\mathbb{Z}})$. Each set G_0 in (9.25) is open and finite dimensional. If G in $\mathscr{M}_s(\Gamma^{\mathbb{Z}})$ is nonempty and open and $G = \cup G_0$, then for each set G_0 contained in G

$$\liminf_{n\to\infty} \frac{1}{n} \log Q_n^{(3)}\{G\} \geq \liminf_{n\to\infty} \frac{1}{n} \log Q_n^{(3)}\{G_0\} \geq -I_\rho^{(3)}\{G_0\}.$$

This follows from the lower large deviation bound for the open, finite-dimensional set G_0. In the last display, we may replace $-I_\rho^{(3)}(G_0)$ by $\sup_{G_0 \subset G}\{-I_\rho^{(3)}(G_0)\}$, and we conclude that

$$\begin{aligned}\liminf_{n\to\infty} \frac{1}{n} \log Q_n^{(3)}\{G\} \geq \sup_{G_0 \subset G} \{-I_\rho^{(3)}(G_0)\} &= -\inf_{G_0 \subset G} I_\rho^{(3)}(G_0) \\ &= -I_\rho^{(3)}(\cup G_0) \\ &= -I_\rho^{(3)}(G).\end{aligned}$$

This is (9.11).

We end by proving the upper large deviation bound (9.10) for any nonempty closed set K in $\mathscr{M}_s(\Gamma^{\mathbb{Z}})$. For the empty set, (9.10) holds trivially. Let $\delta > 0$ be given. We have just seen that any nonempty open set in $\mathscr{M}_s(\Gamma^{\mathbb{Z}})$ contains a finite dimensional set of the form (9.25). Since $I_\rho^{(3)}$ is lower semicontinuous, it follows that for each $P \in K$ there exist cylinder sets $\Sigma_1, \ldots, \Sigma_l$ (depending on P) and a positive real number ε_P such that if $\bar{P}$ is contained in the set

$$K_P = \{\bar{P} \in \mathscr{M}_s(\Gamma^{\mathbb{Z}}): |\bar{P}\{\Sigma_i\} - P\{\Sigma_i\}| \leq \varepsilon_P, i = 1, \ldots, l\},$$

then $I_\rho^{(3)}(\bar{P}) > I_\rho^{(3)}(P) - \delta$.* We have $K \subseteq \bigcup_{P \in K} K_P$, and since $\mathscr{M}_s(\Gamma^{\mathbb{Z}})$ is compact, K is compact [Lemma A.9.2(c)]. Hence there exist finitely many elements $P_1, \ldots, P_s$ of K such that $K \subseteq \bigcup_{i=1}^s K_{P_i}$. Each set K_{P_i} is closed and finite dimensional. By the upper large deviation bound for K_{P_i},

$$\begin{aligned}\limsup_{n\to\infty} \frac{1}{n} \log Q_n^{(3)}\{K\} &\leq \limsup_{n\to\infty} \frac{1}{n} \log\left(\sum_{j=1}^{s} Q_n^{(3)}\{K_{P_i}\}\right) \\ &= \max_{i=1,\ldots,s} \limsup_{n\to\infty} \frac{1}{n} \log Q_n^{(3)}\{K_{P_i}\}\end{aligned}$$

* $I_\rho^{(3)}(P) < \infty$ for $P \in \mathscr{M}_s(\Gamma^{\mathbb{Z}})$ [Theorem IX.2.3].

$$\leq \max_{i=1,\ldots,s} \{-I_\rho^{(3)}(K_{P_i})\}$$

$$\leq \max_{i=1,\ldots,s} \{-I_\rho^{(3)}(P_i) + \delta\}$$

$$\leq -I_\rho^{(3)}(K) + \delta.$$

We obtain (9.10) upon letting δ tend to 0. This completes the proof of the level-3 large deviation property.

IX.5. Notes

1 (page 269). The methods of this chapter can be used to prove large deviation properties for Markov chains with a finite state space. The level-1 property is given as Problem IX.6.6, the level-2 property as Problem IX.6.7, and the level-3 property as Problems IX.6.10–IX.6.15. Donsker and Varadhan (1985) prove a level-3 large deviation property for Gaussian processes. Orey (1985) studies level-3 large deviation problems related to dynamical systems.

2 (page 273) Theorem IX.2.3 showed that $\lim_{\alpha\to\infty} \alpha^{-1} I_{\pi_\alpha P}^{(2)}(\pi_\alpha P)$ exists and equals the quantity $I_\rho^{(3)}(P)$ defined in (9.1). Here is an elementary proof of just the existence of $\lim_{\alpha\to\infty} \alpha^{-1} I_{\pi_\alpha P_\rho}^{(2)}(\pi_\alpha P)$. Using the notation on page 271, we define

$$H_\alpha(P) = \begin{cases} -\sum_{i_1=1}^{r} p(x_{i_1}) \log p(x_{i_1}) & \text{for } \alpha = 1, \\ -\sum p(x_{i_1}, \ldots, x_{i_\alpha}) \log p(x_{i_\alpha} | x_{i_1}, \ldots, x_{i_{\alpha-1}}) & \text{for } \alpha \geq 2. \end{cases}$$

For $\alpha \geq 2$, the sum defining $H_\alpha(P)$ runs over all $i_1, \ldots, i_{\alpha-1}, i_\alpha$ for which $p(x_{i_1}, \ldots, x_{i_{\alpha-1}}) > 0$. $H_\alpha(P)$ is non-negative since $0 \leq p(x_{i_1}) \leq 1$ and $0 \leq p(x_{i_\alpha} | x_{i_1}, \ldots, x_{i_{\alpha-1}}) \leq 1$. $H_\alpha(P)$ is known as the *conditional entropy of* X_α *given* $X_1, \ldots, X_{\alpha-1}$. As in Lemma IX.2.1(b),

$$I_{\pi_\alpha P_\rho}^{(2)}(\pi_\alpha P) = -H_1(P) - H_2(P) - \cdots - H_\alpha(P) - \alpha \sum_{i_1=1}^{r} p(x_{i_1}) \log \rho_{i_1}$$

and $H_1(P) \geq H_2(P) \geq \cdots \geq H_\alpha(P)$. Since $H_\alpha(P)$ is non-negative, the non-increasing sequence $H_1(P), H_2(P), \ldots$ has a limit which we denote by $h(P)$ (the *mean entropy* of P). It follows that $\lim_{\alpha\to\infty} \alpha^{-1} I_{\pi_\alpha P_\rho}^{(2)}(\pi_\alpha P)$ exists and

$$\lim_{\alpha\to\infty} \frac{1}{\alpha} I_{\pi_\alpha P_\rho}^{(2)}(\pi_\alpha P) = -h(P) - \sum_{i_1=1}^{r} p(x_{i_1}) \log \rho_{i_1} < \infty.$$

The existence of $\lim_{\alpha\to\infty} \alpha^{-1} I_{\pi_\alpha P_\rho}^{(2)}(\pi_\alpha P)$ is also a consequence of the super-additivity of the sequence $\{I_{\pi_\alpha P_\rho}^{(2)}(\pi_\alpha P);\ \alpha = 1, 2, \ldots\}$ [page 274] and the bounds

$$0 \leq I_{\pi_\alpha P_\rho}^{(2)}(\pi_\alpha P) \leq \alpha \sum_{i=1}^{r} \log \frac{1}{\rho_i}, \qquad \alpha = 1, 2, \ldots.$$

IX.6. Problems

IX.6.1. Consider the function $I_{\rho,\alpha}^{(3)}(\tau)$, $\alpha \geq 2$, defined in (9.6). Using Theorem IX.3.3 (contraction principle) and the fact that $I_\rho^{(3)}(P)$ is lower semicontinuous and affine [Section IX.2], prove that $I_{\rho,\alpha}^{(3)}(\tau)$ is a closed convex function on $\mathcal{M}_s(\Gamma^\alpha)$.

IX.6.2. Here is another proof of the convexity of $I_{\rho,\alpha}^{(3)}(\tau)$, $\tau \in \mathcal{M}_s(\Gamma^\alpha)$. $\mathcal{M}_s(\Gamma^\alpha)$ is homeomorphic to the compact convex subset $\mathcal{M}_{s,\alpha}$ of $\mathbb{R}^{r^\alpha}$ defined after the proof of Lemma IX.4.1. By (9.13) and (9.19),

$$I_{\rho,\alpha}^{(3)}(\tau) = \bar{I}_{\rho,\alpha}^{(3)}(\tau) = \sup_{t \in \mathbb{R}^{r^\alpha}} \{\langle t, \tau\rangle - \log \lambda(B_\alpha(t))\} \qquad \text{for } \tau \in \mathcal{M}_{s,\alpha}.$$

For $\tau \in \mathbb{R}^{r^\alpha} \backslash \mathcal{M}_{s,\alpha}$, $\bar{I}_{\rho,\alpha}^{(3)}(\tau)$ is defined to be ∞.

(a) Using the last display, prove that $\bar{I}_{\rho,\alpha}^{(3)}$ is a closed convex function on $\mathbb{R}^{r^\alpha}$. It follows that $I_{\rho,\alpha}^{(3)}$ is a closed convex function on $\mathcal{M}_s(\Gamma^\alpha)$.

(b) Prove that $\bar{I}_{\rho,\alpha}^{(3)}$ is essentially strictly convex. [*Hint:* Theorem VI.5.6.]

IX.6.3. (a) Prove that for any $\tau \in \mathcal{M}_s(\Gamma^\alpha)$, $\alpha \geq 2$, $I_{\rho,\alpha}^{(3)}(\tau)$ is non-negative and $I_{\rho,\alpha}^{(3)}(\tau)$ equals 0 if and only if $\tau = \pi_\alpha P_\rho$ (finite product measure).

(b) Let ν be a probability measure on $\mathscr{B}(\Gamma)$. Prove that $I_{\rho,\alpha}^{(3)}(\tau)$ attains its infimum over the set $\{\tau \in \mathcal{M}_s(\Gamma^\alpha) : \pi_1\tau = \nu\}$ at the unique measure $\pi_\alpha P_\nu$ and

$$\inf\{I_{\rho,\alpha}^{(3)}(\tau) : \tau \in \mathcal{M}_s(\Gamma^\alpha), \pi_1\tau = \nu\} = I_{\rho,\alpha}^{(3)}(\pi_\alpha P_\nu) = I_\rho^{(2)}(\nu).$$

IX.6.4. Let $B = \{B_{ij}\}$ be a positive $m \times m$ matrix (some $m \geq 2$). Let u and w be positive left and right eigenvectors associated with $\lambda(B)$ and normalized so that $\langle u, w\rangle = 1$. Denote by $\mathcal{M}_{s,2}$ the set of all $\tau = \{\tau_{ij}; i,j = 1, \ldots, m\}$ which satisfy $\tau_{ij} \geq 0$ for each i and j, $\sum_{i,j=1}^m \tau_{ij} = 1$, and $\sum_{j=1}^m \tau_{ij} = \sum_{k=1}^m \tau_{ki}$ for each i. Let $(\nu_\tau)_i = \sum_{j=1}^m \tau_{ij}$. Prove that

$$\log \lambda(B) = \sup_{\tau \in \mathcal{M}_{s,2}} \sum \tau_{ij} \log \frac{B_{ij}(\nu_\tau)_i}{\tau_{ij}}$$

and that the supremum is attained at the unique point $\tau_{ij}^0 = u_i B_{ij} w_j / \lambda(B)$. The sum in the last display runs over all i and j for which $(\nu_\tau)_i > 0$. [*Hint:* τ^0 is the unique point in $\operatorname{ri} \mathcal{M}_{s,2}$ at which $f(\tau) = \sum_{i,j=1}^m \tau_{ij} \log [B_{ij}(\nu_\tau)_i / \tau_{ij}]$ attains its supremum [Theorem IX.4.4]. If $\rho_1, \ldots, \rho_m$ are positive numbers which sum to 1, then

$$f(\tau) = \sum_{i,j=1}^m \tau_{ij} \log(B_{ij}/\rho_j) - I_{\rho,2}^{(3)}(\tau).$$

Complete the proof using Problem IX.6.1.]

The remaining problems concern finite-state Markov chains. We use the following notation.

Γ a finite set $\{x_1, x_2, \ldots, x_r\}$ with $x_1 < x_2 < \cdots < x_r$.

$\gamma = \{\gamma_{ij}\}$ a positive $r \times r$ stochastic matrix.

$\nu = \{\nu_i; i = 1, \ldots, r\}$ the unique positive solution of the equations $\sum_{i=1}^r \nu_i \gamma_{ij} = \nu_j$, $\sum_{i=1}^r \nu_i = 1$.

P_γ the Markov chain in $\mathcal{M}_s(\Gamma^{\mathbb{Z}})$ with transition matrix γ and invariant measure ν.

$\{X_j; j \in \mathbb{Z}\}$ the coordinate representation process on $\Gamma^{\mathbb{Z}}$.

$\lambda(B)$ the largest positive eigenvalue of a primitive matrix B [Lemma IX.4.1].

$\mathcal{M}$ the set $\{\sigma \in \mathbb{R}^r : \sigma_i \geq 0, \sum_{i=1}^r \sigma_i = 1\}$. $\mathcal{M}$ is homeomorphic to the set $\mathcal{M}(\Gamma)$ of probability measures on $\mathcal{B}(\Gamma)$; $\sigma \in \mathcal{M}$ corresponds to the measure on $\mathcal{B}(\Gamma)$ with $\sigma\{x_i\} = \sigma_i$.

IX.6.5. [Renyi (1970a), Spitzer (1971)]. The purpose of this problem is to prove the limit $\lim_{n\to\infty} \gamma_{ij}^n = \nu_j$ using relative entropy [see Lemma A.9.5.]

(a) Using Lemma IX.4.1, prove that the equations $\sum_{i=1}^r \nu_i \gamma_{ij} = \nu_j$, $\sum_{i=1}^r \nu_i = 1$ have a unique positive solution $\nu = \{\nu_i; i = 1, \ldots, r\}$.

(b) Given $\sigma \in \mathcal{M}$, define $\sigma\gamma^n \in \mathcal{M}$ by $(\sigma\gamma^n)_j = \sum_{i=1}^r \sigma_i \gamma_{ij}^n$. Set $J_n(\sigma) = \sum_{i=1}^r \nu_i \log(\nu_i/(\sigma\gamma^n)_i)$. Prove that for each $j \in \{1, \ldots, r\}$,

$$\log \frac{\nu_j}{(\sigma\gamma^{n+1})_j} \leq \sum_{i=1}^r \nu_i \frac{\gamma_{ij}}{\nu_j} \log \frac{\nu_i}{(\sigma\gamma^n)_i}$$

with equality iff $(\sigma\gamma^n)_i = \nu_i$. Deduce that $J_{n+1}(\sigma) \leq J_n(\sigma)$ with equality iff $\sigma\gamma^n = \nu$.

(c) Let $\{n'\}$ be a subsequence such that $\bar{\nu} = \lim_{n'\to\infty} \sigma\gamma^{n'}$ exists. Prove that $\lim_{n'\to\infty} J_{n'+n}(\nu) = J_n(\bar{\nu})$ for all $n \in \{1, 2, \ldots\}$. Deduce that $\sigma\gamma^n \to \nu$ for any $\sigma \in \mathcal{M}$. It follows that $\lim_{n\to\infty} \gamma_{ij}^n = \nu_j$ for each i and j.

IX.6.6 [Ellis (1981)]. Part (a) proves the level-1 large deviation property.

(a) Let $B(t)$ be the $r \times r$ matrix $\{e^{tx_i}\gamma_{ij}; i, j = 1, \ldots, r\}$, $t \in \mathbb{R}$. Prove that the P_γ-distributions of $S_n/n = \sum_{j=1}^n X_j/n$ have a large deviation property with $a_n = n$ and entropy function

$$I_\gamma^{(1)}(z) = \sup_{t \in \mathbb{R}} \{tz - \log \lambda(B(t))\} \qquad \text{for } z \in \mathbb{R}.$$

(b) Prove that $I_\gamma^{(1)}(z) \geq 0$ with equality if and only if z equals $z_0 = \sum_{i=1}^r x_i \nu_i$.

(c) Prove that $S_n/n \xrightarrow{\exp} \sum_{i=1}^r x_i \nu_i$.

IX.6.7. This problem proves the level-2 large deviation property.

(a) Let $L_n(\omega, \cdot) = n^{-1} \sum_{j=1}^n \delta_{X_j(\omega)}(\cdot)$, $n = 1, 2, \ldots$ (empirical measure). Prove that $L_n(\omega, \cdot) \Rightarrow \nu \quad P_\gamma$-a.s.

(b) [Ellis (1984)]. Let $B(t)$ be the $r \times r$ matrix $\{e^{t_i}\gamma_{ij}; i, j = 1, \ldots, r\}$, $t \in \mathbb{R}^r$. Prove that the P_γ-distributions of $\{L_n; n = 1, 2, \ldots\}$ on $\mathcal{M}(\Gamma)$ have a large deviation property with $a_n = n$ and entropy function

$$I_\gamma^{(2)}(\sigma) = \sup_{t \in \mathbb{R}^r} \{\langle t, \sigma \rangle - \log \lambda(B(t))\} \qquad \text{for } \sigma \in \mathcal{M}(\Gamma).$$

(c) Let $\mathcal{N}^+$ denote the set of vectors $t \in \mathbb{R}^r$ of the form $t_i = \log[u_i/(\gamma u)_i]$ for some $u > 0$ in $\mathbb{R}^r$ ($(\gamma u)_i = \sum_{j=1}^r \gamma_{ij} u_j$). Prove that $\lambda(B(t)) = 1$ if and only if $t \in \mathcal{N}^+$ and that for any $t \in \mathbb{R}^r$

$$h(t) = t - [\log \lambda(B(t))]\mathbf{1} \text{ belongs to } \mathcal{N}^+ \ (\mathbf{1} = (1, \ldots, 1)).$$

(d) [Donsker and Varadhan (1975a)]. Prove that for $\sigma \in \mathcal{M}(\Gamma)$, $I_\gamma^{(2)}(\sigma) = -\inf_{u>0} \sum_{i=1}^r \sigma_i \log[(\gamma u)_i/u_i]$.

(e) Prove that $I_\gamma^{(2)}(\sigma) \geq 0$ with equality if and only if $\sigma = \nu$.

IX.6.8. Let B be a positive $r \times r$ matrix. The purpose of this problem is to prove that

$$\lambda(B) = \sup_{\sigma \in \mathcal{M}} \inf_{u>0} \sum_{i=1}^r \sigma_i \frac{(Bu)_i}{u_i}. \tag{9.26}$$

For generalizations, see Donsker and Varadhan (1975e), Friedland and Karlin (1975), and Friedland (1981).

(a) Let $B(t) = \{e^{t_i}\gamma_{ij}; i, j = 1, \ldots, r\}$, $t \in \mathbb{R}^r$, where $\gamma = \{\gamma_{ij}; i, j = 1, \ldots, r\}$ is a positive stochastic matrix. Prove that

$$\log \lambda(B(t)) = \sup_{\sigma \in \mathcal{M}} \left\{ \langle t, \sigma \rangle + \inf_{u>0} \sum_{i=1}^r \sigma_i \log \frac{(\gamma u)_i}{u_i} \right\} \qquad \text{for } t \in \mathbb{R}^r.$$

(b) Prove that $\log \lambda(B) = \sup_{\sigma \in \mathcal{M}} \inf_{u>0} \sum_{i=1}^r \sigma_i \log[(Bu)_i/u_i]$ by suitable choice of t and γ in part (a). Derive (9.26).

IX.6.9 [Kac (1980)]. Let B be a positive $r \times r$ matrix which is also symmetric ($B_{ij} = B_{ji}$). Show that (9.26) reduces to the Rayleigh–Ritz formula

$$\lambda(B) = \sup \left\{ \sum_{i,j=1}^r B_{ij} v_i v_j : v_i \text{ real}, \sum_{i=1}^r v_i^2 = 1 \right\}.$$

[*Hint*: If $\sigma \in \mathcal{M}$ is positive, then with $w_i = u_i/\sqrt{\sigma_i}$ ($u > 0$)

$$\sum_{i=1}^r \sigma_i \frac{(Bu)_i}{u_i} = \frac{1}{2} \sum_{i,j=1}^r B_{ij} \sqrt{\sigma_i} \sqrt{\sigma_j} \left(\frac{w_i}{w_j} + \frac{w_j}{w_i} \right).\Bigg]$$

The next six problems show how to prove the level-3 large deviation property.

IX.6.10. For $P \in \mathcal{M}_s(\Gamma^{\mathbb{Z}})$, set $I_{\pi_\alpha P_\gamma}^{(2)}(\pi_\alpha P) = \sum_{\omega \in \Gamma^\alpha} \pi_\alpha P\{\omega\} \log[\pi_\alpha P\{\omega\}/\pi_\alpha P_\gamma\{\omega\}]$. Prove that $I_\gamma^{(3)}(P) = \lim_{\alpha \to \infty} \alpha^{-1} I_{\pi_\alpha P_\gamma}^{(2)}(\pi_\alpha P)$ exists and express $I_\gamma^{(3)}(P)$ as in (9.1).

IX.6.11. (a) Prove that $I_\gamma^{(3)}(P)$ is lower semicontinuous, has compact level sets, and is affine.

(b) Prove that $I_\gamma^{(3)}(P) \geq 0$ and that equality holds if and only if $P = P_\gamma$. [*Hint*: Problems IX.6.7(e) and IX.6.13.]

IX.6.12. For $\tau \in \mathcal{M}_s(\Gamma^\alpha)$, $\alpha \geq 2$, set $\tau_{i_1 \cdots i_\alpha} = \tau\{x_{i_1}, \ldots, x_{i_\alpha}\}$ and $(v_\tau)_{i_1, \ldots, i_{\alpha-1}} = \sum_{i_\alpha=1}^r \tau_{i_1 \cdots i_\alpha}$ and define

$$I_{\gamma,\alpha}^{(3)}(\tau) = \sum \tau_{i_1 \cdots i_\alpha} \log \frac{\tau_{i_1 \cdots i_\alpha}}{(v_\tau)_{i_1 \cdots i_{\alpha-1}} \gamma_{i_{\alpha-1} i_\alpha}},$$

where the sum runs over all $i_1, \ldots, i_{\alpha-1}, i_\alpha$ for which $(v_\tau)_{i_1 \cdots i_{\alpha-1}} > 0$. Prove that

$$\inf\{I_\gamma^{(3)}(P): P \in \mathcal{M}_s(\Gamma^{\mathbb{Z}}), \pi_\alpha P = \tau\} = I_{\gamma,\alpha}^{(3)}(\tau) \tag{9.27}$$

and determine the measure(s) P at which the infimum is attained. Formula (9.27) is a contraction principle relating $I_\gamma^{(3)}$ and $I_{\gamma,\alpha}^{(3)}$.

IX.6.13. The level-2 entropy function is given by $I_\gamma^{(2)}(\sigma) = -\inf_{u>0} \sum_{i=1}^r \sigma_i \log[(\gamma u)_i/u_i]$, $\sigma \in \mathcal{M}(\Gamma)$ [Problem IX.6.7]. The contraction principle relating $I_\gamma^{(3)}$ and $I_\gamma^{(2)}$ states that

$$\inf\{I_\gamma^{(3)}(P): P \in \mathcal{M}_s(\Gamma^{\mathbb{Z}}), \pi_1 P = \sigma\} = I_\gamma^{(2)}(\sigma) \qquad \text{for } \sigma \in \mathcal{M}(\Gamma). \tag{9.28}$$

According to (9.27), (9.28) can be proved by showing that

$$\inf\{I_{\gamma,2}^{(3)}(\tau): \tau \in \mathcal{M}_s(\Gamma^2), \pi_1 \tau = \sigma\} = I_\gamma^{(2)}(\sigma) \qquad \text{for } \sigma \in \mathcal{M}(\Gamma). \tag{9.29}$$

Note that $(\pi_1 \tau)_i = (v_\tau)_i = \sum_{j=1}^r \tau_{ij}$. Prove (9.29) and determine the measure(s) P at which the infimum in (9.28) is attained.
[*Hint*: For $\sigma > 0$ and $u > 0$, let $f_\sigma(u) = \sum_{i=1}^r \sigma_i \log[(\gamma u)_i/u_i]$. Show that if $\xi > 0$ is a minimum point of f_σ, then $I_{\gamma,2}^{(3)}(\tau)$ attains its infimum over the set $\{\tau \in \mathcal{M}_s(\Gamma^2): v_\tau = \sigma\}$ at the unique measure $\tau_{ij} = \sigma_i \gamma_{ij} \xi_j/(\gamma\xi)_i$.]

IX.6.14. Let $R_n(\omega, \cdot) = n^{-1} \sum_{k=0}^{n-1} \delta_{T^k X(n,\omega)}(\cdot)$, $n = 1, 2, \ldots$ (empirical process). For $\alpha \geq 2$, prove that the P_γ-distributions of the α-dimensional marginals $\{\pi_\alpha R_n(\omega, \cdot); n = 1, 2, \ldots\}$ on $\mathcal{M}_s(\Gamma^\alpha)$ have a large deviation property with $a_n = n$ and entropy function $I_{\gamma,\alpha}^{(3)}$.

IX.6.15. (a) Prove that $R_n(\omega, \cdot) \Rightarrow P_\gamma$ P_γ-a.s.

(b) Prove that the P_γ-distributions of $\{R_n(\omega, \cdot)\}$ on $\mathcal{M}_s(\Gamma^{\mathbb{Z}})$ have a large deviation property with $a_n = n$ and entropy function $I_\gamma^{(3)}$.

Appendices

Appendix A: Probability

A.1. Introduction

In Sections A.2–A.6 of this appendix we list basic definitions and facts from probability theory which are needed in the text. No effort at completeness has been made. Halmos (1950), Meyer (1966, Chapters I–II), and Breiman (1968) are references for these sections. In the rest of the appendix (Sections A.7–A.11), we discuss the existence of probability measures on various product spaces and study some of the properties of these measures, including weak convergence and ergodicity. For application to Chapters I–III and VII–IX, the space is $(\mathbb{R}^d)^{\mathbb{Z}}$ or $\Gamma^{\mathbb{Z}}$, where Γ is a finite subset of $\mathbb{R}$, and the measures include infinite product measures and Markov chains. The Kolmogorov existence theorem is used to construct these measures. For application to Chapters IV and V, the space is $\{1, -1\}^{\mathbb{Z}^D}$, and the measures are infinite-volume Gibbs states. The Riesz representation theorem is used to show the existence of these measures. In these last five sections, we have aimed at a more self-contained and complete presentation, but the results are tailored specifically to our needs in the text.

A.2. Measurability

Definition A.2.1. Let Ω be a nonempty set and $\mathscr{F}$ a collection of subsets of Ω. $\mathscr{F}$ is said to be a *σ-field* if it contains the empty set and is closed under the formation of complements and countable unions. The pair $(\Omega, \mathscr{F})$ is called a *measurable space*.

Definition A.2.2. Let $(\Omega, \mathscr{F})$ and $(\mathscr{X}, \mathscr{B})$ be two measurable spaces. A mapping X from Ω into $\mathscr{X}$ is said to be *measurable*, or a *random vector*, if $X^{-1}(B) \in \mathscr{F}$ for every $B \in \mathscr{B}$. Such a mapping is also called a *random vector taking values in* $\mathscr{X}$.

Definition A.2.3. Let $\mathscr{C}$ be a collection of subsets of Ω. The *σ-field generated by* $\mathscr{C}$, denoted by $\mathscr{F}(\mathscr{C})$, is defined as the smallest σ-field of Ω which contains $\mathscr{C}$. Let $\mathscr{A}$ be some index set and $\{X_\alpha; \alpha \in \mathscr{A}\}$ a family of mappings from

$(\Omega, \mathscr{F})$ into measurable spaces $\{(\mathscr{X}_\alpha, \mathscr{B}_\alpha); \alpha \in \mathscr{A}\}$. The *σ-field generated by the mappings* $\{X_\alpha; \alpha \in \mathscr{A}\}$, denoted by $\mathscr{F}(X_\alpha; \alpha \in \mathscr{A})$, is defined as the smallest σ-field of Ω with respect to which all the mappings $\{X_\alpha; \alpha \in \mathscr{A}\}$ are measurable.

Definition A.2.4. Let Ω be a topological space. The *Borel σ-field*, denoted by $\mathscr{B}(\Omega)$, is defined as the σ-field generated by the open subsets of Ω. The elements of $\mathscr{B}(\Omega)$ are called *Borel subsets of* Ω or simply *Borel sets.*

Proposition A.2.5. (a) *Let $(\Omega, \mathscr{F})$ be a measurable space, X a mapping from Ω into a set $\mathscr{X}$, and $\mathscr{C}$ a collection of subsets of $\mathscr{X}$. Assume that $\mathscr{X}$ is given the σ-field $\mathscr{F}(\mathscr{C})$. Then X is measurable if and only if $X^{-1}(C) \in \mathscr{F}$ for every $C \in \mathscr{C}$.*

(b) *Let Ω and $\mathscr{X}$ be topological spaces and X a continuous mapping from Ω into $\mathscr{X}$. Then X is a measurable mapping from $(\Omega, \mathscr{B}(\Omega))$ into $(\mathscr{X}, \mathscr{B}(\mathscr{X}))$.*

(c) *Let Ω be a topological space and assume that the topology of Ω admits a countable base (this is the case if, for example, Ω is a separable metric space). Then the σ-field generated by the countable base is the Borel σ-field $\mathscr{B}(\Omega)$.*

A.3. Product Spaces

Definition A.3.1. Let $\{(\Omega_\alpha, \mathscr{F}_\alpha); \alpha \in \mathscr{A}\}$ be a family of measurable spaces and Ω the *product space* $\prod_{\alpha \in \mathscr{A}} \Omega_\alpha$. For $\omega = \{\omega_\alpha; \alpha \in \mathscr{A}\}$ a point in Ω, define $X_\alpha(\omega) = \omega_\alpha$. $\{X_\alpha; \alpha \in \mathscr{A}\}$ are called the *coordinate mappings*. The σ-field $\mathscr{F}(X_\alpha; \alpha \in \mathscr{A})$ is called the *product σ-field* and is denoted by $\prod_{\alpha \in \mathscr{A}} \mathscr{F}_\alpha$. A *cylinder set* is defined to be a set of the form $\{\omega \in \Omega: (\omega_{\alpha_1}, \ldots, \omega_{\alpha_r}) \in F\}$, where $\alpha_1, \ldots, \alpha_r$ are distinct elements of $\mathscr{A}$ and F is a set in the product σ-field $\prod_{i=1}^r \mathscr{F}_{\alpha_i}$. A *product cylinder set* is defined to be a set of the form $\{\omega \in \Omega: \omega_{\alpha_1} \in F_1, \ldots, \omega_{\alpha_r} \in F_r\}$, where $\alpha_1, \ldots, \alpha_r$ are distinct elements of $\mathscr{A}$ and $F_1, \ldots, F_r$ are sets in $\mathscr{F}_{\alpha_1}, \ldots, \mathscr{F}_{\alpha_r}$, respectively.

Proposition A.3.2. *The σ-field generated by the cylinder sets and the σ-field generated by the product cylinder sets both equal the product σ-field.*

Notation A.3.3. If the spaces $\{\Omega_\alpha; \alpha \in \mathscr{A}\}$ are all equal to a space $\tilde{\Omega}$, then write $\tilde{\Omega}^{\mathscr{A}}$ for the product space $\prod_{\alpha \in \mathscr{A}} \tilde{\Omega}$. If $\mathscr{A}$ has finite cardinality k, then write $\tilde{\Omega}^k$ for $\tilde{\Omega}^{\mathscr{A}}$.

Proposition A.3.4. *Let $\{f_\alpha; \alpha \in \mathscr{A}\}$ be a family of measurable mappings from a measurable space $(\mathscr{X}, \mathscr{B})$ into measurable spaces $\{(\Omega_\alpha, \mathscr{F}_\alpha); \alpha \in \mathscr{A}\}$. Define the mapping f from $\mathscr{X}$ into $\Omega = \prod_{\alpha \in \mathscr{A}} \Omega_\alpha$ by $f(x) = \{f_\alpha(x); \alpha \in \mathscr{A}\}$. Then f is measurable when Ω is given the product σ-field.*

If each space Ω_α is a topological space, then we take $\mathscr{F}_\alpha$ to be the Borel σ-field $\mathscr{B}(\Omega_\alpha)$. The product space Ω is a topological space with respect to the product topology. A base for this topology consists of the sets $\prod_{\alpha\in\mathscr{A}} U_\alpha$, where all but a finite number of the U_α equal Ω_α and the remainder, say $U_{\alpha_1}, \ldots, U_{\alpha_k}$, are open sets in Ω_α. Thus Ω has two natural σ-fields: the Borel σ-field $\mathscr{B}(\Omega)$ and the product σ-field $\prod_{\alpha\in\mathscr{A}} \mathscr{B}(\Omega_\alpha)$.

Proposition A.3.5. (a) $\mathscr{B}(\Omega) \supseteq \prod_{\alpha\in\mathscr{A}} \mathscr{B}(\Omega_\alpha)$.

(b) *Assume that each Ω_α is a separable metric space and that $\mathscr{A}$ is countable. Then $\mathscr{B}(\Omega) = \prod_{\alpha\in\mathscr{A}} \mathscr{B}(\Omega_\alpha)$.*

(c) (Tychonoff) *If each topological space Ω_α is compact, then Ω is compact.*

Example A.3.6. (a) $(\mathbb{R}^d)^{\mathbb{Z}}$. For $d \in \{1, 2, \ldots\}$, $\mathbb{R}^d$ denotes d-dimensional Euclidean space. $\mathbb{R}^d$ is a complete separable metric space with respect to the ordinary metric $\|x - y\| = \sum_{i=1}^d [(x_i - y_i)^2]^{1/2}$ $(x = (x_1, \ldots, x_d),\ y = (y_1, \ldots, y_d))$. The sets in $\mathscr{B}(\mathbb{R}^d)$ are called d-dimensional Borel sets.

$(\mathbb{R}^d)^{\mathbb{Z}}$ denotes the product space $\prod_{j\in\mathbb{Z}} \mathbb{R}^d$, where $\mathbb{Z}$ is the set of integers. By Proposition A.3.5(b), the Borel σ-field $\mathscr{B}((\mathbb{R}^d)^{\mathbb{Z}})$ equals the product σ-field $\prod_{j\in\mathbb{Z}} \mathscr{B}(\mathbb{R}^d)$, and the latter coincides with the σ-field generated by the cylinder sets [Proposition A.3.2]. For x and y in $\mathbb{R}^d$, define $b_0(x, y) = \|x - y\|/(1 + \|x - y\|)$; b_0 is a metric on $\mathbb{R}^d$ equivalent to the ordinary metric $\|x - y\|$. For $\omega, \bar{\omega} \in (\mathbb{R}^d)^{\mathbb{Z}}$, the function $b(\omega, \bar{\omega}) = \sum_{j\in\mathbb{Z}} b_0(\omega_j, \bar{\omega}_j)2^{-|j|}$ defines a metric on $(\mathbb{R}^d)^{\mathbb{Z}}$. We have the following facts [Billingsley (1968, pages 218–219)].

(i) $((\mathbb{R}^d)^{\mathbb{Z}}, b)$ is a complete separable metric space.

(ii) The topology on $(\mathbb{R}^d)^{\mathbb{Z}}$ defined by b coincides with the product topology. Convergence in this topology is equivalent to coordinate-wise convergence in $\mathbb{R}^d$.

(iii) A subset K of $(\mathbb{R}^d)^{\mathbb{Z}}$ is compact if and only if the set $\{\omega_j : \omega \in K\}$ is for each $j \in \mathbb{Z}$ a compact set in $\mathbb{R}^d$.

(b) $\Gamma^{\mathbb{Z}}$, $\Gamma \subseteq \mathbb{R}$ *finite*. Let Γ be a finite subset of $\mathbb{R}$ and $\Gamma^{\mathbb{Z}}$ the corresponding infinite product space with metric $b(\omega, \bar{\omega}) = \sum_{j\in\mathbb{Z}} |\omega_j - \bar{\omega}_j| 2^{-|j|}$. Γ is topologized by the discrete topology and $\Gamma^{\mathbb{Z}}$ by the product topology. The topology on $\Gamma^{\mathbb{Z}}$ defined by b coincides with the product topology. Since Γ is compact, $\Gamma^{\mathbb{Z}}$ is a compact metric space. $\mathscr{B}(\Gamma)$ coincides with the set of all subsets of Γ and $\mathscr{B}(\Gamma^{\mathbb{Z}})$ with the σ-field generated by the cylinder sets [Propositions A.3.2 and A.3.5(b)].

A.4. Probability Measures and Expectation

Definition A.4.1. A *probability measure* on a measurable space $(\Omega, \mathscr{F})$ is a positive measure P on $\mathscr{F}$ having total mass 1. The system $(\Omega, \mathscr{F}, P)$ is called a *probability space*. We say that $A \in \mathscr{F}$ occurs *almost surely*, and write A occurs P-a.s., if $P\{A\} = 1$.

A collection of subsets of Ω is called a *field* if it contains the empty set and is closed under the formation of complements and finite unions.

Proposition A.4.2. *Let $\mathscr{F}_0$ be a field of subsets of Ω. If P and Q are probability measures on $(\Omega, \mathscr{F}(\mathscr{F}_0))$ such that $P(F) = Q(F)$ for every $F \in \mathscr{F}_0$, then P equals Q on $\mathscr{F}(\mathscr{F}_0)$.*

Corollary A.4.3. *Let $\{(\Omega_\alpha, \mathscr{F}_\alpha); \alpha \in \mathscr{A}\}$ be a family of measurable spaces, Ω the product space $\prod_{\alpha \in \mathscr{A}} \Omega_\alpha$, and $\mathscr{F}$ the product σ-field $\prod_{\alpha \in \mathscr{A}} \mathscr{F}_\alpha$. A probability measure on $(\Omega, \mathscr{F})$ is uniquely determined by its values on the cylinder sets of Ω.*

Definition A.4.4. A random vector on $(\Omega, \mathscr{F}, P)$ which takes values in $\mathbb{R}$ is called a *random variable*. If X is a random variable on $(\Omega, \mathscr{F}, P)$ which is P-integrable, then $E\{X\} = \int_\Omega X(\omega) P(d\omega)$ is called the *expectation* of X. $L^1(\Omega, P)$ denotes the set of P-integrable random variables on $(\Omega, \mathscr{F}, P)$. If $X = (X_1, \ldots, X_d)$ is a random vector taking values in $\mathbb{R}^d$ and each $X_1, \ldots, X_d$ is integrable, then $E\{X\} = \int_\Omega X(\omega) P(d\omega)$ denotes the vector $(E\{X_1\}, \ldots, E\{X_d\})$.

Definition A.4.5. Let X be a random vector on $(\Omega, \mathscr{F}, P)$ taking values in a measurable space $(\mathscr{X}, \mathscr{B})$. The *distribution* of X is the probability measure P_X on $\mathscr{B}$ defined by $P_X(B) = P\{X^{-1}(B)\}$, $B \in \mathscr{B}$.

Proposition A.4.6. *Let h be a random variable on $(\mathscr{X}, \mathscr{B})$. The function h is P_X-integrable if and only if $h(X(\omega))$ is P-integrable, and then*

$$\int_{\mathscr{X}} h(x) P_X(dx) = \int_\Omega h(X(\omega)) P(d\omega).$$

Definition A.4.7. Let $\{X_\alpha; \alpha \in \mathscr{A}\}$ be a family of random vectors on $(\Omega, \mathscr{F}, P)$ taking values in a measurable space $(\mathscr{X}, \mathscr{B})$. Then $X(\omega) = \{X_\alpha(\omega); \alpha \in \mathscr{A}\}$ is a random vector taking values in $(\mathscr{X}^{\mathscr{A}}, \prod_{\alpha \in \mathscr{A}} \mathscr{B})$. Let P_X be the distribution of X on $\prod_{\alpha \in \mathscr{A}} \mathscr{B}$ and $\hat{X}_\alpha(\omega) = \omega_\alpha$ the coordinate mappings on $\mathscr{X}_{\mathscr{A}}$. The collection $\{\hat{X}_\alpha; \alpha \in \mathscr{A}\}$ is called the *coordinate representation process*. Its distribution on $(\mathscr{X}^{\mathscr{A}}, \prod_{\alpha \in \mathscr{A}} \mathscr{B}, P_X)$ is the distribution P_X of the original collection $\{X_\alpha; \alpha \in \mathscr{A}\}$.

Definition A.4.8. Let $\{X_\alpha; \alpha \in \mathscr{A}\}$ be a family of random vectors on $(\Omega, \mathscr{F}, P)$ taking values in measurable spaces $\{(\mathscr{X}_\alpha, \mathscr{B}_\alpha); \alpha \in \mathscr{A}\}$. The random vectors are said to be *independent* if

$$P\{\bigcap_{\alpha \in \Lambda} X_\alpha^{-1}(B_\alpha)\} = \prod_{\alpha \in \Lambda} P\{X_\alpha^{-1}(B_\alpha)\}$$

for each nonempty finite subset Λ of $\mathscr{A}$ and arbitrary sets $B_\alpha \in \mathscr{B}_\alpha$.

Proposition A.4.9. (Chebyshev's Inequality). *Let X be a random variable and f a non-negative, nondecreasing function on the range of X such that $E\{f(X)\}$ is finite. Then for any real number b for which $f(b)$ is positive,*

$$P\{\omega: X(\omega) \geq b\} \leq \frac{E\{f(X)\}}{f(b)}.$$

A.5. Convergence of Random Vectors

Let $\{X_n; n = 1, 2, \ldots\}$ and X be random vectors defined on a probability space $(\Omega, \mathscr{F}, P)$ and taking values in $\mathbb{R}^d$.

X_n is said to *converge almost surely* to X (written $X_n \xrightarrow{\text{a.s.}} X$ or $X_n \to X$ P-a.s.) if $P\{\omega: \lim_{n\to\infty} X_n(\omega) = X(\omega)\} = 1$.

X_n is said to *converge in probability* to X (written $X_n \xrightarrow{P} X$) if for every $\varepsilon > 0$ $P\{\omega: \|X_n(\omega) - X(\omega)\| \geq \varepsilon\} \to 0$ as $n \to \infty$.

X_n is said to *converge in* L^1 to X (written $X_n \to X$) if $E\{\|X_n - X\|\} \to 0$ as $n \to \infty$.

Proposition A.5.1. (a) *If* $X_n \xrightarrow{\text{a.s.}} X$, *then* $X_n \xrightarrow{P} X$.

(b) *If* $X_n \xrightarrow{L^1} X$, *then* $X_n \xrightarrow{P} X$.

Lemma A.5.2 (Borel–Cantelli). *Let* $\{A_n; n = 1, 2, \ldots\}$ *be subsets of* $\mathscr{F}$ *and define* $B = \bigcap_{m=1}^{\infty} \bigcup_{n=m}^{\infty} A_n$.* *If* $\sum_{n=1}^{\infty} P\{A_n\} < \infty$, *then* $P\{B\} = 0$.

Theorem A.5.3. *If* $\sum_{n=1}^{\infty} P\{\|X_n\| \geq \varepsilon\}$ *is finite for every* $\varepsilon > 0$, *then* $X_n \xrightarrow{\text{a.s.}} 0$.

Proof. For each positive integer k, define the set

$$B(k) = \bigcap_{m=1}^{\infty} \bigcup_{n=m}^{\infty} \{\omega: \|X_n(\omega)\| \geq 1/k\}.$$

By the Borel–Cantelli lemma, $P\{B(k)\} = 0$. Since $\{\omega: X_n(\omega) \to 0 \text{ a.s.}\}^c = \bigcup_{k=1}^{\infty} B(k)$, the proof is done. □

Theorem A.5.4 (Law of Large Numbers). *Let* $\{X_j; j = 1, 2, \ldots\}$ *be a sequence of independent, identically distributed* (i.i.d.) *random vectors and define* $S_n = \sum_{j=1}^{n} X_j$. *If* $E\{\|X_1\|\}$ *is finite, then*

$$\frac{S_n}{n} \xrightarrow{\text{a.s}} E\{X_1\} \text{ (strong law) and therefore } \frac{S_n}{n} \xrightarrow{P} E\{X_1\} \text{ (weak law).}$$

A.6. Conditional Expectation, Conditional Probability, and Regular Conditional Distribution

Let $(\Omega, \mathscr{F}, P)$ be a probability space, Y an integrable random variable on this space, $\mathscr{D}$ a σ-field contained in $\mathscr{F}$, and A a subset of $\mathscr{F}$. The *conditional expectation of* Y *given* $\mathscr{D}$ is defined as any random variable $E\{Y|\mathscr{D}\}$ which

* B is the event that the $\{A_n\}$ occur *infinitely often*. For $P\{B\}$ we may write $P\{A_n \text{ i.o.}\}$.

is measurable with respect to $\mathscr{D}$ and satisfies

$$\text{(A.1)} \qquad \int_D E\{Y|\mathscr{D}\}dP = \int_D YdP \qquad \text{for all subsets } D \in \mathscr{D}.$$

The *conditional probability of A given* $\mathscr{D}$ is defined as any random variable $P\{A|\mathscr{D}\}$ which is measurable with respect to $\mathscr{D}$ and satisfies

$$\text{(A.2)} \qquad \int_D P\{A|\mathscr{D}\}dP = P\{A \cap D\} \qquad \text{for all subsets } D \in \mathscr{D}.$$

Theorem A.6.1. (a) *$E\{Y|\mathscr{D}\}$ exists and is determined up to equivalence.*

(b) *$P\{A|\mathscr{D}\}$ exists and is determined up to equivalence. A version of $P\{A|\mathscr{D}\}$ is $E\{\chi_A|\mathscr{D}\}$.**

If X is a non-negative random variable on $(\Omega, \mathscr{D}, P)$, then for any subset A of $\mathscr{F}$,

$$\text{(A.3)} \qquad \int_\Omega XP\{A|\mathscr{D}\}dP = \int_A XdP.$$

To prove this, approximate X by simple functions on $(\Omega, \mathscr{D}, P)$ and use (A.2) and the monotone convergence theorem.

Let $X_1, X_2, \ldots$ be random vectors on $(\Omega, \mathscr{F}, P)$ taking values in $\mathbb{R}^d$. If $\mathscr{D}$ is the σ-field $\mathscr{F}(X_1, \ldots, X_n)$ or $\mathscr{F}(\{X_j; j \geq 1\})$, then denote $P\{A|\mathscr{D}\}$ by $P\{A|X_1, \ldots, X_n\}$ or $P\{A|\{X_j; j \geq 1\}\}$, respectively. Part (a) of the next theorem is a consequence of the martingale convergence theorem.

Theorem A.6.2. (a) *$P\{A|X_1, \ldots, X_n\}$ converges to $P\{A|\{X_j; j \geq 1\}\}$ almost surely.*

(b) *$P\{A|\{X_j; j \geq 1\}\}$ is a non-negative measurable function of $(X_1(\omega), X_2(\omega), \ldots)$.*

Let X be a random vector on $(\Omega, \mathscr{F}, P)$ taking values in $\mathbb{R}^d$ and let $\mathscr{D}$ be a σ-field contained in $\mathscr{F}$. A function $\tilde{P}(B|\mathscr{D})(\omega)$ from $\mathscr{B}(\mathbb{R}^d) \times \Omega$ into $\mathbb{R}$ is said to be a *regular conditional distribution*, with respect to P, of X given $\mathscr{D}$ if the following hold.

(a) For any $B \in \mathscr{B}(\mathbb{R}^d)$ fixed, $\tilde{P}\{B|\mathscr{D}\}(\omega)$ is a version of the conditional probability $P\{X^{-1}(B)|\mathscr{D}\}$.

(b) For any $\omega \in \Omega$ fixed, $\tilde{P}\{B|\mathscr{D}\}(\omega)$ is a probability measure on $\mathscr{B}(\mathbb{R})^d$.

The following theorem is proved in Breiman (1968, page 79).

Theorem A.6.3. *A regular conditional distribution of X given $\mathscr{D}$ exists.*

Let X, Y, and Z be random vectors on $(\Omega, \mathscr{F}, P)$. We say that X and Y are *conditionally independent given Z* if for every set $A \in \mathscr{F}(X)$ and $B \in \mathscr{F}(Y)$

$$P\{A \cap B|Z\} = P\{A|Z\} \cdot P\{B|Z\} \quad P\text{-a.s.}$$

* χ_A is the characteristic function of A.

Theorem A.6.4 [Loève (1960, page 351)]. *X and Y are conditionally independent given Z if and only if for every set $A \in \mathscr{F}(X)$*

$$P\{A|Y,Z\} = P\{A|Z\} \qquad P\text{-a.s.}$$

Assume that X, Y, and Z have finite state spaces. Then it is easy to show that X and Y are conditionally independent given Z if and only if for all points x, y, and z

$$P\{X = x|Y = y, Z = z\} = P\{X = x|Z = z\} \qquad \text{whenever } P\{Y = y, Z = z\} > 0.$$

A.7. The Kolmogorov Existence Theorem

For proofs of results in this and the next section, we refer to Parthasarathy (1967) and Billingsley (1968) by page number.

Let P be a probability measure on the product space $((\mathbb{R}^d)^{\mathbb{Z}}, \mathscr{B}((\mathbb{R}^d)^{\mathbb{Z}}))$. Given integers m and k with $k \geq 1$, let $\pi_{m,k}$ be the projection from $(\mathbb{R}^d)^{\mathbb{Z}}$ into $(\mathbb{R}^d)^k$ defined by $\pi_{m,k}\omega = (\omega_{m+1}, \ldots, \omega_{m+k})$. Also define projections ψ_k and ϕ_k from $(\mathbb{R}^d)^k$ into $(\mathbb{R}^d)^{k-1}$ $(k \geq 2)$ by

$$\psi_k(\omega_{m+1}, \ldots, \omega_{m+k}) = (\omega_{m+1}, \ldots, \omega_{m+k-1}),$$

$$\phi_k(\omega_{m+1}, \ldots, \omega_{m+k}) = (\omega_{m+2}, \ldots, \omega_{m+k}).$$

Let $\mu_{m,k}$ be the probability measure on $\mathscr{B}((\mathbb{R}^d)^k)$ defined by $\mu_{m,k} = P\pi_{m,k}^{-1}$. The measures $\{\mu_{m,k}; m \in \mathbb{Z}\}$ are called *k-dimensional marginals* of P. Since $\psi_k \pi_{m,k} = \pi_{m,k-1}$ and $\phi_k \pi_{m,k} = \pi_{m+1,k-1}$, the measures $\mu_{m,k}$ satisfy the consistency conditions

$$\text{(A.4)} \qquad \mu_{m,k-1} = \mu_{m,k}\psi_k^{-1}, \qquad \mu_{m+1,k-1} = \mu_{m,k}\phi_k^{-1}.$$

Kolmogorov showed that one can reverse this procedure and construct, from probability measures $\{\mu_{m,k}\}$ satisfying (A.4), a probability measure P on $\mathscr{B}((\mathbb{R}^d)^{\mathbb{Z}})$ with these marginals.

Theorem A.7.1 [Billingsley (pages 228–230)]. *Let $\{\mu_{m,k}\}$ be probability measures on $\mathscr{B}((\mathbb{R}^d)^k)$ which are consistent in the sense that* (A.4) *holds for all integers m and k with $k \geq 1$. Then there exists a unique probability measure P on $\mathscr{B}((\mathbb{R}^d)^{\mathbb{Z}})$ such that $\mu_{m,k} = P\pi_{m,k}^{-1}$ for all integers m and k with $k \geq 1$.*

In applications, one often wants to construct measures P on the product space $(\Gamma^{\mathbb{Z}}, \mathscr{B}(\Gamma^{\mathbb{Z}}))$, where Γ is a finite subset of $\mathbb{R}$. In this case, the consistency conditions are particularly simple.

Corollary A.7.2. *Let Γ be a finite subset of $\mathbb{R}$. Assume that $p(s_1, \ldots, s_k)$ is specified for all finite sequences of elements of Γ and satisfies*

$$p(\cdot) \geq 0, \qquad \sum_{s_{k+1} \in \Gamma} p(s_1, \ldots, s_{k+1}) = p(s_1, \ldots, s_k),$$

$$\sum_{s_1 \in \Gamma} p(s_1, \ldots, s_{k+1}) = p(s_2, \ldots, s_{k+1}), \quad \sum_{s_1 \in \Gamma} p(s_1) = 1.$$

Then there exists a unique probability measure P on $\mathscr{B}(\Gamma^{\mathbb{Z}})$ such that $P\{\omega: \omega_{m+1} = s_1, \ldots, \omega_{m+k} = s_k\} = p(s_1, \ldots, s_k)$ for all integers m and k with $k \geq 1$.

Proof. Let $\mathbf{s}$ denote a k-tuple with coordinates in Γ. For any integer m and any nonempty set $B \in \mathscr{B}(\mathbb{R}^k)$, the function $\mu_{m,k}\{B\} = \sum_{\mathbf{s} \in B \cap \Gamma^k} p(\mathbf{s})$ defines a probability measure on $\mathscr{B}(\mathbb{R}^k)$ ($\mu_{m,k}\{\phi\} = 0$). The measures $\{\mu_{m,k}\}$ are consistent, and so Theorem A.7.1 may be applied. The resulting measure P is easily shown to satisfy $P\{\Gamma^{\mathbb{Z}}\} = 1$ and so is a measure on $\mathscr{B}(\Gamma^{\mathbb{Z}})$. □

Example A.7.3. (a) *Infinite product measure.* Let ρ be a probability measure on $\mathscr{B}(\mathbb{R}^d)$. If $B_1, \ldots, B_k$ are Borel sets, then define for each integer m

$$\mu_{m,k}\{B_1 \times \cdots \times B_k\} = \prod_{i=1}^{k} \rho\{B_i\};$$

$\mu_{m,k}$ extends to a probability measure on $\mathscr{B}((\mathbb{R}^d)^k)$. The measures $\{\mu_{m,k}\}$ are consistent. The corresponding measure P on $\mathscr{B}((\mathbb{R}^d)^{\mathbb{Z}})$ is called the *infinite product measure* with identical one-dimensional marginals ρ and is written P_ρ. The coordinate representation process $X_j(\omega) = \omega_j$ is a sequence of independent random vectors on $((\mathbb{R}^d)^{\mathbb{Z}}, \mathscr{B}((\mathbb{R}^d)^{\mathbb{Z}}), P_\rho)$. Each X_j has P_ρ-distribution ρ.

(b) *Markov chain.* Let Γ be a finite set of distinct real numbers $\{x_1, x_2, \ldots, x_r\}$ with $r \geq 2$ and $v = (v_1, \ldots, v_r)$ an r-tuple of non-negative real numbers which sum to 1. Then v defines the probability measure $\sum_{i=1}^r v_i \delta_{x_i}$ on $\mathscr{B}(\Gamma)$. Let $\gamma = \{\gamma_{ij}\}$ be an $r \times r$ stochastic matrix ($\gamma_{ij} \geq 0$ and $\sum_{j=1}^r \gamma_{ij} = 1$ for each i) and assume that $\sum_{i=1}^r v_i \gamma_{ij} = v_j$ for each i. We define $p(x_i) = v_i$ and for each k-tuple $(x_{i_1}, x_{i_2}, \ldots, x_{i_k})$ of elements in Γ with $k \geq 2$ define

$$p(x_{i_1}, \ldots, x_{i_k}) = v_{i_1} \gamma_{i_1 i_2} \cdots \gamma_{i_{k-1} i_k}.$$

By the assumptions on v and γ, these functions satisfy the hypotheses of Corollary A.7.2. Hence there exists a unique probability measure P on $\mathscr{B}(\Gamma^{\mathbb{Z}})$ such that for all integers m and k with $k \geq 1$

$$P\{\omega : \omega_{m+1} = x_{i_1}, \ldots, \omega_{m+k} = x_{i_k}\} = p(x_{i_1}, \ldots, x_{i_k})$$

Clearly, $P\{\omega : \omega_m = x_i\} = v_i$ and provided $P\{\omega : \omega_{m+1} = x_{i_1}, \ldots, \omega_{m+k-1} = x_{i_{k-1}}\}$ is positive ($k \geq 2$)

$$P\{\omega : \omega_{m+k} = x_{i_k} | \omega_{m+1} = x_{i_1}, \ldots, \omega_{m+k-1} = x_{i_{k-1}}\}$$
$$= P\{\omega : \omega_{m+k} = x_{i_k} | \omega_{m+k-1} = x_{i_{k-1}}\} = \gamma_{i_{k-1} i_k}.$$

The measure P is called the *Markov chain with transition matrix γ and invariant* *measure* ν. The same term refers to the coordinate representation process $X_j(\omega) = \omega_j$ on $(\Gamma^{\mathbb{Z}}, \mathscr{B}(\Gamma^{\mathbb{Z}}), P)$. If γ_{ij} is independent of i, then the condition $\sum_{i=1}^{r} \nu_i \gamma_{ij} = \gamma_j$ implies that $\gamma_j = \nu_j$, and the Markov chain reduces to the infinite product measure on $\mathscr{B}(\Gamma^{\mathbb{Z}})$ with identical one-dimensional marginals ν.

A.8. Weak Convergence of Probability Measures on a Metric Space

Let $\mathscr{X}$ be a metric space, $\mathscr{B}(\mathscr{X})$ the Borel σ-field of $\mathscr{X}$, and $\mathscr{M}(\mathscr{X})$ the set of probability measures on $\mathscr{B}(\mathscr{X})$. Elements of $\mathscr{M}(\mathscr{X})$ are called *Borel probability measures on* $\mathscr{X}$. Let $\mathscr{C}(\mathscr{X})$ be the space of bounded, continuous, real-valued functions f on $\mathscr{X}$ with the supremum norm $\|f\|_\infty = \sup_{x \in \mathscr{X}} |f(x)|$. Each $P \in \mathscr{M}(\mathscr{X})$ is uniquely determined by the integrals $\{\int_{\mathscr{X}} f dP; f \in \mathscr{C}(\mathscr{X})\}$ [Billingsley (page 9)]. We say that a sequence $\{P_n; n = 1, 2, \ldots\}$ in $\mathscr{M}(\mathscr{X})$ *converges weakly* to $P \in \mathscr{M}(\mathscr{X})$, and write $P_n \Rightarrow P$, if $\int_{\mathscr{X}} f dP_n \to \int_{\mathscr{X}} f dP$ for every $f \in \mathscr{C}(\mathscr{X})$. If A is a Borel subset of $\mathscr{X}$, then $\operatorname{int} A$ denotes the interior of A and $\operatorname{cl} A$ the closure of A. A is called a *P-continuity set* if $P\{\operatorname{int} A\} = P\{\operatorname{cl} A\}$.

Theorem A.8.1 [Billingsley (page 11)]. *Let $\{P_n; n = 1, 2, \ldots\}$ and P be probability measures on $(\mathscr{X}, \mathscr{B}(\mathscr{X}))$. The following five conditions are equivalent.*

(a) $P_n \Rightarrow P$.
(b) $\lim_{n\to\infty} \int_{\mathscr{X}} f dP_n = \int_{\mathscr{X}} f dP$ *for all uniformly continuous* $f \in \mathscr{C}(\mathscr{X})$.
(c) $\limsup_{n\to\infty} P_n\{K\} \le P\{K\}$ *for all closed sets* K.
(d) $\liminf_{n\to\infty} P_n\{G\} \ge P\{G\}$ *for all open sets* G.
(e) $\lim_{n\to\infty} P_n\{A\} = P\{A\}$ *for all P-continuity sets* A.

Let $\{P_\beta; \beta > 0\}$ be a subset of $\mathscr{M}(\mathscr{X})$. We say that $\{P_\beta\}$ converges weakly to $P \in \mathscr{M}(\mathscr{X})$ as $\beta \to 0^+$ (resp., $\beta \to \infty$), and write $P_\beta \Rightarrow P$ as $\beta \to 0^+$ (resp., $\beta \to \infty$), if $P_{\beta_n} \Rightarrow P$ for every sequence $\beta_n \to 0^+$ (resp., $\beta_n \to \infty$).

The next theorem states a useful fact for large deviation theory.

Theorem A.8.2. *Let $\{Q_n; n = 1, 2, \ldots\}$ be a sequence in $\mathscr{M}(\mathscr{X})$ and x a point in $\mathscr{X}$. The following statements are equivalent.*

(a) $Q_n \Rightarrow \delta_x$, *where δ_x is the unit point measure at x.*
(b) $Q_n\{A\} \to 0$ *for any Borel subset A of $\mathscr{X}$ whose closure does not contain x.*

Proof. (a) $\Rightarrow$ (b) If the closure of A does not contain x, then A is a δ_x-continuity set and $Q_n\{A\} \to \delta_x\{A\} = 0$.

(b) $\Rightarrow$ (a) Let G be an open set containing x and define $K = G^c$. For any $f \in \mathscr{C}(\mathscr{X})$

$$\left| \int_{\mathscr{X}} f dQ_n - \int_{\mathscr{X}} f d\delta_x \right| \leq \sup_{y \in G} |f(y) - f(x)| + 2\|f\|_\infty Q_n\{K\}.$$

Since f is continuous at x, the first term can be made arbitrarily small by suitable choice of G. The second term converges to 0 as $n \to \infty$ by hypothesis (b). Thus $\int_{\mathscr{X}} f dQ_n \to \int_{\mathscr{X}} f d\delta_x$. □

We make $\mathscr{M}(\mathscr{X})$ into a Hausdorff topological space by taking as the basic neighborhoods of $P \in \mathscr{M}(\mathscr{X})$ all sets of the form

$$\left\{ Q \in \mathscr{M}(\mathscr{X}) : \left| \int_{\mathscr{X}} f_i dQ - \int_{\mathscr{X}} f_i dP \right| < \varepsilon, \, i = 1, \ldots, k \right\},$$

where $f_1, \ldots, f_k$ are elements of $\mathscr{C}(\mathscr{X})$ and ε is positive. The resulting topology is called the *topology of weak convergence*. A sequence $\{P_n; n = 1, 2, \ldots\}$ converges to P in this topology if and only if $P_n \Rightarrow P$. All allusions to $\mathscr{M}(\mathscr{X})$ as a topological space refer to $\mathscr{M}(\mathscr{X})$ with the topology of weak convergence.

Theorem A.8.3 [Parthasarathy (pages 43–46)]. (a) *If $\mathscr{X}$ is a complete separable metric space, then $\mathscr{M}(\mathscr{X})$ is metrizable as a complete separable metric space.*

(b) *If $\mathscr{X}$ is a compact metric space, then $\mathscr{M}(\mathscr{X})$ is a compact metric space.*

Example A.8.4. (a) $\mathscr{X} = \mathbb{R}^d$. Since $\mathbb{R}^d$ is complete and separable, $\mathscr{M}(\mathbb{R}^d)$ is a complete separable metric space. Given points y and z in $\mathbb{R}^d$, we write $y \leq z$ if $y_i \leq z_i$ for each $i \in \{1, \ldots, d\}$. A measure $P \in \mathscr{M}(\mathbb{R}^d)$ has a distribution function (d.f.) defined by $F(y) = P\{z : z \leq y\}$. Let $\{P_n; n = 1, 2, \ldots\}$ and P be measures in $\mathscr{M}(\mathbb{R}^d)$ with corresponding d.f.'s $\{F_n; n = 1, 2, \ldots\}$ and F. Then $P_n \Rightarrow P$ if and only if $F_n(y) \to F(y)$ for all continuity points y of F [Billingsley (page 18)].

(b) $\mathscr{X} = \Gamma^k$. Let $k \geq 1$ be an integer and Γ a finite subset of $\mathbb{R}$ consisting of r numbers. Since Γ^k is compact, $\mathscr{M}(\Gamma^k)$ is a compact metric space. $\mathscr{M}(\Gamma^k)$ is homeomorphic to the compact convex subset of $\mathbb{R}^{r^k}$ consisting of all vectors $v = \{v_{i_1 \cdots i_k}; i_1, \ldots, i_k = 1, \ldots, r\}$ which satisfy $v_{i_1 \cdots i_k} \geq 0$ and $\sum_{i_1, \ldots, i_k = 1}^{r} v_{i_1 \cdots i_k} = 1$. Thus, $v_n \Rightarrow v$ in $\mathscr{M}(\Gamma^k)$ if and only if $(v_n)_{i_1 \cdots i_k} \to v_{i_1 \cdots i_k}$ for each $i_1, \ldots, i_k$.

(c) $\mathscr{X} = (\mathbb{R}^d)^{\mathbb{Z}}$. Since $(\mathbb{R}^d)^{\mathbb{Z}}$ is complete and separable, $\mathscr{M}((\mathbb{R}^d)^{\mathbb{Z}})$ is a complete separable metric space. Given measures $\{P_n; n = 1, 2, \ldots\}$ and P in $\mathscr{M}((\mathbb{R}^d)^{\mathbb{Z}})$, set $(\mu_n)_{m,k} = P_n \pi_{m,k}^{-1}$ and $\mu_{m,k} = P \pi_{m,k}^{-1}$, where m and k are integers with $k \geq 1$ [see page 301]. Then $P_n \Rightarrow P$ if and only if for all m and $k (\mu_n)_{m,k} \Rightarrow \mu_{m,k}$ in $\mathscr{M}((\mathbb{R}^d)^k)$ [Billingsley (page 30)].

(d) $\mathscr{X} = \Gamma^{\mathbb{Z}}$. Let Γ be a finite subset of $\mathbb{R}$. Since $\Gamma^{\mathbb{Z}}$ is a compact metric space, $\mathscr{M}(\Gamma^{\mathbb{Z}})$ is a compact metric space. As in the previous example, a sequence $\{P_n; n = 1, 2, \ldots\}$ converges weakly to P in $\mathscr{M}(\Gamma^{\mathbb{Z}})$ if and only if for all integers m and k with $k \geq 1$, $(\mu_n)_{m,k} \Rightarrow \mu_{m,k}$ in $\mathscr{M}(\Gamma^k)$. Since every subset of Γ^k is a $\mu_{m,k}$-continuity set, $P_n \Rightarrow P$ if and only if $(\mu_n)_{m,k}\{A\} \to \mu_{m,k}\{A\}$

for every subset A of Γ^k [Theorem A.8.1]. Equivalently, $P_n \Rightarrow P$ if and only if $P_n\{\Sigma\} \to P\{\Sigma\}$ for all cylinder sets Σ of $\Gamma^{\mathbb{Z}}$; i.e., for all sets of the form $\Sigma = \{\omega \in \Gamma^{\mathbb{Z}}; (\omega_{m+1}, \ldots, \omega_{m+k}) \in F\}$, where F is a subset of Γ^k.

Let $\{X_n; n = 1, 2, \ldots\}$ be a sequence of random variables on probability space $\{(\Omega_n, \mathscr{F}_n, P_n); n = 1, 2, \ldots\}$ and X a random variable on a probability space $(\Omega, \mathscr{F}, P)$. Denote by P_{X_n} and by P_X the distributions of X_n and of X, respectively. X_n is said to *converge in distribution* to X (written $X_n \to X$) if P_{X_n} converges weakly to P_X. If P is a probability measure on $\mathbb{R}$, then X_n is said to converge in distribution to P (written $X_n \stackrel{\mathscr{D}}{\to} P$) if P_{X_n} converges weakly to P.

Theorem A.8.5 (Central Limit Theorem) [Billingsley (page 45)]. *Let* $\{X_j; j = 1, 2, \ldots\}$ *be a sequence of* i.i.d. *random variables and define* $S_n = \sum_{j=1}^n X_j$. *If* $E\{X_1\} = m$ *and* $\sigma^2 = E\{(X_1 - m)^2\}$ *are finite and* $\sigma^2 > 0$, *then*

$$\frac{S_n - nm}{\sqrt{n}} \stackrel{\mathscr{D}}{\to} N(0, \sigma^2),$$

where $N(0, \sigma^2)$ *is the probability measure on* $\mathbb{R}$ *with density*

$$(2\pi\sigma^2)^{-1/2} \exp[-x^2/2\sigma^2].$$

Random variables $\{X_n; n = 1, 2, \ldots\}$ on probability spaces $\{(\Omega_n, \mathscr{F}_n, P_n); n = 1, 2, \ldots\}$ are said to be *uniformly integrable* if

$$\lim_{\alpha \to \infty} \sup_{n \geq 1} \int_{\{|X_n| \geq \alpha\}} |X_n| dP_n = 0.$$

Theorem A.8.6 [Billingsley (page 32)]. *Suppose that* $X_n \stackrel{\mathscr{D}}{\to} X$ *and that* h *is a continuous function from* $\mathbb{R}$ *into* $\mathbb{R}$ *such that the random variables* $\{h(X_n); n = 1, 2, \ldots\}$ *are uniformly integrable. Then* $E_n\{h(X_n)\} \to E\{h(X)\}$ *as* $n \to \infty$.

□

We also state a continuity theorem for moment generating functions.

Theorem A.8.7. *Let* $\{P_n; n = 1, 2, \ldots\}$ *be a sequence of probability measures in* $\mathscr{M}(\mathbb{R})$. *Assume that the moment generating functions* $g_n(t) = \int_{\mathbb{R}} \exp(tx) P_n(dx)$ *are defined for all* t *in an interval* C *which has nonempty interior and which contains the origin. Assume that for all* $t \in C$ *the limit* $g(t) = \lim_{n \to \infty} g_n(t)$ *exists and that* $g(t) \to g(0) = 1$ *as* $t \in C, t \neq 0$, *converges to* 0. *Then the following conclusions hold.*

(a) $\{P_n; n = 1, 2, \ldots\}$ *converges weakly to some* P *in* $\mathscr{M}(\mathbb{R})$ *and*

$$g(t) = \int_{\mathbb{R}} \exp(tx) P(dx) \qquad \text{for all } t \in C.$$

(b) *If in addition* $0 \in \operatorname{int} C$, *then for any* $j \in \{1, 2, \ldots\}$, $\int_{\mathbb{R}} x^j P_n(dx) \to \int_{\mathbb{R}} x^j P(dx)$ *as* $n \to \infty$.

See, e.g., Martin-Löf (1973, Appendix C) for a proof of part (a). Part (b) follows from part (a) and Theorem A.8.6.

A.9. The Space $\mathcal{M}_s((\mathbb{R}^d)^{\mathbb{Z}})$ and the Ergodic Theorem

We write Ω for the space $(\mathbb{R}^d)^{\mathbb{Z}}$. Let T be the shift mapping on Ω defined by $(T\omega)_j = \omega_{j+1}, j \in \mathbb{Z}$. T is $\mathscr{B}(\Omega)$-measurable. A probability measure P on $\mathscr{B}(\Omega)$ is called *strictly stationary* or *translation invariant* if $P\{T^{-1}B\} = P\{B\}$ for all Borel sets B. By Corollary A.4.3, this condition is equivalent to requiring $P\{T^{-1}\Sigma\} = P\{\Sigma\}$ for all cylinder sets Σ. Let $\mu_{m,k}$ denote the k-dimensional marginal of P obtained by projecting P onto the coordinates $\omega_{m+1}, \ldots, \omega_{m+k}$ [see page 301]. Clearly P is strictly stationary if and only if all the k-dimensional marginals $\{\mu_{m,k}; m \in \mathbb{Z}\}$ are equal for each $k \geq 1$. When P is strictly stationary, the marginals $\{\mu_{m,k}; m \in \mathbb{Z}\}$ will be written as $\pi_k P$. We denote by $\mathcal{M}_s(\Omega)$ the subset of $\mathcal{M}(\Omega)$ consisting of strictly stationary probability measures. If Γ is a finite subset of $\mathbb{R}$, then $\mathcal{M}_s(\Gamma^{\mathbb{Z}})$ denotes the subset of $\mathcal{M}_s(\mathbb{R}^{\mathbb{Z}})$ consisting of probability measures P for which $P\{\Gamma^{\mathbb{Z}}\} = 1$.

Example A.9.1. (a) The infinite product measure P_ρ in $\mathcal{M}(\Omega)$ [Example A.7.3(a)] is strictly stationary.

(b) Any probability measure on $\mathscr{B}(\Gamma^{\mathbb{Z}})$ constructed according to Corollary A.7.2 is strictly stationary. In particular, the Markov chain in Example A.7.3(b) is strictly stationary. The transition probabilities $\gamma_{ij} = P\{\omega : \omega_{m+k} = x_j | \omega_{m+k-1} = x_i\}$ are stationary in the sense that they are independent of m.

The relative topology of $\mathcal{M}_s(\Omega)$ as a subset of $\mathcal{M}(\Omega)$ coincides with the topology or weak convergence. An open base for this topology is the class of sets of the form

$$\left\{Q \in \mathcal{M}_s(\Omega) : \left|\int_\Omega f_i dQ - \int_\Omega f_i dP\right| < \varepsilon, i = 1, \ldots, k\right\},$$

where P is in $\mathcal{M}_s(\Omega)$, $f_1, \ldots, f_k$ are elements of $\mathscr{C}(\Omega)$, and ε is positive.

$\mathcal{M}_s(\Omega)$ is a closed subset of $\mathcal{M}(\Omega)$. Indeed, if $\{P_n; n = 1, 2, \ldots\}$ in $\mathcal{M}_s(\Omega)$ converges weakly to a measure P in $\mathcal{M}(\Omega)$, then for all functions f in $\mathscr{C}(\Omega)$

$$\int_\Omega f dP = \lim_{n\to\infty} \int_\Omega f dP_n = \lim_{n\to\infty} \int_\Omega f \circ T dP_n = \int_\Omega f \circ T dP.*$$

This implies that P is strictly stationary and thus that $\mathcal{M}_s(\Omega)$ is closed. The

* For $\omega \in \Omega$, $f \circ T(\omega) = f(T(\omega))$.

next result follows from properties of $\mathcal{M}(\Omega)$ and $\mathcal{M}(\Gamma^{\mathbb{Z}})$ [Example A.8.4(c)–(d)].

Theorem A.9.2. (a) *$\mathcal{M}_s(\Omega)$ is a closed convex subspace of $\mathcal{M}(\Omega)$. In particular, $\mathcal{M}_s(\Omega)$ is a complete separable metric space.*

(b) *$P_n \Rightarrow P$ in $\mathcal{M}_s(\Omega)$ if and only if for all positive integers k $\pi_k P_n \Rightarrow \pi_k P$ in $\mathcal{M}((\mathbb{R}^d)^k)$.*

(c) *If Γ is a finite subset of $\mathbb{R}$, then $\mathcal{M}_s(\Gamma^{\mathbb{Z}})$ is a compact convex subspace of $\mathcal{M}_s(\mathbb{R}^{\mathbb{Z}})$. In particular, $\mathcal{M}_s(\Gamma^{\mathbb{Z}})$ is a compact metric space. $P_n \Rightarrow P$ in $\mathcal{M}_s(\Gamma^{\mathbb{Z}})$ if and only if $P_n\{\Sigma\} \to P\{\Sigma\}$ for all cylinder sets Σ of Ω.*

A Borel subset A of Ω is said to be *invariant* if $T^{-1}A = A$. The class of invariant sets forms a σ-field $\mathcal{I}$. A measure $P \in \mathcal{M}_s(\Omega)$ is said to be *ergodic* if for every $A \in \mathcal{I}$, $P\{A\}$ equals 0 or 1. A measure $P \in \mathcal{M}_s(\Omega)$ is said to be *mixing* if for any Borel sets A and B

$$\lim_{n\to\infty} P\{A \cap T^{-n}B\} = P\{A\} \cdot P\{B\}.$$

Birkhoff's ergodic theorem is stated next.

Theorem A.9.3. (a) *Let P be a measure in $\mathcal{M}_s(\Omega)$. Then for any function $f \in L^1(\Omega, P)$*

$$\lim_{n\to\infty} \frac{1}{n} \sum_{i=0}^{n-1} f(T^i\omega) = E\{f \mid \mathcal{I}\} \qquad P\text{-a.s. and in } L^1.$$

(b) *$P \in \mathcal{M}_s(\Omega)$ is ergodic if and only if for every $f \in L^1(\Omega, P)$ $E\{f \mid \mathcal{I}\} = E\{f\}$ P-a.s.*

(c) *If $P \in \mathcal{M}_s(\Omega)$ is mixing, then P is ergodic.*

Parts (a) and (b) are proved in Breiman (1968, pages 112–115). To prove part (c), note that if $P \in \mathcal{M}_s(\Omega)$ is mixing, then $P\{A\} = [P\{A\}]^2$ for any invariant set A. Thus P is ergodic.

If a measure $P \in \mathcal{M}_s(\Omega)$ satisfies

$$\lim_{n\to\infty} P\{\Sigma_1 \cap T^{-n}\Sigma_2\} = P\{\Sigma_1\}P\{\Sigma_2\}$$

for all cylinder sets Σ_1 and Σ_2, then a standard approximation argument [Halmos (1950, page 56)] shows that P is mixing.

Example A.9.4. (a) Let $P_\rho \in \mathcal{M}_s(\Omega)$ be the infinite product measure with identical one-dimensional marginals $\rho \in \mathcal{M}(\mathbb{R}^d)$ [Example A.7.3(a)]. If Σ_1 and Σ_2 are cylinder sets, then $P_\rho\{\Sigma_1 \cap T^{-n}\Sigma_2\} = P_\rho\{\Sigma_1\}P_\rho\{\Sigma_2\}$ for all sufficiently large n. Thus P_ρ is mixing.

(b) We find conditions on the Markov chain in Example A.7.3(b) which ensure that it is mixing. Let Γ be a finite set of distinct real numbers $\{x_1, x_2, \ldots, x_r\}$ with $r \geq 2$ and $\gamma = \{\gamma_{ij}\}$ an $r \times r$ stochastic matrix. The matrix γ

is called *irreducible* if for each pair $x_i \neq x_j$ either $\gamma_{ij} > 0$ or there exists a sequence $x_{i_1} = x_i, x_{i_2}, \ldots, x_{i_k} = x_j$ such that $\gamma_{i_\alpha i_{\alpha+1}} > 0$ for $\alpha \in \{1, \ldots, k-1\}$. The matrix γ is called *aperiodic* if for each $i \in \{1, \ldots, r\}$ the set of positive integers n for which γ_{ii}^n is positive is nonempty and has greatest common divisor 1. Clearly both properties hold if γ is a positive matrix (each $\gamma_{ij} > 0$).

Lemma A.9.5 [Feller (1957, page 356)]. (a) *The stochastic matrix γ is irreducible and aperiodic if and only if there exists an integer $N \geq 1$ such that each $\gamma_{ij}^N > 0$.**

(b) *If γ is irreducible and aperiodic, then the limit $\lim_{m\to\infty} \gamma_{ij}^m = v_j$ exists and is independent of i, each v_j is positive, and $\{v_j; j = 1, \ldots, r\}$ is the unique solution of the equations $\sum_{i=1}^r v_i\gamma_{ij} = v_j$, $\sum_{i=1}^r v_i = 1$. We call $v = \sum_{i=1}^r v_i\delta_{x_i}$ the invariant measure corresponding to γ.*

Theorem A.9.6. *Let γ be an $r \times r$ irreducible, aperiodic stochastic matrix and v the corresponding invariant measure. Then the Markov chain P in $\mathcal{M}_s(\Gamma^{\mathbb{Z}})$ with transition matrix γ and invariant measure v [Example A.7.3(b)] is mixing and thus is ergodic.*

Proof. Consider cylinder sets of the form

$$A_1 = \{\omega \in \Gamma^{\mathbb{Z}} : \omega_j = x_{i_j}, \omega_{j+1} = x_{i_{j+1}}, \ldots, \omega_{j+m} = x_{i_{j+m}}\},$$

$$A_2 = \{\omega \in \Gamma^{\mathbb{Z}} : \omega_k = x_{i_k}, \omega_{k+1} = x_{i_{k+1}}, \ldots, \omega_{k+l} = x_{i_{k+l}}\}.$$

Whenever $n + k > j + m$,

$$P\{A_1 \cap T^{-n}A_2\} = P\{A_1\}\gamma_{i_{j+m}i_k}^{n+k-j-m}\gamma_{i_ki_{k+1}} \cdots \gamma_{i_{k+l-1}i_{k+l}}.$$

By Lemma A.9.5(b), $\gamma_{i_{j+m}i_k}^{n+k-j-m} \to v_{i_k}$ as $n \to \infty$, and so $P\{A_1 \cap T^{-n}A_2\} \to P\{A_1\}P\{A_2\}$. It follows that P is mixing. □

Let $X_j(\omega) = \omega_j$ denote the coordinate functions on $\Omega = (\mathbb{R}^d)^{\mathbb{Z}}$ and $X(n, \omega)$ the periodic point in Ω obtained by repeating $(X_1(\omega), X_2(\omega), \ldots, X_n(\omega))$ periodically. The empirical measure is defined as $L_n(\omega, \cdot) = n^{-1}\sum_{j=1}^n \delta_{X_j(\omega)}\{\cdot\}$. The empirical process is defined as $R_n(\omega, \cdot) = n^{-1}\sum_{k=0}^{n-1} \delta_{T^kX(n,\omega)}\{\cdot\}$. For each $\omega \in \Omega$, $L_n(\omega, \cdot)$ is a probability measure on $\mathscr{B}(\mathbb{R}^d)$ and $R_n(\omega, \cdot)$ is a strictly stationary probability measure on $\mathscr{B}(\Omega)$.

Theorem A.9.7. (a) *If $P \in \mathcal{M}_s(\Omega)$ is ergodic, then $L_n(\omega, \cdot) \Rightarrow \pi_1 P$ P-a.s.*
(b) *If $P \in \mathcal{M}_s(\Omega)$ is ergodic, then $R_n(\omega, \cdot) \Rightarrow P$ P-a.s.*

Proof. (a) Since $L_n(\omega, \cdot) = \pi_1 R_n(\omega, \cdot)$, this follows from part (b) and Theorem A.9.2(b).

*Such a matrix is called primitive [page 279].

(b) There exists a metric b_k on $(\mathbb{R}^d)^k$ with the following properties: b_k is equivalent to the Euclidean metric; the set $U((\mathbb{R}^d)^k, b_k)$ of bounded, uniformly continuous, real-valued functions on $((\mathbb{R}^d)^k, b_k)$ is a separable Banach space with respect to the supremum norm [Parthasarathy (page 43)]. Let A_k be a countable dense subset of $U((\mathbb{R}^d)^k, b_k)$. The k-dimensional marginal of $R_n(\omega, \cdot)$, $\pi_k R_n(\omega, \cdot)$, is a probability measure on $\mathscr{B}((\mathbb{R}^d)^k)$. For $g \in A_k$,

$$\int_{(\mathbb{R}^d)^k} g(\bar{\omega})\pi_k R_n(\omega, d\bar{\omega}) = \frac{1}{n}\sum_{j=1}^{n} g(Y_{j,k}(\omega)) + O\left(\frac{1}{n}\right),$$

where $Y_{j,k}(\omega) = (X_j(\omega), X_{j+1}(\omega), \ldots, X_{j+k-1}(\omega))$. By the ergodic theorem, there exists a P-null set N_g such that

$$\lim_{n\to\infty} \frac{1}{n}\sum_{j=1}^{n} g(Y_{j,k}(\omega)) = \int_{(\mathbb{R}^d)^k} g(\bar{\omega})\pi_k P(d\bar{\omega}) \qquad \text{for all } \omega \notin N_g.$$

Hence for each $f \in U((\mathbb{R}^d)^k, b_k)$,

$$\lim_{n\to\infty} \int_{(\mathbb{R}^d)^k} f(\bar{\omega})\pi_k R_n(\omega, d\bar{\omega}) = \int_{(\mathbb{R}^d)^k} f(\bar{\omega})\pi_k P(d\bar{\omega}) \quad \text{for } \omega \notin \bigcup_{g \in A_k} N_g.$$

It follows from Theorem A.8.1 that $\pi_k R_n(\omega, \cdot) \Rightarrow \pi_k P$ for $\omega \notin \bigcup_{g \in A_k} N_g$ and from Theorem A.9.2(b) that $R_n(\omega, \cdot) \Rightarrow P$ for $\omega \notin \bigcup_{k=1}^{\infty} \bigcup_{g \in A_k} N_g$. □

We now restate Theorem A.9.7 for an arbitrary sequence $X_1, X_2, \ldots$ of i.i.d. random vectors defined on some probability space $(\Omega_0, \mathscr{F}_0, P_0)$ and taking values in $\mathbb{R}^d$. Let ρ be the distribution of X_1 and P_ρ the corresponding infinite product measure on $\mathscr{B}((\mathbb{R}^d)^{\mathbb{Z}})$. Define $L_n(\omega, \cdot)$ and $R_n(\omega, \cdot)$ to be, respectively, the empirical measure and the empirical process corresponding to $X_1, X_2, \ldots, X_n$.

Corollary A.9.8. $L_n(\omega, \cdot) \Rightarrow \rho \quad P_0$-a.s. *and* $R_n(\omega, \cdot) \Rightarrow P_\rho \quad P_0$-a.s.

Proof. P_ρ is ergodic [Example A.9.4(a)]. The a.s. convergence of $L_n(\omega, \cdot)$ and of $R_n(\omega, \cdot)$ depends only on the distributions of the respective quantities. Apply Theorem A.9.7. □

Let P be a measure in $\mathscr{M}_s(\Omega)$ and $X_j(\omega) = \omega_j$ the coordinate functions on Ω. We denote by $\tilde{P}_\omega$ a regular conditional distribution, with respect to P, of X_1 given the σ-field $\mathscr{F}(\{X_j; j \le 0\})$ [see page 300]. The next theorem shows that the level-3 entropy function $I_\rho^{(3)}$ in Theorem II.4.4 is well defined.

Theorem A.9.9. *Let P be a measure in $\mathscr{M}_s(\Omega)$ and ρ a measure in $\mathscr{M}(\mathbb{R}^d)$. Define*

$$h_P(\omega, x) = \begin{cases} (d\tilde{P}_\omega/d\rho)(x) & \text{if } \tilde{P}_\omega \ll \rho, \\ \infty & \textit{otherwise}. \end{cases}$$

Then there exists a version of $h_P(\omega, x)$ which

is a non-negative measurable function of $(\bar{\omega}, x) \in (\prod_{j=0}^{\infty} \mathbb{R}^d) \times \mathbb{R}^d$, *where* $\bar{\omega} = (\ldots, X_{-2}(\omega), X_{-1}(\omega), X_0(\omega))$. *The relative entropy*

$$I_\rho^{(2)}(\tilde{P}_\omega) = \int_{\mathbb{R}^d} \log h_P(\omega, x) \tilde{P}_\omega(dx)$$

is a non-negative measurable function of $\bar{\omega}$. *The level-3 entropy function*

$$I_\rho^{(3)}(P) = \int_\Omega I_\rho^{(2)}(\tilde{P}_\omega) P(d\omega)$$

is well defined and non-negative.

For each set $B \in \mathscr{B}(\mathbb{R}^d)$, $\tilde{P}_\omega(B)$ is a measurable function of $\bar{\omega}$ [Theorem A.6.2(b)]. A theorem of Doob [see Dellacherie and Meyer (1982, page 52)] implies the statements about $h_P(\omega, x)$, which in turn yield the rest of theorem.

A measure $P \in \mathscr{M}_s(\Omega)$ is called an *extremal point* of $\mathscr{M}_s(\Omega)$ if $P = \lambda P_1 + (1 - \lambda) P_2$ for $\lambda \in (0, 1)$ and $P_1, P_2 \in \mathscr{M}_s(\Omega)$ implies $P_1 = P_2 = P$. Let $\mathscr{E}(\Omega)$ be the set of ergodic measures in $\mathscr{M}_s(\Omega)$.

Theorem A.9.10. $\mathscr{E}(\Omega)$ *is the set of extremal points of* $\mathscr{M}_s(\Omega)$.

Proof. If P is in $\mathscr{M}_s(\Omega)$ but not in $\mathscr{E}(\Omega)$, then there exists an invariant set A such that $0 < P\{A\} < 1$. Define measures P_1 and P_2 in $\mathscr{M}_s(\Omega)$ by

$$P_1\{B\} = \frac{P\{B \cap A\}}{P\{A\}}, \qquad P_2\{B\} = \frac{P\{B \cap A^c\}}{P\{A^c\}}, \qquad B \in \mathscr{B}(\Omega).$$

Then $P_1 \neq P_2$ and $P = \lambda P_1 + (1 - \lambda) P_2$, where $\lambda = P\{A\} \in (0, 1)$. Thus P is not an extremal point. We have shown that every extremal point of $\mathscr{M}_s(\Omega)$ is ergodic. Conversely, suppose that $P \in \mathscr{M}_s(\Omega)$ is not an extremal point. Then there exist distinct measures P_1 and P_2 in $\mathscr{M}_s(\Omega)$ and a number $\lambda \in (0, 1)$ such that $P = \lambda P_1 + (1 - \lambda) P_2$. Suppose that P is ergodic. Then P_1 and P_2 are ergodic. Choose a Borel set A such that $P_1\{A\} \neq P_2\{A\}$ and for $i = 1, 2$ define sets

$$C_i = \left\{ \omega : \lim_{n \to \infty} \frac{1}{n} \sum_{j=1}^{n} \chi_A\{T^j \omega\} = P_i\{A\} \right\}.$$

By Theorem A.9.3(a), $P_1\{C_1\} = P_2\{C_2\} = 1$. Since C_1 and C_2 are invariant and disjoint, the ergodicity of P implies that either $P\{C_1\}$ or $P\{C_2\}$ equals 0. But this is impossible because $P = \lambda P_1 + (1 - \lambda) P_2$ and $0 < \lambda < 1$. We have shown that every ergodic measure is an extremal point of $\mathscr{M}_s(\Omega)$. □

The proof of Theorem A.9.10 yields the following corollary.

Corollary A.9.11. *Let* $P_1 \neq P_2$ *be ergodic measures in* $\mathscr{M}_s(\Omega)$. *Then there exist disjoint Borel sets* C_1 *and* C_2 *such that* $P_1\{C_1\} = P_2\{C_2\} = 1$ *(i.e.,* P_1 *and* P_2 *are singular).*

A.10. n-Dependent Markov Chains

These processes are discussed in Doob (1953, Section V.3) under the name multiple Markov chains. They occur in the present book in Section IX.3. Let Γ be a finite set of distinct real numbers $\{x_1, x_2, \dots, x_r\}$ with $r \geq 2$ and define $\Omega = \Gamma^{\mathbb{Z}}$. In a Markov chain, the conditional probability that ω_{m+k} equals x_{i_k}, given the values $\omega_{m+k-j} = x_{i_{k-j}}, j \geq 1$, depends only on $x_{i_{k-1}}$ [Example A.7.3(b)]. By contrast, in an n-dependent Markov chain ($n \geq 2$), the same conditional probability depends on the n values $\{x_{i_{k-j}}; j = 1, 2, \dots, n\}$.

Let $\{v_{i_1 \dots i_n}; i_1, \dots, i_n = 1, \dots, r\}$ be an array of r^n real numbers which satisfy

$$
\begin{aligned}
& v_{i_1 \dots i_n} \geq 0 \qquad \text{for each } i_1, \dots, i_n, \\
\text{(A.5)} \qquad & \sum_{i_1, \dots, i_n = 1}^{r} v_{i_1 \dots i_n} = 1, \\
& \sum_{j=1}^{r} v_{i_1 \dots i_{n-1} j} = \sum_{k=1}^{r} v_{k i_1 \dots i_{n-1}} \qquad \text{for each } i_1, \dots, i_{n-1}.
\end{aligned}
$$

The measure $v = \sum_{i_1, \dots, i_n = 1}^{r} v_{i_1 \dots i_n} \delta_{\{x_{i_1}, \dots, x_{i_n}\}}$ will play the role of invariant measure for the n-dependent chain. Let $\gamma = \{\gamma_{i_1 \dots i_{n+1}}; i_1, \dots, i_{n+1} = 1, \dots, r\}$ be an array of r^{n+1} real numbers which satisfy

$$
\begin{aligned}
& \gamma_{i_1 \dots i_{n+1}} \geq 0 \qquad \text{for each } i_1, \dots, i_{n+1}, \\
\text{(A.6)} \qquad & \sum_{i_{n+1} = 1}^{r} \gamma_{i_1 \dots i_{n+1}} = 1 \qquad \text{for each } i_1, \dots, i_n, \\
& \sum_{i_1 = 1}^{r} v_{i_1 \dots i_n} \gamma_{i_1 \dots i_{n+1}} = v_{i_2 \dots i_{n+1}} \qquad \text{for each } i_2, \dots, i_{n+1}.
\end{aligned}
$$

The number $\gamma_{i_1 \dots i_{n+1}}$ will be identified with the conditional probability that ω_{n+1} equals $x_{i_{n+1}}$ given the values $\omega_{n+1-j} = x_{i_{n+1-j}}$, $1 \leq j \leq n$.

We define $p(x_{i_1}, \dots, x_{i_n}) = v_{i_1 \dots i_n}$; for each k-tuple $(x_{i_1}, \dots, x_{i_k})$ with $1 \leq k \leq n-1$

$$
p(x_{i_1}, \dots, x_{i_k}) = \sum_{i_{k+1}, \dots, i_n = 1}^{r} v_{i_1 \dots i_n};
$$

and for each k-tuple $(x_{i_1}, \dots, x_{i_k})$ with $k \geq n+1$,

$$
p(x_{i_1}, \dots, x_{i_k}) = v_{i_1 \dots i_n} \gamma_{i_1 \dots i_{n+1}} \cdots \gamma_{i_{k-n} \dots i_k}.
$$

By (A.5) and (A.6), the hypotheses of Corollary A.7.2 are satisfied. Hence we have the following result.

Theorem A.10.1. *There exists a unique measure P in $\mathcal{M}_s(\Gamma^{\mathbb{Z}})$ such that for all integers m and k with $k \geq 1$*

$$P\{\omega: \omega_{m+1} = x_{i_1}, \ldots, \omega_{m+k} = x_{i_k}\} = p(x_{i_1}, \ldots, x_{i_k}).$$

P satisfies $P\{\omega: \omega_{m+1} = x_{i_1}, \ldots, \omega_{m+n} = x_{i_n}\} = \nu_{i_1 \ldots i_n}$ *and provided* $P\{\omega: \omega_{m+1} = x_{i_1}, \ldots, \omega_{m+n} = x_{i_n}\}$ *is positive,*

$$P\{\omega: \omega_{m+n+1} = x_{i_{n+1}} | \omega_{m+1} = x_{i_1}, \ldots, \omega_{m+n} = x_{i_n}\} = \gamma_{i_1 \ldots i_{n+1}}.$$

P is called the n-dependent Markov chain with transition array γ *and invariant measure* ν.

If $\gamma_{i_1 i_2 \ldots i_{n+1}}$ is independent of i_1, then the n-dependent Markov chain reduces to the $(n-1)$-dependent Markov chain with transition array $\{\gamma_{i_2 \ldots i_{n+1}}; i_2, \ldots, i_{n+1} = 1, \ldots, r\}$ and invariant measure $\sum_{i_2, \ldots, i_n}^r \sum_{i_1=1}^r \nu_{i_1 \ldots i_n} \delta_{\{x_{i_2}, \ldots, x_{i_n}\}}$. If $n = 1$, then the 1-dependent Markov chain is the Markov chain in Example A.7.3(b).

Define $\tilde{X}_j(\omega)$ to be the n-dimensional random vector $(X_{j-n+1}(\omega), X_{j-n+2}(\omega), \ldots, X_j(\omega))$, where $X_j(\omega) = \omega_j$ are the coordinate functions on $\Gamma^{\mathbb{Z}}$. If P is an n-dependent Markov chain, then with respect to P, $\{\tilde{X}_j; j \in \mathbb{Z}\}$ is a Markov chain. It follows from Theorem A.9.6 that if the transition array γ and invariant measure ν are both positive, then the n-dependent Markov chain P is mixing.

A.11. Probability Measures on the Space $\{1, -1\}^{\mathbb{Z}^D}$

Given a positive integer D, $\mathbb{Z}^D$ denotes the D-dimensional integer lattice. Define $\Omega = \{1, -1\}^{\mathbb{Z}^D}$ to be the product space $\prod_{j \in \mathbb{Z}^D} \{1, -1\}$ consisting of all sequences $\omega = \{\omega_j; j \in \mathbb{Z}^D\}$ with each $\omega_j \in \{1, -1\}$. Ω is the configuration space for ferromagnetic spin systems on $\mathbb{Z}^D$, which are studied in Chapters IV and V. The results of this section are tailored specifically to Ω. With minor modifications, the results generalize to the space $\Omega_0 = \mathscr{X}^{\mathbb{Z}^D}$, where $\mathscr{X}$ is a compact metric space. Ω_0 serves as the configuration space for lattice gases and other lattice systems (see the references in Note 2a of Chapter IV).

The set $\Gamma = \{1, -1\}$ is topologized by the discrete topology and the set Ω by the product topology. The σ-field generated by the open sets of the product topology is called the Borel σ-field of Ω and is denoted by $\mathscr{B}(\Omega)$. A cylinder set is defined to be a set of the form $\{\omega \in \Omega: (\omega_{i_1}, \ldots, \omega_{i_r}) \in F\}$, where $i_1, \ldots, i_r$ are distinct elements of $\mathbb{Z}^D$ and F is a subset of $\{1, -1\}^r$. $\mathscr{B}(\Omega)$ coincides with the σ-field generated by the cylinder sets [Propositions A.3.2 and A.3.5(b)]. We define a metric on Ω by the formula $b(\omega, \bar{\omega}) = \sum_{j \in \mathbb{Z}^D} |\omega_j - \bar{\omega}_j| 2^{-\|j\|}$, where $\|-\|$ denotes the Euclidean norm on $\mathbb{Z}^D$. The topology on Ω defined by b coincides with the product topology, and the space (Ω, b) is a compact metric space.

Let $\mathscr{M}(\Omega)$ denote the set of probability measures on $\mathscr{B}(\Omega)$ and $\mathscr{C}(\Omega)$ the space of bounded, continuous, real-valued functions on Ω with the supremum norm. We say that a sequence $\{P_n; n = 1, 2, \ldots\}$ in $\mathscr{M}(\Omega)$ *converges weakly* to $P \in \mathcal{M}(\Omega)$, and write $P_n \Rightarrow P$ or $P = w\text{-lim}_{n \to \infty} P_n$, if $\int_\Omega f dP_n \to \int_\Omega f dP$ for every $f \in \mathscr{C}(\Omega)$. $\mathcal{M}(\Omega)$ is a Hausdorff topological space with respect to the topology of weak convergence. An open base for this topology consists of all sets of the form

$$\left\{ Q \in \mathcal{M}(\Omega): \left| \int_\Omega f_i dQ - \int_\Omega f_i dP \right| < \varepsilon, \quad i = 1, \ldots, k \right\},$$

where P is in $\mathcal{M}(\Omega)$, $f_1, \ldots, f_k$ are elements of $\mathscr{C}(\Omega)$, and ε is positive.

In order to prove several results in this section, we need the *Stone–Weierstrass theorem* for the compact metric space Ω.

Theorem A.11.1 [Dunford and Schwartz (1958, page 272)]. *Let $\mathscr{A}$ be a set of continuous functions in $\mathscr{C}(\Omega)$ with the following properties.*

(a) *The constant function $f(\omega) = 1$, $\omega \in \Omega$, belongs to $\mathscr{A}$.*

(b) *If f and g belong to $\mathscr{A}$, then $\alpha f + \beta g$ belong to $\mathscr{A}$ for any α and β real.*

(c) *If f and g belong to $\mathscr{A}$, then fg belongs to $\mathscr{A}$.*

(d) *If $\omega \neq \bar{\omega}$ are two points of Ω, then there exists a function f in $\mathscr{A}$ such that $f(\omega) \neq f(\bar{\omega})$.*

It then follows that $\mathscr{A}$ is dense in $\mathscr{C}(\Omega)$; that is, given any function f in $\mathscr{C}(\Omega)$ and any $\varepsilon > 0$, there exists a function g in $\mathscr{A}$ such that $\sup_{x \in \Omega} |f(x) - g(x)| < \varepsilon$.

We next prove some facts about the space $\mathcal{M}(\Omega)$.

Theorem A.11.2. (a) *$\mathcal{M}(\Omega)$ is compact with respect to weak convergence.*

(b) *$\mathcal{M}(\Omega)$ is metrizable.*

(c) *$P_n \Rightarrow P$ in $\mathcal{M}(\Omega)$ if and only if $P_n\{\Sigma\} \to P\{\Sigma\}$ for all cylinder sets Σ.*

Proof. (a) This follows from Theorem A.8.3(b), but we prefer a direct proof. Let $\{P_n; n = 1, 2, \ldots\}$ be any sequence in $\mathcal{M}(\Omega)$. Since there are only countably many cylinder sets, we can find an infinite subsequence $\{P_{n'}\}$ such that the limit $\lim_{n' \to \infty} P_{n'}\{\Sigma\}$ exists for each cylinder set Σ. Let $\mathscr{A}$ be the subset of $\mathscr{C}(\Omega)$ consisting of all finite, real, linear combinations $\sum a_i \chi_{\Sigma i}$, where $\{\Sigma_i\}$ are cylinder sets. By the Stone–Weierstrass theorem, $\mathscr{A}$ is dense in $\mathscr{C}(\Omega)$. Hence the limit $\Lambda(f) = \lim_{n' \to \infty} \int_\Omega f dP_{n'}$ exists for every $f \in \mathscr{C}(\Omega)$. Λ defines a nonnegative linear functional on $\mathscr{C}(\Omega)$ and $\Lambda(1) = 1$. The Riesz representation theorem guarantees the existence of a measure $P \in \mathcal{M}(\Omega)$ such that $\Lambda(f) = \int_\Omega f dP$. We conclude that $P_{n'} \Rightarrow P$. It follows that $\mathcal{M}(\Omega)$ is compact.

(b) Since Ω is compact, Ω is complete and separable. Hence part (b) follows from Theorem A.8.3(a).

(c) While this can be proved as in Example A.8.4(d), we prefer a direct proof. If Σ is a cylinder set, then the function $f(\omega) = \chi_\Sigma(\omega)$ is continuous, and so $P_n \Rightarrow P$ implies

$$P_n\{\Sigma\} = \int_\Omega \chi_\Sigma(\omega) P_n(d\omega) \to \int_\Omega \chi_\Sigma(\omega) P(d\omega) = P\{\Sigma\}.$$

For the converse, we apply the Stone–Weierstrass theorem as in part (a). $P_n\{\Sigma\} \to P\{\Sigma\}$ for all cylinder sets Σ implies $P_n \Rightarrow P$. □

Let $\{P_n; n = 1, 2, \dots\}$ be an arbitrary sequence in $\mathscr{M}(\Omega)$. A subset $\mathscr{S}$ of $\mathscr{C}(\Omega)$ is called a *convergence-determining class* if the existence of the limit $\lim_{n\to\infty} \int_\Omega f dP_n$ for each $f \in \mathscr{S}$ implies that $\{P_n; n = 1, 2, \dots\}$ converges weakly to some probability measure P. $\mathscr{C}(\Omega)$ itself is a convergence-determining class because of the Riesz representation theorem [see the proof of Theorem A.11.2(a)].

Theorem A.11.3. *The following subsets of $\mathscr{C}(\Omega)$ are convergence-determining classes.*

(a) *The set $\mathscr{S}_1$ consisting of all functions $f(\omega) = \prod_{i\in B} \omega_i$ for B* a *finite subset of $\mathbb{Z}^D$ (define $f(\omega) = 1$ if B is empty).*

(b) *The set $\mathscr{S}_2$ consisting of all functions $f(\omega) = \chi_\Sigma(\omega)$ for Σ a cylinder set.*

Proof. By the Stone–Weierstrass theorem, the set consisting of all finite, real, linear combinations of functions in $\mathscr{S}_1$ (resp., $\mathscr{S}_2$) is dense in $\mathscr{C}(\Omega)$. Hence $\mathscr{S}_1$ (resp., $\mathscr{S}_2$) is a convergence-determining class since $\mathscr{C}(\Omega)$ is a convergence determining class. □

For each $\alpha \in \{1, \dots, D\}$ let T_α denote the shift mapping on Ω defined by $(T_\alpha\omega)_j = \omega_{j+u_\alpha}$, where u_α is the unit coordinate vector in the αth direction. The shifts $\{T_\alpha; \alpha = 1, \dots, D\}$ commute. A measure $P \in \mathscr{M}(\Omega)$ is said to be *translation invariant* if for each Borel set B and index α $P\{T_\alpha^{-1}B\} = P\{B\}$. This is equivalent to requiring $P\{T_\alpha^{-1}\Sigma\} = P\{\Sigma\}$ for every cylinder set Σ and index α. Let $\mathscr{M}_s(\Omega)$ denote the subset of $\mathscr{M}(\Omega)$ consisting of translation invariant probability measures. The next theorem restates Theorem A.9.2.

Theorem A.11.4. (a) *$\mathscr{M}_s(\Omega)$ is a compact convex subspace of $\mathscr{M}(\Omega)$. In particular, $\mathscr{M}_s(\Omega)$ is a compact metric space.*

(b) *$P_n \Rightarrow P$ in $\mathscr{M}_s(\Omega)$ if and only if $P_n\{\Sigma\} \to P\{\Sigma\}$ for all cylinder sets Σ.*

A Borel subset A of Ω is said to be invariant if $T_\alpha^{-1}A = A$ for each index α. The class of invariant sets forms a σ-field $\mathscr{I}$. A measure $P \in \mathscr{M}_s(\Omega)$ is said to be *ergodic* if for every $A \in \mathscr{I}$, $P\{A\}$ equals 0 or 1. Part (a) of the next theorem states that for functions $f \in L^1(\Omega, P)$, ergodic averages with respect to symmetric hypercubes in $\mathbb{Z}^D$ converge almost surely and in L^1. The theorem follows from Theorem 4.1 in Brunel (1973). For $j \in \mathbb{Z}^D$, T^j denotes the product $T_1^{j_1} \cdots T_D^{j_D}$.

Theorem A.11.5. (a) *Let P be a measure in $\mathscr{M}_s(\Omega)$ and $\{\Lambda\}$ an increasing sequence of symmetric hypercubes whose union is $\mathbb{Z}^D$. Then for any function $f \in L^1(\Omega, P)$,*

$$\lim_{\Lambda \uparrow \mathbb{Z}^D} \frac{1}{|\Lambda|} \sum_{j \in \Lambda} f(T^j \omega) = E\{f|\mathscr{I}\} \qquad P\text{-a.s. and in } L^1.$$

(b) *$P \in \mathscr{M}_s(\Omega)$ is ergodic if and only if for every $f \in L^1(\Omega, P)$, $E\{f|\mathscr{I}\} = E\{f\}$ P-a.s.*

An ergodic theorem holds for increasing sequences of rectangles in $\mathbb{Z}^D$ if f satisfies $\int_\Omega |f|(\log^+ f)^{D-1} dP < \infty$ [Fava (1972)]. Ergodic theorems have been proved by many other people, including Wiener (1939), Dunford and Schwartz (1958), Tempel'man (1972), Nguyen and Zessin (1979), and Sucheston (1983). Also see Krengel (1985).

If $\mathscr{D}$ is a nonempty subset of $\mathscr{M}(\Omega)$, then a measure P in $\mathscr{D}$ is called an *extremal point* of $\mathscr{D}$ if $P = \lambda P_1 + (1 - \lambda) P_2$ for $\lambda \in (0, 1)$ and P_1, $P_2 \in \mathscr{D}$ implies $P_1 = P_2 = P$. The *closed convex hull* of $\mathscr{D}$ is defined as the intersection of all closed convex subsets of $\mathscr{M}(\Omega)$ containing $\mathscr{D}$.

Theorem A.11.6 [Dunford and Schwartz (1958, pages 439–440)]. (a) *If $\mathscr{D}$ is a nonempty compact subset of $\mathscr{M}(\Omega)$, then $\mathscr{D}$ has extremal points.*

(b) (Krein–Milman) *If $\mathscr{D}$ is a nonempty compact convex subset of $\mathscr{M}(\Omega)$, then $\mathscr{D}$ equals the closed convex hull of its extremal points.*

Let $\mathscr{E}(\Omega)$ be the set of ergodic measures in $\mathscr{M}_s(\Omega)$. The next theorem is proved exactly like Theorem A.9.10 and Corollary A.9.11.

Theorem A.11.7. (a) *$\mathscr{E}(\Omega)$ is the set of extremal points of $\mathscr{M}_s(\Omega)$.*

(b) *If $P_1 \neq P_2$ are ergodic measures in $\mathscr{M}_s(\Omega)$, then there exist disjoint Borel sets C_1 and C_2 such that $P_1\{C_1\} = P_2\{C_2\} = 1$ (i.e., P_1 and P_2 are singular).*

Let J be a summable ferromagnetic interaction on $\mathbb{Z}^D$. Given $\beta > 0$ and h real, we define $\mathscr{G}_{\beta,h}$ to be the closed convex hull of the set of weak limits $P = \text{w-lim}_{\Lambda' \uparrow \mathbb{Z}^D} P_{\Lambda',\beta,h,\tilde{\omega}(\Lambda')}$, where $\{\Lambda'\}$ is any increasing sequence of symmetric hypercubes whose union is $\mathbb{Z}^D$, $\tilde{\omega}(\Lambda')$ is any external condition for Λ', and $P_{\Lambda',\beta,h,\tilde{\omega}(\Lambda')}$ is the finite-volume Gibbs state on Λ' corresponding to the interaction J. The elements of $\mathscr{G}_{\beta,h}$ are called *infinite-volume Gibbs states*. $\mathscr{G}_{\beta,h}$ and $\mathscr{G}_{\beta,h} \cap \mathscr{M}_s(\Omega)$ are nonempty. In fact, for any $\beta > 0$ and h real, both sets contain the measure $P_{\beta,h,+} = \text{w-lim}_{\Lambda \uparrow \mathbb{Z}^D} P_{\Lambda,\beta,h,+}$ [Theorems IV.6.5(a) and V.6.1(a)].*

Theorem A.11.8. (a) *$\mathscr{G}_{\beta,h} \cap \mathscr{M}_s(\Omega)$ is a nonempty compact convex subspace of $\mathscr{M}(\Omega)$. In particular, $\mathscr{G}_{\beta,h} \cap \mathscr{M}_s(\Omega)$ is a compact metric space.*

* Λ denotes the hypercube $\{j \in \mathbb{Z}^D : |j_\alpha| \le N, \alpha = 1, \ldots, D\}$, where N is a non-negative integer.

(b) *The set of extremal points of* $\mathscr{G}_{\beta,h} \cap \mathscr{M}_s(\Omega)$ *is nonempty and equals* $\mathscr{G}_{\beta,h} \cap \mathscr{E}(\Omega)$.

Proof. (a) Since $\mathscr{G}_{\beta,h}$ is nonempty, closed, and convex, part (a) follows from Theorem A.11.4(a).

(b) According to Theorem A.11.6(a), the set $\mathscr{G}_{\beta,h} \cap \mathscr{M}_s(\Omega)$ has extremal points. Suppose that P belongs to $\mathscr{G}_{\beta,h} \cap \mathscr{E}(\Omega)$. Then P is an extremal point of $\mathscr{M}_s(\Omega)$ [Theorem A.11.7(a)], and so P is an extremal point of $\mathscr{G}_{\beta,h} \cap \mathscr{M}_s(\Omega)$. The converse uses the Gibbs variational principle, which states that the supremum in the formula

$$-\beta\psi(\beta,h) = \sup_{P \in \mathscr{M}_s(\Omega)} \{-\beta u(h;P) - I_\rho^{(3)}(P)\}$$

is attained precisely on the set $\mathscr{G}_{\beta,h} \cap \mathscr{M}_s(\Omega)$ [Theorems IV.7.3(b) and V.6.3(b)]. Suppose that P is an extremal point of $\mathscr{G}_{\beta,h} \cap \mathscr{M}_s(\Omega)$. If we can show that P is also an extremal point of $\mathscr{M}_s(\Omega)$, then by Theorem A.11.7(a), P must be ergodic and thus belong to the set $\mathscr{G}_{\beta,h} \cap \mathscr{E}(\Omega)$. This will complete the proof of the theorem. We prove that if $P = \lambda P_1 + (1-\lambda)P_2$ for $\lambda \in (0,1)$ and $P_1, P_2 \in \mathscr{M}_s(\Omega)$, then $P_1 = P_2 = P$. Since P is in $\mathscr{G}_{\beta,h} \cap \mathscr{M}_s(\Omega)$ and $u(h;\cdot)$ and $I_\rho^{(3)}$ are affine,

$$\begin{aligned} -\beta\psi(\beta,h) &= -\beta u(h;P) - I_\rho^{(3)}(P) \\ &= \lambda[-\beta u(h;P_1) - I_\rho^{(3)}(P_1)] + (1-\lambda)[-\beta u(h;P_2) - I_\rho^{(3)}(P_2)]. \end{aligned} \tag{A.7}$$

We claim that the measures P_1 and P_2 both belong to $\mathscr{G}_{\beta,h} \cap \mathscr{M}_s(\Omega)$. Otherwise by the Gibbs variational principle, the last term in (A.7) would be strictly less than $-\beta\psi(\beta,h)$, and this would give a contradiction. We have shown that if $P \in \mathscr{G}_{\beta,h} \cap \mathscr{M}_s(\Omega)$ can be written as $\lambda P_1 + (1-\lambda)P_2$ for $\lambda \in (0,1)$ and $P_1, P_2 \in \mathscr{M}_s(\Omega)$, then P_1 and P_2 are also in $\mathscr{G}_{\beta,h} \cap \mathscr{M}_s(\Omega)$. Since P is an extremal point of $\mathscr{G}_{\beta,h} \cap \mathscr{M}_s(\Omega)$, we conclude that $P_1 = P_2 = P$. Thus, P is an extremal point of $\mathscr{M}_s(\Omega)$. □

Corollary A.11.9. *For each* $\beta > 0$, *h real, and choice of sign, the measure* $P_{\beta,h,\pm} = w\text{-}\lim_{\Lambda \uparrow \mathbb{Z}^D} P_{\Lambda,\beta,h,\pm}$ *in* $\mathscr{G}_{\beta,h} \cap \mathscr{M}_s(\Omega)$ *is ergodic.*

Proof. The FKG inequality implies that the measures $P_{\beta,h,+}$ and $P_{\beta,h,-}$ are extremal points of $\mathscr{G}_{\beta,h}$ [see page 123]. Therefore, these measures are extremal points of $\mathscr{G}_{\beta,h} \cap \mathscr{M}_s(\Omega)$. □

For an exposition of the material in the remainder of this appendix, see Preston (1976, Section 2). These results generalize to the many-body interactions introduced in Appendix C.3. A measure $P \in \mathscr{M}_s(\Omega)$ is said to be *mixing* if for all Borel subsets A and B of Ω

$$\lim_{\|j\| \to \infty} P\{A \cap T^{-j}B\} = P\{A\}P\{B\}.$$

If Λ is a finite subset of $\mathbb{Z}^D$, then $\mathscr{F}_{\Lambda^c}$ denotes the σ-field generated by the coordinate functions $Y_j(\omega) = \omega_j$ for $j \in \Lambda^c$. A measure $P \in \mathscr{M}(\Omega)$ is said to have *short-range correlations* if for any Borel subset A of Ω and any $\varepsilon > 0$, there exists a finite subset Λ of $\mathbb{Z}^D$ such that for all sets B in $\mathscr{F}_{\Lambda^c}$

$$|P\{A \cap B\} - P\{A\}P\{B\}| < \varepsilon.$$

Define $\mathscr{F}_\infty = \bigcap_\Lambda \mathscr{F}_{\Lambda^c}$, where the intersection runs over the finite subsets of $\mathbb{Z}^D$.

Proposition A.11.10. (a) *If $P \in \mathscr{M}_s(\Omega)$ is* mixing, *then P is ergodic.*

(b) *If $P \in \mathscr{M}_s(\Omega)$ has short-range correlations, then P is mixing.*

(c) *$P \in \mathscr{M}(\Omega)$ has short-range correlations if and only if for every $A \in \mathscr{F}_\infty$, $P\{A\}$ equals 0 or 1.*

Proof. (a) If $P \in \mathscr{M}_s(\Omega)$ is mixing, then $P\{A\} = [P\{A\}]^2$ for any invariant set A. Thus P is ergodic.

(b) If P in $\mathscr{M}_s(\Omega)$ has short-range correlations, then for any cylinder sets Σ_1 and Σ_2 and any $\varepsilon > 0$, there exists a positive real number N such that

$$|P\{\Sigma_1 \cap T^{-j}\Sigma_2\} - P\{\Sigma_1\}P\{\Sigma_2\}| < \varepsilon \tag{A.8}$$

whenever $\|j\| \geq N$. The proof of (A.8) for arbitrary Borel sets A and B follows from a standard approximation argument [Halmos (1950, page 56)].

(c) See Lanford and Ruelle (1969) or Preston (1976, page 23). □

Theorem A.11.11. *Let J be a summable ferromagnetic interaction on $\mathbb{Z}^D$ and for $\beta > 0$ and h real, let $\mathscr{G}_{\beta,h}$ be the set of corresponding infinite-volume Gibbs states. Then the following conclusions hold.*

(a) *$\mathscr{G}_{\beta,h}$ is a nonempty compact convex subspace of $\mathscr{M}(\Omega)$. In particular, $\mathscr{G}_{\beta,h}$ is a compact metric space.*

(b) *The set of extremal points of $\mathscr{G}_{\beta,h}$ is nonempty* [*Theorem* A.11.6(a)] *and equals the set of infinite-volume Gibbs states with short-range correlations.*

Proof. (a) $\mathscr{G}_{\beta,h}$ contains $P_{\beta,h,+}$ and is a closed convex subset of $\mathscr{M}(\Omega)$. Hence part (a) follows from Theorem A.11.2.

(b) See Lanford and Ruelle (1969) or Preston (1976, page 21). □

The following corollary sharpens Corollary A.11.9.

Corollary A.11.12. (a) *For each $\beta > 0$, h real, and choice of sign, the measure $P_{\beta,h,\pm} = w\text{-}\lim_{\Lambda \uparrow \mathbb{Z}^D} P_{\Lambda,\beta,h,\pm}$* in *$\mathscr{G}_{\beta,h} \cap \mathscr{M}_s(\Omega)$ has short-range correlations.*

(b) *For each choice of sign, let $\langle Y_0; Y_k \rangle_{\beta,h,\pm}$ denote the pair correlation*

$$\langle Y_0; Y_k \rangle_{\beta,h,\pm} = \int_\Omega Y_0 Y_k dP_{\beta,h,\pm} - \int_\Omega Y_0 dP_{\beta,h,\pm} \int_\Omega Y_k dP_{\beta,h,\pm}.$$

Then for each $\beta > 0$, h real, and choice of sign, $\langle Y_0; Y_k \rangle_{\beta,h,\pm} \to 0$ as $\|k\| \to \infty$.

Proof. (a) The FKG inequality implies that the measures $P_{\beta,h,+}$ and $P_{\beta,h,-}$ are extremal points of $\mathscr{G}_{\beta,h}$. Hence these measures have short-range correlations.

(b) Define the cylinder sets $\Sigma_+ = \{\omega : \omega_0 = 1\}$ and $\Sigma_- = \{\omega : \omega_0 = -1\}$ and write P for $P_{\beta,h,+}$. Then

$$\begin{aligned}\langle Y_0 ; Y_k\rangle_{\beta,h,+} &= P\{\Sigma_+ \cap T^{-k}\Sigma_+\} + P\{\Sigma_- \cap T^{-k}\Sigma_-\} \\ &\quad - P\{\Sigma_+ \cap T^{-k}\Sigma_-\} - P\{\Sigma_- \cap T^{-k}\Sigma_+\} \\ &\quad - [P\{\Sigma_+\} - P\{\Sigma_-\}]\cdot[P\{T^{-k}\Sigma_+\} - P\{T^{-k}\Sigma_-\}].\end{aligned}$$

Since P has short-range correlations, P is mixing, and thus $\langle Y_0; Y_k\rangle_{\beta,h,+} \to 0$ as $\|k\| \to \infty$. The same proof shows that $\langle Y_0; Y_k\rangle_{\beta,h,-} \to 0$ as $\|k\| \to \infty$. □

For finite-range interactions, Martin-Löf (1972) has a direct proof of part (b) of the corollary.

Appendix B: Proofs of Two Theorems in Section II.7

B.1. Proof of Theorem II.7.1

(a) Assume that F is a continuous real-valued function on a complete separable metric space $\mathscr{X}$ and that $\sup_{x\in\mathscr{X}} F(x)$ is finite. Clearly $\sup_{x\in\mathscr{X}} \{F(x) - I(x)\} > -\infty$, and since I is non-negative, $\sup_{x\in\mathscr{X}} \{F(x) - I(x)\} < \infty$. We prove that

$$\lim_{n\to\infty} \frac{1}{a_n} \log \int_{\mathscr{X}} \exp(a_n F(x)) Q_n(dx) = \sup_{x\in\mathscr{X}} \{F(x) - I(x)\}.$$

For A a Borel subset of $\mathscr{X}$, define

$$I(A) = \inf_{x\in A} I(x) \qquad \text{and} \qquad J_n(A) = \int_A \exp(a_n F(x)) Q_n(dx)$$

and write J_n for $J_n(\mathscr{X})$. $I(\phi)$ equals ∞. We suppose that for all $x \in \mathscr{X}$, $F(x) \leq L < \infty$. Choose a number $M < \min(L, \sup_{x\in\mathscr{X}}\{F(x) - I(x)\})$. For N a positive integer and $j = 1, \ldots, N$, define closed sets

$$\text{(B.1)} \qquad K_{N,j} = \left\{x \in \mathscr{X} : M + \frac{j-1}{N}(L - M) \leq F(x) \leq M + \frac{j}{N}(L - M)\right\},$$

We have $\bigcup_{j=1}^{N} K_{N,j} = \{x \in \mathscr{X} : F(x) \geq M\}$ and the upper large deviation bounds

$$\limsup_{n\to\infty} \frac{1}{a_n} \log Q_n\{K_{N,j}\} \leq -I(K_{N,j}).$$

Therefore,

$$\begin{aligned}
&\limsup_{n\to\infty} \frac{1}{a_n} \log J_n(x \in \mathscr{X} : F(x) \geq M\}) \\
&\qquad \leq \max_{j=1,\ldots,N} \left\{M + \frac{j}{N}(L - M) - I(K_{N,j})\right\} \\
&\qquad \leq \max_{j=1,\ldots,N} \sup_{x\in K_{N,j}} \{F(x) - I(x)\} + \frac{L - M}{N}. \\
&\qquad \leq \sup_{x\in\mathscr{X}} \{F(x) - I(x)\} + \frac{L - M}{N}.
\end{aligned}$$

Taking $N \to \infty$ yields

$$\limsup_{n\to\infty} \frac{1}{a_n} \log J_n(\{x \in \mathscr{X} : F(x) \geq M\}) \leq \sup_{x \in \mathscr{X}} \{F(x) - I(x)\}.$$

Since $J_n(\{x \in \mathscr{X} : F(x) \leq M\}) \leq e^{a_n M}$, it follows that

$$\limsup_{n\to\infty} \frac{1}{a_n} \log J_n \leq \max\left(M, \sup_{x \in \mathscr{X}} \{F(x) - I(x)\}\right) = \sup_{x \in \mathscr{X}} \{F(x) - I(x)\}.$$

We now prove that

$$\text{(B.2)} \qquad \liminf_{n\to\infty} \frac{1}{a_n} \log J_n \geq \sup_{x \in \mathscr{X}} \{F(x) - I(x)\}.$$

Let x_0 be an arbitrary point in $\mathscr{X}$ and ε an arbitrary positive number. If G denotes the open set $\{x \in \mathscr{X} : F(x) > F(x_0) - \varepsilon\}$, then the lower large deviation bound

$$\liminf_{n\to\infty} \frac{1}{a_n} \log Q_n\{G\} \geq -I(G)$$

implies

$$\liminf_{n\to\infty} \frac{1}{a_n} \log J_n \geq \liminf_{n\to\infty} \frac{1}{a_n} \log J_n(G)$$

$$\geq F(x_0) - \varepsilon - I(G) \geq F(x_0) - I(x_0) - \varepsilon.$$

Since x_0 and ε are arbitrary, (B.2) follows. This completes the proof of part (a) of Theorem II.7.1. □

(b) We prove the limit (2.37) for a continuous real-valued function F which satisfies (2.38). As in the proof of part (a),

$$\text{(B.3)} \qquad \liminf_{n\to\infty} \frac{1}{a_n} \log J_n \geq \sup_{x \in \mathscr{X}} \{F(x) - I(x)\}.$$

Hypothesis (2.38) guarantees that $\limsup_{n\to\infty} a_n^{-1} \log J_n$ is finite. Hence by (B.3), $\sup_{x \in \mathscr{X}} \{F(x) - I(x)\}$ is finite. By (2.38) there exists a number $L > 0$ such that

$$\limsup_{n\to\infty} \frac{1}{a_n} \log J_n(\{x \in \mathscr{X} : F(x) > L\}) \leq \sup_{x \in \mathscr{X}} \{F(x) - I(x)\}.$$

Define $\bar{F}(x) = \min[F(x), L]$ and $\bar{J}_n = \int_{\mathscr{X}} \exp(a_n \bar{F}(x)) Q_n(dx)$. $\bar{F}$ satisfies the hypotheses of part (a). We have

$$J_n = J_n(\{x \in \mathscr{X} : F(x) \leq L\}) + J_n(\{x \in \mathscr{X} : F(x) > L\})$$
$$= \bar{J}_n + J_n(\{x \in \mathscr{X} : F(x) > L\}),$$

$$\limsup_{n\to\infty} \frac{1}{a_n} \log J_n \le \max\left[\sup_{x\in\mathscr{X}} \{\bar{F}(x) - I(x)\}, \sup_{x\in\mathscr{X}} \{F(x) - I(x)\}\right]$$
$$= \sup_{x\in\mathscr{X}} \{F(x) - I(x)\}.$$

This completes the proof of part (b) of Theorem II.7.1.

B.2. Proof of Theorem II.7.2

A proof of Theorem II.7.2 was sketched in Section II.7. In order to complete the proof, we show the following facts. For A a Borel subset of $\mathscr{X}$, let $J_n(A) = \int_A \exp(a_n F(x)) Q_n(dx)$.

(a) $\limsup_{n\to\infty} a_n^{-1} \log J_n(K) \le \sup_{x\in K} \{F(x) - I(x)\}$ for each nonempty closed set K in $\mathscr{X}$.

(b) $\liminf_{n\to\infty} a_n^{-1} \log J_n(G) \ge \sup_{x\in G} \{F(x) - I(x)\}$ for each nonempty open set G in $\mathscr{X}$.

(c) $I_F(x) = I(x) - F(x) - \inf_{x\in\mathscr{X}} \{I(x) - F(x)\}$ has compact level sets.

Proof. (a) First assume that for all $x \in \mathscr{X}$, $F(x) \le L < \infty$. Choose a number $M < \min(L, \sup_{x\in K} \{F(x) - I(x)\})$. For N a positive integer and $j = 1, \ldots, N$, define closed sets $\tilde{K}_{N,j} = K \cap K_{N,j}$, where $K_{N,j}$ is defined in (B.1). As in the proof of part (a) of Theorem II.7.1,

$$\limsup_{n\to\infty} \frac{1}{a_n} \log J_n(\{x \in K : F(x) \ge M\})$$
$$\le \max_{j=1,\ldots,N} \sup_{x\in\tilde{K}_{N,j}} \{F(x) - I(x)\} + \frac{L-M}{N}$$
$$\le \sup_{x\in K} \{F(x) - I(x)\} + \frac{L-M}{N}.$$

Taking $N \to \infty$ yields

$$\limsup_{n\to\infty} \frac{1}{a_n} \log J_n(\{x \in K : F(x) \ge M\}) \le \sup_{x\in K} \{F(x) - I(x)\}.$$

Since $J_n(\{x \in K : F(x) \le M\}) \le e^{a_n M}$, it follows that

$$\limsup_{n\to\infty} \frac{1}{a_n} \log J_n(K) \le \max\left(M, \sup_{x\in K} \{F(x) - I(x)\}\right) = \sup_{x\in K} \{F(x) - I(x)\}.$$

This proves the upper bound (a) for a continuous real-valued function F which is bounded above. If F is a continuous real-valued function which is unbounded above and satisfies (2.38), then the proof of the upper bound (a) proceeds exactly as in the proof of part (b) of Theorem II.7.1.

(b) Let x_0 be an arbitrary point in $\mathscr{X}$ and ε an arbitrary positive number. If $\tilde{G}$ denotes the open set $\{x \in G: F(x) > F(x_0) - \varepsilon\}$, then

$$\begin{aligned}\liminf_{n\to\infty} \frac{1}{a_n} \log J_n(G) &\geq \liminf_{n\to\infty} \frac{1}{a_n} \log J_n(\tilde{G}) \\ &\geq F(x_0) - \varepsilon - I(\tilde{G}) \geq F(x_0) - I(x_0) - \varepsilon.\end{aligned}$$

Since x_0 and ε are arbitrary, the lower bound (b) follows.

(c) If F is bounded above and continuous, then I_F has compact level sets since I has compact level sets. So we assume that F is unbounded above and continuous and satisfies (2.38). By Theorem II.7.1, $\inf_{x \in \mathscr{X}} \{I(x) - F(x)\}$ is finite. By the lower bound (b) which has just been proved,

$$\begin{aligned}-\infty &= \lim_{L\to\infty} \limsup_{n\to\infty} \frac{1}{a_n} \log J_n(\{x \in \mathscr{X} : F(x) > L\}) \\ &\geq \lim_{L\to\infty} \sup_{\{x \in \mathscr{X} : F(x) > L\}} \{F(x) - I(x)\}.\end{aligned}$$

Thus

$$\lim_{L\to\infty} \inf\{I(x) - F(x) : x \in \mathscr{X}, F(x) > L\} = \infty. \tag{B.4}$$

Consider the level set $K_b = \{x \in \mathscr{X} : I_F(x) \leq b\}$, b real. We must show that for any sequence $\{x_n; n = 1, 2, \ldots\}$ in K_b, there exists a subsequence $\{x_{n'}\}$ converging to some element x in K_b. First suppose that $\sup_{n=1,2,\ldots} F(x_n) \leq c < \infty$. Then

$$\sup_{n=1,2,\ldots} I(x_n) \leq b + c + \inf_{x \in \mathscr{X}} \{I(x) - F(x)\} < \infty.$$

Since I has compact level sets, there exists a subsequence $\{x_{n'}\}$ and an element x such that $x_{n'} \to x$; x belongs to K_b since I is lower semicontinuous and F is continuous. Now suppose that $\sup_{n=1,2,\ldots} F(x_n) = \infty$. Then by (B.4) $\sup_{n=1,2,\ldots} \{I(x_n) - F(x_n)\} = \infty$. Since the latter contradicts $I_F(x_n) \leq b$, the proof is done. $\square$

Appendix C: Equivalent Notions of Infinite-Volume Measures for Spin Systems

C.1. Introduction

In Chapters IV and V, we proved the existence and properties of infinite-volume Gibbs states for ferromagnetic models on $\mathbb{Z}^D$. In this appendix, we generalize the definition of infinite-volume Gibbs states to include many-body interactions. We also discuss the equivalence between Gibbs states and other notions of infinite-volume measures that are standard in the literature. In Appendix C.5, a proof of the Gibbs variational formula and principle is sketched. In the last section of this appendix, we solve the Gibbs variational formula for finite-range interactions on $\mathbb{Z}$.

C.2. Two-Body Interactions and Infinite-Volume Gibbs States

Let us first recall some definitions from Chapters IV and V. J is a non-negative, symmetric, summable function on $\mathbb{Z}^D$; β a positive real number; h a real number; Λ a symmetric hypercube in $\mathbb{Z}^D$; Ω_Λ the set $\{1,-1\}^\Lambda$; $\mathscr{B}(\Omega_\Lambda)$ the set of all subsets of Ω_Λ; $\pi_\Lambda P_\rho$ the product measure on $\mathscr{B}(\Omega_\Lambda)$ with identical one-dimensional marginals $\rho = \frac{1}{2}\delta_1 + \frac{1}{2}\delta_{-1}$; and $\tilde{\omega} = \tilde{\omega}(\Lambda)$ a point in $\Omega_{\Lambda^c} = \{1,-1\}^{\Lambda^c}$. We defined the finite-volume Gibbs state on Λ with external condition $\tilde{\omega}$ to be the probability measure $P_{\Lambda,\beta,h,\tilde{\omega}}$ on $\mathscr{B}(\Omega_\Lambda)$ which assigns to each $\{\omega\}$, $\omega \in \Omega_\Lambda$, the probability

$$\text{(C.1)} \quad P_{\Lambda,\beta,h,\tilde{\omega}}\{\omega\} = \exp[-\beta H_{\Lambda,h,\tilde{\omega}}(\omega)]\pi_\Lambda P_\rho(\omega) \cdot \frac{1}{Z(\Lambda,\beta,h,\tilde{\omega})}.$$

$H_{\Lambda,h,\tilde{\omega}}(\omega)$ is the Hamiltonian

$$-\frac{1}{2}\sum_{i,j\in\Lambda} J(i-j)\omega_i\omega_j - h\sum_{i\in\Lambda}\omega_i - \sum_{i\in\Lambda}\sum_{j\in\Lambda^c} J(i-j)\tilde{\omega}_j\omega_i, \qquad \omega \in \Omega_\Lambda,$$

and $Z(\Lambda,\beta,h,\tilde{\omega})$ is the normalization $\int_{\Omega_\Lambda} \exp[-\beta H_{\Lambda,h,\tilde{\omega}}(\omega)]\pi_\Lambda P_\rho(d\omega)$. In (5.24), $P_{\Lambda,\beta,h,\tilde{\omega}}$ was extended to a probability measure on $\mathscr{B}(\Omega)$, which is the Borel σ-field of the space $\Omega = \{1,-1\}^{\mathbb{Z}^D}$. $\mathscr{M}(\Omega)$ denotes the set of probability measures on $\mathscr{B}(\Omega)$. $\mathscr{M}_s(\Omega)$ denotes the subset of $\mathscr{M}(\Omega)$ consisting of translation invariant probability measures. $\mathscr{G}_{\beta,h}$ denotes the subset of $\mathscr{M}(\Omega)$ consisting of infinite-volume Gibbs states. $\mathscr{G}_{\beta,h}$ was defined to be the closed convex hull of the set of weak limits

$$P = w\text{-}\lim_{\Lambda' \uparrow \mathbb{Z}^D} P_{\Lambda',\beta,h,\tilde{\omega}(\Lambda')},$$

where $\{\Lambda'\}$ is any increasing sequence of symmetric hypercubes whose union is $\mathbb{Z}^D$ and $\tilde{\omega}(\Lambda')$ is any external condition for Λ'. The reader is referred to Appendix A.11 for information about the sets $\mathscr{M}(\Omega)$, $\mathscr{M}_s(\Omega)$, and $\mathscr{G}_{\beta,h}$.

Since $\omega_i^2 = 1$, the Hamiltonian can be written as

$$\begin{aligned} H_{\Lambda,h,\tilde{\omega}}(\omega) = &-\frac{N}{2}J(0) - \sum_{\{i,j\}} J(\{i,j\})\omega_i\omega_j - \sum_{i \in \Lambda} J(\{i\})\omega_i \\ &- \sum_{i \in \Lambda}\sum_{j \in \Lambda^c} J(\{i,j\})\tilde{\omega}_j\omega_i. \end{aligned} \tag{C.2}$$

The first sum runs over all unordered pairs $\{i,j\}$ in $\Lambda \times \Lambda$ with $i \neq j$, $J(\{i,j\})$ equals $J(i-j)$, and $J(\{i\})$ equals h. The constant $-\frac{1}{2}NJ(0)$ cancels in the numerator and denominator of (C.1). It was useful to allow nonzero $J(0)$ since in the Curie–Weiss model $J(i-j)$ equals $\mathscr{J}_0/|\Lambda|$ for all i and j in Λ. The terms $J(\{i,j\})$ in (C.2) are two-body interactions. The terms $J(\{i\})$ are one-body interactions. Hamiltonians with many-body interactions will be considered in the next section.

In the above definitions, the non-negativity of J is not required. J need only be real-valued, symmetric, and summable. For any such J, $P_{\Lambda,\beta,h,\tilde{\omega}}$ and $\mathscr{G}_{\beta,h}$ are well-defined, and since $\mathscr{M}(\Omega)$ is compact, $\mathscr{G}_{\beta,h}$ is nonempty. The assumption that J is non-negative was needed in order to analyze the structure of $\mathscr{G}_{\beta,h}$ for different values of β and h [Theorems IV.6.5 and V.6.1]. This analysis depended on the FKG inequality and on the moment inequalities proved in Section V.3.

We also comment on the choice of subsets of $\mathbb{Z}^D$ which define the infinite-volume limit. A number of proofs in Chapters IV and V involved the specific Gibbs free energy

$$\psi(\beta,h) = -\beta^{-1} \lim_{\Lambda \uparrow \mathbb{Z}^D} \frac{1}{|\Lambda|} \log Z(\Lambda,\beta,h,\tilde{\omega}(\Lambda)).$$

The limit exists and is independent of the choice of external conditions $\{\tilde{\omega}(\Lambda)\}$ if the sets $\{\Lambda\}$ are increasing symmetric hypercubes whose union is $\mathbb{Z}^D$. Clearly, since the interaction defined by J is translation invariant, the hypercubes need not be symmetric. According to Appendix D.2, other sequences of subsets are allowed. The definition of the set of infinite-volume Gibbs states may also be modified. We may consider the closed convex hull of the set of weak limits of finite-volume Gibbs states $\{P_{\Lambda',\beta,h,\tilde{\omega}(\Lambda')}\}$, where $\{\Lambda'\}$ is any increasing sequence of nonempty finite sets whose union is $\mathbb{Z}^D$ (not necessarily symmetric hypercubes). In the next few sections, we explore these and related matters.

C.3. Many-Body Interactions and Infinite-Volume Gibbs States

A (many-body) *interaction* is a function J which assigns to each nonempty finite subset A of $\mathbb{Z}^D$ a real number $J(A)$. The interaction is assumed to be translation invariant in the sense that $J(A+i) = J(A)$ for all A and all $i \in \mathbb{Z}^D$. We denote by $\mathscr{J}$ the set of interactions for which $|||J||| = \sum_{A \ni 0} |J(A)|/|A|$ is finite* and by $\tilde{\mathscr{J}}$ the set of interactions for which $\|J\|_{\sim} = \sum_{A \ni 0} |J(A)|$ is finite. $\tilde{\mathscr{J}}$ is a subset of $\mathscr{J}$. If $J(A)$ is non-negative for each A, then the interaction is called *ferromagnetic*. In this appendix, we do not restrict to such interactions.

Let Λ be a nonempty finite subset of $\mathbb{Z}^D$, J an interaction in $\tilde{\mathscr{J}}$, and $\tilde{\omega} = \tilde{\omega}(\Lambda)$ a point in Ω_{Λ^c} (external condition). The Hamiltonian is defined by the formula

$$\text{(C.3)} \quad H_{\Lambda,J,\tilde{\omega}}(\omega) = -\sum_{A \subset \Lambda} J(A)\omega_A - \sum_{\substack{A \subset \Lambda, B \subset \Lambda^c \\ A, B \neq \varnothing}} J(A \cup B)\omega_A \tilde{\omega}_B, \qquad \omega \in \Omega_\Lambda,$$

where $\omega_A = \prod_{i \in A} \tilde{\omega}_i$ and $\tilde{\omega}_B = \prod_{i \in A} \tilde{\omega}_i$. The sums converge since J is in $\tilde{\mathscr{J}}$.† We define the finite-volume Gibbs state on Λ with external condition $\tilde{\omega}$ to be the probability measure $P_{\Lambda,J,\tilde{\omega}}$ on $\mathscr{B}(\Omega_\Lambda)$ which assigns to each $\{\omega\}$, $\omega \in \Omega_\Lambda$, the probability

$$\text{(C.4)} \qquad P_{\Lambda,J,\tilde{\omega}}\{\omega\} = \exp[-H_{\Lambda,J,\tilde{\omega}}(\omega)]\pi_\Lambda P_\rho(\omega) \cdot \frac{1}{Z(\Lambda, J, \tilde{\omega})}.$$

$Z(\Lambda, J, \tilde{\omega})$ is the normalization $\int_{\Omega_\Lambda} \exp[-H_{\Lambda,J,\tilde{\omega}}(\omega)]\pi_\Lambda P_\rho(d\omega)$. We have absorbed the temperature dependence into J. As in (5.24), $P_{\Lambda,J,\tilde{\omega}}$ is extended to a probability measure on $\mathscr{B}(\Omega)$ (denoted by the same symbol $P_{\Lambda,J,\tilde{\omega}}$).

We consider two definitions of infinite-volume Gibbs states. Define $\mathscr{G}_J$ to be the closed convex hull of the set of weak limits

$$P = \underset{\Lambda' \uparrow \mathbb{Z}^D}{w\text{-lim}}\, P_{\Lambda',J,\tilde{\omega}(\Lambda')},$$

where $\{\Lambda'\}$ is any increasing sequence of symmetric hypercubes whose union is $\mathbb{Z}^D$ and $\tilde{\omega}(\Lambda')$ is any external condition for Λ'. Define $\bar{\mathscr{G}}_J$ to be the closed convex hull of the set of weak limits

$$P = \underset{\Lambda' \uparrow \mathbb{Z}^D}{w\text{-lim}}\, P_{\Lambda',J,\tilde{\omega}(\Lambda')},$$

where $\{\Lambda'\}$ is any increasing sequence of nonempty finite sets whose union is $\mathbb{Z}^D$ and $\tilde{\omega}(\Lambda')$ is any external condition for Λ'. Since $\mathscr{M}(\Omega)$ is compact, both $\mathscr{G}_J$ and $\bar{\mathscr{G}}_J$ are nonempty. By definition $\mathscr{G}_J$ is a subset of $\bar{\mathscr{G}}_J$; according to Theorem C.4.1 below, $\mathscr{G}_J$ equals $\bar{\mathscr{G}}_J$. The structure of the set of infinite-

* $|A|$ is the number of sites in A.

† $\max_{\omega \in \Omega_\Lambda} |H_{\Lambda,J,\tilde{\omega}}(\omega) \leq |\Lambda| \cdot \|J\|_{\sim}$.

volume Gibbs states has been analyzed by S. A. Pirogov and Ya. G. Sinai (see Sinai (1982, Chapter II)) and by Holsztynski and Slawny (1979).

C.4. DLR States

Let J be an interaction in $\tilde{\mathscr{J}}$. Another formulation of infinite-volume measures for spin systems is due to Dobrushin (1968a, 1968b, 1968c, 1969, 1970) and to Lanford and Ruelle (1969). In order to motivate this formulation, consider a finite-volume Gibbs state $P_{\bar{\Lambda},J,\bar{\omega}}$ on a nonempty subset $\bar{\Lambda}$ of $\mathbb{Z}^D$ with external condition $\bar{\omega}$. Let Λ be a nonempty subset of $\bar{\Lambda}$, ω_0 a point in Ω_Λ, and ω_1 a point in $\Omega_{\bar{\Lambda}\setminus\Lambda}$. A straightforward calculation shows that the conditional probability

$$P_{\bar{\Lambda},J,\bar{\omega}}\{\omega\in\Omega_{\bar{\Lambda}}:\omega=\omega_0 \text{ on } \Lambda|\omega=\omega_1 \text{ on } \bar{\Lambda}\setminus\Lambda\} \quad \text{equals} \quad P_{\Lambda,J,\tilde{\omega}}\{\omega_0\},$$

where the external condition $\tilde{\omega}$ equals ω_1 on $\bar{\Lambda}\setminus\Lambda$ and $\bar{\omega}$ on $\bar{\Lambda}^c$. If we formally pass to the limit $\bar{\Lambda}\uparrow\mathbb{Z}^D$ with Λ held fixed, then we are led to the definition of DLR state given below.

We need some notation. Let Λ be a nonempty finite subset of $\mathbb{Z}^D$; $\mathscr{F}_{\Lambda^c}$ the σ-field of subsets of Ω generated by the coordinate mappings $Y_j(\omega)=\omega_j$, $j\in\Lambda^c$; π_Λ the projection of Ω onto Ω_Λ defined by $(\pi_\Lambda\omega)_i=\omega_i$, $i\in\Lambda$; and π_{Λ^c} the corresponding projection of Ω onto Ω_{Λ^c}. For each point $\omega\in\Omega$, $\pi_{\Lambda^c}\omega$ defines an external condition for Λ. If P is a probability measure on $\mathscr{B}(\Omega)$, then define a probability measure $\pi_\Lambda P$ on $\mathscr{B}(\Omega_\Lambda)$ by requiring

$$\pi_\Lambda P\{F\}=P\{\pi_\Lambda^{-1}F\} \qquad \text{for subsets } F \text{ of } \Omega_\Lambda.$$

A probability measure P on $\mathscr{B}(\Omega)$ is called a *DLR state* for the interaction J if for each Λ and each point $\omega_0\in\Omega_\Lambda$

$$\text{(C.5)} \qquad P\{\pi_\Lambda^{-1}\{\omega_0\}|\mathscr{F}_{\Lambda^c}\}(\omega)=P_{\Lambda,J,\pi_{\Lambda^c}\omega}\{\omega_0\} \qquad P\text{-a.s.}$$

By the definition of conditional probability, this is equivalent to requiring

$$\text{(C.6)} \qquad P\{\pi_\Lambda^{-1}\{\omega_0\}\cap G\}=\int_G P_{\Lambda,J,\pi_{\Lambda^c}\omega}\{\omega_0\}P(d\omega)$$

for each Λ, each point $\omega_0\in\Omega_\Lambda$, and each set G in $\mathscr{F}_{\Lambda^c}$. One may show that the set of DLR states for J is a nonempty compact convex subset of $\mathscr{M}(\Omega)$.

According to the next theorem, the notions of infinite-volume Gibbs state and DLR state are equivalent.

Theorem C.4.1. *Let J be an interaction in $\tilde{\mathscr{J}}$. The following are equivalent.* (a) *P is in $\mathscr{G}_J$.* (b) *P is in $\bar{\mathscr{G}}_J$.* (c) *P is a DLR state for J.*

Comments on Proof. (a) $\Rightarrow$ (b) By definition, $\mathscr{G}_J$ is a subset of $\bar{\mathscr{G}}_J$.

(b) $\Rightarrow$ (c) See Dobrushin (1968a, page 206), Lanford (1973, page 104), or Ruelle (1978, page 18).

(c) $\Rightarrow$ (a) Georgii (1973) notes that if P is an extremal DLR state, then for P-almost every ω, the sequence $\{P_{\Lambda, J, \pi_{\Lambda^c}\omega}; \Lambda$ symmetric hypercubes$\}$ converges weakly to P as $\Lambda \uparrow \mathbb{Z}^D$. This is an immediate consequence of the martingale convergence theorem. Each DLR state for J belongs to the closed convex hull of the extremal DLR states for J [Theorem A.11.6(b)]. It follows that if P is a DLR state for J, then P belongs to $\mathscr{G}_J$. □

C.5. The Gibbs Variational Formula and Principle

Throughout this section Λ denotes a symmetric hypercube in $\mathbb{Z}^D$. Let J be an interaction in $\mathscr{J}$, $H_{\Lambda,J}(\omega)$ the Hamiltonian $-\sum_{A \subset \Lambda} J(A)\omega_A$ $(\omega \in \Omega_\Lambda)$, and $Z(\Lambda, J)$ the partition function

$$\text{(C.7)} \qquad Z(\Lambda, J) = \int_{\Omega_\Lambda} \exp[-H_{\Lambda,J}(\omega)]\pi_\Lambda P_\rho(d\omega).$$

According to Theorem D.I.I, the limit

$$\text{(C.8)} \qquad p(J) = \lim_{\Lambda \uparrow \mathbb{Z}^D} \frac{1}{|\Lambda|} \log Z(\Lambda, J)$$

exists. The functional $p(J)$ is called the *pressure*. It equals negative the specific Gibbs free energy (for $\beta = 1$). Let P belong to the set $\mathscr{M}_s(\Omega)$ of translation invariant probability measures on Ω. The limit

$$\text{(C.9)} \qquad \langle J, P \rangle = -\lim_{\Lambda \uparrow \mathbb{Z}^D} \frac{1}{|\Lambda|} \int_{\Omega_\Lambda} H_{\Lambda,J}(\omega)\pi_\Lambda P(d\omega)$$

exists and

$$\text{(C.10)} \qquad \langle J, P \rangle = \sum_{A \ni 0} (J(A)/|A|) \int_{\Omega} \omega_A P(d\omega).$$

Also the limit

$$\text{(C.11)} \qquad I_\rho^{(3)}(P) = \lim_{\Lambda \uparrow \mathbb{Z}^D} \frac{1}{|\Lambda|} I_{\pi_\Lambda P_\rho}^{(2)}(\pi_\Lambda P)$$

exists, where $I_{\pi_\Lambda P_\rho}^{(2)}(\pi_\Lambda P)$ denote the relative entropy of $\pi_\Lambda P$ with respect to $\pi_\Lambda P_\rho$. See, for example, Israel (1979, Chapter II) for a discussion of the limits $\langle J, P \rangle$ and $I_\rho^{(3)}(P)$.

The next theorem states the Gibbs variational formula and Gibbs variational principle (parts (a) and (b), respectively). The theorem is due to Ruelle (1967) and Lanford and Ruelle (1969). If $J(A)$ equals 0 whenever $|A| > 2$, then this theorem reduces to Theorems IV.7.3 and V.6.3.

Theorem C.5.1. *Let J be an interaction. Then the following conclusions hold.*

(a) *For* $J \in \mathscr{J}$, $p(J) = \sup_{P \in \mathscr{M}_s(\Omega)} \{\langle J, P\rangle - I_\rho^{(3)}(P)\}$.

(b) *For* $J \in \tilde{\mathscr{J}}$, *the set of* $P \in \mathscr{M}_s(\Omega)$, *at which the supremum in part* (a) *is attained equals* $\mathscr{G}_J \cap \mathscr{M}_s(\Omega)$, *the set of translation invariant infinite-volume Gibbs states.*

We sketch a proof of the theorem for interactions J in $\tilde{\mathscr{J}}$. This proof is due to Föllmer (1973) and Preston (1976). See Pirlot (1980) and Künsch (1981) for generalizations. Given measures P and Q in $\mathscr{M}_s(\Omega)$, consider the relative entropy $I_{\pi_\Lambda Q}^{(2)}(\pi_\Lambda P)$ of $\pi_\Lambda P$ with respect to $\pi_\Lambda Q$. If the limit

$$\lim_{\Lambda \uparrow \mathbb{Z}^D} \frac{1}{|\Lambda|} I_{\pi_\Lambda Q}^{(2)}(\pi_\Lambda P)$$

exists, then we denote the limit by $h(P|Q)$ and call it the *specific information gain* of P with respect to Q. Clearly $h(P|Q)$ is non-negative and if Q equals the infinite product measure P_ρ, then $h(P|Q)$ equals $I_\rho^{(3)}(P)$. Theorem C.5.1 for $J \in \tilde{\mathscr{J}}$ is a consequence of the following result and Theorem C.4.1.

Theorem C.5.2. *Let J be an interaction in $\tilde{\mathscr{J}}$ and let Q be a DLR state for J. Then for any $P \in \mathscr{M}_s(\Omega)$ the following conclusions hold.*

(a) *$h(P|Q)$ exists and satisfies*

$$0 \le h(P|Q) = p(J) - \langle J, P\rangle + I_\rho^{(3)}(P).$$

(b) *$h(P|Q)$ equals* 0 *if and only if P is a DLR state for J.*

Sketch of Proof. (a) Consider the probability measure $P_{\Lambda,J}$ on $\mathscr{B}(\Omega_\Lambda)$ defined by

$$P_{\Lambda,J}\{\omega_0\} = \exp[-H_{\Lambda,J}(\omega_0)]\pi_\Lambda P_\rho\{\omega_0\} \cdot \frac{1}{Z(\Lambda, J)} \qquad \text{for } \omega_0 \in \Omega_\Lambda.$$

We have

$$I_{P_{\Lambda,J}}^{(2)}(\pi_\Lambda P) - I_{\pi_\Lambda Q}^{(2)}(\pi_\Lambda P) = \sum_{\omega_0 \in \Omega_\Lambda} \log \frac{\pi_\Lambda Q\{\omega_0\}}{P_{\Lambda,J}\{\omega_0\}} \cdot \pi_\Lambda P\{\omega_0\}.$$

Since Q is a DLR state for J,

$$\frac{\pi_\Lambda Q\{\omega_0\}}{P_{\Lambda,J}\{\omega_0\}} = \int_\Omega \frac{P_{\Lambda,J,\pi_{\Lambda^c}\omega}\{\omega_0\}}{P_{\Lambda,J}\{\omega_0\}} Q(d\omega).$$

By elementary estimates, one shows that for any $\omega_0 \in \Omega_\Lambda$ and $\omega \in \Omega$

$$e^{-a(\Lambda)} \le \frac{P_{\Lambda,J,\pi_{\Lambda^c}\omega}\{\omega_0\}}{P_{\Lambda,J}\{\omega_0\}} \le e^{a(\Lambda)},$$

where $a(\Lambda)$ does not depend on ω_0 and ω and satisfies $a(\Lambda)/|\Lambda| \to 0$ as $\Lambda \uparrow \mathbb{Z}^D$. It follows that

(C.12) $$\lim_{\Lambda \uparrow \mathbb{Z}^D} \frac{1}{|\Lambda|} [I^{(2)}_{\pi_\Lambda Q}(\pi_\Lambda P) - I^{(2)}_{P_{\Lambda,J}}(\pi_\Lambda P)] = 0.$$

Since

$$I^{(2)}_{P_{\Lambda,J}}(\pi_\Lambda P) = \log Z(\Lambda, J) + \int_{\Omega_\Lambda} H_{\Lambda,J}(\omega_0)\pi_\Lambda P_\rho(d\omega_0) + I^{(2)}_{\pi_\Lambda P_\rho}(\pi_\Lambda P),$$

we conclude that $h(P|Q)$ exists and satisfies

$$0 \le h(P|Q) = p(J) - \langle J, P\rangle + I^{(3)}_\rho(P).$$

(b) If P is a DLR state for J, then (C.12) with $Q = P$ shows that $h(P|Q)$ equals 0. The most difficult part of the proof is the converse: $h(P|Q) = 0$ implies that P is a DLR state for J. We omit this proof, referring the reader to Preston (1976, pages 115–122) for details. □

We end this section by stating another characterization of the measures in $\mathscr{G}_J \cap \mathscr{M}_s(\Omega)$. Heuristically, the Gibbs variational formula is a Legendre–Fenchel transform relating $I^{(3)}_\rho(P)$ and $p(J)$. By Theorem D.1.1, $p(J)$ is a convex function of $J \in \mathscr{J}$. Let J_0 be an interaction in $\mathscr{J}$. A measure $P_0 \in \mathscr{M}_s(\Omega)$ is said to be in the *subdifferential of p at* J_0 if

(C.13) $$p(J_0 + J) \ge p(J_0) + \langle J, P_0\rangle \qquad \text{for all } J \in \mathscr{J}.$$

This equation means that the functional $J \to \langle J, P_0\rangle$ is tangent to the graph of p at $(J_0, p(J_0))$. A measure $P_0 \in \mathscr{M}_s(\Omega)$ satisfying (C.13) is called a *translation invariant equilibrium state for* J_0.

Theorem C.5.3. (a) *Let J_0 be an interaction in $\mathscr{J}$. Then P_0 is a translation invariant equilibrium state for J_0 if and only if P_0 gives the supremum in the Gibbs variational formula*

$$p(J_0) = \sup_{P \in \mathscr{M}_s(\Omega)} \{\langle J_0, P\rangle - I^{(3)}_\rho(P)\}.$$

(b) *Let J_0 be an interaction in $\tilde{\mathscr{J}}$. Then the following are equivalent.*

(i) *P_0 is a translation invariant equilibrium state for J_0.*
(ii) *P_0 is a translation invariant DLR state for J_0.*
(iii) *P_0 belongs to the set $\mathscr{G}_{J_0} \cap \mathscr{M}_s(\Omega)$.*

Comments on Proof. (a) This follows from Theorem C.5.1 and the inverse formula $I^{(3)}_\rho(P_0) = \sup_{J \in \mathscr{J}}\{\langle J, P_0\rangle - p(J)\}$. See Ruelle (1978, page 48) or Israel (1979, page 49).

(b) (i) ⇔ (ii) Part (a) and Theorem C.5.2.
(ii) ⇔ (iii) Theorem C.4.1. □

Lanford (1973, Chapters B–C), Preston (1976), and Israel (1979) discuss in detail the material contained in the previous three sections.

C.6. The Solution of the Gibbs Variational Formula for Finite-Range Interactions on $\mathbb{Z}$

Let J be a many-body interaction on $\mathbb{Z}$ which has finite range. Thus we suppose there exists an integer $\alpha \geq 2$ such that if $n_1 < \cdots < n_k$ are integers and $n_k - n_1 \geq \alpha$, then $J(\{n_1, \ldots, n_k\}) = 0$. We assume that α is the smallest integer with this property. Then the range of the interaction is defined to be $\alpha - 1$. As in Sections C.3–C.5, we absorb the temperature dependence into J. We sketch a proof that the supremum in the Gibbs variational formula is attained at a unique measure which is a Markov chain if $\alpha = 2$ and is an $(\alpha - 1)$-dependent Markov chain if $\alpha \geq 3$. The proof depends on results developed in Chapter IX. For two-body ferromagnetic interactions which are of finite-range, it follows from Theorems IV.6.5 and C.5.1 that the critical inverse temperature β_c equals ∞. Another proof is given at the end of this section.

Given $P \in \mathcal{M}_s(\Omega)$ $(\Omega = \{1, -1\}^{\mathbb{Z}})$, set $\tau = \pi_\alpha P$. Because of translation invariance, the functional $\langle J, P\rangle$ in (C.10) equals

$$\sum_{1 \in A \subseteq \{1,2,\ldots,\alpha\}} (J(A)/|A|) \int_\Omega \omega_A P(d\omega) = \sum_{1 \in A \subseteq \{1,2,\ldots,\alpha\}} (J(A)/|A|) \int_{\Omega_\alpha} \omega_A \tau(d\omega),$$

where $\Omega_\alpha = \{1, -1\}^\alpha$. Given $(x_{i_1}, \ldots, x_{i_\alpha}) \in \Omega_\alpha$, set $\tau_{i_1 \cdots i_\alpha} = \tau\{x_{i_1}, \ldots, x_{i_\alpha}\}$ and define a point $t \in \mathbb{R}^{2^\alpha}$ by

$$t_{i_1 \ldots i_\alpha} = \sum_{1 \in A \subseteq \{1,2,\ldots,\alpha\}} (J(A)/|A|) \prod_{j \in A} x_{i_j}.$$

By Theorem IX.3.3 (contraction principle) and (9.13), we can write the Gibbs variational formula as

$$\begin{aligned} p(J) &= \sup_{P \in \mathcal{M}_s(\Omega)} \{\langle J, P\rangle - I_\rho^{(3)}(P)\} = \sup_{\tau \in \mathcal{M}_s(\Omega_\alpha)} \{\langle t, \tau\rangle - I_{\rho,\alpha}^{(3)}(\tau)\} \\ &= \sup_{\tau \in \mathcal{M}_{s,\alpha}} \{\langle t, \tau\rangle - \bar{I}_{\rho,\alpha}^{(3)}(\tau)\}. \end{aligned}$$

According to (9.20),

$$\sup_{\tau \in \operatorname{ri} \mathcal{M}_{s,\alpha}} \{\langle t, \tau\rangle - \bar{I}_{\rho,\alpha}^{(3)}(\tau)\} = \log \lambda(B_\alpha(t)),$$

where the matrix $B_\alpha(t)$ is defined in (9.14). As in the proof of Theorem IX.4.4, the supremum in the last display is attained at the unique point

$$\tau^0_{i_1 \ldots i_\alpha} = u_{i_1 \ldots i_{\alpha-1}} B_\alpha(t)_{i_1 \ldots i_{\alpha-1}, i_2 \ldots i_\alpha} w_{i_2 \ldots i_\alpha} / \lambda(B_\alpha(t)),$$

where u and w are positive left and right eigenvectors associated with $\lambda(B_\alpha(t))$ and normalized so that $\langle u, w\rangle = 1$. Since $f(\tau) = \langle t, \tau\rangle - \bar{I}_{\rho,\alpha}^{(3)}(\tau)$ is concave [Problem IX.6.1 or IX.6.2], $f(\tau)$ attains its supremum over all of $\mathcal{M}_{s,\alpha}$ at the unique point $\tau^0 = \{\tau^0_{i_1 \cdots i_\alpha}\}$. We conclude that

$$p(J) = \sup_{\tau \in \mathcal{M}_{s,\alpha}} \{\langle t, \tau\rangle - \bar{I}_{\rho,\alpha}^{(3)}(\tau)\} = \langle t, \tau^0\rangle - \bar{I}_{\rho,\alpha}^{(3)}(\tau^0) = \log \lambda(B_\alpha(t))$$

and that the supremum in the Gibbs variational formula is attained at a unique measure P. If $\alpha = 2$, then P is the Markov chain satisfying $\pi_2 P = \tau^0$; if $\alpha \geq 3$, then P is the $(\alpha - 1)$-dependent Markov chain satisfying $\pi_\alpha P = \tau^0$ [Theorem IX.3.3]. $B_\alpha(t)$ plays the role of transfer matrix [Example IV.5.4].

The analytic implicit function theorem [Bochner and Martin (1948, page 39)] implies that the function $\log \lambda(B_\alpha(t))$ is a real analytic function of the quantities $\{J(A)\}$. Now assume that J is a finite-range ferromagnetic interaction as we considered in Chapter IV: $J(\{i,j\}) = \beta J(i-j) \geq 0$ $(i \neq j)$, $J(\{i\}) = h$, and $J(A) = 0$ whenever $|A| > 2$. It follows that the specific Gibbs free energy is a real analytic function of h and thus that the critical inverse temperature β_c equals ∞.

Appendix D: Existence of the Specific Gibbs Free Energy*

D.1. Existence Along Hypercubes

Let J be a non-negative, symmetric, summable function on $\mathbb{Z}^D$; Λ a symmetric hypercube in $\mathbb{Z}^D$; $\beta > 0$; and h real. In Chapters IV and V, the specific Gibbs free energy was defined by the formula

$$\psi(\beta, h)$$

$$= -\beta^{-1} \lim_{\Lambda \uparrow \mathbb{Z}^D} \frac{1}{|\Lambda|} \log \int_{\Omega_\Lambda} \exp\left[\frac{\beta}{2} \sum_{i,j \in \Lambda} J(i-j)\omega_i\omega_j + \beta h \sum_{i \in \Lambda} \omega_i\right] \pi_\Lambda P_\rho(d\omega),$$

(D.1)

where $\Omega_\Lambda = \{1, -1\}^\Lambda$. We now show the existence of $\psi(\beta, h)$ by proving the existence of a more general quantity.

We use the notation and terminology of Section C.3. A (many-body) interaction J on $\mathbb{Z}^D$ is said to have *finite-range* if $J(A)$ equals 0 for all A of sufficiently large diameter. The linear space of finite-range interactions is denoted by $\mathscr{J}_0$. We denote by $\mathscr{J}$ the space of interactions for which

$$|||J||| = \sum_{A \ni 0} |J(A)|/|A|$$

is finite. $\mathscr{J}$ is a separable Banach space and $\mathscr{J}_0$ is a dense subset of $\mathscr{J}$. For Λ a finite subset of $\mathbb{Z}^D$, the Hamiltonian is defined by the formula

$$H_{\Lambda,J}(\omega) = -\sum_{A \subset \Lambda} J(A)\omega_A, \qquad \omega \in \Omega_\Lambda.$$

The function $p_\Lambda(J) = |\Lambda|^{-1} \log \int_{\Omega_\Lambda} \exp[-H_{\Lambda,J}(\omega)]\pi_\Lambda P_\rho(d\omega)$ is called the *pressure for* Λ. Note that

(D.2) $$\|H_{\Lambda,J}\|_\infty \le \sum_{A \subset \Lambda} |J(A)| = \sum_{i \in \Lambda} \sum_{\{A \subset \Lambda : i \in A\}} |J(A)|/|A| = |\Lambda| \cdot |||J|||,$$

(D.3) $$|p_\Lambda(J)| \le \frac{1}{|\Lambda|} \|H_{\Lambda,J}\|_\infty \le |||J|||,$$

(D.4) $$|p_\Lambda(J) - p_\Lambda(J_0)| \le \frac{1}{|\Lambda|} \|H_{\Lambda,J} - H_{\Lambda,J_0}\|_\infty \le |||J - J_0|||,$$

*This appendix follows Israel (1979, Section I.2).

where $\|-\|_\infty$ denotes the maximum over Ω_Λ. The first inequality in (D.4) follows from the comparison lemma, Theorem II.7.4.

Theorem D.1.1. *For positive integers b, let $\Lambda(b)$ be a hypercube of side length b. Then for $J \in \mathscr{J}$, $p(J) = \lim_{b\to\infty} p_{\Lambda(b)}(J)$ exists and is a convex, Lipschitz continuous function. The function $p(J)$ is called the pressure.*

Proof. By (D.4), it suffices to prove the theorem for $J \in \mathscr{J}_0$. Let m be a positive integer such that $J(A)$ equals 0 whenever A has diameter at least m. Let a and b be positive integers such that $b > a + m$. Thus $b = n(m + a) + c$ for some $n \geq 1$ and $0 \leq c < m + a$. We write the hypercube $\Lambda(b)$ as the union of two sets $\Lambda' \cup \Lambda''$; Λ' equals the union of n^D hypercubes of side length a which are separated by corridors of width m; Λ'' equals $\Lambda\backslash\Lambda'$ [see Israel (1979, page 11)]. Since the n^D hypercubes in Λ' do not interact with each other, $H_{\Lambda',J}$ is the sum of n^D copies of $H_{\Lambda(a),J}$ which are independent with respect to $\pi_{\Lambda'}P_\rho$. Therefore

$$p_{\Lambda'}(J) = \frac{1}{n^D a^D}\log\int_{\Omega_{\Lambda'}} \exp[-H_{\Lambda',J}(\omega)]\pi_{\Lambda'}P_\rho(d\omega)$$
$$= \frac{1}{a^D}\log\int_{\Omega_{\Lambda(a)}} \exp[-H_{\Lambda(a),J}(\omega)]\pi_{\Lambda(a)}P_\rho(d\omega) = p_{\Lambda(a)}(J).$$

We have

$$\big|\,|\Lambda'|p_{\Lambda'}(J) - |\Lambda(b)|\,p_{\Lambda(b)}(J)\big| \leq \|H_{\Lambda',J} - H_{\Lambda(b),J}\|_\infty$$
$$\leq \sum_{A\subset\Lambda(b),\,A\not\subset\Lambda'} |J(A)| \leq |\Lambda(b)\backslash\Lambda'|\cdot\|J\|_\sim,$$

where $\|J\|_\sim = \sum_{A\ni 0}|J(A)|$. Dividing through by $|\Lambda(b)| = b^D$ yields

$$|(na/b)^D p_{\Lambda(a)}(J) - p_{\Lambda(b)}(J)| \leq (1 - (na/b)^D)\|J\|_\sim.$$

For each fixed a, as $b\to\infty$, $n(a+m)/b \to 1$ and so $na/b = a/(a+m) + o(1)$. Thus

(D.5)
$$\left|\left(\frac{a}{a+m} + o(1)\right)^D p_{\Lambda(a)}(J) - p_{\Lambda(b)}(J)\right|$$
$$\leq \left(1 - \left(\frac{a}{a+m} + o(1)\right)^D\right)\|J\|_\sim \qquad \text{as } b\to\infty,$$

(D.6)
$$\limsup_{b,b'\to\infty} |p_{\Lambda(b)}(J) - p_{\Lambda(b')}(J)| \leq 2\left(1 - \left(\frac{a}{a+m}\right)^D\right)\|J\|_\sim.$$

Taking $a\to\infty$ in (D.6), we see that $\{p_{\Lambda(b)}(J)\}$ forms a Cauchy sequence with limit $p(J)$. The convexity of $p_{\Lambda(b)}(J)$, and thus of $p(J)$, follows from Hölder's inequality. The Lipschitz continuity of $p(J)$ is a consequence of (D.4). □

Corollary D.1.2. *Let J be a symmetric, summable function on* $\mathbb{Z}^D$. *Then the specific Gibbs free energy* $\psi(\beta, h)$ *defined in* (D.1) *exists.*

Proof. Let $\bar{J}$ be the interaction defined by $\bar{J}(\{i,j\}) = \beta J(i-j)$ $(i \neq j)$, $\bar{J}(\{i\}) = \beta h$, and $\bar{J}(A) = 0$ whenever $|A| > 2$. $\bar{J}$ is in $\mathscr{J}$ since J is summable. We have

$$\psi(\beta, h) = -J(0) - \beta^{-1} \lim_{\Lambda \uparrow \mathbb{Z}^D} p_\Lambda(\bar{J}) = -J(0) - \beta^{-1} p(\bar{J}). \qquad \square$$

D.2. An Extension

For each positive integer a and point $n \in \mathbb{Z}^D$, let $\Lambda_n(a)$ be the hypercube $\{j \in \mathbb{Z}^D : n_\alpha a \leq j < (n_\alpha + 1)a, \alpha = 1, \ldots, D\}$. For fixed a and Λ a finite subset of $\mathbb{Z}^D$, we define $N_a^+(\Lambda)$ as the number of sets $\Lambda_n(a)$ which intersect Λ and $N_a^-(\Lambda)$ as the number of sets $\Lambda_n(a)$ which are contained in Λ. A sequence of finite subsets $\{\Lambda\}$ of $\mathbb{Z}^D$ is said to *converge to infinity in the sense of van Hove* if for all a, $N_a^-(\Lambda) \to \infty$ and $N_a^-(\Lambda)/N_a^+(\Lambda) \to 1$. Roughly, this means that the boundaries of $\{\Lambda\}$ become negligible in the limit as compared to $\{\Lambda\}$.

Theorem D.2.1. *If* $\{\Lambda\}$ *converges to infinite in the sense of van Hove and if* $J \in \mathscr{J}$, *then* $p_\Lambda(J) \to p(J)$.

A proof is given in Israel (1979, page 12).

List of Frequently Used Symbols

Symbol	Meaning	Section
$\operatorname{aff} C$	Affine hull of the convex set C	VI.3
a.s.	Almost surely	A.4
$\xrightarrow{\text{a.s.}}$	Almost sure convergence	A.4
$\operatorname{bd} C$	Boundary of the convex set C	VI.3
$\mathscr{B}(\Omega)$	Borel σ-field of Ω	A.2
$\mathbb{C}$	Set of complex numbers	VII.5
$\mathbb{C}^d$	Complex Euclidean space	VII.5
$\operatorname{cc} S_\rho$	Closed convex hull of the support of ρ	VIII.4
$\operatorname{cl} A$	Closure of the set A	A.8
$\operatorname{conv} S_\rho$	Convex hull of the support of ρ	VIII.3
$c_{\mathscr{W}}(t)$	Free energy function of the sequence $\mathscr{W}$	II.6
$\mathscr{C}(\mathscr{X})$	Space of bounded continuous real-valued functions on the metric space $\mathscr{X}$	A.8
$c_\rho(t)$	Free energy function of ρ	II.4
$\xrightarrow{\mathscr{D}}$	Convergence in distribution	A.8
$\operatorname{dom} f$	Effective domain of f	VI.2
$d\nu/d\rho$	Radon–Nikodym derivative of ν with respect to ρ	II.4
$E\{\ \}$	Expectation	A.4
$\operatorname{epi} f$	Epigraph of f	VI.2
$\xrightarrow{\text{exp}}$	Exponential convergence	II.6
$E\{Y \mid \mathscr{D}\}$	Conditional expectation of Y given $\mathscr{D}$	A.6
$\mathscr{F}$	σ-field	A.2
$\mathscr{F}(X_\alpha; \alpha \in \mathscr{A})$	σ-field generated by $\{X_\alpha; \alpha \in \mathscr{A}\}$	A.2
$f^*(y)$	Legendre–Fenchel transform of f	VI.5
$f'_+(x), f'_-(x)$	Right-hand derivative and left-hand derivative of f at x	VI.3
$\partial f(y)$	Subdifferential of f at y	VI.3
$\mathscr{G}_{\beta,h}$	Set of infinite-volume Gibbs states	IV.6
$h(P)$	Mean entropy of P	I.6
$H_{\Lambda,h}(\omega)$, $H_{\Lambda,h,\tilde{\omega}}(\omega)$	Hamiltonian or interaction energy	IV.3, IV.6
$H(\nu)$	Shannon entropy of ν	I.4

int A	Interior of the set A	A.8
$I_{\mathscr{W}}(z)$	Legendre–Fenchel transform of $c_{\mathscr{W}}(t)$	II.6
$I_\rho^{(1)}(z)$	Level-1 entropy function	II.4
$I_\rho^{(2)}(v)$	Relative entropy of v with respect to ρ (level-2 entropy function)	I.4, II.4
$I_\rho^{(3)}(P)$	Mean relative entropy of P with respect to P_ρ (level-3 entropy function)	II.4
L_n, $L_n(\omega, \cdot)$	Empirical measure	I.2
log	Natural logarithm	I.1
$\mathscr{M}(\mathscr{X})$	Set of probability measures on the metric space $\mathscr{X}$	A.8
$\mathscr{M}_s(\Omega)$	Set of strictly stationary probability measures on the space Ω	A.9
$m(\beta, h)$	Specific magnetization	VI.3
$m(\beta, +)$, $m(\beta, -)$	Limits of $m(\beta, h)$ as $h \to 0^+$ and $h \to 0^-$	VI.5
m_ρ	Mean of ρ	II.4
$m_\rho(t)$	Moment generating function of ρ	VII.5
$P\{A \mid \mathscr{D}\}$	Conditional probability of A given $\mathscr{D}$	A.6
$P_{n,\Delta,A}$	Microcanonical ensemble of the discrete ideal gas	III.3
$P_{n,\Delta,\beta}$	Canonical ensemble of the discrete ideal gas	III.6
$P_{\beta,h,+}$, $P_{\beta,h,-}$, $P_{\beta,h}$, $P_{\beta,h}^{(\lambda)}$	Infinite-volume Gibbs state	IV.6
$P_{\Lambda,\beta,h}$, $P_{\Lambda,\beta,h,\tilde{\omega}}$, $P_{\Lambda,\beta,h,+}$, $P_{\Lambda,\beta,h,-}$	Finite-volume Gibbs state on Λ	IV.3, IV.6
P_ρ	Infinite-product measure with identical one-dimensional marginals ρ	A.7
$\tilde{P}_\omega$, $\tilde{P}\{B \mid \mathscr{D}\}(\omega)$	Regular conditional distribution	II.4, A.6
$Q_n^{(1)}$	Distribution of S_n/n	I.3
$Q_n^{(2)}$	Distribution of L_n	I.4
$Q_n^{(3)}$	Distribution of R_n	I.6
rbd C	Relative boundary of the convex set C	VI.3
$\mathbb{R}^d$	Real Euclidean space	A.3
$(\mathbb{R}^d)^{\mathbb{Z}}$	$\prod_{\alpha \in \mathbb{Z}} \mathbb{R}^d$	A.3
ri C	Relative interior of the convex set C	VI.3
R_n, $R_n(\omega, \cdot)$	Empirical process	I.6
S_n, $S_n(\omega)$	nth partial sum of i.i.d. random vectors	I.2
$S_\Lambda(\omega)$	Total spin in Λ	IV.3
S_ρ	Support of ρ	VIII.3
T	Shift mapping	I.6
$\mathscr{U}$	$\{(\beta, h) \colon \beta > 0, h \neq 0 \text{ and } 0 < \beta < \beta_c, h = 0\}$	V.7
$\mathbb{Z}$, $\mathbb{Z}^+$	Set of integers, set of positive integers	I.1, III.8
$\mathbb{Z}^D$	D-dimensional integer lattice	V.2

$Z(\Lambda, \beta, h)$, $Z(\Lambda, \beta, h, \tilde{\omega})$	Partition function	IV.3, IV.6
β_c	Critical inverse temperature	IV.5
δ_x	Unit point measure at x	I.2
$\lambda(B)$	Largest positive eigenvalue of the primitive matrix B	IX.4
$\nu \ll \rho$	ν is absolutely continuous with respect to ρ	II.4
$\pi_n P_\rho$	Finite product measure with identical one-dimensional marginals ρ	I.3
$\pi_\alpha P$	α-dimensional marginal of P	I.6
$\pi_\Lambda P_\rho$	Product measure on Ω_Λ with identical one-dimensional marginals $\rho = \frac{1}{2}\delta_1 + \frac{1}{2}\delta_{-1}$	IV.3
$\bar{\rho}_\beta$	Maxwell–Boltzmann distribution	III.5
$\phi(\beta)$	Specific free energy of the discrete ideal gas	III.8
χ_C	Characteristic function of the set C	III.4
$\chi(\beta, h)$	Specific magnetic susceptibility	V.7
$\psi(\beta, h)$	Specific Gibbs free energy	IV.5
$(\Omega, \mathscr{F}, P)$	Probability space	A.4
Ω_Λ	$\{1, -1\}^\Lambda$	IV.3
$\langle\ ,\ \rangle$	Euclidean inner product on $\mathbb{R}^d$	II.4
$\|\ \|$	Euclidean norm on $\mathbb{R}^d$	II.4
$\|\ \|_\infty$	Supremum norm	A.8
$\Rightarrow$	Weak convergence of probability measures	A.8

References

D. B. ABRAHAM (1978). Susceptibility of the rectangular Ising ferromagnet. J. Statist. Phys. *19*, 349–358.

D. B. ABRAHAM AND A. MARTIN-LÖF (1973). The transfer matrix for a pure phase in the two-dimensional Ising model. Commun. Math. Phys. *32*, 245–268.

D. B. ABRAHAM AND P. REED (1976). Interface profile of the Ising ferromagnet in two dimensions. Commun. Math. Phys. *49*, 35–46.

C. J. ADKINS (1975). *Equilibrium Thermodynamics*. Second edition. McGraw-Hill, London.

M. AIZENMAN (1979). Instability of phase coexistence and translation invariance in two dimensions. Phys. Rev. Lett. *43*, 407–409.

M. AIZENMAN (1980). Translation invariance and instability of phase coexistence in the two-dimensional Ising system. Commun. Math. Phys. *73*, 83–94.

M. AIZENMAN (1981). Proof of the triviality of φ_d^4 field theory and some mean-field features of Ising models for $d > 4$. Phys. Rev. Lett. *47*, 1–4.

M. AIZENMAN (1982). Geometric analysis of φ^4 fields and Ising models. Parts I and II. Commun. Math. Phys. *86*, 1–48.

M. AIZENMAN (1983). Rigorous results on the critical behavior in statistical mechanics. In: *Scaling and Self-Similarity in Physics*, pp. 181–201. Edited by J. Fröhlich. Progress in Physics 7. Birkhäuser, Boston.

M. AIZENMAN (1984). Absence of an intermediate phase in one-component ferromagnetic systems (in preparation).

M. AIZENMAN (1985). Rigorous studies of critical behavior II. In: *Statistical Physics and Dynamical Systems-Rigorous Results-Proceedings 1984*. Edited by D. Petz and D. Szász. Birkhäuser, Boston (in press).

M. AIZENMAN AND R. GRAHAM (1983). On the renormalized coupling constant and the susceptibility in φ_4^4 field theory and the Ising model in four dimensions. Nuclear Phys. B *225* [FS9], 261–288.

M. AIZENMAN AND E. H. LIEB (1981). The third law of thermodynamics and the degeneracy of the ground state for lattice systems. J. Statist. Phys. *24*, 279–297.

M. AIZENMAN AND C. M. NEWMAN (1984). Tree graph inequalities and critical behavior in percolation models. J. Statist. Phys. *36*, 107–143.

M. AIZENMAN, S. GOLDSTEIN, AND J. L. LEBOWITZ (1978). Conditional equilibrium and the equivalence of microcanonical and grandcanonical ensembles in the thermodynamic limit. Commun. Math. Phys. *62*, 279–302.

C. ARAGÃO de CARVALHO, S. CARACCIOLO, AND J. FRÖHLICH (1983). Polymers and $g|\varphi|^4$ theory in 4 dimensions. Nuclear Phys. B *215* [FS7], 209–248.

R. AZENCOTT (1980). Grandes deviations et applications. In: *Ecole d'Eté de Probabilités de Saint-Flour VIII-1978*, pp. 1–176. Edited by P. L. Hennequin. Lecture Notes in Mathematics *774*, Springer, Berlin.

R. R. BAHADUR (1967). Rates of convergence of estimates and test statistics. Ann. Math. Statist. *38*, 303–324.

R. R. BAHADUR (1971). *Some Limit Theorems in Statistics*. SIAM, Philadelphia.

R. R. BAHADUR AND R. RANGA RAO (1960). On deviations of the sample mean. Ann. Math. Statist. *31*, 1015–1027.

R. R. BAHADUR AND S. L. ZABELL (1979). Large deviations of the sample mean in general vector spaces. Ann. Probab. *7*, 587–621.

G. A. BAKER, JR. (1977). Analysis of hyperscaling in the Ising model by the high-temperature series method. Phys. Rev. B *15*, 1552–1559.

G. A. BAKER, JR. and J. M. KINCAID (1981). The continuous spin Ising model, $g_0 : \phi :_d$ field theory, and the renormalization group. J. Statist. Phys. *24*, 469–528.

O. BARNDORFF-NIELSEN (1970). *Exponential Families: Exact Theory*. Various Publication Series, Number 19, Institute of Mathematics, Aarhus Univ.

O. BARNDORFF-NIELSEN (1978). *Information and Exponential Families in Statistical Theory*. Wiley, Chichester.

G. A. BATTLE AND L. ROSEN (1980). The FKG inequality for the Yukawa_2 quantum field theory. J. Statist. Phys. *22*, 123–192.

L. E. BAUM, M. KATZ, AND R. R. READ (1962). Exponential convergence rates for the law of large numbers. Trans. Amer. Math. Soc. *102*, 187–199.

R. J. BAXTER (1982). *Exactly Solved Models in Statistical Mechanics*. Academic, New York.

G. BENETTIN, G. GALLAVOTTI, G. JONA-LASINIO, AND A. L. STELLA (1973). On the Onsager-Yang value of the spontaneous magnetization. Commun. Math. Phys. *30*, 45–54.

P. BILLINGSLEY (1965). *Ergodic Theory and Information*. Wiley, New York.

P. BILLINGSLEY (1968). *Convergence of Probability Measures*. Wiley, New York.

D. BLACKWELL AND J. L. HODGES (1959). The probability in the extreme tail of a convolution. Ann. Math. Statist. *30*, 1113–1120.

A BLANC-LAPIERRE AND A. TORTRAT (1956). Statistical mechanics and probability theory. In: *Proceedings of the Third Berkeley Symposium on Mathematical Statistics and Probability*, Volume III, pp. 145–170. Edited by J. Neyman. University of California Press, Berkeley.

S. BOCHNER AND W. T. MARTIN (1948). *Several Complex Variables*. Princeton Univ. Press, Princeton.

E. BOLTHAUSEN (1984a). Laplace approximation for sums of independent random vectors. Z. Wahrsch. verw. Geb. (submitted).

E. BOLTHAUSEN (1984b). On the probability of large deviations in Banach spaces. Ann. Probab. *12*, 427–435.

L. BOLTZMANN (1872). Weitere Studien über das Wärmegleichgewicht unter Gasmolekulen, Wien. Akad. Sitz. *66*, 275–370.

L. BOLTZMANN (1877). Über die Beziehung zwischen dem zweiten Hauptsatze der mechanischen Wärmetheorie und der Wahrscheinlichkeitsrechnung respektive den Sätzen über das Wärmegleichgewicht. Wien. Akad. Sitz. *76*, 373–435.

L. BREIMAN (1968). *Probability*. Addison-Wesley, Reading.

J. BRETAGNOLLE (1979). Formule de Chernoff pour les lois empiriques de variables a valeurs dans des espaces généraux. In: *Grandes Déviations et Applications Statistiques*, pp. 33–52. Astérisque *68*, Société Mathématique de France, Paris.

E. BRÉZIN, J. C. le GUILLOU, AND J. ZINN-JUSTIN (1973). Approach to scaling in normalized perturbation theory. Phys. Rev. D *8*, 2418–2430.

E. BRÉZIN, J. C. le GUILLOU, AND J. ZINN-JUSTIN (1976). Field theoretical approach to critical phenomena. In: *Phase Transitions and Critical Phenomena*, Vol. 6, pp. 127–247. Edited by C. Domb and M. S. Green. Academic, New York.

J. BRICMONT AND J.-R. FONTAINE (1982). Perturbation about the mean field critical point. Commun. Math. Phys. *86*, 337–362.

J. BRICMONT, J. L. LEBOWITZ, AND C.-E. PFISTER (1979). On the equivalence of boundary conditions. J. Statist. Phys. *21*, 573–582.

A. BRØNDSTED (1964). Conjugate convex functions in topological vector spaces. Mat.-Fys. Medd. Danske Vid. Selsk. *34*, 1–26.

L. D. BROWN (1985). *Foundations of Exponential Families*. Institute of Mathematical Statistics Lecture Notes—Monograph Series (in preparation).

A. BRUNEL (1973). Théorème ergodique ponctuel pour un semi-groupe commutatif finiment engendré de contractions de L^1. Ann. Inst. H. Poincaré *9*, Section B, 327–343.

S. BRUSH (1967). History of the Lenz-Ising model. Rev. Modern Phys. *39*, 883–893.

D. BRYDGES (1982). Field theories and Symanzik's polymer representation. In: *Gauge Theories: Fundamental Interactions and Rigorous Results*, pp. 311–337. 1981 Brasov lectures. Edited by P. Dita, V. Georgescu, and R. Purice. Birkhäuser, Boston.

D. C. BRYDGES, J. FRÖHLICH, AND A. D. SOKAL (1983). The random walk representation of classical spin systems and correlation inequalities, II. Commun. Math. Phys. *91*, 117–139.

D. BRYDGES, J. FRÖHLICH, AND T. SPENCER (1982). The random walk representation of classical spin systems and correlation inequalities. Commun. Math. Phys. *83*, 123–150.

M. J. Buckingham and J. D. Gunton (1969). Correlations at the critical point of the Ising model. Phys. Rev. *178*, 848–853.

H. B. Callen (1960). *Thermodynamics*. Wiley, New York.

M. Cassandro and G. Jona-Lasinio (1978). Critical point behavior and probability theory. Adv. in Phys. *27*, 913–941.

M. Cassandro and E. Olivieri (1981). Renormalization group and analyticity in one dimension: a proof of Dobrushin's theorem. Commun. Math. Phys. *80*, 255–269.

M. Cassandro, E. Olivieri, A. Pellegrinotti, and E. Presutti (1978). Existence and uniqueness of DLR measures for unbounded spin systems. Z. Wahrsch. verw. Geb. *41*, 313–334.

C. Cercignani (1975). *Theory and Application of the Boltzmann Equation*. Elsevier, New York.

N. R. Chaganty and J. Sethuraman (1982). Central limit theorems in the area of large deviations for some dependent random variables. Old Dominion Univ.–Florida State Univ. preprint.

N. R. Chaganty and J. Sethuraman (1984). Multidimensional large deviation local limit theorems. Old Dominion Univ.–Florida State Univ. preprint.

N. R. Chaganty and J. Sethuraman (1985). Large deviation local limit theorems for arbitrary sequences of random variables. Ann. Probab. *13* (in press).

H. Chernoff (1952). A measure of asymptotic efficiency for tests of a hypothesis based on the sum of observations. Ann. Math. Statist. *23*, 493–507.

H. Chernoff (1972). *Sequential Analysis and Optimal Design*. SIAM, Philadelphia.

S. Chevet (1983). Gaussian measures and large deviations. In: *Probability in Banach Spaces IV-Proceedings 1982*, pp. 30–46. Edited by A. Beck and K. Jacobs. Lecture Notes in Mathematics *990*, Springer, Berlin.

T.-S. Chiang (1982). Large deviations of some Markov processes on compact Polish spaces. Z. Wahrsch. verw. Geb. *61*, 271–281.

K. L. Chung (1968). *A Course in Probability Theory*. Harcourt, Brace and World, New York.

R. J. E. Clausius (1850). Über die bewegende Kraft der Wärme und die Gesetze die sich daraus für die Wärmelehre selbst ableiten lassen. Ann. Phys. *79*, 368–397, 500–524. (English translation in Philos. Mag. *2*, 1-21, 102-119 (1851).)

R. J. E. Clausius (1865). Über verschiedene für die Anwendung bequeme Formen der Hauptgleichungen der mechanischen Wärmetheorie. Ann. Phys. Chem. *125*, 353–400.

P. Collet and J.-P. Eckmann (1978). *A Renormalization Group Analysis of the Hierarchical Model in Statistical Mechanics*. Lecture Notes in Physics *74*, Springer, Berlin.

R. Courant and D. Hilbert (1962). *Methods of Mathematical Physics*, Volume II. Interscience, New York.

H. Cramér (1938). Sur un nouveaux theorème-limite da la theorie des probabilités. Actualités Scientifiques et Industrielles, No. 736, pp. 5–23. Colloque consacré à la théorie des probabilités (held in October, 1937), Vol. 3. Hermann, Paris.

I. Csiszár (1975). I-divergence geometry of probability distributions and minimization problems. Ann. Probab. *3*, 146–158.

I. Csiszár (1984). Sanov property, generalized I-projection, and a conditional limit theorem. Ann. Probab. *12*, 768–793.

D. Dacunha-Castelle (1979). Formule de Chernoff pour une suite de variables réelles. In: *Grandes Déviations et Applications Statistiques*, pp. 19–24. Astérisque *68*, Société Mathématique de France, Paris.

I. Davies and A. Truman (1982). Laplace expansions of conditional Wiener integrals and applications to quantum physics, In: *Stochastic Processes in Quantum Theory and Statistical Physics—Proceedings 1981*, pp. 40–55. Edited by S. Albeverio, Ph. Combe, and M. Sirugue-Collin. Lecture Notes in Physics *173*, Springer, Berlin.

D. A. Dawson (1983). Critical dynamics and fluctuations for a mean field model of cooperative behavior. J. Statist. Phys. *31*, 29–85.

A. de Acosta (1984). Upper bounds for large deviations of dependent random vectors. Z. Wahrsch. verw. Geb. (in press).

A. de Acosta (1985). On large deviations of sums of independent random vectors. Case Western Reserve University preprint.

J. De Coninck (1984). Scaling limit of the energy variable for the two-dimensional Ising ferromagnet. Commun. Math. Phys. *95*, 53–59.

C. Dellacherie and P.-A. Meyer (1982). *Probabilities and Potential B*. Translated by J. P. Wilson. North-Holland, Amsterdam.

R. L. Dobrushin (1965). The existence of a phase transition in the two- and three-dimensional Ising models. Theory Probab. Appl. *10*, 193–213.

R. L. DOBRUSHIN (1968a). The description of a random field by means of conditional probabilities and conditions of its regularity. Theory Probab. Appl. *13*, 197–224.

R. L. DOBRUSHIN (1968b). Gibbsian random fields for lattice systems with pairwise interactions. Functional Anal. Appl. *2*, 292–301.

R. L. DOBRUSHIN (1968c). The problem of uniqueness of a Gibbsian random field and the problem of phase transitions. Functional Anal. Appl. *2*, 302–312.

R. L. DOBRUSHIN (1969). Gibbsian random fields. The general case. Functional Anal. Appl. *3*, 22–28.

R. L. DOBRUSHIN (1970). Prescribing a system of random variables by conditional distributions. Theory Probab. Appl. *15*, 458–486.

R. L. DOBRUSHIN (1972). Gibbs state describing coexistence of phases for a three-dimensional Ising model. Theory Probab. Appl. *17*, 582–601.

R. L. DOBRUSHIN (1973). Analyticity of correlation functions in one-dimensional classical systems with slowly decreasing potentials. Commun. Math. Phys. *32*, 269–289.

R. L. DOBRUSHIN (1980). Automodel generalized random fields and their renorm-group. In: *Multicomponent Random Systems*, pp. 153–198. Edited by R. L. Dobrushin and Ya. G. Sinai. Marcel Dekker, New York.

R. L. DOBRUSHIN and B. TIROZZI (1977). The central limit theorem and the problem of equivalence of ensembles. Commun. Math. Phys. *54*, 173–192.

M. D. DONSKER AND S. R. S. VARADHAN (1975a). Asymptotic evaluation of certain Markov process expectations for large time, I. Comm. Pure Appl. Math. *28*, 1–47.

M. D. DONSKER AND S. R. S. VARADHAN (1975b). Asymptotic evaluation of certain Markov process expectations for large time, II. Comm. Pure Appl. Math. *28*, 279–301.

M. D. DONSKER AND S. R. S. VARADHAN (1975c). Asymptotic evaluation of certain Wiener integrals for large time. In: *Functional Integration and Its Applications*, pp. 15–33. Proceedings of the International Conference Held at Cumberland Lodge, Windsor Great Park, London, April 1974. Edited by A. M. Arthurs. Clarendon, Oxford.

M. D. DONSKER AND S. R. S. VARADHAN (1975d). Asymptotics for the Wiener sausage. Comm. Pure Appl. Math. *28*, 525–565.

M. D. DONSKER AND S. R. S. VARADHAN (1975e). On a variational formula for the principal eigenvalue for operators with a maximum principle. Proc. Nat. Acad. Sci. U.S.A. *72*, 780–783.

M. D. DONSKER AND S. R. S. VARADHAN (1976a). Asymptotic evaluation of certain Markov process expectations for large time, III. Comm. Pure Appl. Math. *29*, 389–461.

M. D. DONSKER AND S. R. S. VARADHAN (1976b). On the principal eigenvalue of second-order elliptic differential operators. Comm. Pure Appl. Math. *29*, 595–621.

M. D. DONSKER AND S. R. S. VARADHAN (1977a). Some problems of large deviations. Symposia Mathematica *21*, 313–318. Academic, New York.

M. D. DONSKER AND S. R. S. VARADHAN (1977b). On laws of the iterated logarithm for local times. Comm. Pure Appl. Math. *30*, 707–753.

M. D. DONSKER AND S. R. S. VARADHAN (1979). On the number of distinct sites visited by a random walk. Comm. Pure Appl. Math. *32*, 721–747.

M. D. DONSKER AND S. R. S. VARADHAN (1983a). Asymptotic evaluation of certain Markov process expectations for large time, IV. Comm. Pure Appl. Math. *36*, 183–212.

M. D. DONSKER AND S. R. S. VARADHAN (1983b). Asymptotics for the polaron. Comm. Pure Appl. Math. *36*, 505–528.

M. D. DONSKER AND S. R. S. VARADHAN (1985). Large deviations for stationary Gaussian processes. Commum. Math. Phys. *97*.

J. L. DOOB (1953). *Stochastic Processes*. Wiley, New York.

M. DUNEAU, B. SOUILLARD, AND D. IAGOLNITZER (1975). Decay of correlations for infinite-range interactions. J. Math. Phys. *16*, 1662–1666.

N. DUNFORD AND J. T. SCHWARTZ (1958). *Linear Operators*, Part I. Interscience, New York.

F. DUNLOP (1977). Zeros of partition functions via correlation inequalities. J. Statist. Phys. *17*, 215–228.

F. DUNLOP AND C. M. NEWMAN (1975). Multicomponent field theories and classical rotators. Commun. Math. Phys. *44*, 223–235.

R. DURRETT (1981). An introduction to infinite particle systems. Stochastic Process. and Appl. *11*, 109–150.

R. DURRETT (1984a). Oriented percolation in two dimensions. Ann. Probab. *12*, 999–1040.

R. DURRETT (1984b). Some general results concerning the critical exponent of percolation processes. Z. Wahrsch. verw. Geb. (in press).

F. J. DYSON (1969a). Existence of a phase transition in a one-dimensional Ising ferromagnet. Commun. Math. Phys. *12*, 91–107.

F. J. DYSON (1969b). Non-existence of spontaneous magnetization in a one-dimensional Ising ferromagnet. Commun. Math. Phys. *12*, 212–215.

F. J. DYSON (1971). An Ising ferromagnet with discontinuous long-range order. Commun. Math. Phys. *21*, 269–283.

F. J. DYSON (1972). Existence and nature of phase transitions in one-dimensional Ising ferromagnets. In: *Mathematical Aspects of Statistical Mechanics*, pp. 1–11. Edited by J. C. T. Pool. Amer. Math. Soc., Providence.

M. L. EATON (1982). A review of selected topics in multivariate probability inequalities. Ann. Statist. *10*, 11–43.

B. EFRON AND D. TRUAX (1968). Large deviation theory in exponential families. Ann. Math. Statist. *39*, 1402–1424.

P. EHRENFEST AND T. EHRENFEST (1959). *The Conceptual Foundations of the Statistical Approach in Mechanics.* Translated by M. J. Moravcsik. Cornell Univ. Press, Ithaca. (Translation of 1912 article in Encyklopädie der mathematischen Wissenschaften.)

T. EISELE AND R. S. ELLIS (1983). Symmetry breaking and random waves for magnetic systems on a circle. Z. Wahrsch. verw. Geb. *63*, 297–348.

I. EKELAND AND R. TEMAN (1976). *Convex Analysis and Variational Problems.* Translated by Minerva Translations, Ltd. North Holland, Amsterdam.

R. S. ELLIS (1981). Large deviations and other limit theorems for a class of dependent random variables with applications to statistical mechanics. Univ. of Massachusetts preprint (unpublished).

R. S. ELLIS (1984). Large deviations for a general class of random vectors. Ann. Probab. *12*, 1–12.

R. S. ELLIS AND J. L. MONROE (1975). A simple proof of the GHS and further inequalities. Commun. Math. Phys. *41*, 33–38.

R. S. ELLIS AND C. M. NEWMAN (1976). Quantum mechanical soft springs and reverse correlation inequalities. J. Math. Phys. *17*, 1682–1683.

R. S. ELLIS AND C. M. NEWMAN (1978a). Fluctuationes in Curie–Weiss exemplis. In: *Mathematical Problems in Theoretical Physics-Proceedings 1977*, pp. 312–324. Edited by G. Dell'Antonio, S. Doplicher, and G. Jona-Lasinio. Lecture Notes in Physics *80*, Springer, Berlin.

R. S. ELLIS AND C. M. NEWMAN (1978b). Limit theorems for sums of dependent random variables occurring in statistical mechanics. Z. Wahrsch. verw. Geb. *44*, 117–139.

R. S. ELLIS AND C. M. NEWMAN (1978c). Necessary and sufficient conditions for the GHS inequality with applications to probability and analysis. Trans. Amer. Math. Soc. *237*, 83–99.

R. S. ELLIS AND C. M. NEWMAN (1978d). The statistics of Curie-Weiss models, J. Statist. Phys. *19*, 149–161.

R. S. ELLIS, C. M. NEWMAN, AND J. S. ROSEN (1980). Limit theorems for sums of dependent random variables occurring in statistical mechanics, II: conditioning, multiple phases, and metastability. Z. Wahrsch. verw. Geb. *51*, 153–169.

R. S. ELLIS AND J. S. ROSEN (1981). Asymptotic analysis of Gaussian integrals, II: manifold of minimum points. Commun. Math. Phys. *82*, 153–181.

R. S. ELLIS AND J. S. ROSEN (1982a). Asymptotic analysis of Gaussian integrals, I: isolated minimum points. Trans. Amer. Math. Soc. *273*, 447–481.

R. S. ELLIS AND J. S. ROSEN (1982b). Laplace's method for Gaussian integrals. Ann. Probab. *10*, 47–66. (Correction: Ann. Probab. *11*, 246 (1983).)

R. S. ELLIS, J. L. MONROE, AND C. M. NEWMAN (1976). The GHS and other correlation inequalities for a class of even ferromagnets. Commun. Math. Phys. *46*, 167–182.

A. ERDÉLYI (1956). *Asymptotic Expansions.* Dover, New York.

M. FANNES, P. VANHEUVERZWIJN, AND W. VERBEURE (1982). Energy-entropy inequalities for classical lattice systems. J. Statist. Phys. *29*, 547–560.

I. E. FARQUHAR (1964). *Ergodic Theory in Statistical Mechanics.* Interscience, London.

N. A. FAVA (1972). Weak inequalities for product operators. Studia Math. *42*, 271–288.

W. FELLER (1957). *An Introduction to Probability Theory and Its Applications*, Vol. I, Second edition. Wiley, New York.

W. FELLER (1966). *An Introduction to Probability Theory and Its Applications*, Vol. II. Wiley, New York.

W. FENCHEL (1949). On conjugate convex functions. Canad. J. Math. *1*, 73–77.

M. E. FISHER (1965). The nature of critical points. In: *Lectures in Theoretical Physics*. Vol. VIIC, pp. 1–159. Edited by W. E. Britten. Univ. of Colorado Press, Boulder.

M. E. FISHER (1967a). Critical temperatures of anisotropic Ising lattices. II. General upper bounds. Phys. Rev. *162*, 480–485.

M. E. FISHER (1967b). The theory of equilibrium critical phenomena. Rep. Progr. Phys. *30*, 617–730.

M. E. FISHER (1969). Rigorous inequalities for critical-point correlation exponents. Phys. Rev. *180*, 594–600.

M. E. FISHER (1973). General scaling theory for critical points. In: *Collective Properties of Physical Systems-Proceedings of the Twenty-Fourth Nobel Symposium 1973*, pp. 16–37. Edited by B. Lundqvist and S. Lundqvist. Academic, New York.

M. E. FISHER (1974). The renormalization group in the theory of critical behavior. Rev. Modern Phys. *46*, 597–616.

M. E. FISHER (1983). Scaling, universality, and renormalization group theory. In: *Critical Phenomena*, pp. 1–139. Edited by F. J. W. Hahne. Lecture Notes in Physics *186*, Springer, Berlin.

H. FÖLLMER (1973). On entropy and information gain in random fields. Z. Wahrsch. verw. Geb. *26*, 207–217.

C. FORTUIN, P. KASTELYN, AND J. GINIBRE (1971). Correlation inequalities on some partially ordered sets. Commun. Math. Phys. *22*, 89–103.

M. I. FREIDLIN AND A. D. WENTZELL (1984). *Random Perturbations of Dynamical Systems*. Translated by J. Szücs. Springer, New York.

S. FRIEDLAND (1981). Convex spectral functions. J. Linear and Multilinear Algebra *9*, 299–316.

S. FRIEDLAND AND S. KARLIN (1975). Some inequalities for the spectral radius of non-negative matrices and applications. Duke Math. J. *42*, 459–490.

J. FRÖHLICH (1978). The pure phases (harmonic functions) of generalized processes. Bull. Amer. Math. Soc. *84*, 165–193.

J. FRÖHLICH (1980). On the mathematics of phase transitions and critical phenomena. In: *Proceedings of the International Congress of Mathematicians, Helsinki 1978*, pp. 896–904.

J. FRÖHLICH (1982). On the triviality of $\lambda\phi_d^4$ theories and the approach to the critical point in $d \underset{(-)}{>} 4$ dimensions. Nuclear Phys. B *200* [FS4], 281–296.

J. FRÖHLICH AND T. SPENCER (1982a). The phase transition in the one-dimensional Ising model with $1/r^2$ interaction energy. Commun. Math. Phys. *84*, 87–101.

J. FRÖHLICH AND T. SPENCER (1982b). Some recent rigorous results in the theory of phase transitions and critical phenomena In: *Séminaire Bourbaki—volume 1981/1982—Exposés 579–596*, pp. 159–200. Astérisque *92–93*, Sociéte Mathématique de France, Paris.

J. FRÖHLICH, B. SIMON, AND T. SPENCER (1976). Infrared bounds, phase transitions, and continuous symmetry breaking. Commun. Math. Phys. *50*, 79–85.

J. FRÖHLICH, R. ISRAEL, E. H. LIEB, AND B. SIMON (1978). Phase transitions and reflection positivity. I. Commun. Math. Phys. *62*, 1–34.

J. GÄRTNER (1977). On large deviations from the invariant measure. Theory Probab. Appl. *22*, 24–39.

R. G. GALLAGER (1968). *Information Theory and Reliable Communication*. Wiley, New York.

G. GALLAVOTTI (1972a). Instabilities and phase transitions in the Ising model. Riv. Nuovo Cimento *2*, 133–169.

G. GALLAVOTTI (1972b). The phase separation line in the two-dimensional Ising model. Comm. Math. Phys. *27*, 103–136.

K. GAWĘDZKI AND A. KUPIAINEN (1982). Triviality of φ_4^4 and all that in a hierarchical model approximation. J. Statist. Phys. *29*, 683–698.

K. GAWĘDZKI AND A. KUPIAINEN (1983). Block spin renormalization group for dipole gas and $(\nabla\phi)^4$. Ann. Phys. *147*, 198–243.

H.-O. GEORGII (1972). *Phasenübergang 1. Art bei Gittergasmodellen*. Lecture Notes in Physics *16*. Springer, Berlin.

H.-O. GEORGII (1973). Two remarks on extremal equilibrium states. Commun. Math. Phys. *32*, 107–118.

J. W. GIBBS (1873a). Graphical methods in the thermodynamics of fluids. Trans. of the Connecticut Acad. *2*, 309–342.

J. W. GIBBS (1873b). A method of geometrical representation of the thermodynamic properties of substances by means of surfaces. Trans. of the Connecticut Acad. *2*, 382–404.

J. W. GIBBS (1875–1878). On the equilibrium of heterogeneous substances. Trans. of the Connecticut Acad. *3*, 108–248 (1875–1876), 343–524 (1877–1878).

J. W. GIBBS (1960). *Elementary Principles of Statistical Mechanics*. Dover, New York. (Republication of 1902 work published by Yale Univ. Press.)

J. GINIBRE (1970). General formulation of Griffiths' inequalities. Commun. Math. Phys. *16*, 310–328.

J. GINIBRE (1972). Correlation inequalities in statistical mechanics. In: *Mathematical Aspects of Statistical Mechanics*, pp. 27–45. Edited by J. C. T. Pool. Amer. Math. Soc., Providence.

J. GLIMM AND A. JAFFE (1981). *Quantum Physics: a Functional Integral Point of View*. Springer, New York.

H. GOLDSTEIN (1959). *Classical Mechanics*. Addison-Wesley, Reading.

R. L. GRAHAM (1983). Applications of the FKG inequality and its relatives. In: *Mathematical Programming—Bonn 1982*, pages 116–131. Edited by A. Bachem, M. Grötschel, and B. Korte. Springer, New York.

D. GRIFFEATH (1978). *Additive and Cancellative Interacting Particle Systems*. Lecture Notes in Mathematics *724*, Springer, Berlin.

R. B. GRIFFITHS (1964). Peierls proof of spontaneous magnetization in a two-dimensional Ising ferromagnet. Phys. Rev. *136A*, 437–439.

R. B. GRIFFITHS (1967a). Correlations in Ising ferromagnets. I. J. Math. Phys. *8*, 478–483.

R. B. GRIFFITHS (1967b). Correlations in Ising ferromagnets. II. J. Math. Phys. *8*, 484–489.

R. B. GRIFFITHS (1967c). Correlations in Ising ferromagnets. III. Commun. Math. Phys. *6*, 121–127.

R. B. GRIFFITHS (1969). Rigorous results for Ising ferromagnets of arbitrary spin. J. Math. Phys. *10*, 1559–1565.

R. B. GRIFFITHS (1971). Phase transitions. In: *Statistical Mechanics and Quantum Field Theory*, pp. 241–279. Summer School of Theoretical Physics, Les Houches 1970. Edited by C. de Witt and R. Stora. Gordon and Breach, New York.

R. B. GRIFFITHS (1972). Rigorous results and theorems. In: *Phase Transitions and Critical Phenomena*, Vol. 1, pp. 7–109. Edited by C. Domb and M. S. Green. Academic, New York.

R. B. GRIFFITHS, C. A. HURST, AND S. SHERMAN (1970). Concavity of magnetization of an Ising ferromagnet in a positive external field. J. Math. Phys. *11*, 790–795.

P. GROENEBOOM, J. OOSTERHOFF, AND F. H. RUYMGAART (1979). Large deviation theorems for empirical probability measures. Ann. Probab. 7, 553–586.

L. GROSS (1979). Decay of correlations in classical lattice models at high temperature. Commun. Math. Phys. *68*, 9–27.

L. GROSS (1982). Thermodynamics, Statistical Mechanics, and Random Fields. In: *Ecole d'Eté de Probabilités de Saint-Flour X-1980*, pp. 101–204. Edited by P. L. Hennequin. Lecture Notes in Mathematics *929*, Springer, Berlin.

F. GUERRA, L. ROSEN, AND B. SIMON (1975). The $P(\phi)_2$ Euclidean quantum field theory as classical statistical mechanics. Ann. of Math. *101*, 111–259.

P. R. HALMOS (1950). *Measure Theory*. Van Nostrand, Princeton.

T. E. HARRIS (1960). A lower bound for the critical probability in a certain percolation process. Proc. Camb. Phil. Soc. *59*, 13–20.

P. C. HEMMER AND J. L. LEBOWITZ (1976). Systems with weak long-range potentials. In: *Phase Transitions and Critical Phenomena*, Volume 5b, pp. 107–203. Edited by C. Domb and M. S. Green. Academic, London.

P. HENRICI (1977). *Applied and Computational Complex Analysis*, Volume 2. Wiley, New York.

E. HEWITT AND K. STROMBERG (1965). *Real and Abstract Analysis*. Springer, Berlin.

Y. HIGUCHI (1982). On the absence of non-translationally invariant Gibbs states for the two-dimensional Ising system. In: *Random Fields: Rigorous Results in Statistical Mechanics and Quantum Field Theory–Proceedings 1979*, pp. 517–534. Edited by J. J. Fritz, J. L. Lebowitz, and D. Szász. Colloquia Mathematica Societatis Janos Bolyai *27*. North Holland, Amsterdam.

T. HÖGLUND (1979). A unified formulation of the central limit theorem for small and large deviations from the mean. Z. Wahrsch. verw. Gebiete *49*, 105–117.

R. A. HOLLEY AND D. W. STROOCK (1976a). Applications of the stochastic Ising model to the Gibbs states. Commun. Math. Phys. *48*, 249–265.

R. A. HOLLEY AND D. W. STROOCK (1976b). L_2 theory for the stochastic Ising model. Z. Wahrsch. verw. Geb. *35*, 87–101.

R. A. HOLLEY AND D. W. STROOCK (1977). In one and two dimensions, every stationary measure

for a stochastic Ising model is a Gibbs state. Commun. Math. Phys. *55*, 37–45.
W. HOLSZTYNSKI AND J. SLAWNY (1979). Phase transitions in ferromagnetic spin systems at low temperatures. Commun. Math. Phys. *66*, 147–166.
K. HUANG (1963). *Statistical Mechanics*. Wiley, New York.
K. HUSIMI (1953). Statistical mechanics of condensation. In: *Proceedings of the International Conference of Theoretical Physics*, pp. 531–533. Edited by H. Yukawa. Science Council of Japan, Tokyo.
D. IAGOLNITZER AND B. SOUILLARD (1977). Decay of correlations for slowly decreasing potentials. Phys. Rev. *16A*, 1700–1704.
D. IAGOLNITZER AND B. SOUILLARD (1979). Lee-Yang theory and normal fluctuations. Phys. Rev. *19B*, 1515–1518.
J. Z. IMBRIE (1982). Decay of correlations in the one-dimensional Ising model with $J_{ij} = |i-j|^{-2}$. Commun. Math. Phys. *85*, 491–515.
A. IOFFE AND V. TIKHOMIROV (1968). Duality of convex functions and extremum problems. Russian Math. Surveys *23*, Number 6, 53–124.
E. ISING (1925). Beitrag zur Theorie des Ferromagnetismus. Z. Phys. *31*, 253–258.
R. B. ISRAEL (1979). *Convexity in the Theory of Lattice Gases*. Princeton Univ. Press, Princeton.
N. C. JAIN (1982). A Donsker–Varadhan type of invariance principle. Z. Wahrsch. verw. Gebiete *59*, 117–138.
E. T. JAYNES (1957a). Information theory and statistical mechanics. Phys. Rev. *106*, 620–630.
E. T. JAYNES (1957b). Information theory and statistical mechanics, II. Phys. Rev. *108*, 171–190.
E. T. JAYNES (1963). Information theory and statistical mechanics. In: *Statistical Physics*, Vol. 3, pp. 181–218. Brandeis University Summer Institute in Theoretical Physics 1962. Edited by K. W. Ford. Benjamin, New York.
E. T. JAYNES (1968). Prior probabilities. IEEE Trans. Syst. Sci. Cybernetics. *4*, 227–241.
E. T. JAYNES (1982). On the rationale of maximum-entropy methods. Proc. IEEE *70*, 939–952.
J.-W. JEON (1983). Weak convergence of processes occurring in statistical mechanics. J. Korean Statist. Soc. *12*, 10–17.
S. JOHANSEN (1979). *Introduction to the Theory of Regular Exponential Families*. Lecture Notes, Institute of Mathematical Statistics, Univ. of Copenhagen.
G. JONA-LASINIO (1975). The renormalization equation: a probabilistic view. Nuovo Cimento *26B*, 99–119.
G. JONA-LASINIO (1983). Large fluctuations of random fields and renormalization group: some perspectives. In: *Scaling and Self-Similarity in Physics*, pp. 11–28. Edited by J. Fröhlich. Birkhäuser, Boston.
M. KAC (1959a). On the partition function of a non-dimensional gas. Phys. Fluids *2*, 8–12.
M. KAC (1959b). *Probability and Related Topics in Physical Sciences*. Interscience, New York.
M. KAC (1968). Mathematical mechanisms of phase transitions. In: *Statistical Physics: Phase Transitions and Superfluidity*, Vol. 1, pp. 241–305. Brandeis University Summer Institute in Theoretical Physics 1966. Edited by M. Chretien, E. P. Gross, and S. Deser. Gordon and Breach, New York.
M. KAC (1980). *Integration in Function Spaces*. Fermi Lectures. Academia Nazionale dei Lincei Scuola Normale Superiore, Pisa.
A. M. KAGAN, YU V. LINNIK, AND C. R. RAO (1973). *Characterization Problems in Mathematical Statistics*. Translated by B. Ramachandran. Wiley, New York.
E. KAMKE (1930). *Differentialgleichungen Reeler Funktionen*. Akademische Verlagsgesellschaft, Leipzig.
E. KAMKE (1974). *Differentialgleichungen Lösungsmethoden und Lösungen*. Third unaltered edition. Chelsea, Bronx.
D. G. KELLY AND S. SHERMAN (1968). General Griffiths' inequalities on correlations in Ising ferromagnets. J. Math. Phys. *9*, 466–484.
H. KESTEN (1982). *Percolation Theory for Mathematicians*. Birkhäuser, Boston.
A. KHINCHIN (1929). Über einen neuen Grenzwertsatz der Wahrscheinlichkeitsrechnung. Math. Annalen *101*, 745–752.
A. I. KHINCHIN (1949). *Mathematical Foundations of Statistical Mechanics*. Translated by G. Gamow. Dover, New York.
A. I. KHINCHIN (1957). *Mathematical Foundations of Information Theory*. Translated by R. A. Silverman and M. D. Friedman. Dover, New York.
R. KINDERMANN AND J. L. SNELL (1980). *Markov Random Fields and Their Applications*. American Mathematical Society, Providence.

M. KLEIN (1970). Maxwell, his demon, and the second law of thermodynamics. American Scientist *58*, 84–97.

A. KOLMOGOROV (1930). Sur la loi forte des grandes nombres. C. R. Acad. Sci. *191*, 910–912.

V. I. KOLOMYTSEV AND A. V. ROKHLENKO (1978). Sufficient condition for ordering of an Ising ferromagnet. Theoret. and Math. Phys. *35*, 487–493.

V. I. KOLOMYSTEV AND A. V. ROKHLENKO (1979). A new sufficient condition for the existence of a phase transition in a classical system with long-range interaction. Soviet. Phys. Dokl. *24*, 902–904.

H. A. KRAMERS AND G. H. WANNIER (1941). Statistics of the two-dimensional ferromagnet, I and II. Phys. Rev. *60*, 252–276.

M. A. KRASNOSEL'SKIĬ AND Y. B. RUTICKIĬ (1961). *Convex Functions and Orlicz Spaces*. Translated by L. F. Boron. Noordhoff, Groningen.

U. KRENGEL (1985). *Ergodic Theorems*. De Gruyter, Berlin (in press).

H. KÜNSCH (1981). Almost sure entropy and the variational principle for random fields with unbounded state space. Z. Wahrsch. verw. Gebiete *58*, 69–85.

H. KÜNSCH (1982). Decay of correlations under Dobrushin's uniqueness condition and its applications. Commun. Math. Phys. *84*, 207–222.

S. KULLBACK (1959). *Information Theory and Statistics*. Wiley, New York.

S. KULLBACK AND R. A. LEIBLER (1951). On information and sufficiency. Ann. Math. Statist. *22*, 79–86.

S. KUSUOKA AND Y. TAMURA (1984). Gibbs measures for mean field potentials. Journal of the Faculty of Science, Univ. of Tokyo, Section IA, Vol. 31, 223–245.

J. LAMPERTI (1966). *Probability*. Benjamin, New York.

O. E. LANFORD (1973). Entropy and equilibrium states in classical statistical mechanics. In: *Statistical Mechanics and Mathematical Problems*, pp. 1–113. Edited by A. Lenard. Lecture Notes in Physics *20*. Springer, Berlin.

O. E. LANFORD (1975). Time evolution of large classical systems. In: *Dynamical Systems, Theory and Applications-Proceedings 1975*, pp. 1–111. Edited by J. Moser. Lecture Notes in Physics *38*. Springer, Berlin.

O. E. LANFORD (1976). On a derivation of the Boltzmann equation. In: *International Conference on Dynamical Systems in Mathematical Physics, Rennes, 1975, Sept. 14–21*, pp. 117–137. Astérisque *40*, Société Mathématique de France, Paris.

O. E. LANFORD AND D. RUELLE (1969). Observables at infinity and states with short range correlations in statistical mechanics. Commun. Math. Phys. *13*, 194–215.

R. LANG (1981). An elementary introduction to the mathematical problems of renormalization group. Lecture notes, Institut für Angewandte Mathematik, Univ. Heidelberg (unpublished).

A. I. LARKIN AND D. E. KHMEL'NITSKII (1969). Phase transition in uniaxial ferroelectrics. Soviet Phys. JETP *29*, 1123–1128.

J. L. LEBOWITZ (1972). Bounds on the correlations and analyticity properties of ferromagnetic Ising spin systems. Commun. Math. Phys. *28*, 313–321.

J. L. LEBOWITZ (1974). The GHS and other inequalities. Commun. Math. Phys. *35*, 87–92.

J. L. LEBOWITZ (1975). Uniqueness, analyticity, and decay properties of correlations in equilibrium systems. In: *International Symposium on Mathematical Problems in Theoretical Physics-Proceedings 1975*, pp. 370–379. Edited by H. Araki. Lecture Notes in Physics *39*. Springer, Berlin.

J. L. LEBOWITZ (1977). Coexistence of phases in Ising ferromagnets. J. Statist. Phys. *16*, 463–476.

J. L. LEBOWITZ (1978). Exact results in nonequilibrium statistical mechanics: where do we stand? Progress of Theoretical Physics, Supplement No. 64, 35–49.

J. L. LEBOWITZ AND A. MARTIN-LÖF (1972). On the uniqueness of the equilibrium state for Ising spin systems. Commun. Math. Phys. *25*, 276–282.

J. L. LEBOWITZ AND O. PENROSE (1966). Rigorous treatment of the van der Waals-Maxwell theory of the liquid-vapor transition. J. Math. Phys. *7*, 98–113.

J. L. LEBOWITZ AND O. PENROSE (1968). Analytic and clustering properties of thermodynamic functions and distribution functions for classical lattice and continuum systems. Commun. Math. Phys. *7*, 99–124.

J. L. LEBOWITZ AND O. PENROSE (1973). Modern ergodic theory. Physics Today *26*, Feb. 1973, 23–29.

J. L. LEBOWITZ AND E. PRESUTTI (1976). Statistical mechanics of systems of unbounded spins. (Commun. Math. Phys. *50*, 195–218. (Erratum: Commun. Math. Phys. *78*, 151 (1980).)

T. D. LEE AND C. N. YANG (1952). Statistical theory of equations of state and phase transitions,

II. Lattice gas and Ising model. Phys. Rev. *87*, 410–419.
A. LENARD (1978). Thermodynamical proof of the the Gibbs formula for elementary quantum systems. J. Statist. Phys. *19*, 575–586.
W. LENZ (1920). Beitrag zum Verständnis der magnetischen Erscheinungen in festen Körpern. Z. Physik *21*, 613–615.
E. H. LIEB (1978). New proofs of long range order. In: *Mathematical Problems in Theoretical Physics-Proceedings 1977*, pp. 59–67. Edited by G. Dell'Antonio, S. Doplicher, and G. Jona-Lasinio. Lecture Notes in Physics *80*, Springer, Berlin.
E. H. LIEB (1980). A refinement of Simon's correlation inequality. Commun. Math. Phys. *77*, 127–135.
E. H. LIEB AND A. D. SOKAL (1981). A general Lee-Yang theorem for one-component and multi-component ferromagnets. Commun. Math. Phys. *80*, 153–179.
T. M. LIGGETT (1977). The stochastic evolution of infinite systems of interacting particles. In: *Ecole d'Eté de Probabilités de Saint-Flour VI-1976*, pp. 187–248. Edited by P. L. Hennequin. Lecture Notes in Mathematics *598*, Springer, Berlin.
T. M. LIGGETT (1985). *Interacting Particle Systems*. Springer, New York (in press).
YU. V. LINNIK (1961). On the probability of large deviations for the sums of independent variables. In: *Proceedings of the Fourth Berkeley Symposium on Mathematical Statistics and Probability*, Vol. II, pp. 289–306. Edited by J. Neyman. Univ. of California Press, Berkeley.
M. LOÈVE (1960). *Probability Theory*. Second edition. Van Nostrand, Princeton.
J. M. LUTTINGER (1982). A new method for the asymptotic evaluation of a class of path integrals. J. Math. Phys. *23*, 1011–1016.
S.-K. MA (1976). *Modern Theory of Critical Phenomena*. Benjamin, Reading.
G. W. MACKEY (1974). Ergodic theory and its significance for statistical mechanics and probability theory. Adv. Math. *12*, 178–268.
N. F. G. MARTIN AND J. W. ENGLAND (1981). *Mathematical Theory of Entropy*. Addison-Wesley, Reading.
A. MARTIN-LÖF (1972). On the spontaneous magnetization in the Ising model. Commun. Math. Phys. *24*, 253–259.
A. MARTIN-LÖF (1973). Mixing properties, differentiability of the free energy, and the central limit theorem for a pure phase in the Ising model at low temperature. Commun. Math. Phys. *32*, 75–92.
A. MARTIN-LÖF (1979). *Statistical Mechanics and the Foundations of Thermodynamics*. Lecture Notes in Physics 101. Springer, Berlin.
A. MARTIN-LÖF (1982). A Laplace approximation for sums of independent random variables. Z. Wahrsch. verw. Geb. *59*, 101–116.
J. C. MAXWELL (1860). Illustrations of the dynamical theory of gases. Philos. Mag. Ser. 4, *19*, 19–32; *20*, 21–37.
J. C. MAXWELL (1867). On the dynamical theory of gases. Philos. Trans. Roy. Soc. London *157*, 49–88.
B. M. MCCOY AND T. T. WU (1973). *The Two-Dimensional Ising Model*. Harvard Univ. Press, Cambridge.
R. J. MCELIECE (1977). *Information Theory*. Addison-Wesley, Reading.
K. MENDELSSOHN (1966). *The Quest for Absolute Zero*. Weidenfeld and Nicolson, London.
A. MESSAGER AND S. MIRACLE-SOLE (1977). Correlation functions and boundary conditions in the Ising ferromagnet. J. Statist. Phys. *77*, 245–262.
J. MESSER AND H. SPOHN (1982). Statistical mechanics of the isothermal Lane–Emden equation. J. Statist. Phys. *29*, 561–578.
P.-A. MEYER (1966). *Probability and Potentials*. Blaisdell, Waltham.
J. L. MONROE AND P. A. PEARCE (1979). Correlation inequalities for vector spin models. J. Statist. Phys. *21*, 615–633.
E. W. MONTROLL, R. B. POTTS, AND J. C. WARD (1963). Correlations and spontaneous magnetization in the two-dimensional Ising model. J. Math. Phys. *4*, 308–322.
J.-J. MOREAU (1962). Fonctions convexes en dualité (multigraph). Faculté des Sciences, Séminaires de Mathématiques, Univ. de Montpellier, Montpellier.
J.-J. MOREAU (1966). *Fonctionelles Convexes*. Lecture Notes, Séminaire "Equations aux dérivées partielles," Collège de France.
C. M. NEWMAN (1974). Zeros of the partition function for generalized Ising systems. Comm. Pure Appl. Math. *27*, 143–159.

C. M. NEWMAN (1975a) Inequalities for Ising models and field theories which obey the Lee-Yang theorem. Commun. Math. Phys. *41*, 1–9.

C. M. NEWMAN (1975b). Moment inequalities for ferromagnetic Gibbs distributions. J. Math. Phys. *16*, 1956–1959.

C. M. NEWMAN (1975c). Gaussian correlation inequalities for ferromagnets. Z. Wahrsch. verw. Geb. *33*, 75–93.

C. M. NEWMAN (1976a). Classifying general Ising models. In: *Les Méthodes Mathématiques de la Théorie Quantique des Champs*, pp. 273–288. Editions du Centre National de la Recherche Scientifique, Paris.

C. M. NEWMAN (1976b). Fourier transforms with only real zeros. Proc. Amer. Math. Soc. *61*, 245–251.

C. M. NEWMAN (1976c). Rigorous results for general Ising ferromagnets. J. Statist. Phys. *15*, 399–406.

C. M. NEWMAN (1979). Critical point inequalities and scaling limits. Commun. Math. Phys. *66*, 181–196.

C. M. NEWMAN (1980). Normal fluctuations and the FKG inequalities. Commun. Math. Phys. *74*, 119–128.

C. M. NEWMAN (1981a). Self-similar random fields in mathematical physics. In: *Measure Theory and Its Applications, Proceedings of the 1980 Conference*, pp. 93–119. Edited by G. A. Goldin and R. F. Wheeler. Northern Illinois Univ. Press, De Kalb.

C. M. NEWMAN (1981b). Shock waves and mean field bounds—concavity and analyticity of the magnetization at low temperature. Univ. of Arizona preprint. Also talk given at 46th Statistical Mechanics Meeting, Rutgers Univ., Dec. 1981. Title published in J. Statist. Phys. *27*(1982), 83.

C. M. NEWMAN (1983). A general central limit theorem for FKG systems. Commun. Math. Phys. *91*, 75–80.

C. M. NEWMAN (1984). Asymptotic independence and limit theorems for positively and negatively dependent random variables. In: *Inequalities in Statistics and Probability*, pp. 127–140. Institute of Mathematical Statistics Lecture Notes—Monograph Series, Volume 5. Edited by Y. L. Tong.

P. NEY (1983). Dominating points and the asymptotics of large deviations for random walk on $\mathbb{R}^d$. Ann. Probab. *11*, 158–167.

X. X. NGUYEN AND H. ZESSIN (1979). Ergodic theorems for spatial processes. Z. Wahrsch. verw. Geb. *48*, 133–158.

B. G. NICKEL (1981). Hyperscaling and universality in three dimensions. Physica 106A, 48–58.

B. G. NICKEL (1982). The problem of confluent singularities. In: *Phase Transitions—Cargèse 1980*, pp. 291–329. Edited by M. Lévy, J.-C. Le Guillou, and J. Zinn-Justin. Plenum, New York.

B. G. NICKEL AND B. SHARPE (1979). On hyperscaling in the Ising model in three dimensions. J. Phys. A *12*, 1819–1834.

L. ONSAGER (1944). Crystal statistics I. A two-dimensional model with an order-disorder transitions. Phys. Rev. *65*, 117–149.

L. ONSAGER (1949). Discussion remark (spontaneous magnetization of the two-dimensional Ising model). Nuovo Cimento (Supplement) *6*, 261.

S. OREY (1985). Large deviations in ergodic theory. In: *Seminar on Stochastic Processes, 1984*. Edited by E. Çinlar, K. L. Chung, and R. K. Getoor. Birkhäuser, Boston (in press).

A. PAIS (1982). *'Subtle Is the Lord...' The Science and the Life of Albert Einstein*. Oxford Univ. Press, New York.

J. PALMER AND C. TRACY (1981). Two-dimensional Ising correlations: convergence of the scaling limit. Adv. Appl. Math. *2*, 329–388.

K. R. PARTHASARATHY (1967). *Probability Measures on Metric Spaces*. Academic, New York.

P. A. PEARCE (1981). Mean field bounds on the magnetization for ferromagnetic spin models. J. Statist. Phys. *25*, 309–320.

R. F. PEIERLS (1936). On Ising's model of ferromagnetism. Proc. Cambridge Philos. Soc. *32*, 477–481.

O. PENROSE (1979). Foundations of statistical mechanics. Rep. Progr. Phys. *42*, 1939–2006.

O. PENROSE AND J. L. LEBOWITZ (1974). On the exponential decay of correlation functions. Commun. Math. Phys. *39*, 165–184.

J. K. PERCUS (1975). Correlation inequalities for Ising spin lattices. Commun. Math. Phys. *40*, 283–308.

V. V. PETROV (1975). *Sums of Independent Random Variables.* Translated by A. A. Brown. Springer, Berlin.

C.-E. PFISTER (1983). Interface and surface tension in the Ising model. In: *Scaling and Self-Similarity in Physics*, pp. 139–161. Edited by J. Fröhlich. Progress in Physics 7. Birkhäuser, Boston.

M. PINCUS (1968). Gaussian processes and Hammerstein integral equations. Trans. Amer. Math. Soc. *134*, 193–216.

R. PINSKY (1985). On evaluating the Donsker-Varadhan I-function. Ann. Probab. *13* (in press).

A. B. PIPPARD (1957). *Elements of Classical Thermodynamics.* Cambridge Univ. Press, New York.

M. PIRLOT (1980). A strong variational principle for continuous spin systems. J. Appl. Probab. *17*, 47–58.

C. J. PRESTON (1974a). An application of the GHS inequalities to show the absence of phase transitions for Ising spin systems. Commun. Math. Phys. *35*, 253–255.

C. J. PRESTON (1974b). *Gibbs States on Countable Sets.* Cambridge Univ. Press, Cambridge.

C. J. PRESTON (1976). *Random Fields.* Lecture Notes in Mathematics *534*. Springer, Berlin.

I. PRIGOGINE (1973). Microscopic aspects of entropy and the statistical foundations of non-equilibrium thermodynamics. In: *Foundations of Continuum Thermodynamics*, pp. 81–112. Proceedings of an International Symposium held at Bussaco, Portugal. Edited by J. J. D. Domingos, M. N. R. Nina, and J. H. Whitelaw. Wiley, New York.

I. PRIGOGINE (1978). Time, structure, and fluctuations. Science *201*, 777–785.

W. G. PROCTOR (1978). Negative absolute temperatures. Scientific American *239*, August 1978, 90–99.

L. E. REICHL (1980). *A Modern Course in Statistical Physics.* Univ. of Texas Press, Austin.

A. RÉNYI (1970a). *Foundations of Probability.* Holden-Day, San Francisco.

A. RÉNYI (1970b). *Probability Theory.* North-Holland, Amsterdam.

V. RICHTER (1957). Local limit theorems for large deviations. Theory Probab. Appl. *2*, 206–220.

V. RICHTER (1958). Multidimensional local limit theorems for large deviations. Theory Probab. Appl. *3*, 100–106.

A. W. ROBERTS AND D. E. VARBERG (1973). *Convex Functions.* Academic, New York.

D. W. ROBINSON AND D. RUELLE (1967). Mean entropy of states in classical statistical mechanics. Commun. Math. Phys. *5*, 288–300.

R. T. ROCKAFELLAR (1970). *Convex Analysis.* Princeton Univ. Press, Princeton.

J. B. ROGERS AND C. J. THOMPSON (1981). Absence of long-range order in one-dimensional spin systems. J. Statist. Phys. *25*, 669–678.

J. ROSEN (1977). The Ising model limit of ϕ^4 lattice fields. Proc. Amer. Math. Soc. *66*, 114–118.

W. RUDIN (1974). *Real and Complex Analysis.* Second edition. McGraw-Hill, New York.

D. RUELLE (1967). A variational formulation of equilibrium statistical mechanics and the Gibbs phase rule. Commun. Math. Phys. *5*, 324–329.

D. RUELLE (1968). Statistical mechanics of a one-dimensional lattice gas. Commun. Math. Phys. *9*, 267–278.

D. RUELLE (1969). *Statistical Mechanics: Rigorous Results.* Benjamn, Reading.

D. RUELLE (1976). Probability estimates for continuous spin systems. Commun. Math. Phys. *50*, 189–194.

D. RUELLE (1978). *Thermodynamic Formalism.* Addison-Wesley, Reading.

S. SAKAI (1976). On commutative normal $*$-derivations, II. J. Funct. Anal. *21*, 203–208.

I. N. SANOV (1957). On the probability of large deviations of random variables (in Russian). Mat. Sb. *42*, 11–44. (English translation in *Selected Translations in Mathematical Statistics and Probability I*, pp. 213–244 (1961).)

M. SCHILDER (1966). Some asymptotic formulae for Wiener integrals. Trans. Amer. Math. Soc. *125*, 63–85.

E. SENETA (1981). *Non-Negative Matrices and Markov Chains.* Second edition. Springer, New York.

J. SETHURAMAN (1964). On the probability of large deviations of families of sample means. Ann. Math. Statist. *35*, 1304–1316. (Corrections: Ann. Math. Statist. *41*, 1376–1380 (1970)).

C. E. SHANNON (1948). A mathematical theory of communication. Bell System Tech. J. *27*, 379–423, 623–656.

B. SIMON (1974). *The $P(\phi)_2$ Euclidean (Quantum) Field Theory.* Princeton Univ. Press, Princeton.

B. SIMON (1979a). *Functional Integration and Quantum Physics.* Academic, New York.

B. SIMON (1979b). A remark on Dobrushin's uniqueness theorem. Commun. Math. Phys. *68*, 183–185.

B. SIMON (1980a). Correlation inequalities and the decay of correlations in ferromagnets. Commun. Math. Phys. *77*, 111–126.

B. SIMON (1980b). Mean field upper bound on the transition temperature in multicomponent ferromagnets. J. Statist. Phys. *22*, 491–493.

B. SIMON (1981). Absence of continuous symmetry breaking in a one-dimensional n^{-2} model. J. Statist. Phys. *26*, 307–311.

B. SIMON (1983). Instantons, double wells, and large deviations. Bull. (New Series) Amer. Math. Soc. *8*, 323–326.

B. SIMON (1985). *The Statistical Mechanics of Lattice Gases*. Volume 1. Princeton Univ. Press, Princeton (in preparation).

B. SIMON AND R. B. GRIFFITHS (1973). The $(\phi)_2^4$ field theory as a classical Ising model. Commun. Math. Phys. *33*, 145–164.

B. SIMON AND A. D. SOKAL (1981). Rigorous entropy-energy arguments. J. Statist. Phys. *25*, 679–694.

YA. G. SINAI (1977). *Introduction to Ergodic Theory*. Translated by V. Scheffer. Princeton Univ. Press, Princeton.

YA. G. SINAI (1982). *Theory of Phase Transitions: Rigorous Results*. Pergamon, Oxford.

J. SLAWNY (1974). A family of equilibrium states relevant to low temperature behavior of spin −1/2 classical ferromagnets. Breaking of translation symmetry. Commun. Math. Phys. *35*, 297–305.

J. SLAWNY (1983). On the mean field theory bound on the magnetization. J. Statist. Phys. *32*, 375–388.

A. D. SOKAL (1979). A rigorous inequality for the specific heat of an Ising or ϕ^4 ferromagnet. Phys. Lett. *71*A, 451–453.

A. D. SOKAL (1981). More inequalities for critical exponents. J. Statist. Phys. *25*, 25–50.

A. D. SOKAL (1982a). Mean field bounds and correlation inequalities. J. Statist. Phys. *28*, 431–439.

A. D. SOKAL (1982b). More surprises in the general theory of lattice systems. Commun. Math. Phys. *86*, 327–336.

F. L. SPITZER (1970). Interaction of Markov processes. Adv. Math. *5*, 246–290.

F. L. SPITZER (1971). *Random Fields and Interacting Particle Systems*. Math. Assoc. of America, Washington.

F. L. SPITZER (1974). Introduction aux processus de Markov à parametre dans $\mathbb{Z}_r$. In: *Ecole d'Eté de probabilités de Saint Flour III-1973*, pages 115–189. Edited by A. Badrikian and P. L. Hennequin. Lecture Notes in Mathematics *390*, Springer, Berlin.

H. E. STANLEY (1971). *Introduction to Phase Transitions and Critical Phenomena*. First edition. Oxford Univ. Press, New York.

H. E. STANLEY (1985). *Introduction to Phase Transitions and Critical Phenomena*. Second edition. Oxford Univ. Press, New York (in preparation).

J. STEINEBACH (1978). Convergence rates of large deviation probabilities in the multidimensional case. Ann. Probab. *6*, 751–759.

D. W. STROOCK (1984). *An Introduction to the Theory of Large Deviations*. Springer, New York.

L. SUCHESTON (1983). On one-parameter proofs of almost sure convergence of multiparameter processes. Z. Wahrsch. verw. Geb. *63*, 43–49.

G. S. SYLVESTER (1976a). Continuous-spin inequalities for Ising ferromagnets. J. Statist. Phys. *15*, 327–342.

G. S. SYLVESTER (1976b). Continuous spin Ising ferromagnets. MIT dissertation.

J. SZARSKI (1965). *Differential Inequalities*. PWN-Polish Scientific Publishers, Warsaw.

A. A. TEMPEL'MAN (1972). Ergodic theorems for general dynamical systems. Trans. Moscow Math. Soc. *26*, 94–132.

H. N. V. TEMPERLEY (1954). The Mayer theory of condensation tested against a simple model of the imperfect gas. Proc. Phys. Soc. A *67*, 233–238.

C. J. THOMPSON (1971). Upper bounds for Ising model correlation functions. Commun. Math. Phys. *24*, 61–66.

C. J. THOMPSON (1972). *Mathematical Statistical Mechanics*. Macmillan, New York.

C. J. THOMPSON AND H. SILVER (1973). The classical limit of n-vector spin models. Commun. Math. Phys. *33*, 53–60.

D. J. THOULESS (1969). Long-range order in one-dimensional Ising systems. Phys. Rev. *187*, 732–733.

C. TRUESDELL (1961). Ergodic theory in classical statistical mechanics. In: *Ergodic Theories*, pp. 21–56. Proceedings of the International School of Physics "Enrico Fermi," XIV Course, Varenna, Italy, May 23–31, 1960. Edited by P. Caldirola. Academic Press, New York.

G. E. Uhlenbeck and G. W. Ford (1963). *Lectures in Statistical Mechanics.* American Mathematical Society, Providence.

H. van Beijeren (1975). Interface sharpness in the Ising system. Commun. Math. Phys. *40*, 1–6.

H. van Beijeren and G. S. Sylvester (1978). Phase transitions for continuous-spin Ising ferromagnets. J. Funct. Anal. *28*, 145–167.

J. van Campenhout and T. M. Cover (1981). Maximum entropy and conditional probability. IEEE Trans. Inform. Theory, IT-*27*, 483–489.

S. R. S. Varadhan (1966). Asymptotic probabilities and differential equations. Comm. Pure Appl. Math. *19*, 261–286.

S. R. S. Varadhan (1980a). Lectures on large deviations at Osaka Univ, Osaka, Japan, Summer 1980 (unpublished).

S. R. S. Varadhan (1980b). Some problems of large deviations. In: *Proceedings of the International Congress of Mathematicians, Helsinki 1978*, pp. 755–762.

S. R. S. Varadhan (1984). *Large Deviations and Applications.* SIAM, Philadephia.

J. Voight (1981). Stochastic operators, information, and entropy. Commun. Math. Phys. *81*, 31–38.

D. J. Wallace and R. K. P. Zia (1978). The renormalization group approach to scaling in physics. Rep. Progr. Phys. *41*, 1–85.

F. J. Wegner (1975). Critical phenomena and the renormalization group. In: *Trends in Elementary Particle Theory—Proceedings 1974*, pp. 171–196. Lecture Notes in Physics *37*. Springer, Berlin.

F. J. Wegner and E. K. Riedel (1973). Logarithmic corrections to the molecular-field behavior of critical and tricritical systems. Phys. Rev. B *7*, 248–256.

A. Wehrl (1978). General Properties of entropy. Rev. Mod. Phys. *50*, 221–260.

R. J. B. Wets (1980). Convergence of convex functions, variational inequalities, and convex optimization problems. In: *Variational Problems and Complementarity Problems*, Chapter 25. Edited by R. W. Cottle, F. Giannessi, and J.-L. Lions. Wiley, Chichester.

N. Wiener (1939). The ergodic theorem. Duke Math. J. *5*, 1–18.

N. Wiener (1948). *Cybernetics.* Wiley, New York.

A. S. Wightman (1979). Convexity and the notion of equilibrium state in thermodynamics and statistical mechanics. Introduction to R. B. Israel (1979), op. cit.

R. A. Wijsman (1966). Convergence of sequences of convex sets, cones, and functions, II. Trans. Amer. Math. Soc. *123*, 32–45.

K. G. Wilson (1974). Critical phenomena in 3.99 dimensions. Physica *73*, 119–128.

K. G. Wilson (1979). Problems in physics with many scales of length. Scientific American *241*, August 1979, 158–179.

K. G. Wilson (1983). The renormalization group and critical phenomena. Rev. Modern Phys. *55*, 583–600.

K. G. Wilson and J. Kogut (1974). The renormalization group and the ε expansion. Phys. Reports *12C*, 75–200.

J. M. Wozencraft and I. M. Jacobs (1965). *Principles of Communication Engineering.* Wiley, New York.

C. N. Yang (1952). The spontaneous magnetization of a two-dimensional Ising model. Phys. Rev. *85*, 808–816.

C. N. Yang and T. D. Lee (1952). Statistical theory of equations of state and phase transitions, I. Theory of condensation. Phys. Rev. *87*, 404–409.

S. L. Zabell (1980). Rates of convergence for conditional expectations. Ann. Probab. *8*, 928–941.

M. W. Zemansky (1968). *Heat and Thermodynamics.* Fifth edition. McGraw-Hill, New York.

Author Index

Subject Index

Printing and Binding: Strauss GmbH, Mörlenbach